General Analytical Chemistry
Separation and Spectral Methods

Gwenola Burgot and Jean-Louis Burgot

Faculty of Pharmaceutical and Biological Sciences
University of Rennes 1
Rennes, France

CRC Press
Taylor & Francis Group
Boca Raton London New York

CRC Press is an imprint of the
Taylor & Francis Group, an **informa** business

A SCIENCE PUBLISHERS BOOK

Cover illustration provided by the authors.

First edition published 2024
by CRC Press
2385 NW, Executive Center Drive, Suite 320, Boca Raton, FL 33431

and by CRC Press
4 Park Square, Milton Park, Abingdon, Oxon, OX14 4RN

Library of Congress Cataloging-in-Publication Data (applied for)

ISBN: 978-1-032-03914-5 (hbk)
ISBN: 978-1-032-03915-2 (pbk)
ISBN: 978-1-003-18968-8 (ebk)

DOI: 10.1201/9781003189688

Typeset in Times New Roman
by Radiant Productions

Preface

Analytical chemistry can be defined as the study of the physical and chemical methods of chemical analysis. Hence, analytical chemistry and chemical analysis must not be confused. Besides, the *Encyclopédie Universalis* defines analytical chemistry as follows:

> Analytical chemistry is the branch of chemistry whose goal is the identification, the characterization and the quantification of chemical substances together with the development of the methods necessary for this analysis. It is also devoted to the understanding of the phenomena being to play in the processes and techniques of analysis in order to ameliorate them (http://www.universalis-edu.com.passerelle.univ-rennes1.fr/encyclopedie/chimie-analytique/ consulted on 20 november 2022).

Our aim is to comply with these precepts. Besides, it would be impossible to give the very numerous examples of applications to the chemical analysis of the separation, spectral and others methods studied here. We confine ourselves to give a few. Among them, in case of need, we shall choose examples coming from bioanalytical chemistry. Perhaps, in this connection, let us already say that the greater difference which exists between chemical analysis and bioanalysis is the high molecular weight of biomolecules.

This first book is a translation, and an extension of our book has been written in French, titled "Chimie analytique, méthodes de separation, méthodes spectrales", and edited in 2017 by Lavoisier (Paris). It will be prolonged by a volume devoted to the electrochemical methods of analysis.

Acknowledgments

We thank Mr. Arsène Lancien who, as usual, has corrected our imperfect English. Also special thanks to our family who have been supportive throughout writing of this book.

Contents

Glossary

Mathematical

C_n^p number of combination of n things taken p to p

f( ) or ϕ( ) or φ( ) or ψ( ) function of

I imaginary symbol

log decimal or common logarithm and Ln or Log = Napierian or natural logarithm

y'(x) derivative of y(x) with respect to variable x

dy, dx…differentials of y, x…

Δy, Δx changes of y and x

dy/dx idem

$(\partial M/\partial N)_{ij}$ partial derivatives of functions M with respect to variable N, other variables i and j being constants

(*Remark*: the highlighted quantity means a partial derivative.)

Physico-chemical

α angle/relative volatility/algebraic intermediary/corrected partition coefficient

A amplitude of the sinusoid wave/absorbance/mass number/surface/Ampere

Å angstrom unit

a_i activity of i

β electronic Bohr magneton

β_N nuclear Bohr magneton

B(x) magnetic field

B_0 magnetic field of induction

C or c velocity of the propagation

C number of compounds

C_i or c_i concentration of i

Cte integration constant

C(z,t) concentration at z coordinate and at the time t

δ chemical shift in N.M.R.

ξ electro-kinetic potential

d distance

D distribution ratio/diffusion coefficient

Dp diameter of a particle

D' number of moles collected or distillate

ç factor of proportionality

ε dielectric constant (relative permittivity)/ε_i molecular energy level of i/ε_i^{tr} molecular energy of translation level of i/ε_i^{int} molecular energy level internal of i/molar absorption coefficient

ε^{rot} molecular energy of rotation level

ε^{vib} molecular energy of vibration level

ε^{electr} molecular energy of electronic level

ε^n molecular energy of nuclear level

η viscosity

E electrical potential/ratio relative error about concentration/electric energy

E average efficacy of the plate

Em interaction energy of the center and the magnetic field (E.S.R.)

E(x) electric field

eV electron-volt

φ number of phases of a system/proportionality factor quantum yield

ϕ radiative power Fi quantum yield

F or ν frequency or *f*

F force

F module/ratio of concentrations/fraction or fluorescence intensity/factor of enrichment of the extraction

f molecular partition function or fugacity/frequency (Fourier transform)

γ gravity coefficient/composition of azeotropes/magnetogyric factor/ratio

γ_i activity coefficient of *i*

G Gibbs energy

G_{ibar} Partial molal Gibbs energy (formerly free enthalpy)

G(t) Fourier transform of the frequency intensity g(t)

g multiplier coefficient of gravity

H demand/rate of flow/enthalpy/plate height

h Planck's constant

I luminous intensity/ionic strength/intensity of current/quantum number of a proton (and neutron)

I(w) frequency intensity

i electric current intensity (in R.M.N.)

J flux of a species

χ–1 layer of ions not bonded to the first layer

K proportionality constant/distribution coefficient/equilibrium constant

K° permeability constant of the column

Ka, pKa acidity constant

k proportionality constant

ki-1 radius of the diffuse layer

k_{ij} kinetic constant

λ wavelength/partition coefficient

L vaporization latent heat/length/angular momentum of an isolated mobile/kinetic moment

L Number of moles liquid given by a theoretical plate in distillation

l length/quantum number

μ_i chemical potential of *i*, $\mu_{i°}$ standard chemical potential of I/μep electrophoretic mobility/μ0 electroosmotic mobility/μ_0 osmosis mobility

$\boldsymbol{\mu}$ magnetic moment (vector)

M mass molar

m mass/quantum number

m_e mass of an electron

m.p. melting point

v frequency of the radiation

N number of molecules of the system/number of plates

n number of moles/number of solutes

n number of turns of circumference per second

n vector (solid angle)

ψ electrical potential at the surface of a conducting sphere

π bond/osmotic pressure

ω degeneracy of the ith molecular energy level/angular rate

P pressure/thermodynamic partition coefficient/power (or intensity)/power of radiation

p partial pressure/probability of process

q electrical charge

Q number of moles/partition function of a thermodynamic macroscopic system

ρ yield

R radius of a circumference/R perfect gas constant

R distance in N.M.R (vector)

R_f retention factor

R_D ratio of the reflux to the overhead product

R_V ratio of the reflux to the vapor

r radius

Σ intersection area

σ wavenumber/σ bond/standard deviation/screening constant

S solubility/substance/rate of flow (demand)/surface/internal surface of the circuit/spin total

s num of plate/bond/spin of the electron

θ angle

τ time in N.M.R./one scale in N.M.R./Turbidimetry coefficient/lifetime of the excited state

T absolute temperature/period/transmittance

t time

U internal energy

u vector (solid angle)

u instantaneous flow

V linear velocity of the mobile phase

V number of moles on a vapor received and given by a theoretical plate (distillation)

V volume/potential voltage

∇ Laplacian operator

$\bar{v}$ wave number

v linear rate/variance

v_i velocity of the propagation

υ efficient volume

W mass of the adsorbent in a packed column

w width of a chromatographic peak

x molar fraction

xD instantaneous composition of the distillate

Y amplitude of beating of FN.M.R.

y molar fraction in gaseous phase/fraction

Z atomic number

z fraction

Acronyms

cAMP	Adenosine 3'5'-cyclic monophosphate
BTEX	Benzene, Toluene, Ethylbenzene and Xylene
Cap-LC	Capillary Liquid Chromatography
CE	Capillary Electrophoresis
CEC	Capillary Electro Chromatography
CIEF	Capillary Iso Electric Focusing
CITP	Capillary Isotachophoresis
DNA	Deoxyribo Nucleic Acid
DVD	DiVinyl Benzene
E.D.T.A.	Ethylene Diamine Tetracetic Acid
EKC	Electrokinetic Chromatography
ELISA	Enzyme-Linked Immuno Sorbent Assay
ELSD	Evaporating Light Scattering Detector
ELONA	Enzyme-Linked Oligo Nucleotide Assay
FAAS	Flame Absorption Atomic Spectrometry
FID	Flame Ionization Detector
FSOT	Fused Silica Open Tubular
FT-NMR	Fourier Transforms Nuclear Magnetic Resonance
GC	Gas Chromatography
GCxGC GC by GC	Comprehensive Two-Dimensional Gas Chromatography
GFAA	Graphite Furnace Absorption Atomic
HPLC	High Performance Liquid Chromatography
HPTLC	High Performance Thin Layer Chromatography
HTLC	High Temperature Liquid Chromatography
ICP	Inductively Coupled Plasma
ICP-MS	Inductively Coupled Plasma-Mass Spectrometry
ICP-OES	Inductively Coupled Plasma-Optical Emission Spectroscopy
IT-SPME	In-tube Solid-phase Microextraction
IUPAC	International Union of Pure and Applied Chemistry
LC	Liquid Chromatography
LC-UV	Liquid Chromatography-UV
LC-MS	Liquid Chromatography-Mass spectrometry
MALDI-MS	Matrix Assisted Laser Desorption Ionization coupled with Mass Spectrometer
MALLS	Multi-Angle Laser Light Scattering
MIP	Molecular Imprinted Polymer
MS	Mass Spectrometry

N.M.R.	Nuclear Magnetic Resonance
NOE	Nuclear Overhauser Enhancement
PA	Poly-Acrylate
PC	Paramagnetic Center
PDMS	PolyDiMethylSiloxane
PLOT	Porous Layer Open-Tubular
PMT	Photo Multiplier Tube
QuEChERS	Quick, Easy, Cheap, Efficient, Rugged, Safe
RAM	Restricted Access Material
RNA	Ribo Nucleic Acid
SBSE	Stir Bar Sorptive Extraction
SCOT	Support Coated Open-Tubular
SDS	Sodium Dodecyl Sulfate
SDS-PAGE	Sodium Dodecyl Sulfate-Poly Acrylamide Gel Electrophoresis
SEC	Size Exclusion Chromatography
SPE	Solid-Phase Extraction
SPME	Solid-Phase MicroExtraction
TFME	Thin Film MicroExtraction
TLC	Thin Layer Chromatography
TOF	Time of Flight
TSH	Thyroid Stimulating Factor
U-HPLC	Ultra-High Performance Liquid Chromatography
UV-visible	Ultra-Violet–visible
WCOT	Wall-Coated Open Tubular

Part I
Separation Methods

Introduction to the Methods of Separation

Separation methods are methods that have been grouped formerly in the methods of *proximate analysis*.

It is a usual fact that the sample to analyze is a mixture and that the substance under study is one of its components. Most often, the mixture arises as only one phase which can be solid, liquid, or gaseous (one phase being one ensemble that appears homogenous to the eye). More rarely, the mixture is heterogenous and involves several phases.

Today, one can witness a true revival of the analysis. It is notably due to the revival of the proximate analysis, the methods that essentially have the goal to extract without adulterating the substance to analyze. The deep insight into the knowledge we have now, such as about vitamins, hormones, antibiotics, and alkaloids, is due in great part to the progress of the proximate analysis.

The separation methods are numerous, and their choice depends on the nature and the composition of the mixture to analyze. One can distinguish either simple operations, such as filtration and centrifugation or more complicated methods, which can be successful even for the resolution of homogeneous phases into their components. Therefore, one can classify the separation methods under the following headings:

- separation of a "mixture" of two phases;
- resolution of a solid phase into its constituents by:
  - fractionated dissolution;
  - sublimation;
- resolution of a liquid phase into its constituents by:
  - rupture of phase (concentration and precipitation);
  - phases transfer or extraction;
  - distillations;
  - chromatography;
  - electrophoresis.

They will be studied now. Some of these subjects are the matter of a chapter, sometimes brief. Others are the matter of several chapters, given the fact that they deal with very important methods, such as those of chromatography.

CHAPTER 1

Separation of a Mixture of Distinct Phases

Mixture of Solid Phases

The separation of a "mixture" of several solid phases is seldom carried out. One can mention the historical separation of tartrates left and right by Pasteur with the help of pliers and a lens as a celebrated example.

Levigation is a physical means of separation widely used in the mining industry. It consists of carrying away the less dense particles by a stream of water. More generally, such a mixture is mixed until it becomes as homogeneous as possible through a mechanical means; after which it considers only one substance and treats it as one solid.

Mixture of Solid and Liquid Phases

Filtration

Filtering a mixture is to force it to traverse a porous membrane, permeable to liquids and impermeable to the solids which are in suspension in order to separate both phases. The size of the particles must be in ratio to the diameter of the pores. However, one can retain particles, the diameter of which is slightly inferior to that of the pores. The first passing liquid causes trouble. Then, some of them are retained by adsorption. The filter clogs, and it becomes effective.

Porous membranes must exhibit some qualities, which are often difficult to join together such as mechanical resistance, effectiveness, and speed. One can use organic and inorganic substances.

Organic Substances

The filter paper is the most used. It is constituted by the paper that is not glued. It is prepared by starting from non-purified cellulose. It is attacked by very acidic and very basic solutions. It can retain some substances by adsorption, such as colorings. Some filters are constituted by thin disks of cellulose, the pores of which are very thin and regular. They have the disadvantage to be dissolved by some solvents such as ethyl acetate.

Inorganic Substances

Among them, formerly, one used asbestos which is now known to be carcinogenic. One also uses glass wool and cotton wool (it has the disadvantage to give up noteworthy quantities of alkali ions to the solution), sand, porous porcelain, and sintered glass from numbers 0 to 5:

N°0: pores very large (diameter about 200 μ);

N°5: pores very thin (diameter about 1 μ).

Their efficacity increases with the number, but the rate of filtration decreases with it. From a technical standpoint, it is superfluous to recall the different modes of filtration here.

Centrifugation

When a mass m turns on a circumference of radius R, according to a uniform circular motion, it is attracted by a force F called centripetal force. The uniform circular motion implicates that the mass is displacing on a circle with a rate of constant intensity. Given the fact that the intensity of the rate is constant and taking into account the expression of the angular rate of the mass, it is easy to demonstrate that the modulus of F is proportional to m, to the square of the linear rate v and inversely proportional to the radius R:

$$F = mv^2/R = 4\pi^2\, mRn^2 \qquad \text{since v} = 2\pi Rn$$

where n = number of turns of circumference per second.

But the centripetal Force F is counterbalanced by the force of inertia or centrifugal F', which is equal and opposite to the former. Therefore, we can write:

$$F' = mv^2/R \qquad \text{(in modules)}$$

The densest parts of the mixture are submitted to the strongest forces and settle toward the exterior by forming a dottle. The apparatus intervenes in the efficacy of the operation. The latter is proportional to the developed centrifuge force, which itself functions as the characteristics of the apparatus and its form is:

$$F = KRn^2$$

From a practical viewpoint, R comprises between 10 and 30 cm and it cannot be increased. In contrast, the factor n may considerably increase. This fact enlarges the efficacy very quickly as this factor intervenes through its square.

According to a general rule, the efficacy is expressed by a multiplier of the gravity coefficient and the number g. It is the ratio γ/g:

$$\text{centrifuge force/force of the gravity} = m\gamma/mg = \gamma/g$$

According to the apparatus, the number of g obtained is:

- 1,500 g with an apparatus turning to 3,600 t/min;
- 10,000 g with an electric apparatus turning to 9,000 t/min;
- 40,000 g with a super-centrifuge turning to 21,000 t/min;
- 260,000 g with an ultra-centrifuge turning to 51,000 t/min.

Some ultra-centrifuges can reach 1,000,000 g.

The analytical applications of centrifugation are numerous. The method is rapid and effective in separating small precipitates. It has the advantage to destroy the emulsions formed between two liquids of different densities. As a general rule, it is limited to volumes that are relatively small, except when they are working continuously. By ultra-centrifugation, the ion iodide I^- (which is a particularly heavy ion) has been partially separated from the ion sodium Na^+ (lighter) by starting from a solution of sodium iodide.

Mixture of Liquid Phases

As a general rule, the separation is mechanical. One lets the mixture repose in a decantation funnel or in an industrial decanter until the separation of the two phases takes place. These are then successively taken away. This method is generally impeded by the occurrence of emulsions. The latter ones can be destroyed by as following:

- by rotation around the axis of the funnel;
- by heating;
- by change of the pH value (as a rule, the emulsions are more stable in alkaline media);
- by addition of a surfactant agent;
- by centrifugation.

CHAPTER 2

Resolution of a Solid Phase in its Constituents

We consider two methods: fractionated dissolution and sublimation.

Fractionated Dissolution

Principle

The goal of the method is to extract one or several constituents from the solid mixture quantitatively and, if it is possible, selectively. In order to do that, the mixture is extracted by one or several solvents that are successively used. From the quantitative standpoint, one operation with one solvent may be sufficient to extract the whole of the substance, but this case is rare. Usually, it is necessary to carry out several successive extractions with the same solvent and eventually recommence with another solvent.

From the strict selectivity point of view, it is reached with a given solvent only in favorable cases. Very often, it is difficult to reach a total separation of the constituents since:

- The insolubility is never absolute in a given solvent. The extraction has for a consequence not only preferentially extracts one or several constituents but also extracts some quantities of non-wished others.

- A solvent does not act on a mixture of constituents as it does on an isolated one. The solution of the first constituent generally possesses a solvating power upon the other constituents higher than that the pure solvent possesses.

After each operation, therefore, one obtains:

- a solid phase that still contains a part of the compound that we want to extract;
- a liquid phase that may contain several dissolved compounds.

Therefore, it is essential:

- to separate both phases;
- to perfect the dissolution by starting from the solid phase by carrying out a series of discontinuous extractions or a continuous extraction;
- to resolve the liquid phase into its constituents.

The Solvents

Only a few theoretical and practical considerations upon solvents are recalled here.

Theoretical Considerations

They are related to the dissolving power of the considered liquid. This fact prompts us to investigate its origin.

A solvent product acts upon a substance by separating the molecules which are then dispersed; the solvent associates with the dissolved compound and this phenomenon is called "solvation". In some pure compounds, such as pure liquid acetic acid, the molecules are not free but are associated between them per group of two. In general, the molecules dissolved in water are not associated between them, but they are associated with water molecules. Today, the classification of solvents takes place on a property that is called *polarity*. This notion was first introduced in 1912 by Debye.[1] The molecules are named "Polar" when the center of action of their intramolecular charges (+) is distinct from that of charges (−).

These molecules are dipoles that possess a permanent electrical dipole. An electrical dipole is a vector of length; $q.d$ where q is the charge of each pole and d is their distance. If the molecule possesses one or several dipoles, its electrical momentum is expressed in units C.G.S. or debyes (1 debye = 10^{-18} units U.E.S. C.G.S.). Beyond that, there exists a relation between this electrical momentum and the permittivity relative constant (dielectric constant) (Table 1).

Table 1: Relative Permittivity (Dielectric Constants) of Some Solvents.

Non-Polar Molecules

Ethane-hexane	2
Benzene	3
Chloroform	5

Polar Molecules		**Polar Molecules**	
Water	81	Formic acid	58
Methanol	33	Nitrobenzene	35
Ethanol	24	Formamide	110

The association between the solvent and the dissolved compound may be accounted for by some structural analogy between them. The dissolution takes rise when the solvation energy is higher than the grating-reticular energy, which will vanish (reticular energy). Now, in one crystal, the force of attraction between opposed charges is proportional to $1/\varepsilon$ (ε; dielectric constant). It is the reason why inorganic salts are, as a general rule, insoluble in benzene as the necessary energy for the separation of charges is too high.

Among the polar solvents, water occupies the first place. Due to the hydroxyl group, it dissolves alcohols and carbohydrates and yet it must be pointed out that the solubility of alcohols in hot water decreases when the molecular weight increases. Therefore, ethylic alcohol CH_3-CH_2-OH is miscible in water, even though it possesses a methyl group that is non-polar and hydrophobic. The lengthening of the chain diminishes the polar character, and finally the alcohols in C_7 and upper are insoluble in water. Moreover, water dissolves ionized compounds. Most often, it does not dissolve organic compounds except those that are polar.

Inversely, non-polar solvents dissolve non-polar compounds. As a result of the preceding considerations, it can be concluded that a product of an organic homologous series dissolves

---

[1] Debye Petrus, Dutch physicist and chemist, naturalyzed american, 1884–1966, Nobel Prize in Chemistry (1936).

others of the same series. Therefore, petroleum dissolves vaselin® and benzene dissolves aromatic compounds. Finally, when two compounds have neighboring structures, their solubility in the same solvent is a function of their vapor tension and their fusion temperature. For example:

- phenanthrene   m.p.: 103°C; is soluble in ether;
- anthracene   m.p.: 207°C; is poorly soluble in ether.

Practical Considerations

It is desirable to use solvents:

- of moderate prices
- easy, to distill
- non-toxicological

Theoretical Study

The theoretical study of the fractionated dissolution, above all, presents one interest in the industrial domain. This is the reason why we limit ourselves to mentioning its principal conclusion. It justifies the modalities of applications, which are as follows:

- The yield of the dissolution, which is the ratio of the extracted matter and the total quantity to extract, is all the better as the solubility of the compound to be extracted is higher in the chosen solvent and the solvent volume is higher as well.

- For the same volume of solvent, the yield is better when this volume is fractionated into several portions in order to extract the same quantity of initial mixture in a single period only. One consequence of this fact is that an identical yield is obtained with several successive extractions with a total volume less than that used when there is only one operation of extraction.

- It needs to be recalled that the solubility also varies with the temperature. Generally, it increases when the temperature itself increases.

- Lastly, the rate of dissolution is all the more rapid as the surface of the solid in direct contact with the solvent is greater and the surfaces in contact are renewed. It follows that effective agitation brings an advantage.

Techniques of Dissolution

There are first preliminary techniques. They consist of reducing the sample to very thin particles and then, in case the mixture is particularly complex, they also establish the order according to which the solvents will be successively used; beginning with the less good solvent and ending with the best one.

In practice, it is preferable to stir the mixture and, when it is possible from the standpoint of the stability of the mixture, to carry out the dissolution in warm conditions. When it is necessary to carry out several successive extractions with the same solvent, it is possible to work discontinuously or continuously.

Discontinuous Extraction

After the first extraction, the liquid and solid phases are separated. The solid phase is extracted a second time with a new volume of solvent and so forth. The several fractions of the liquid phase are finally joined together.

Continuous Extraction

It is more advantageous to process by continuous extraction in order to reach the same result. A suitable apparatus is necessary. It always contains:

- one balloon in which the solvent is heated and vaporized;
- one device which permits the condensation of the vapors of solvent;
- one device in which is introduced a filtering refill (the extractor) containing the solid mixture to be extracted and in which the condensed solvent falls back. After filtration by the refill, the extracting solution is periodically eliminated from the extractor with the help of a siphon and falls back into the balloon.

The principal types of apparatus are given below:

- The Soxhlet apparatus: The extraction is carried out by the condensed and chilled solvent (Figures 1 and 2).
- The Kumagawa apparatus: The extractor is placed in the middle of the current of the stream. Hence, the extraction is carried out at a temperature close to that of the ebullition of the solvent. As a rule, it is more effective than the preceding one (Figure 3).

Figure 1: Soxhlet Apparatus.

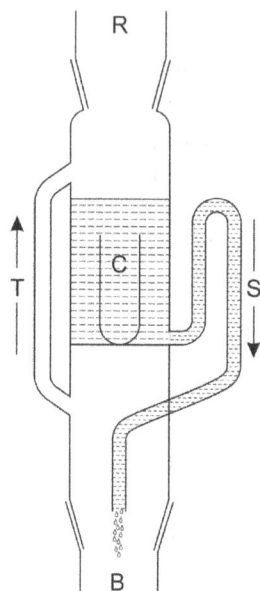

Figure 2: Apparatus of Soxhlet B (Detailed View); (R = Refrigerating Device; C = Cartridge; S = Siphon; T = Tube; B = Balloon).

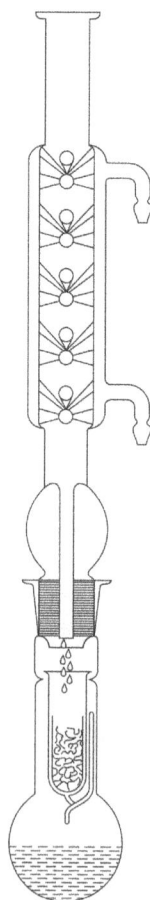

Figure 3: KUMAGAWA Apparatus.

The advantage of this apparatus is to allow carrying out the number of extractions we want with a minimal volume of solvent. Their major drawback is that one needs to heat the extracted substances for a long time. There is, therefore, a risk of altering them.

Sublimation

Principle

To sublime a volatile compound, it requires making it pass from the solid to the gaseous state and at a new time, once again, returning it to the solid state without passing through the liquid state. As shown in Figure 4, which is the diagram of changes of states of a pure compound, the operation of sublimation is facilitated by raising the temperature and decreasing the pressure (Figure 4).

Studying this diagram permits an understanding of how any changes in pressure or temperature may enable the phenomenon of sublimation. For example, a compound located at point A (Figure 4) by a decrease in the pressure can pass directly from the solid to the gaseous state.

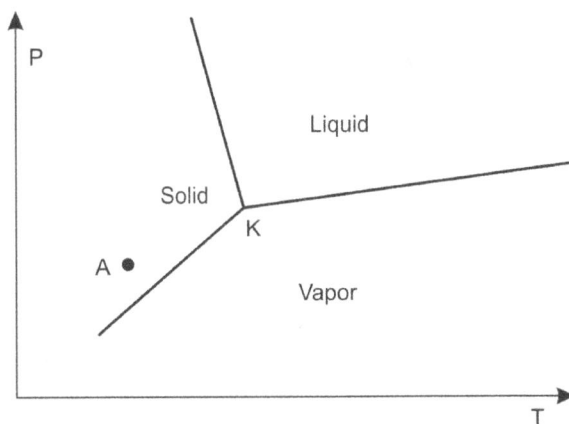

Figure 4: Diagram of Change of State of a Pure Compound (Sublimation); (P = Pressure; T = Temperature; K = triple point).

Modalities

• Under ordinary pressure: There exists one very simple device to carry out a sublimation in this case. The substance is placed in a porcelain dish. It is gently heated and the vapor is condensed upon a cool surface.

• Under vacuum: This technique is called molecular distillation. This case is well illustrated by the methodology called *molecular distillation* (refer to Chapter 12). The operation is accepted under a high vacuum of the order of 10^{-3} 10^{-5} mmHg (1 mmHg = 133.322 Pa), and it is necessary to exclude every vapor (water or organic solvent) from the system. In order to reach this state, the molecules to distillate must have a sufficient speed in order to escape from the surface of the liquid and reach the surface of condensation. Therefore, the distillation depends no longer on the vapor tension of the liquid. The molecular distillation permits to carry out of sublimations at not too high temperatures and hence with a loss of risks to adulterating the product.

The process of sublimation is employed to separate volatile substances from non-volatile impurities. It is a very convenient process. It can apply to some organic compounds and some inorganic ones as well. Unfortunately, it cannot apply (by far) to all organic and inorganic products.

CHAPTER 3

Separation by Rupture of Phase

In this group of methods, the goal is to reduce a liquid phase into its constituents where one provokes a rupture of phase, that is to say, the apparition of a solid phase in the bulk solution. Let us consider a constituent dissolved in any liquid phase. This phase is homogeneous by definition of a phase, and the concentration C of the constituent is lower than its solubility S. On the contrary case, it is no longer homogeneous since its concentration C is higher than its solubility (at the same temperature). There is a rupture of phase. It can be provoked either by increasing C or by lowering S. Hence, there is precipitation. Finally, separation is also possible by fractioned crystallization.

Concentration

Principle

One seeks to increase C by eliminating partially or wholly solvent. Any liquid, at each temperature, corresponds to vapor pressure (vapor tension), which also increases with temperature. Otherwise, a liquid product boils when its vapor tension is equal to the pressure that surmounts it. In order to eliminate one solvent, we can act either on pressure, temperature or both simultaneously.

Techniques

Elimination of the Solvent by Ebullition

The simplest methodology consists in boiling the solvent at the ambient pressure. According to the working modalities, one can distinguish the operation in open media and the *distillation*. In an open medium, the vapor spreads in the atmosphere. That may cause some risks (poisoning, inflammation, etc.). In both cases, the dissolved substance is heated. As a result, there is a risk of adulteration. One limits the rising of temperature by operating under reduced pressure.

Elimination of the Solvent by Evaporation at Ambient Temperature

When the volume of the solution is very small and also when the solvents are very volatile (such as ethyl ether), the vapor pressure at ambient temperature is sufficiently strong so that the evaporation could be complete even without enhancing the temperature.

Elimination of the Last Traces of Solvent

There is a difficulty. The residue obtained at this point always contains traces of solvent; the elimination of which is difficult, particularly when it is water. It is eliminated with the help of a device named 'desiccator'. The elimination of the last traces of solvent can also be carried out with vacuum ovens.

(*Remark*: The desiccation under reduced pressure may provoke some chemical modifications.)

Precipitation

The goal is now to decrease the solubility by physical action.

Precipitation by Difference in Temperature

Most often, the solubility changes and increases with temperature. By cooling, it is possible to precipitate a dissolved compound, the concentration of which is at the beginning of the process near its saturation.

Precipitation by a Mixture of Solvents

One adds up a second liquid miscible to the first which is a very bad solvent of the considered substance. The concentration decreases but less than its solubility, and thus the precipitation begins. For example, an alcoholic solution of salicylic acid precipitates after the addition of water, and likewise a solution of digitalis into chloroform by addition of ligroine. However, this method is brutal and may also precipitate impurities.

Precipitation by Salting-Out

If to a solution of a Compound A in a solvent, one adds a Compound B that modifies the solubility of A, the latter can become isolated; this phenomenon is called the salting-out. Most often, it is observed in aqueous solutions. The salted-out compound can be often organic, that is it can be in the state of solid or liquid. The compound provoking the salting-out is often an inorganic compound. For example:

- It is the case of ammonium sulfate which is soluble in water and provokes the salting-out. The acetone, poorly soluble in an aqueous solution of ammonium sulfate, separates into a distinct phase out.

Compounds That Can Be Salted-Out

They are often organic substances. In their structures, one distinguishes polar functional groups that account for their solubility in water and non-polar functional groups, bringing their organic character and their poor solubility in water. The substance can be all the more non-soluble as the organic character is more pronounced; that is to say, the hydrophobic groups dominate.

Compounds Provoking the Salting-Out

They are essentially inorganic salts. Their choice is directed by their solubility in water and also by their price. One principally uses ammonium sulfate and sodium chloride. The compounds provoking the salting-out can be ranked by decreasing order of efficacy (Hofmeister's series):

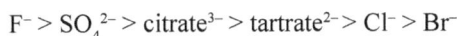

$$F^- > SO_4^{2-} > citrate^{3-} > tartrate^{2-} > Cl^- > Br^-$$

$$Li^+ > Na^+ > K^+ > NH_4^+ > Mg^{2+}$$

Advantages of the Salting-Out

It permits the use of solvents that are miscible to water for the extractions of aqueous solutions. It is carried out with less chance of adulteration of the salting-out substances. It avoids some chemical and physical processes, which with other methodologies can destroy the product to be recovered. The salting-out is used in the industry of soaps. It is also used in the separation of the benzenesulfonic acid during the synthesis of phenol, in the separation of colorings as well as in biochemistry in the separation of proteins and biliary pigments.

Isolation by Crystallization

Crystallization is the sudden spontaneous apparition of a solid body in the bulk solution. The maximal quantity that can remain dissolved in the solution is constant at a determined temperature. This constant is called the *solubility* of the body. The maximal concentration can be exceeded in two cases:

- Some bodies have a marked tendency toward saturation. It is precluded by priming.
- The body which appears first is not always the most stable crystalline form. It turns by degrees into a crystalline form more stable that is always less soluble.

In favorable cases, the body which crystallizes is single. One obtains it in a pure state after a series of crystallizations. For example, stearic acid crystallizes in ethanol. One can also find oleic acid with it. It is obtained in a pure state after 7 or 8 crystallizations. In the cases of mixtures of bodies, their solubility may be sufficiently close to provoking a simultaneous crystallization. One can even try to separate them by using the differences in crystallization rates or the differences in rates of "redissolution" or by using another solvent that is more favorable to simple crystallization.

In some cases, the dissolved bodies form mixed crystals in which they are strictly associated and where they can form solid solutions that are homogeneous from the macroscopic standpoint. In other words, it can undergo *syn-crystallization*. In order to separate products that syn-crystallize, one can try *fractioned crystallization*. A good example is provided by the acids α and β thiophene carboxylic acids. They form an inseparable mixture.

The fractioned crystallization applies to a mixture of solids, the properties of which are very close. Let us suppose two bodies A and B, where A is a little less soluble than B. One makes a solution of A and B in one solvent. One concentrates and crystallizes (Figure 1).

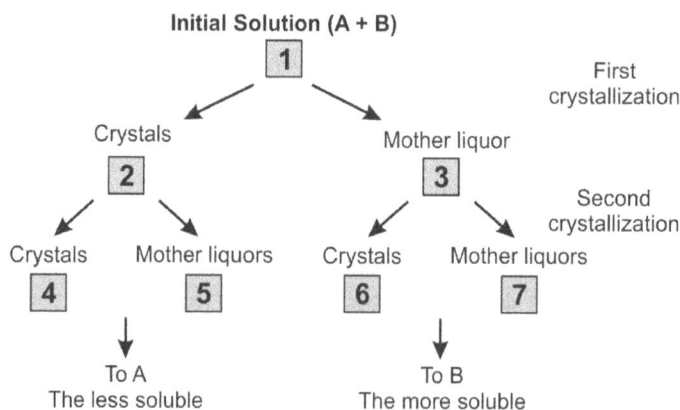

Figure 1: An Example of Fractioned Crystallization.

Crystals 2 contain more Compound A, which is less soluble than Compound B. Mother-Liquors 3 contain more Compound B than A. One takes Crystals 3, dissolves them and crystallizes them (second crystallization), which gives Crystals 4 and the Mother-Liquors 5. Crystals 4 is still richer in A. It remains so until compounds A and B are very pure obtained.

Isolation by Filtering Membranes

A filtering membrane is difficult to define. In their most general meaning, there is one interface that separates two media or two phases. They are also defined by the part they play as being a selective barrier permitting or not passing some components of the media from one medium to the other under the influence of a transfer force. Generally, their separative power results from the difference in rates of transfer of the components through the membrane. (This paragraph is more thoroughly developed in Chapter 5, which is devoted to osmose and dialysis).

The semi-permeable membranes are artificial membranes, such as the tannate of gelatin. The copper ferrocyanide deposited in the canals of a porous vessel only permits water to traverse them.

Other membranes permit water and crystalloids to traverse these colloids. They are permeable membranes.

Diffusion

Two important displacement phenomena are related to isolation by filtering membranes.

- **Diffusion:** The phenomenon of diffusion can be illustrated by the displacement of particles that spontaneously go from a region of higher concentration to a region of lower one until the concentration becomes uniform throughout. This definition is incomplete and will be completed in the chapters devoted to the electrochemical methods of analysis.

- **Osmosis:** Osmosis is the phenomenon of diffusion of one solvent through a selective membrane. It corresponds to a gradient of concentration which pushes the solvent to leave the solute less concentrated, to traverse the membrane as well as dilute the more concentrated solute. (More will be said in the next chapter). The membrane plays a significant part in the exchanges.

Other Applications of Filtering Membranes

Three applications of filtering membranes are:

- the dialysis
- the ultrafiltration
- the electrodialysis

(More details will be given in Chapter 5).

CHAPTER 4

Partition Coefficient
Thermodynamic Basis

After having studied this chapter, we shall essentially consider the dialysis, osmose and extractions of liquid-liquid. These points are the first steps before studying the distillation, chromatography and electrophoresis at some length. It must be well understood that all these phenomena are governed and quantified by the notion of partition coefficient. It is important to state that it is precisely from the standpoint of thermodynamics.

This chapter may be considered the theoretical introduction of all the following chapters concerning the methods of separation.

Extraction and Separation

Usually, one distinguishes the two phenomena called *extraction* and *separation* from one another based on this occurrence that we are considering two analytical methods, which are founded on the extraction of one or several solutes from a solution by a non-miscible or a poorly miscible solvent to it. One attributes:

- the term "extraction" to the study of the general method; the goal of which is to make the solute(s), present in the initial solution, pass selectively and quantitatively into the extractive solvent;

- whereas, the term "separation" which is less general than the preceding, refers to the method whose goal is to separately isolate each substance constituting the initial mixture. It is also founded on the different extractions of the solutes that are initially present in the mixture. Hence, the separation concerns initial mixtures of at least two solutes. In the French chemical literature, the term "fractionation", which is sometimes encountered, is used to denote separation.

Actually, these definitions can be considered as qualifying quasi-identical methods since extraction and separation are often processes that are quasi-identical and occur simultaneously. Therefore, it is not rare that a single extraction leads to a beginning of separation.

Partial Molal Gibbs Energy

Let us consider a liquid phase constituted by the solvent (compound 1) and by several solutes (compounds 2,3,4...i). By definition, one calls the *partial molal Gibbs energy* $\overline{G}_i$ of component i, the quantity:

$$(\partial G/\partial n_i)_{T,P,nj} = \overline{G}_i$$

Here, G is the Gibbs[2] energy of the whole system, T is its absolute temperature, P is its pressure and n_j is the number of moles of the components of the system others from the compound *i* but including the solvent. Hence, the partial molal Gibbs energy is a partial derivative from the mathematical standpoint (Figure 1).

(*Remark*: The Gibbs energy had still recently the name of free enthalpy.)

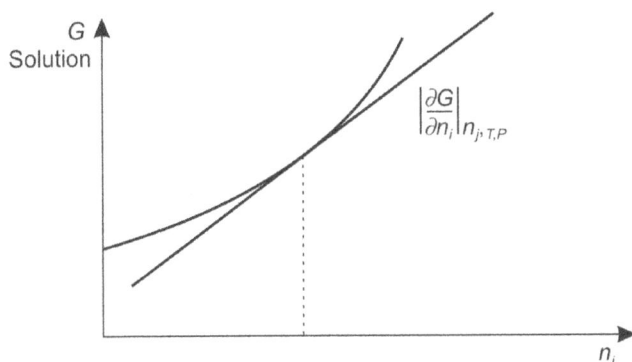

Figure 1: Partial Molal Gibbs Energy (Reprinted with Permission from Gwenola and Jean-Louis Burgot) (Paris, Lavoisier, Tec et Doc 2017, 46).

The physical significance of this quantity can be obtained by the following reasoning. Let us take an example of the aqueous solution of the sole Solute M. By definition, it is:

$$\bar{G}_M = (\partial G/\partial n_M)_{T,P,nH_2O}$$

According to the theory of partial derivatives, one can say that $\bar{G}_M$ is the change in the Gibbs energy of the *whole* solution (taken as the whole system) where one mole of Solute M is added to an infinitely great amount of solution in such a manner that this addition does not markedly change the concentration of M in the solution. Another way to grasp the significance of $\bar{G}_M$, equivalent to the preceding, is to consider that $\bar{G}_M$ is the change in the Gibbs energy of the whole system when one adds an infinitesimally small number (a differential) of moles, dn_M moles, in such a way that the concentration of M should be considered as being constant.

Otherwise, one proves that the whole Gibbs energy of the system G_{syst} is given by the relation:

$$G_{syst} = n_1 \bar{G}_1 + n_2 \bar{G}_2 + \ldots\ldots$$

Clearly, there are as many partial molal Gibbs energies as components. The examination of this relationship shows that $\bar{G}_i$ effectively plays the part of a *molar* Gibbs energy. However, it is important to notice that, most often, the partial molal Gibbs energies $\bar{G}_i$ change with the composition of the solution, whence the notion of partial molal Gibbs energies defined with the help of partial derivatives is as given above.

Chemical Potential

The partial molal Gibbs energy has another name; it is the *chemical potential*. It has been introduced by J.W. Gibbs under the vocable "escaping tendency". It is symbolized by μ. In the preceding case of the solute M in water, one immediately notices there are two chemical potentials to consider, those of M and water, i.e., μ_M and μ_{H2O}. Contrary to the Gibbs energy of a system which is an extensive quantity, the chemical potential of a substance is an intensive quantity.

---

[2] Gibbs, Josiah Willard, American physicist-chemist, (1839–1903).

Let us recall that the chemical potential of a substance is its tendency to leave its current thermodynamic state either by a physical, chemical or biological process at constant pressure and temperature.

Voltage and Electric Current

We know that when two compartments of an electrochemical cell, possess different electric potentials, are connected by a conducting wire, one observes an electric current (Figure 2).

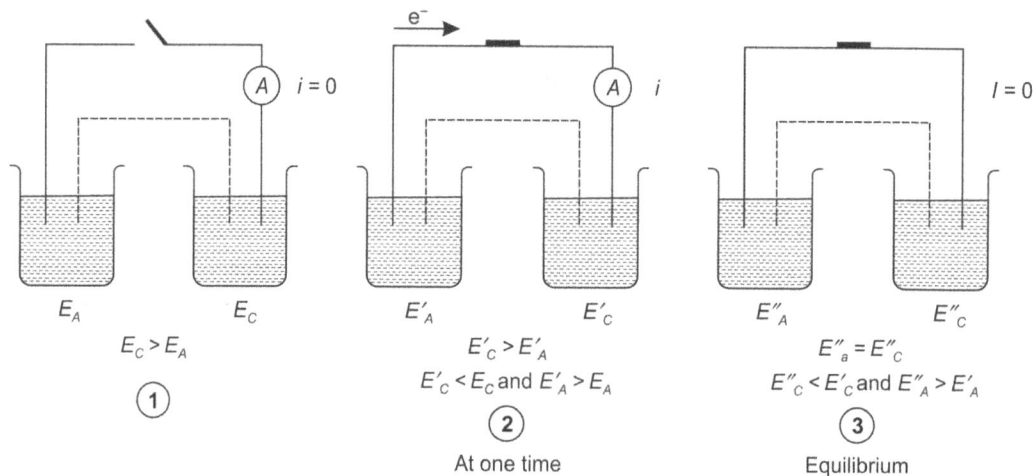

Figure 2: Difference Between Potential and Electric Current (E_A, E'_A, E''_A Anode Potentials – E_C, E'_C, E''_C Cathode Potentials) (Reprinted with Permission from Gwenola and Jean-Louis Burgot, Paris, Lavoisier, Tec et Doc, 2017, 47).

The electrons go from the electrode from a less positive to a more positive one (anode → cathode). The spontaneous process stops when, under the effect of the current, both electrodes possess the same electric potential. That is to say when the equilibrium state is reached.

Therefore, it is the difference of potential or a voltage which is at the origin of the current, that says of the displacement of the electrons.

Equilibrium of One Solute Between Two Liquid Phases

Let us again take the example of the preceding aqueous solution containing the solute M. Let us put it in presence of an organic solvent immiscible in water (they can be the toluene, benzene and carbon tetrachloride. They can be considered as wholly immiscible in water at a first approximation). One conceives that the Solute M partitions in both solvents. As in the case of the above electrochemical cell, one can imagine three times in the process of equilibrating:

- At time 1, all the solute is still in an aqueous solution;
- At time 2, one sees the passing of M into the organic solvent;
- At time 3, the solute is partitioned. One does not witness any longer partitioning of the solute into the organic solvent. The partitioning equilibrium is reached (Figure 3).

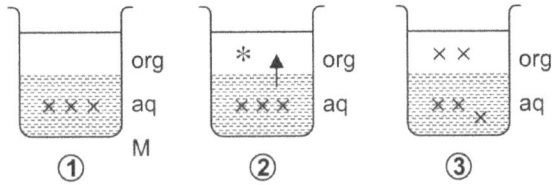

Figure 3: Establishing the Partitioning Equilibrium of a Solute M Between Two Phases (Reprinted with Permission from Gwenola and Jean-Louis Burgot, Paris, Lavoisier, Tec et Doc, 2017, 48).

The setting up of the state of equilibrium 3 of the solute with respect to the two phases is figured by the relation:

$$\overline{G}_{Maq} = \overline{G}_{Morg}$$

or equivalently;

$$\mu_{Maq} = \mu_{Morg}$$

The chemical potential of the solute is the same in both phases at equilibrium. In contrast, at the times 1 and 2, the equilibrium is not still established and:

$$\overline{G}_{Maq} \neq \overline{G}_{Morg}$$

More precisely:

$$\overline{G}_{Maq} > \overline{G}_{Morg}$$

In the initial state, in this case, the chemical potential is greater in the aqueous phase. It is the transfer of the solute that leads to the equality of the chemical potentials of the solute in both phases.

The partial molal Gibbs energy plays an identical part as the electrode potential. The analogy difference in partial molal Gibbs energy and difference in electrode potentials is perfect. There are these differences that spontaneously impose the displacements of the solutes and electrons. Hence, the name chemical potential. The greater they are, the greater tendency of the solute to go into another solvent. It is the same for its tendency to react chemically.

Thermodynamic Partition Coefficient

In thermodynamics—in order to reason with non-ideal solutions as with the ideal solutions or as with perfect gases—one introduces the notion of activity of each of the present species in the system. For species i, its activity is a_i and its chemical potential is μ_i. The latter is related to the activity by the relation:

$$\mu_i = \mu_i^{\circ} + RT\ln a_i$$

In the above equation, μ_i° is the chemical potential of i in its standard state, R is the constant of perfect gases and T is the absolute temperature of the system. The standard chemical potential is a constant for a species in a given solution, given temperature and pressure. The activity a_i is related to the concentration C_i of the species (whatever the scale chosen to express its concentration is) by a relation of the kind:

$$a_i = \gamma_i\, C_i$$

Here, γ_i is the activity coefficient of the species i. One must know that:

- $\gamma_i = 1$ for the ideal solutions. This case is rare. It occurs when the components of the solution are very close from a structural standpoint.

- $\gamma_i = 1$ for the very diluted solutions. Otherwise, γ_i changes with the concentration C_i of the species. The linear relationship between γ_i and C_i is, hence, fallacious.

Let us return to the preceding Step 3 when the partitioning equilibrium of M between water and the organic solvent is reached.

$$\mu_{Morg} = \mu_{Maq}$$

$$\mu_{M\,org}^{\circ} + RT\ln a_{Morg} = \mu_{M\,aq}^{\circ} + RT \ln a_{Maq}$$

$$\mu_{M\,aq}^{\circ} - \mu_{M\,org}^{\circ} = RT \ln (a_{Morg}/a_{Maq})$$

Finally, according to what has been said about the standard potentials:

$$a_{Morg}/a_{Maq} = \text{constant (equilibrium)}$$

$$a_{Morg}/a_{Maq} = P$$

P is the thermodynamic partition coefficient of M (between water and the organic solvent).

(*Remark*: It seems that there exist no normalized symbols for this quantity. One finds the symbols K, K_D, K', λ ... and so forth. According to the way it is followed to introduce it, one notices that P depends only on temperature, the pressure and the nature of the solute and the solvents.)

Partition Coefficient

The thermodynamic partition coefficient can also be written:

$$P = \gamma_{Morg} C_{Morg}/\gamma_{Maq} C_{Maq}$$

$$C_{Morg}/C_{Maq} = P \gamma_{Maq}/\gamma_{Morg}$$

One introduces the ratio K written in concentration terms:

$$C_{Morg}/C_{Maq} = K$$

K is simply called the partition coefficient or distribution coefficient. It is a constant, at a given temperature and pressure, but when the two phases are ideal solutions (that case is very rare) or are very diluted solutions. The activity coefficients in both phases then become equal to unity and:

$$K = P; \quad \text{ideal and very diluted solutions}$$

This relation is expressed graphically by a rectilinear line (Figure 4). Otherwise, it is not true.

It is clear that beyond some limits of concentrations, the partitioning isotherm becomes a curve. When the isotherm is a straight line, the distribution is said to be *regular*. The partition coefficients K and P take different values according to the scales of concentrations, molarities, molalities and molar fractions that are taken in order to quantify the activities.

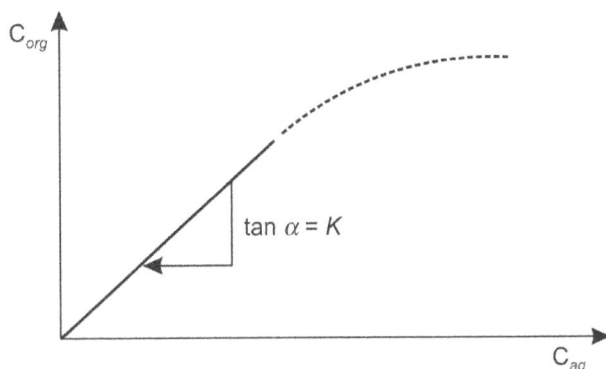

Figure 4: Partition Coefficient (Partition Isotherm) (Reprinted with Permission from Gwenola and Jean-Louis Burgot, Paris, Lavoisier, Tec et Doc, 2017, 51).

Partition Coefficient and Solubility

At a given temperature and pressure, when the solubility of the solute in both solvents is reached at partitioning equilibrium, their ratio is equal to the partitioning coefficient. When the solubilities are weak, it is then considered a thermodynamic coefficient. But the solubilities of the solute in both solvents, when the partition equilibrium is reached, are not obligatorily equal to their activities. In that case, it is when either the two solutions must be ideal or very dilute. This point must be still discussed. It is reconsidered in Chapter 7.

Berthelot- Jungfleich's Law

The relation of C_{Morg}/C_{Maq} = K brings the name of law of Berthelot[3]-Jungfleich[4] (1872). It has been established on a different basis than those given here. It has been established by Nernst (1891).[5]

It is exposed as follows:

whatever the solubility of a compound to partition, whatever the volumes of solvent used, at equilibrium the compound splits up in such a way that its concentrations in both solvents are in a constant ratio (Nernst (1891)).

Non-Validity of Berthelot-Jungfleich's Law

The respect of this law suffers a lot of vicissitudes, which can be easily explained by what has been said previously, that is:

- Expressed in terms of concentrations, it is not rigorous in thermodynamic terms. It might be expressed in terms of activities. Expressed in terms of concentrations, the two phases at equilibrium must be ideal or very dilute solutions.
- So for the law to be verified, the temperature and pressure must be constant.
- Also, the solute must be in the same physical and chemical states in the two solvents at equilibrium. This condition is by far less harmless than we can imagine.

---

[3] Marcellin Berthellot, French chemist, 1807–1907.
[4] Emile Jungfleich, French chemist, 1839–1916.
[5] Walther Nernst, German physical chemist, 1864–1941, nicknamed the father of the modern electrochemistry, Nobel prize in chemistry in 1920.

For example, the carboxylic acids, when they are dissolved in an organic solvent of weak relative permittivity, are in the form of dimers; whereas, in an aqueous solution, they are monomers (Figure 5).

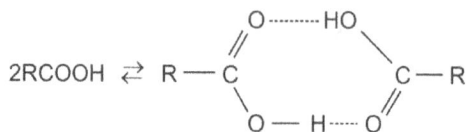

Figure 5: Carboxylic Acid and Dimer in a Solvent of Weak Relative Permittivity.

These are, of course, different states of carboxylic acids. To state another example as a dimer, let us mention that carboxylic acids may be ionized in the aqueous phase. This is another case in which the carboxylic acid is not the same in water as in the organic phase:

- There is no emulsion at the interface. The emulsion, due to surface phenomena, which must not be miscible, has the behavior of a third phase from which nothing is taken into account.

- The solvent must not be miscible. If it is the case, one must consider that one is in the presence of a third phase.

The occurrence of these deviations from the law results in an *irregular* distribution whose consequence is an isothermal partition curve that is no longer a straight line.

Case in Which Both Solvents Are Partially Miscible—The Ternary Diagram

The simplest case that we can encounter in the phenomena of liquid-liquid extraction is that in which:

- there is only one solute to extract;
- the solvent of the solution to be extracted, sometimes called the diluting solvent, is non-miscible with the extracting solvent.

Although the following studies are essentially situated within the frame of the immiscibility of both solvents, it is interesting to somewhat consider the case of partial miscibility of the diluting solvent and the extracting solvent and its resulting consequence. This study is all the more interesting as the partitioning of the solute between both liquids (partially miscible) has the consequence that leads it to appear in phases of variable compositions. The result is that the calculation of the yield of extraction, which is essentially the goal of the following studies devoted to the various types of extractions, becomes difficult. It is, however, facilitated by the examination of the ternary diagram described in Chapter 7.

CHAPTER 5

Dialysis
Osmosis and Reverse Osmosis

In dialysis, osmosis and reverse osmosis (and also in the liquid-liquid extractions) are one or several dissolved components that are transferred into a different liquid phase from which they are more easily recoverable. These methods are osmosis, dialysis and electrodialysis and these are used for the extractions with non-miscible solvents. In this chapter, we only consider osmosis, reverse osmosis, dialysis and electrodialysis.

Osmosis and Reverse Osmosis

Osmosis and reverse osmosis are methods of transferring one solute from a solution to another and both are separated by a membrane. These methods are governed by differences of chemical potentials that solute possesses in each solution. Osmosis leads directly to the notion of osmotic pressure, which is one of the colligative properties of solutions. A solution is defined as being colligative when its properties depend on the quantities of matter which constitute it rather than its nature.

Description of the Phenomenon

Let us consider the apparatus shown in Figure 1:

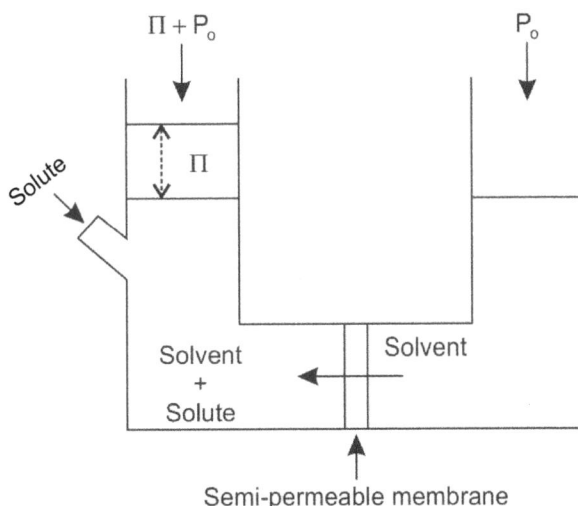

Figure 1: Apparatus Displaying the Occurrence of an Osmotic Pressure.

The apparatus has a U-shaped form. It is made up of two compartments that are separated by Membrane M. At the beginning of the experience, both compartments are filled with pure liquid 1, which we name the solvent. The solvent can traverse the membrane. One of the compartments (on the left side, for example) is equipped with an aperture that allows the addition of a solute. The solute cannot pass through the membrane. It stays confined in the left compartment all along the experience.

In the beginning, the solute is not added. The level of the solvent (in each compartment) is the same because they are under the same pressure P_o (viz., the Appendix 1). When the solute is added through the aperture, we notice an enhancement of the level of solution in the left column. There has been a moving of the solvent from the right to the left through the membrane. It is the phenomenon of osmosis. When the addition of the solute is finished, the level of the highest point of the solvent of the left column is no longer enhanced. The osmotic equilibrium that existed before the addition of solute and was broken during the latter is one more time established. At this point, the difference of height of both columns is equal to the hydrostatic pressure or *osmotic pressure* π. This phenomenon is reversible. It is sufficient to apply a mechanical pressure stronger than the osmotic pressure to the solution, which is most concentrated to reverse the sense of the flux of solvent. This is the phenomenon of reverse osmosis.

The Osmotic Pressure

From the quantitative point of view, the osmotic pressure π is given by the relation of van't Hoff:[6]

$$\pi = n_2 (RT/V)$$

where, n_2 is the number of moles of solutes at equilibrium, R is the constant of perfect gases, T is the absolute temperature of the system and V is the molar volume of the solvent in the pure state at this temperature. It must be clear that π is the supplement of pressure, which must be superimposed to the left compartment so that the whole system can be again at the initial pressure of Po. In other words, it is the pressure that must be exerted on the left compartment in reverse osmosis.

We can notice that π is in a linear relation with n_2. This is a colligative property. Finally, we also notice that the relation of van't Hoff is wholly comparable to that of the perfect gases, but this analogy is purely formal.

This mode of separation is strongly membrane-dependent. Generally, small molecules diffuse easily through membranes and can be separated without any great difficulty from colloids. Colloids are dispersions of small particles of one material into another. Here, small means less than about 500 nm diameter. They cannot pass through a membrane of cellulose. On the contrary, crystalloids can pass through this membrane.

It is interesting to consider how van't Hoff has obtained its relation. Its demonstration is purely thermodynamic by essence. It is mentioned in Appendix 1.

Dialysis

In this method, exchanges of matter are carried out through a semi-permeable membrane, which is traversed by the solvents and the small ions and molecules but not by the great species. The

---

[6] Van't Hoff, J.H., Dutch chemist, 1852–1911, Nobel prize for Physics, 1933.

crystalloids are separable from the colloids by these membranes. The dialyzers are constituted of two concentric Crystallizers A and B. The crystallizer B is without a bottom (Figure 2):

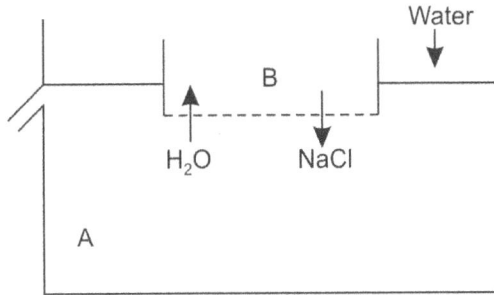

Figure 2: Scheme of a Dialyzer Apparatus for Dialysis.

Crystallizer B without a bottom is closed by a porous membrane (cellophane or collodion). It contains the liquid which must be dialyzed that usually contains the studied product, proteins and NaCl. It is plunged into Crystallizer A through which a stream of pure water circulates. The crystallizer A charges sodium chloride and proteins.

A drawback of dialysis is the fact that this method is very slow. Its duration may last several days. The solution to be dialyzed is diluted by water that traverses the membrane as the non-colloidal species do but in an opposite sense.

Electrodialysis

In this methodology, a potential difference is applied to two points of a solution that for effect directs the migration of ions, the anions towards the anode and the cations towards the cathode. The presence of a membrane permeable to water and ions permits the phenomenon to persist. It also permits it to accelerate the extraction of electrolytes.

The solution to dialyze is placed between the anodic and cathodic compartments from which they are separated by a permeable membrane (cellophane) (Figure 3).

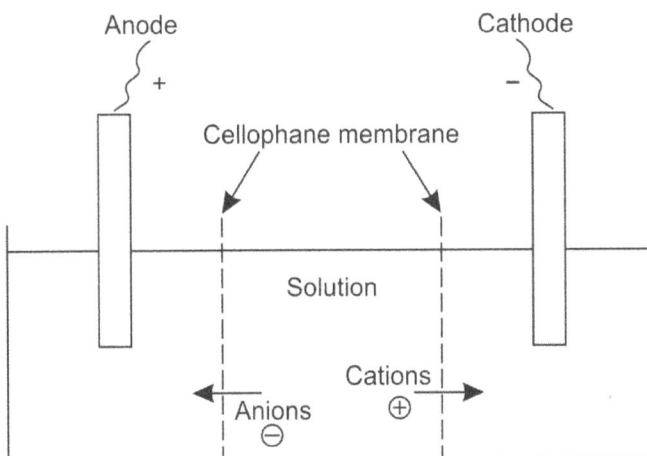

Figure 3: Apparatus for Electrodialysis.

One operates with a direct current of 120 V under an intensity of about one hundred milliamperes. The colloidal solution must be electrolyzed to a pH close to the isoelectric point so that the rate of dialysis could be maximal. Electrodialysis is used in toxicology to quickly extract some toxic ions from complex biological mixtures and to obtain several colloidal substances such as some proteins.

CHAPTER 6

Generalities Upon the Liquid-Liquid Extractions

Some constituents can be extracted from their mixtures with the help of a solvent. The operation is named liquid-liquid extraction when the mixture to be treated is a solution forming one phase. We have already given some explanations concerning solid-liquid extraction. The phenomenon has been referred to as the extraction by a liquid of a mixture containing solid (Chapters 1 and 2). We shall not study further this kind of extraction. From a general standpoint, we only consider the extraction of a solute S from its initial homogeneous solution in solvent A by solvent B (Figure 1).

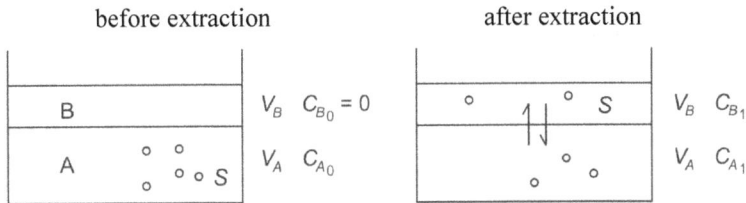

Figure 1: One Simple Extraction and the Position Before the Extraction.

There are several modalities to carry out liquid-liquid extraction. In this chapter, we devote ourselves to:

- general considerations concerning the liquid-liquid extraction;
- we still introduce the concept of partition coefficient by starting from the solubilities of the solute to partition in the two solvents;
- we consider the liquid-liquid extraction from the viewpoint of the phase rule;
- Finally, we pay attention to the treatment of ternary systems.

The detailed studies of one or several stages of partition equilibria, together with those of the different modalities are studied in the next chapters.

General Considerations

From now on, we can emit the following commentaries:

- The added solvent B must provoke the formation of a new phase. The passing of the solute S from the initial mixture into this new phase constitutes the transfer of matter.
- The solubility of solvent B in the initial solution A (to be extracted), together with that of A in which the solute S was initially dissolved, are two important parameters for the success of the

extraction. Evidently, the operation is interesting in the case where A and B are poorly mutually soluble and when the separation of the solute from solvent B is markedly easier than that of the solute from A.

- If there are several solutes, the solvent must be as selective as possible. That is to say, the partition coefficient of the solute between the two phases must be in favor of solvent B rather than in favor of A.

- The extraction can be the result of one or two processes. One is a purely physical one. It is based on the dissolution. The second is grounded on some chemical reactions such as those of the solute with the solvent. For several extractions, the two processes may play a part simultaneously.

- One extraction by solvent entails two operations:

  - Switching of the mixture to the solvent. One must keep their contact as long as possible in order to obtain an equilibrium state between the two phases.

- The separation of the two phases gives:

  - The solution that contains the solute and is rich in the solvent. It is called the *extractum*.

  - The residue is another liquid coat (which may also be a solid residue). It is called the *refined*.

The whole *two* operations constitute *one unit* or *one stage of extraction*. This stage is said to be ideal when the partitioning equilibrium is reached during it. The preceding operations must be completed by the recovery of the solvent.

Its setting to work implicates the borrowing of judicious modalities according to the investigated partitioning. Among these modalities, perhaps the most important is the pH value; above all when one of both solvents is water and the solute does possess an acid-basic character. These conditions are equally studied later.

More About the Thermodynamic Partitioning Coefficient—Theoretical Considerations

General Study

Thermodynamic Partitioning Coefficient

According to some authors, the thermodynamic partitioning coefficient is equal to the ratio of its Solubilities S_1 and S_2 in both solvents (the diluent and the extractor):

$$K = S_1/S_2$$

This assertion is not always right. In principle, even, it is false since the question here is to consider the thermodynamic partitioning coefficient. It is not false when, at the partitioning equilibrium, both solvents are ideal solutions or very diluted solutions; in other words, when the solubilities are equal to the activities. Let us recall that the solubility of one solute in one solvent is the maximal concentration and it can reach it at a given temperature:

$$S = C_{max}$$

One activity is equal to the product of its concentration by its activity coefficient γ for each solvent. Since solubility is a concentration, there exists the following relation between both quantities:

$$S = a/\gamma$$

One sees, therefore, that the previous sentence is only correct when both activity coefficients (the same solute has one different activity coefficient per solvent) are equal to unity; that is to say, at equilibrium partitioning, both solvents are ideal solutions or are very dilute solutions. This is unusual. Besides, it is sufficient that one of the phases at equilibrium is not ideal for the preceding proposition concerning the solubilities to be false.

General Theoretical Study

It is interesting to situate the principle of the method of *batch extraction* with respect to the phase rule. Batch means that all the substance to extract is together in the same volume of solvent at the beginning of the experiment (more will be explained later). The variance v (or the number of degrees of liberty of the system) is given by the relation:

$$v = C + 2 - \varphi$$

where φ is the number of phases and C is the number of components. In the present case, $\varphi = 2$ and $C = 3$. Three is the number of factors that can play a part in the equilibrium. They can be, for example, the concentration of solute, the pressure and temperature.

Other Systems

A more complicated system is that constituted by the ternary mixture. This case can be studied with the help of *triangular diagrams* also called *ternary diagrams*. They also give the composition of each of the constituents. For this reason, they are named the *miscibility diagrams*.

The system we are studying is constituted by the ternary mixture A, B and C and is represented by the equilateral triangle in Figure 2. The A and C represent the two solvents (diluting and extracting) mutually miscible (in this general study), and B is the solute:

- Each of the three apexes of the triangle represents one of the components pure under the form of 100% of the total weight of the whole constituents. As a result, it represents 0% of the two other components.
- The three sides of the triangle represent the three binary mixtures:

  Diluting solvent plus solute; AB

  Extractive solvent plus solute; BC

  Extractive and diluting solvent; CA

By dividing each of these sides into 100 units, one directly reads the proportions of each of the two members of the binary mixture:

- Each point M inside the triangle represents all the possibilities of concentrations of the three components. Each horizontal line parallel to AC corresponds to 0% of compound B (straight line AC itself) until 100% at point B itself. In the same manner, at point M, the system contains 60% of compound B. It is the same for the parallels to the sides AB and BC. At point M, there exists 10% of C and 30% of A.
- On each line coming from each of the summits until any point of the opposite side, all the points correspond to identical percentages of both opposite compounds at the origin of the line (see the line DC for a proportion: 1/3).
- Each point belonging to a parallel linear of a side corresponds to the same percentage of the component located on the opposite.

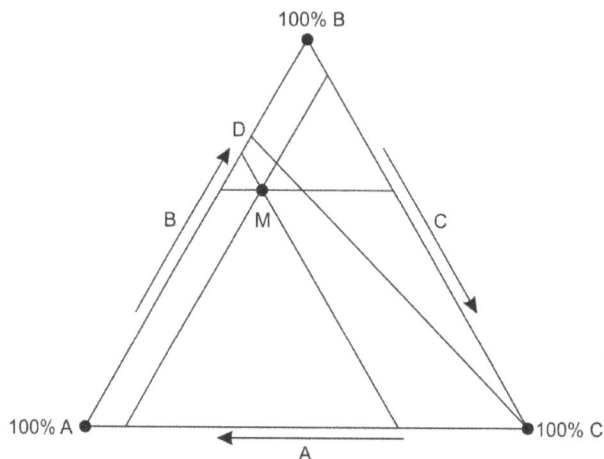

Figure 2: Properties of a Ternary Mixture Represented by a Triangular Diagram.

The interesting case concerning us is that in which two of the three components of the system form two phases of different compositions, which are only partially miscible. The classical case is that of the ternary, water A, ethanol B and benzene C (Figure 3). In our case, one knows that water and benzene are only poorly miscible. The system constituted by these two components leads to two phases. One is solvent benzene C saturated in diluent water A and the other by diluent water A saturated by solvent benzene C. The line AC describes binary mixtures of solvent benzene and diluent Water A. The a and c are respectively within the limit solubilities of solvent S in diluent A and that of diluent A in solvent S (at the working temperature) (Figure 3).

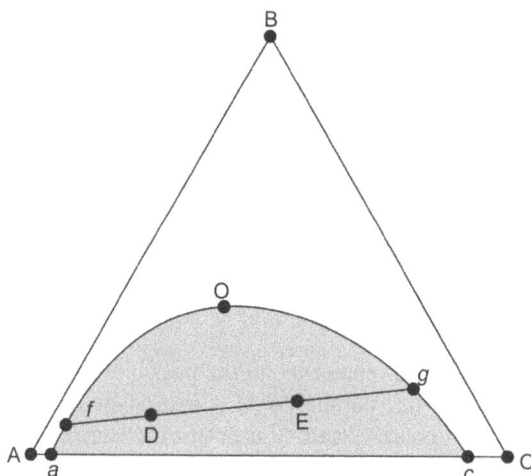

Figure 3: The Ternary System: Solvent C (benzene) – Solute B (Ethanol)-Diluent A (Water).

The dark area of Figure 3 represents the zone of non-miscibility or zone of heterogeneity. The curve which separates the two zones (miscibility and non-miscibility) is called the *binodal curve*. Let us consider the two systems represented by points D and E. They give rise to two phases non-miscible of composition *f* and *g*. More precisely, the systems D and E give rise to two phases which are in the ratio Dg/Df in weight. We can notice that the lines of the type fg (lines called lines of the link) are not necessarily parallel to a side. They are also called *saturation lines* or *partition lines*. The points *f* and *g* are said conjugated. Point O is located at the summit of the binodal curve and is conjugated with itself. It is called the *critical point*.

It is evident that it is obligatory to be placed in the zone of non-miscibility. If not, the extraction is not possible. It is clear that after extraction, the extractive phase still contains some diluting solvent.

CHAPTER 7

One Stage of Liquid-Liquid Extraction
Batch Extraction

In this chapter, we only treat the case of one simple extraction. This method is also called *batch extraction*. The whole process may be symbolized by Figure 1.

1 = Initial Equilibrium   2 = Partitioning   3 = Final Equilibrium

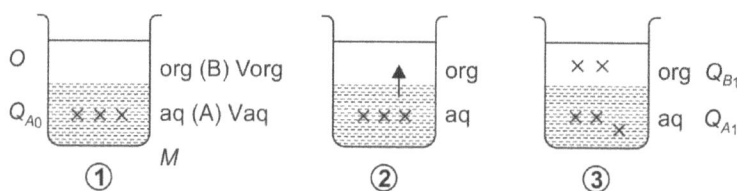

Figure 1: The Setting up of the Partition Equilibrium of a Solute Between Two Phases.

The initial phase is a solution containing the substance S present in solvent A. The studied process is that of the extraction of S by solvent B in principle which is fully immiscible in A. After an energic stirring period followed by a time-interval during which the system remains at rest so that the decantation can be realized, one obtains:

- Phase A which is poorer in S than before;
- Phase B enriched with S.

The whole of both operations (stirring and separation) is called one theoretical stage.

Study in the Case of One Stage of Extraction

In principle, pressure and temperature are factors of equilibrium of partitioning. However, the partitioning coefficient depends weakly on the pressure. One can also investigate the case in which the partitioning is carried out at a constant temperature. In these conditions, the system becomes univariant. If, for example, the concentration of the solute to be extracted is known, the partitioning curve is easily obtained. It can be obtained graphically or by calculation:

Let us call it Q_{Ao}, which is the total number of moles of substance S, V_B and V_A the volumes of the two phases. The parameters to calculate are:

- The concentrations C_{A1} and C_{B1} after extraction
- The quantities of solute Q_{A1} and Q_{B1} after the extraction

- The yield ρ of the extraction defined by the ratio:

ρ = quantity of extracted substance/quantity of substance initially present

$\rho = Q_{B1}/Q_{AO}$

Evidently, $0 \leq \rho \leq 1$.

The known parameters permitting the calculation are Q_{Ao}, V_{A1}, V_{B1}, and K.

Calculation for a Regular Distribution

From a general standpoint, to realize the calculations in this domain of chemistry and particularly that of chemical engineering, one systematically and simultaneously uses the equations expressing the conservation of the matter and the equilibrium partitioning, that is to say:

$$C_{B1}/C_{A1} = K \text{ (equilibrium)}$$

$$Q_{AO} = Q_{A1} + Q_{B1} \text{ (conservation of the matter)} \tag{1}$$

It is important to understand that it is at the level of this second equation that occurs the principal *simplification* of this calculation. Indeed, the partition of the solute is taken into account. But often solvent A, an association A-solute, may also undergo partition. Another group of equations must be handled in this case. By definition of a molar concentration, one finds:

$$C_{B1} = Q_{B1}/V_B \text{ and } C_{A1} = Q_{A1}/V_A$$

As a result:

$$Q_{B1}/Q_{A1} = K \, V_B/V_A \tag{2}$$

(For more, see upcoming chapters)

$$Q_B/Q_A = \alpha$$

Here, K (V_B/V_A) is a new constant known according to the data of the experience. It is symbolized by α. It is called the *corrected partition coefficient*. This is a notion that we can also find in chromatography (the term 'corrected' implies corrected from volumes). Equations (1) and (2) constitute a system of two equations of two unknowns, Q_B and Q_A, the resolution of which gives:

$$C_{A1} = C_{Ao}[1/(1 + \alpha)]$$

$$C_{B1} = C_{Ao}[K/(1 + \alpha)]$$

$$\rho = \alpha/(1 + \alpha) \text{ or } \rho = 1 - 1/(1 + \alpha)$$

The second formula giving the yield shows that it can attain 1 when $\alpha \to \infty$. This means that:

- $K \to \infty$. It is difficult to modify its value at our will since this parameter depends only on the conditions of the experience.
- $V_B \to \infty$. This is a condition quite impossible to reach for reasons of volume and cost.
- $V_A = 0$. This value is located out of the values experimentally possible.

In the same kind of reasoning, one notices that the first relation expressing the yield shows that the greater is α, the closer from 1 is the yield. This means the simplest way to increase the yield is to increase the ratio V_B/V_A. By modulating these two parameters, one can obtain a large interval of changes of α. If one corrects the value (only slightly modifiable) of K by varying the values of this ratio, then the name of "corrected partitioning coefficient" is given to α.

It is interesting to notice that when α increases, the extracted quantity Q_B increases and hence the yield. But at the same time, the concentration of the extractum C_{B1} decreases. It is impossible to simultaneously increase the quantity of extracted substance and its concentration in the extractive phase.

(*Remark*: This result may at first appear to be paradoxical. Actually, the reason for this apparent contradiction lies in the fact that a concentration has not the same definition as a quantity of matter. A concentration is a quantity of matter related to its volume.)

Determination for an Irregular Distribution

The simplest way to determine the same parameters as previously is to proceed graphically. The equation of conservation of the solute is:

$$Q_{Ao} = Q_{A1} + Q_{B1}$$

or

$$C_{A1}V_A + C_{B1}V_B = C_{Ao}V_A$$

$$C_{B1}V_B = -C_{A1}V_A + C_{Ao}V_A$$

The point of coordinates C_{A1} and C_{B1} can be considered as a point of the straight line, the equation of which is:

$$C_B = -(V_A/V_B)\, C_A + (V_A/V_B)\, C_{Ao}$$

It is the equation of a straight line, the slope of which is $-(V_A/V_B)$ and the intercept is (V_A/V_B) C_{Ao}. We notice in passing that $C_B = 0$ for $C_A = C_{Ao}$.

This result is in accordance with the initial data. Hence, on a diagram C_A/C_B, one draws:

- The isothermal curve C_B/C_A after having determined experimentally some concentrations C_{Bi} as a function of C_{Ai}; *i* being the index of the chosen points. It is the curve that represents the *partitioning equilibrium*.

- The straight line of the conservation of matter. It is drawn very easily since one knows C_{Ao} and $-(V_A/V_B)$. The two curves are represented (Figure 2):

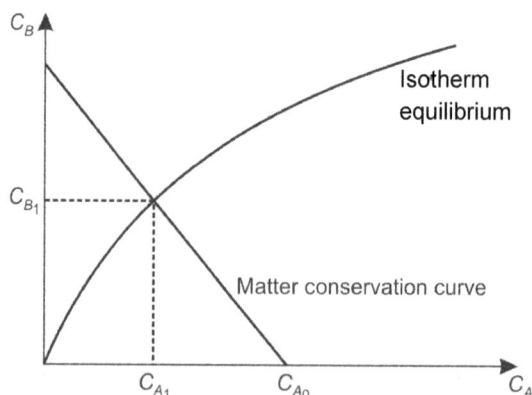

Figure 2: Graphical Determination of the Concentrations in the Two Phases After a Simple Extraction in the Case of an Irregular Distribution. (Reprinted with Permission from Gwenola and Jean-Louis Burgot, Paris, Lavoisier, Tec et Doc, 2017, 59).

We notice that the curve of equilibrium is not a straight line. This is in agreement with the hypothesis. Since the equilibrium condition and the equations of conservation must be verified simultaneously, the concentrations C_{A1} and C_{B1} must be obligatorily and be at the cutting point of the partitioning equilibrium curve and of the line of conservation of matter. Once the concentrations are determined as explained, it is easy to calculate Q_{B1} and Q_{A1} and ρ as we know V_A, V_B and Q_{Ao}.

Example

It is an example in which the extraction is regular. It concerns the pharmaceutical industry. The matter is the extraction of an antibiotic in a culture medium at pH = 2.2 by amyl acetate. At this pH, the partition coefficient is:

$$K(\text{org/aq}) = 15$$

The concentration of the phase which must be extracted is $C_{Ao} = 1{,}000$ units/cm^3. To carry out the experience, one chooses the ratio $V_B/V_A = 1/5$.

(*Remark*: The word unity is not well chosen. Here, it does not designate a unity of concentration but a quantity of matter.)

According to the preceding equations and data, one calculates:

$$\alpha = 15 \times 1/5 = 3$$

$$C_{B1} = 1{,}000 \times [15(1 + 3)]$$

$$C_{B1} = 3{,}750 \text{ units/cm}^3$$

$$C_{A1} = 1000 \times [1(1 + 3)]$$

$$C_{A1} = 250 \text{ units/cm}^3$$

$$\rho = 3/4 = 0.75$$

These results call for the following comments:

- The interest in the extraction appears clearly. One obtains one solution more concentrated than the initial. $C_{B1} = 3{,}750$ units/cm^3 for $C_{Ao} = 1{,}000$ units/cm^3. The interest is located in the fact that numerous apparatuses of instrumental quantitative analysis respond to the concentrations.
- The question which might be asked is the following: what would be the obtained results if we had used three times the preceding extractive volume? With $V_B/V_A = 3/5$, we could obtain $\alpha = 9$, $C_{B1} = 1{,}500$ units/cm^3, $C_{A1} = 100$ units/cm^3, and $\rho = 0.90$.

As we have already seen in the quantitative study, the yield is frankly better by augmenting α. However, the concentration of the extractum is weaker than previously $C_{B1} = 1{,}500$ units/cm^3 for 3,750 units/cm^3. By multiplying by three the volume of the extractive solvent, the yield is enhanced but it is not triplicated. Because of this result, some authors infer that the extractive power of the solvent is less strong when it is used a volume more and more important.

Practical Methods

In the laboratory, the simple extraction is usually carried out with separating funnels. In industry, the agitation is done in a mixer and the separation in a decanter which is distinct from the preceding device. Some apparatuses carry out the two functions.

Separation and Fractionation

Let us recall that one separation is an operation in which the goal is to isolate each substance constituting the initial mixture into distinct fractions of solvent. We shall see that one simple extraction permits in some cases to achieve a beginning of separation. One also says a beginning of fractionation (mentioned below).

Let us consider the case in which there exist two extractible substances. Since they have different chemical structures, their partitioning coefficients are different except for a numerical accident which is hardly probable. Hence, one can conceive that in the extractum, the concentrations of both solutes are no longer the same as in the initial solution. The extractum is enriched with one of the two initial substances, whereas the refined is enriched with the other. One has achieved a beginning of separation.

As an example, let us suppose that the initial solution contains antibiotics I and II of an initial concentration of 1,000 units and respective partitioning coefficients 15 and 5. The ratio of volumes used is V_{org}/V_{aq} which is 1/5. Calculations analogous to the previous ones give results as below:

- for Antibiotic I:

  $\alpha = 15 \times 1/5 = 3$; $C_{B1\,I} = 3{,}750$ units/cm^3;

  $C_{A1\,I} = 250$ units/cm^3

- for Antibiotic II:

  $\alpha = 1$; $C_{B1\,II} = 2{,}500$ units/cm^3;

  $C_{A1\,II} = 500$ units/cm^3

With respect to the initial time, phase I is enriched with compound I. Concerning now the refined, it is enriched with compound II. However, these are relative enrichments expressed in terms of concentrations. A beginning of separation has been achieved.

We shall see later that there exist separation processes more effective (Chapter 9) devoted to the counter-current separation.

Terms Used in Extraction

For several authors, the term fractionation is synonymous with separation. It is the meaning we are keeping. For other authors, the term possesses a more precise meaning. Let us consider a mixture of two solutes *a* and *b* in the initial solvent A that we extract with solvent B. Let us also suppose that *a* is insoluble in B, whereas *b* is soluble. The operation of extraction permits to obtain *b* in its pure state. This is referred to as *extraction with fractionation*. When several constituents of the initial solution are extracted, they are in ratios different from those found in the initial solution after extraction because their partitioning coefficients are different. The result is an enrichment of the phases with one product or with another. This enrichment can be besides pronounced by the beginning again of the extraction operation (Chapter 8). Therefore, one speaks of extraction with fractionation.

Distribution Ratio

Another term encountered in the domain of extractions is that of *distribution ratio D*. It is encountered when the distributing species get chemical interactions with the other components in each phase. We shall not see further this notion since we have decided to treat this point in another part of the syllabus of analytical chemistry that is devoted to ionic equilibria. Nevertheless, these chemical interactions are important in analytical chemistry since they can profoundly affect the concentrations (and the affinities) of the distributing species. Hence, it became necessary to introduce a more general quantity than the partitioning coefficient to describe the extraction. It is called the distribution ratio, and it is defined by the relation:

D_r = Total concentration in organic phase/Total concentration in aqueous phase at partition equilibrium.

This is a stoichiometric ratio including all species of the same component in the respective phases. (The nature "aqueous" is one of the two immiscible phases that is specified because an ionizing solvent must be present in order for ions to be formed).

Interest in the Liquid-Liquid Extraction

The liquid-liquid extraction replaces the distillation (see Chapter12) when:

- The less volatile compound is in small quantities in the mixture submitted to the distillation.
- The compound to extract is thermosensitive. It deteriorates under the action of heat.
- The relative volatility of the product submitted to the distillation is close to unity. In this case, the separation is impossible.
- The initial mixture gives an azeotrope. That is to say, the separation by distillation is impossible.

It must not be forgotten that after the extraction, the obtained mixture must be most often distilled. This can be a drawback. The choice of the extractive solvent is particularly important.

CHAPTER 8

Crosscurrent Extraction or Repetitive Batchwise Extraction by an Immiscible Solvent

Often, only one operation extraction is not sufficient. Moreover, we know that after one extraction, some quantity of solute remains in the refined since the yield of the extraction can never reach unity. Finally, from the economic standpoint, it must not be forgotten that if the yield is better when the volume of extractive solvent used is greater, its solving property decreases and it may become prohibitive to use great quantities of solvent. Hence, the idea is to repeat the simple extraction several times, and each time new parts of fresh solvent are used. This is referred to as *repetitive batchwise extraction* of a solute by an immiscible solvent. The method is also called *Crosscurrent extraction*.

In this chapter, we only study this methodology from a simplified viewpoint. That is to say, we consider that diluting and extracting solvents are immiscible.

Principle and Notations

As previously, one carries out a first extraction of the Volume V_A of Phase A to be extracted with a Volume V_B of pure Solvent B. After separation of the refined in which it remains the quantity Q_{A1} of substance and from the extractum that contains Q_{B1}, the latter is kept apart (Figure 1):

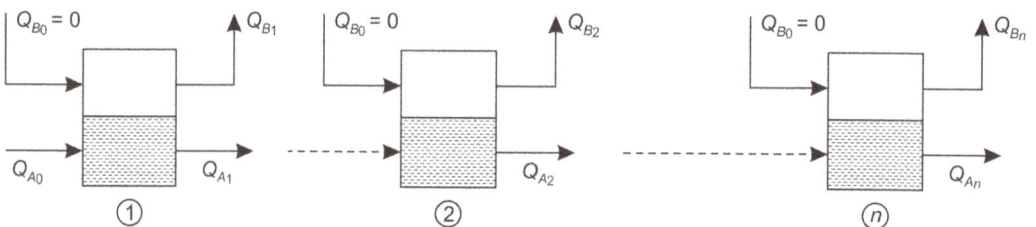

Figure 1: Principle of Crosscurrent Extraction. (Reprinted with Permission from Gwenola and Jean-Louis Burgot, Paris, Lavoisier, Tec et Doc, 2017, 71).

The whole refined comes from the first stage onward. Namely, Volume V_A containing still Q_{A1} is once more extracted by Volume V_B of the extractive pure solvent in the second stage. (Usually, $V_{B2} = V_{B1}$ and .... $V_{Bn} = V_{Bn-1}$). After the separation of phases, Extractum 2 is added to Extractum 1. Then, one carries out the third stage of extraction that applies to the whole refined of Stage 2 with the same Volume V_B of pure extractive solvent and the operation is renewed until the nth stage. Finally, one obtains an nth refined, which is probably still containing a weak quantity of non-extracted solute and a total extractum ΣQ_{Bn} that is constituted by the whole of the n extracta.

From the standpoint of the notations, C_{An} is the remaining concentration in the refined after the nth operation of extraction and C_{Bn} that of the corresponding extractum. It is important to remember that phases that are mutually in partitioning equilibrium possess the same index n.

Quantitative Study

It must be permitted to calculate the yield of the extraction as a function of the initial data and also the final quantities and concentrations in the different phases.

Parameters to be Calculated

- The yield ρ defined by the relation:

  ρ = total quantity extracted (number of moles)/number of moles initially present

  $\rho = \Sigma Q_{Bn}/Q_{Ao}$
- Q_{Bn}, which is the quantity of a substance that is not extracted and remaining in the last refined
- The concentrations in the total extractum $\Sigma Q_{Bn}/\Sigma V_B$ ($\Sigma V_B = n V_B$)
- Accessorily, the different C_{An} and C_{Bn}.

The *initial* data permitting the calculations are C_{Ao}, Q_{Bo}, V_A, V_B, K, and n. Eventually, the coefficient K can be replaced by the distribution ratio D_r, provided it is constant in the case of this study.

Calculations for a Regular Distribution

Again, in this case, the problem is solved by drawing simultaneously the equations that express the partition equilibria and the conservation of the matter.

Conservation of the Matter

Evidently, each extraction corresponds to one stage of extraction. The basic principle is that there exists no accumulation of matter in each stage. The number of moles (of the compound to extract) that enters a stage is equal to the number of moles coming from it. We can write:

$$Q_{Ao} = Q_{B1} + Q_{A1}$$

For the second stage:

$$Q_{A1} = Q_{A2} + Q_{B2}$$

For the nth stage:

$$Q_{An-1} = Q_{An} + Q_{Bn}$$

By calculating the sums of these equations corresponding to the result of all the different stages, we find

$$Q_{Ao} = Q_{An} + \Sigma Q_{Bn} \tag{1}$$

This is the *equation of conservation* of matter.

Condition of the Equilibrium

Since by hypothesis, the distribution is regular and at each stage, the volumes V_D and V_L are always the same by the hypothesis, then:

$$K = C_{A1}/C_{B1} = C_{A2}/C_{B2} = \ldots\ldots C_{An}/C_{Bn}$$

and

$$C_{B1}V_B/C_{A1}V_A = C_{B2}V_B/C_{A2}V_A = C_{Bn}V_B/C_{An}V_A = \alpha$$

$$\alpha = Q_{B1}/Q_{A1} = Q_{B2}/Q_{A2} = \ldots\ldots\ldots\ldots Q_{Bn}/Q_{An}$$

In order to establish the condition of equilibrium, it is sufficient to consider that each stage is equivalent to a simple extraction. Only the initial quantities are different from one stage to the other. According to the results of the preceding chapters, the values are found:

- After the first extraction:

$$Q_{A1} = Q_{Ao}[1/(1 + \alpha)]$$

- After the second extraction:

$$Q_{A2} = Q_{A1}[1/(1 + \alpha)]$$

The quantity that enters the second stage is no longer Q_{Ao} but Q_{A1}. Taking into account the preceding expression of Q_{A1}, we find:

$$Q_{A2} = Q_{Ao}[1/(1 + \alpha)]^2$$

- After the nth extraction, it is evident that:

$$Q_{An} = Q_{Ao}[1/(1 + \alpha)]^n \tag{2}$$

It is the *condition of equilibrium*. Starting from relations (3) and (4), we obtain:

$$\Sigma Q_B = Q_{Do}\{1 - [1/(1 + \alpha)]^n\}$$

$$\rho = \{1 - [1/(1 + \alpha)]^n\} \tag{3}$$

Now, concerning the calculation of concentrations, we obtain immediately:

$$C_{An} = C_{Ao}(1/(1 + \alpha)^n)$$

$$C_{Bn} = C_{Ao}K(1 + \alpha)^n \tag{4}$$

The quantity of substance in the nth extractum as $Q_{Bn}/Q_{An} = \alpha$ is:

$$Q_{Bn} = Q_{Ao}\alpha/(1 + \alpha)^n$$

$$C_\Sigma = \Sigma Q_B/\Sigma V_B$$

$$C_\Sigma = Q_{Ao}[1 - (1/(1 + \alpha)^n]/nV_B \tag{5}$$

- Equation (5) permits the calculation of the yield that shows that the latter increases with the number of n extractions. The term $[1/(1 + \alpha)]^n$ becomes indeed weaker as the number of n increases.
- There is a very interesting comparison to do. It consists in comparing the obtained yields. On one hand, it carries out one simple extraction with a volume V_B. On the other, we carry

out extraction at n stages with a total volume of solution V_B. That is to say by small portions of volumes of V_B/n. So, the comparison makes sense as one must start from the same initial quantity of solute and the same volume V_A. In the simple extraction, the yield is:

$$\rho = \alpha/(1 + \alpha)$$

$$\rho = K(V_B/V_A)/[1 + K(V_B/V_A)]$$

$$\rho = 1 - 1/(1 + KV_B/V_A)$$

For the extraction at n stages:

$$\rho' = 1 - 1/[1 + K(V_B/n)/V_A]^n$$

Let us study the term $[1 + K(V_B/n)/V_A]^n$. By developing it according to the formula of Newton's binomial relation, one obtains an expression of the kind:

$$1 + KV_B/V_A + ...\text{positive terms}$$

It follows that:

$$[1 + K(V_B/n)/V_A]^n > 1 + K(V_B/V_A)$$

and:

$$\rho' > \rho \qquad (6)$$

Moreover, the inequality is all the more effective as n is higher. Hence, the very important conclusion is that if we want to extract a substance by a given fixed volume of extractive solvent, there is an interest to divide this volume into n portions, where n is as great as possible in order to enhance it. As the volume of extractive solvent is limited, even if we choose n $\rightarrow \infty$, the yield cannot be equal to unity. It tends toward the limit $[1 - \exp(-KV_B/V_A)]$. In order to demonstrate this point, it is sufficient:

- To develop according to McLaurin $(1 + x)^n$, to replace x by $K(V_B/n/V_A)$ and make the obtained expression tend toward infinity. We obtain:

$$(1 + K(V_B/V_A))^n = 1 + KV_B/V_A + (K^2/2) V_B^2/V_D^2 +$$

$$(K^3/3!)V_L^3/V_D^3 + ...$$

- On the other hand, to develop in series e^x, $x = KV_B/V_A$. One obtains:

$$e^{KV_B/V_A} = 1 + KV_B/V_A + (K^2/2) V_B^2/V_A^2 + (K^3/3!) V_B^3/V_A^3 +...$$

- Considering the relation $Q_{Bn} = Q_{Ao}[\alpha/(1 + \alpha)^n]$, we can clearly see that the extracted quantity becomes weaker when n increases since $1 + \alpha > 1$. An equivalent behavior should be obtained if, at each stage, the solvent would lose its solvating power.

- There exists in the literature some charts that for a given constant K and ratio V_{Btotal}/V_A, they immediately give the yield ρ as a function of the number n of stages.

Irregular Distribution

Here also, the problem is solved geometrically. One draws the equilibrium curve after having experimentally determined several couples x_i–y_i that are points of this curve. One also draws the equation of conservation of the matter. One proceeds as in the case of extraction at one stage (Figure 2):

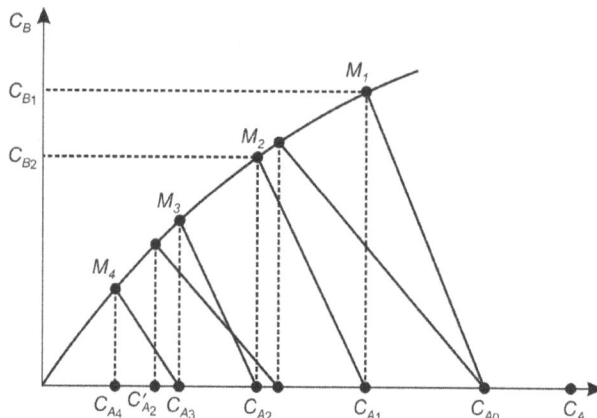

Figure 2: Multiple Extractions and Irregular Distribution. (Reprinted with Permission from Gwenola and Jean-Louis Burgot, Paris, Lavoisier, Tec et Doc, 2017, 76).

Then, one has the curves corresponding to both relations:

$$C_{B1}V_B + C_{A1}V_A = C_{Ao}V_A$$

$$C_B = -(V_A/V_B)C_A + C_{Ao}(V_A/V_B)$$

One determines the point M_1, the coordinates of which are C_{A1} and C_{B1}. Then, one traces the straight line of the conservation of the matter applied to the second stage:

$$C_{B2}V_B + C_{A2}V_A = C_{A1}V_A$$

$$C_B = -(V_A/V_B)C_A + C_{A1}(V_A/V_B)$$

It is parallel to the preceding straight line since their slopes $-V_A/V_B$ are the same. One immediately sees that for $C_B = 0$, $C_A = C_{Ao}$. This line is easily drawn as one knows one of its points and its slope. It intersects the equilibrium curve at point M_2, the coordinates of which are C_{A2} and C_{B2}. This construction is repeated as many times at these stages. The last point M_n has for coordinates C_{An} and C_{Bn}. One easily deduces Q_{An}, ΣQ_B and ρ. In order to do that, one must refer to the definitions of these parameters and not to the preceding calculations as K is varying.

This kind of graphical construction shows that for the same total volume of the extractive phase, it is preferable to work with as many stages as possible. For the preceding construction, there were four stages. For the same global volume of the extractive phase, if one extracts with a volume that is twofold greater, then there are only two stages left. Whereas the equilibrium curve would be the same as before, the straight lines in dotted points express the conservation of the matter that would present a slope that is two times weaker (refer to Figure 2). After two stages, one notices that it remains one concentration C'_{A2} markedly higher than the preceding concentration C'_{A4} that would be. Less solute has been extracted.

(*Remark*: The preceding quantitative studies have been carried out in the case where for each stage, the volume of extractive solvent used does possess the same value V_B. It is the same for the value of the refined volume V_A. This is the most common case. If this does not happen, that is to say, if V_A and V_B are different at each stage, the corrected partitioning coefficient does possess a different value $\alpha_1, \alpha_2 \ldots \alpha_n$ at each stage. Therefore, one must proceed to its study step by step by using equations of the simple extraction, that is to say, Q_{A1} and Q_{B1} with α_1, Q_{A2} and Q_{B2} with α_2, etc.

Finally, one obtains ΣQ_B and ρ with the help of equations:

$$\Sigma Q_B = Q_{Ao} - Q_A \text{ and } \rho = \Sigma Q_B/Q_{Ao}$$

With the graphical methodology, one observes that the straight lines representing the conservation of matter are no longer parallel since the slopes $-V_B/V_A$ are not constant from one stage to another.)

Example of Calculation

Let us again look at the example of the study of the extraction of antibiotics. Let us compare the obtained yields by carrying out, in the first case, the three successive extractions with, for each extraction, one volume of solvent that is equal to 1/5 of the volume of the phase, which must be extracted. In another methodology by proceeding to only one extraction with a ratio $V_B/V_A = 3/5$. That is to say, with the same total volume V_B that has been used previously. One recalls $K = 15$.

In the first case:

$$V_B/V_A = 1/5: \alpha = 15/5 = 3: n = 3$$

$$\rho = 1 - 1/(1 + 3)^3$$

$$\rho = 0.984$$

(*Remark*: Previously, for only one extraction with a ratio $V_B/V_A = 1/5$, the yield was 0.75. One notices a very strong enhancement of the yield, but it is at the expense of a three times larger consumption of solvent. Previously, for only one extraction with the ratio $V_B/V_A = 3/5$, that is to say with the same total volume V_B, one should find:

$$\rho = 0,90$$

These results illustrated nicely what we have said previously.)

Practical Points

In the Laboratory

Currently, one uses separating funnels. One can also operate according to a continual manner. The action of the extractive solvent is ceaseless. Then, an infinity of micro-pumping takes to rise by division of the extractive solvent into droplets, which traverse the solvent to be extracted and are recovered. In France, the most used apparatus is called the *perforateurs of Jalade*. The word *perforateur* has its origin in the fact that the droplets of the extractive solvent go through (perforate) the diluting solvent. These apparatuses are named Wehrli, Kutscher or Streudel's extractors, which is based on the fact that the extractive solvent is heavier or lighter than that to perforate. When it is denser, it traverses the solution from the top to the bottom to low. The *perforateur* is adapted to a balloon and is prolonged by a refluxing refrigerator (see Figure 3).

The solution under study is introduced into the *perforateur*, the extractive solvent into the balloon, and the latter is heated. The solvent distills, and its vapors are condensed in the refrigerating device that falls in fine droplets into the *perforateur*. They traverse the solution and then through a siphon, it falls again into the balloon.

The method has the advantage of reducing the quantity of solvent to use and the time necessary for the operation. The yield is often very good. It sometimes has the drawback to submit the extracted substances after it faces a long time of exposure to the heating. This can adulterate them.

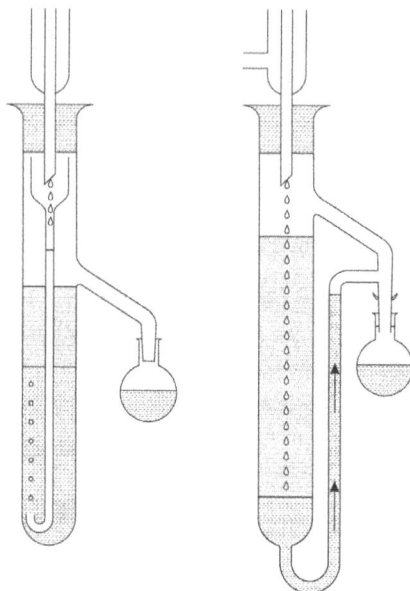

Figure 3: "Perforateurs of Jalade". (Reprinted with Permission from Gwenola and Jean-Louis Burgot, Paris, Lavoisier, Tec et Doc, 2017, 78).

In the Industry

One uses a set of "extractors-detectors". They are analogous to those used for a single extraction.

Extraction with Fractionation

Quite evidently, if there exist several solutes where the partitioning coefficients are different, one carries out one more or less important fractionation. This is because of the occurrence of the multiple successive steps, which is by far put forward than when there exists only one stage. Let us mention the example of both stereoisomers, the maleic and fumaric acids (Figure 4)

Fumaric acid   Maleic acid

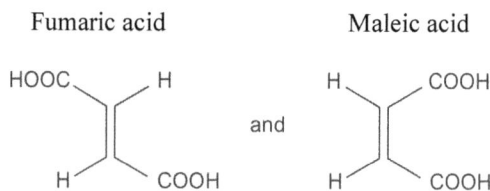

Figure 4: Structures of Fumaric and Maleic Acids. (Reprinted with Permission from Gwenola and Jean-Louis Burgot, Paris, Lavoisier, Tec et Doc, 2017, 79).

The partitioning coefficients of these compounds between water and ether are:

$$K_{mal} = 9.65 \quad \text{and} \quad K_{fum} = 0.90$$

These values are found for equal volumes of solvents. One shows in Table 1 gives the ratios at equilibrium C_{mal}/C_{fum} after n extractions.

Table 1: Ratios of the Concentrations of the Two Acids in Water After *n* Extractions.

n	$(C_{mal}/C_{fum})_{aq}$
0	1.0
1	5.6
2	31.6
3	178.0
4	1,000.0

Conclusion

Carrying out successive extractions improves the yield with respect to one simple extraction. However, the greater the number of stages, the less powerful is solvating of the extractive solvent. As a result of this drawback, there is an important dilution of the extracted substance. Otherwise, the fractionation by successive extractions may become interesting when the number of stages becomes important.

CHAPTER 9

Countercurrent Extraction

The crosscurrent extraction or extractions by multiple contacts (refer to the previous chapters) necessitates the handling of important volumes of solvents of which the first parts are only used with efficacy. It is more rational to choose a methodical extraction process enriching a current of solvent by degrees. The extractive solvent encounters the extracted one at countercurrent, which then becomes poorer according to the same rhythm. The devices used for this goal are only industrial ones. Hence, it is not astonishing that only great volumes of phases that are to be extracted are treated according to these conditions. This method is the best method of extraction for a given expense of extractive solvent to yield upper than those obtained by crosscurrent extraction and evidently by simple extraction.

This method must not be confounded with another method called *Separation at countercurrent* or *Craig's method*, which is not a method of extraction but a separation method of several solutes that are initially in the same solvent (see next chapter). The symbolism we use in this chapter is the same as that used in the preceding chapters.

Principle

Let us suppose, as previously, that the phase to be extracted is a solution of the substance S in Solvent A and that the extractive phase is Solvent B that is immiscible to A. The countercurrent method is a method functioning continuously with the help of an extraction column (Figure 1).

Phase A to be extracted as well as the extractive Solvent B are respectively introduced under pressure at each extremity of the column with a determined outflow. The two phases are collected at the opposite extremity. Thus, both phases run through the column according to an inverse direction. It is the origin of the name countercurrent of the method. The phase to be extracted becomes progressively poorer in Solute S and inversely the extractive phase becomes progressively richer. An appropriate device permits decanting of both phases just before they undergo outflow of the column. One obtains the final *refined* and one final *extractum*. It can be said that the crosscurrent only implies the renewing of one phase, whereas the methodology proposed now involves renewing the two phases that play a part in the extraction.

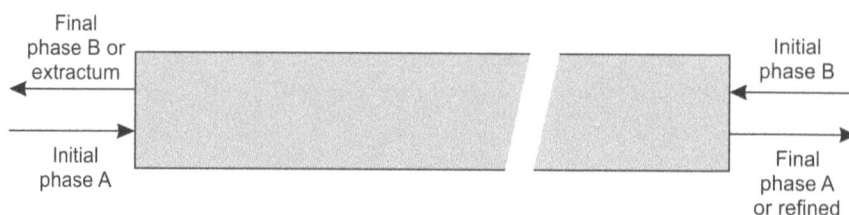

Figure 1: Column for Extraction (Reprinted with Permission from Gwenola and Jean-Louis Burgot, Paris, Lavoisier, Tec et Doc, 2017, 82).

One regularly analyses both phases at the comings out of the column after the beginning of the experiment. One notices that after a certain time interval, their composition no longer changes. One can then say that the *stationary rate* (of extraction) has reached.

Assimilation of an Extraction Column to a Series of N Theoretical Stages

Once the stationary time is reached, one can consider that the column is composed of a series of *n* batches of length 1 (or dl), which is called *theoretical stages.* In Chapters 12, 14 and 15, a strong analogy presents this notion with that of formerly *theoretical plates* that are encountered in distillation and chromatography. The length *l* or *dl* is defined as the length of the column that is necessary for the condition of equilibrium to be satisfied (Figure 2). That is to say, for the following relation at the going out of the stage *n* is:

$$C_{Bn}/C_{An} = K$$

The notion of theoretical stage is typically a notion, even if it is rational to consider that a certain interval of time (that is to say a certain length of column) is necessary for the partitioning equilibrium between the solute and the two solvents to be reached.

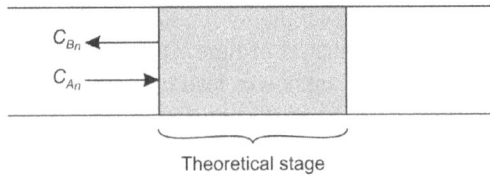

Theoretical stage

Figure 2: Theoretical Stage (Reprinted with Permission from Gwenola and Jean-Louis Burgot, Paris, Lavoisier, Tec et Doc, 2017, 82).

Quantitative Study in the Case of a Regular Distribution

Notations

Each batch (corresponding to a theoretical stage of the column) can be assimilated to a decantation funnel that would be fed by a volume V_A of phase A, which is coming from the stage located at its left, and by volume V_B of phase B, which is coming from the stage located on its right. The phases are stirred up and since this stage is theoretical by definition, they go out of this stage at equilibrium.

The numbering of the stages is done in the sense of the following route of one of the two phases. The quantities and concentrations have the index of the stage from which the phases at equilibrium go out. They are numbered when they enter the column. This corresponds to the initial phase A and the pure solvent B (Figure 3).

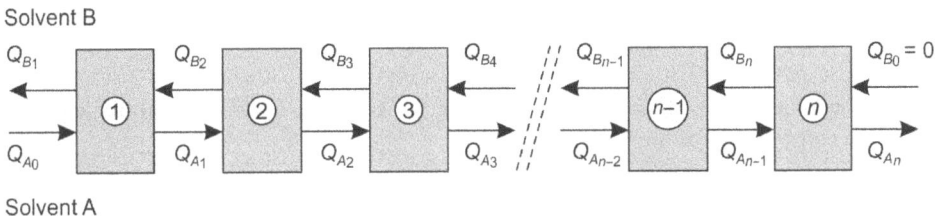

Figure 3: Followed Notation in the Countercurrent Extraction (Reprinted with Permission from Gwenola and Jean-Louis Burgot, Paris, Lavoisier, Tec et Doc, 2017, 83).

As a result, the final refined has the index Q_{An} and the final extract Q_{B1}.

Algebraic Study

When the distribution is regular, one demonstrates that each stage contains the same Volume V_A of solution of refined solvent and the same Volume V_B of extractive solvent, whatever the number of the stage is:

Calculation of Q_{B1}, C_{B1}, ρ, Q_{Ao}, Q_{An}

This is the calculation of the quantities of matter and the concentrations of solute in the phases going out, that is to say, after the whole process has taken place, they are still in questions. As we shall see, to perform these calculations one needs to start from the knowledge of the total volume V of the column, the total number n theoretical stages, the demands H and S (rate of flow) of the extract and extractive phases and the partitioning coefficient K (or of the distribution ratio D_r or of the number of moles Q_{Ao} of the solute). These calculations correspond to the first type of those we can do. One can extract the values of V_A and V_B with the help of the two following relations:

$$V_A/V_B = S/H$$

$$V_A + V_B = V/n$$

(*Remark*: In order to be convinced by the validity of the first relation, it is sufficient to multiply the two symbols S and H by an identical interval of time. One effectively obtains quantities that are volumes. These two equations with two unknowns lead to the two constant volumes V_A and V_B at each stage.)

As previously mentioned, we shall express the equilibrium condition and the conservation of the matter. (Recall that is a convenient way to be grounded upon these two types of relations to solve these kinds of problems)

• The partition equations are:

$$C_{B1}/C_{A1} = K \ldots\ldots\ldots\ldots C_{Bn}/C_{An} = K$$

It must be understood that K may be replaced by the distribution ratio D_r, provided it is constant. In quantities of matter, one can write as previously:

$$Q_{B1}/Q_{A1} = \ldots\ldots\ldots\ldots Q_{Bn}/Q_{An} = K(V_B/V_A) = \alpha$$

• The equations of conservation of matter are:

$$\text{Stage 1: } Q_{Ao} + Q_{B2} = Q_{A1} + Q_{B1}$$

$$\text{Stage 2: } Q_{A1} + Q_{B3} = Q_{A2} + Q_{B2}$$

$$\text{Stage n: } Q_{An-1} + Q_{Bo} = Q_{An} + Q_n \text{ with } Q_{Bo} = 0$$

One seeks to use two equations with two unknowns. The first one is the conservation of the matter for the n stages:

$$Q_{Ao} = Q_{B1} + Q_{An} \tag{1}$$

The second equation is obtained by introducing the coefficient α in the multiple equations. It expresses the conservation of the matter:

$$Q_{Ao} + Q_{B2} = Q_{A1} + Q_{B1}$$

$$Q_{Ao} + Q_{B2} = Q_{B1}/\alpha + \alpha Q_{A1} \text{ (1st stage for example)}$$

By writing that is present at the left of the sign equal and what is present on the right, one obtains:

$$Q_{Ao} - Q_{B1}/\alpha = \alpha Q_{A1} - Q_{B2}$$

$$1/\alpha(\alpha Q_{Ao} - Q_{B1}) = \alpha Q_{A1} - Q_{B2}$$

$$Q_{An-1} = Q_{Bn}/\alpha + \alpha Q_{An}$$

$$Q_{An-1} - Q_{Bn}/\alpha = \alpha Q_{An}$$

$$1/\alpha(\alpha Q_{An-1} - Q_{Bn}) = \alpha Q_{An}$$

The expression $(\alpha Q_{An-1} - Q_{Bn})$ has already been obtained for the (n–1)th stage, from which:

$$(1/\alpha^n)(\alpha Q_{Ao} - Q_{B1}) = \alpha Q_{An}$$

$$\alpha Q_{Ao} - Q_{B1} = \alpha^{n+1} Q_{An} \tag{2}$$

From relations (7) and (8), we obtain immediately:

$$Q_{An} = Q_{Ao}[(\alpha - 1)/(\alpha^{n+1} - 1)]$$

$$Q_{B1} = Q_{Ao}[1 - (\alpha - 1)/(\alpha^{n+1} - 1)]$$

For the yield:

$$\rho = Q_{B1}/Q_{Ao}$$

$$\rho = 1 - (\alpha - 1)/(\alpha^{n+1} - 1)$$

Let us recall that these formulas are only valid in a stationary system. An elementary calculus of the derivative $d\rho/dn$ would show that ρ is always a growing function of n and that the yield is optimal for $n \to \infty$ ($\rho = 1$) since n is positive. As previously, some diagrams permit immediately knowing ρ as a function of n for some values of α. (Pocket-calculators solve the problem of calculation quickly).

Calculation of the Efficiency of a Column

Actually, the problem which has just been studied is rarely put in. What is rather sought is the number of theoretical stages of the column. This number permits the calculation of its efficacy. It is even the mark of its efficiency. The known values which permit this calculation are C_{Ao}, Q_{Ao}, Q_{B1}, C_{B1}, Q_{An} and C_{Bn}. They are easily accessible by chemical analysis. Also, α is known via:

$$\alpha = K(V_B/V_A)$$

$$\alpha = K \, S/H$$

In order to obtain these values, one uses the same equations as previously but the unknown is now n.

Geometrical Study

On one hand, the most frequent problem is knowing C_{Ao} and Q_{B1}, and C_{An} and all these data that are accessible by chemical analysis, the following question is raising what is the number of theoretical stages of the column. In short, what is its efficiency?

On a diagram C_A/C_B, we shall draw the two curves that represent the conservation of the matter and the partitioning equilibrium.

- Equilibrium

The curve is a straight line, the slope of which is $K = C_B/C_A$. It passes through the origin.

- Conservation of the matter

Let us express the conservation of the matter for the different stages by calculating the quantities of matter by the unit of time (mol/time). One can write:

$$HC_{Ao} + SC_{B2} = HC_{A1} + SC_{B1} \quad \text{(1st stage)}$$

$$HC_{A1} + SC_{B3} = HC_{A2} + SC_{B2} \quad \text{(2nd stage)}$$

We can write these two equations differently:

$$SC_{B1} = HC_{Ao} + SC_{B2} - HC_{A1}$$

$$SC_{B2} = HC_{A1} + SC_{B3} - HC_{A2}$$

Let us consider the terms $SC_{B2} - HC_{A1}$ and $SC_{B3} - HC_{A2}$ of the two last equations. Let us compare them. The question is whether these terms are equal.

$$SC_{B2} - HC_{A1} = SC_{B3} - HC_{A2}$$

If, it is the case, then:

$$SC_{B2} + HC_{A2} = SC_{B3} + HC_{A1}$$

This relation is *satisfied* since, in a stationary system, there is no accumulation of matter in each stage. The above equations can be written:

$$C_{B1} = (H/S)C_{Ao} + b$$

$$C_{B2} = (H/S)C_{A1} + b \quad \text{etc...}$$

Also, by setting the following equality:

$$(S_{CB2} - HC_{A1})/S = (SC_{B3} - HC_{A2})/S = \ldots\ldots = b$$

It appears that b is evidently a constant.

The points corresponding to the concentrations (C_{Ao} and C_{B1}), (C_{A1} and C_{B2}), etc., are all located on the same straight line of slope H/S. This line is called an *operational line*. If we trace the straight lines of equilibrium and operational line (Figure 4), we notice that each theoretical stage is represented by three points where two are on the operational line and one is on the equilibrium line.

The equilibrium straight line can be drawn without any problem and it is the same for the operational line. For the latter, indeed, one knows its slope and it passes through the point of coordinates (C_{Ao} and C_{B1}) as has been just demonstrated above. One knows that these points are obtained experimentally by chemical analysis. It is the point A. Let us consider the first theoretical stage:

- Point A is in this first stage since the concentration C_{Ao} enters it and the concentration C_{B1} goes out of it.

- Point B is also in this first stage. The concentrations C_{B1} are naturally on the line of equilibrium. The concentration C_{A1} is given by the abscissa of point B.

- Point C corresponds to the concentration going out C_{A1} and also at the concentration arriving C_{B2}. Point C is on the operational line. It is also common to the second theoretical stage defined

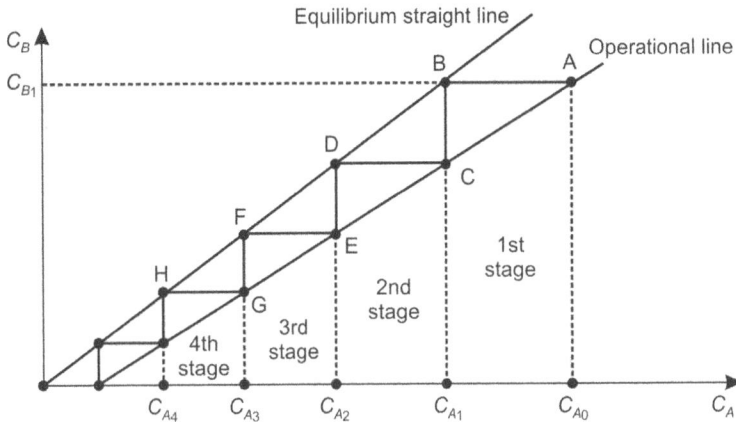

Figure 4: Representation of the Theoretical Stages. (Reprinted with Permission from Gwenola and Jean-Louis Burgot, Paris, Lavoisier, Tec et Doc, 2017, 87).

by Points C, D and E. Therefore, one builds the different theoretical stages by degrees. In the example given above, the apparatus is constituted by four theoretical stages. The concentration C_{An} (here C_{B1}) is given by the abscissa of point H. The number of stairs is the number of theoretical stages. (This geometrical representation of the theoretical stages is due to two authors McCabe and Thiele. We shall find it again about distillation where it would be somewhat more studied).

There is an important point to mention here. The number of theoretical stages is a function of the partitioning coefficient K, which says about the nature of the solute and both solvents. Each solute corresponds to a proper number of theoretical stages. One finds the same phenomenon in chromatography and distillation.

Quantitative Study in the Case of an Irregular Distribution

In this case, it is no longer possible to carry out the above algebraic calculation. Since the distribution is irregular, the stages differ from the others by their volumes of V_A and V_B. To solve the problem, one proceeds usually by seeking a graphical means (Figure 5). If the equilibrium curve is no longer a straight line by definition, the operational straight line remains the same as previously stated.

The equilibrium graph is, as it has been said earlier, a characteristic curve of the irregular distribution. Its graph is purely and simply obtained from experimental data after a study of the partitioning of the solute in different experimental conditions.

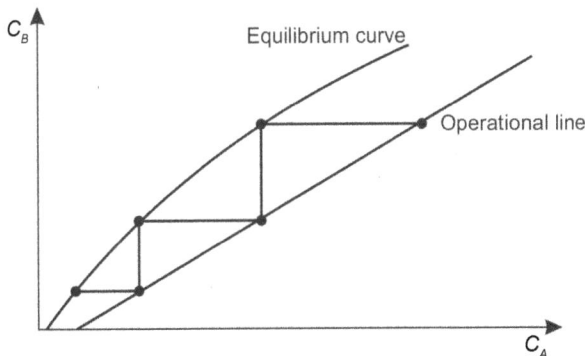

Figure 5: Representation of the Theoretical Stages (Irregular Distribution). (Reprinted with Permission from Gwenola and Jean-Louis Burgot, Paris, Lavoisier, Tec et Doc, 2017, 88).

Example of Calculation for a Regular Distribution

Let us again consider the example of the extraction of the antibiotic for an identical expense of solvent as for the simple and repeated extractions. Let V_B/V_A be 1/5 or S/H = 1/5. We know that there are three theoretical extraction stages. What is the ρ value?

$$\alpha = KV_B/V_A = 15 \times (1/5) = 3$$

$$\rho = 1 - (3 - 1)/(3^4 - 1)$$

$$\rho = 0.975$$

We notice the yield improvement with respect to one extraction ($\rho = 0.875$) or with respect to three successive extractions for an identical expense of Solvent B.

Practical Modalities

Let us recall that countercurrent extraction is an industrial method. The apparatus the most frequently used are:

- The vertical columns or extraction towers. The lightest solvent is introduced at the basis of the tower and the heavier solvent at the top part by pulverizing them in order to increase the surfaces of contact. An example is Scheibel's tower (Figure 6). In this case, the two liquids can self-inter penetrate under the action of the differences in densities.

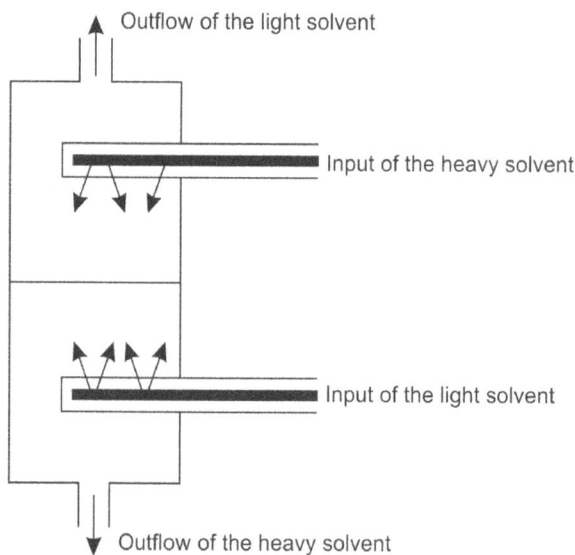

Figure 6: Scheibel's Tower. (Reprinted with Permission from Gwenola and Jean-Louis Burgot, Paris, Lavoisier, Tec et Doc, 2017, 89).

- The apparatus of Podbielnak. It is an apparatus that permits good decantation by increasing the field of gravity by centrifugation. The apparatus (Figure 7) is a kind of rotor. The horizontal shaft is hollow and conducts the solvents. The lighter solvent is admitted at the periphery of the rotor, and the solution to be extracted must be heavier in the axial region. Both arrive by conduits that are parallel to the axis of the rotor. Because of the rotation and hence the centrifugal force, both solvents enter contact on the helicoidal circuit of the rotor.

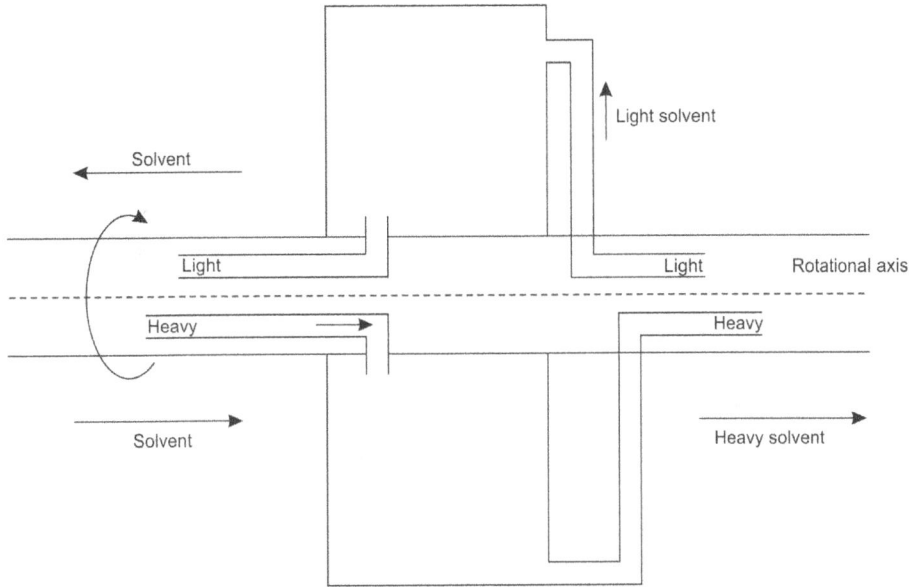

Figure 7: Podbielnak's Apparatus. (Reprinted with Permission from Gwenola and Jean-Louis Burgot, Paris, Lavoisier, Tec et Doc, 2017, 90).

With this kind of apparatus, one can obtain a reduction of the volumes of solvent, the contact time (of the order of 20s) and the losses by volatilization (they are weaker than 2%) with rotations that are of the order of 1,000 rpm.

Applications

For example, let us mention:

- The extraction of antibiotics. It has been possible to carry out good fractionations of isomers of penicillin.
- The concentration and purification:
  - of hormones
  - of vitamins
  - of alkaloids
- The extraction and the fractionation of uranium and thorium.

CHAPTER 10

Countercurrent Separation or Discontinuous Liquid-Liquid Countercurrent Separation

As it has been already recalled (refer to Chapter 4), the goal of these methods called *countercurrent separation* is to separate compounds dissolved in an initial liquid phase by addition of an immiscible solvent and by assembling them in distinct fractions of the two solvents. Hence, the goal of these methods is not to extract a particular compound from the initial mixture by a non-miscible solvent, even if the separation is based on the differential extraction of the components of a mixture from one phase to another. Besides, the separation is only possible when, for the chosen solvents, the substances composing the initial mixture do possess different partitioning coefficients.

This method consists in carrying out multiple liquid-liquid extractions at countercurrent, like the preceding method we have described. As it is also the case for the latter, it also implies a renewed countercurrent. It is due to Craig of the Rockfeller Institute (1949) that it is sometimes called a *discontinuous liquid-liquid countercurrent separation*.

Description of the Apparatus

The Craig's apparatus which permits such operations is quite characteristic. It entails two blocks:

- An inferior block is hollowed out by some number of cavities that are numbered. For example, from 0 to 19 in which there are disposed tubes that open at their superior extremity and are calibrated in order to contain the same volume of liquid when they are filled. They are intended to receive the *heavy solvating phase* (Figure 1).

- A superior block that fits the preceding. It is also hollowed by equidistant cavities numbered from 0 to 19 in the opposite sense of that existing in the inferior block. These cavities are intended to receive tubes that are open at their two extremities. They are calibrated in order to receive the same volume of *light solvent*.

- The cavities of both blocks are matched so that when the apparatus is in a correct position, each inferior tube is in communication with a superior tube. This ensemble of two tubes constitutes a column. The superior block is mobile around a vertical axis and can be displaced toward the left or the right in order to change coincidences between the inferior and superior blocks. Finally, a cover supplied with a joint can fit the superior block, and a handful permits to agitate the whole.

Preparation of the Experience

Let us suppose that the solutes to be separated are contained in the heavy phase. Only, tube 0 of the inferior block is filled with this heavy phase (and with the solutes as well). Other tubes of the inferior block are filled with the same volume of heavy phase, but they are pure. All the tubes of the superior block are filled with the same volume of the light phase. The whole structure is in the position as it is indicated in Figure 1, and the cover is kept closed.

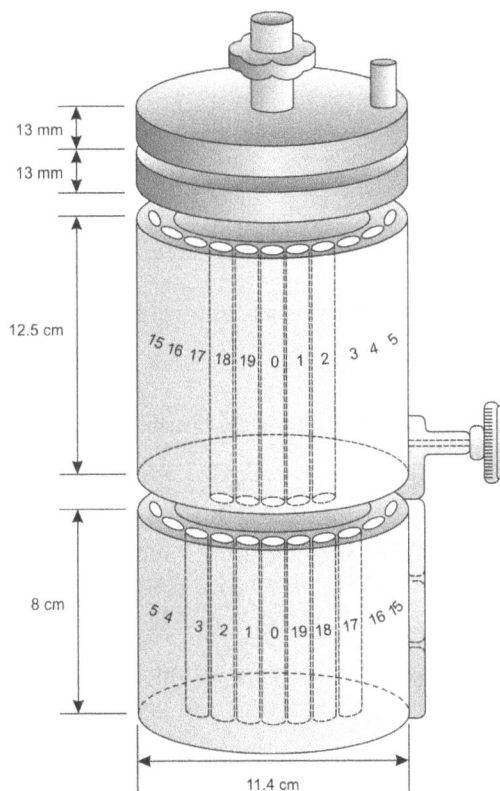

Figure 1: Craig's Apparatus. (Reprinted with Permission from Gwenola and Jean-Louis Burgot, Paris, Lavoisier, Tec et Doc, 2017, 92).

Working of the Apparatus-Case of One Solute: A Qualitative and Quantitative Study

Study of One Operation

The apparatus is energetically agitated by turning it upside down. Then, it is abandoned until the phases are separated and their limits of separation coincide exactly with the plane of rotation separating the two blocks. The superior block is then moved from one space to the right with respect to the heavy phase (Phase 2) by a simple rotation and without stirring. During this rotation, the inferior block remains motionless. In each column, the whole heavy phase has kept its position whereas the whole light phase has moved one step. The set *stirring-rotation* of one step constitutes one *operation*.

Let us examine what is passing during one operation.

- Qualitatively, one notices that after one separation, the substance is present in two columns (Figure 2):

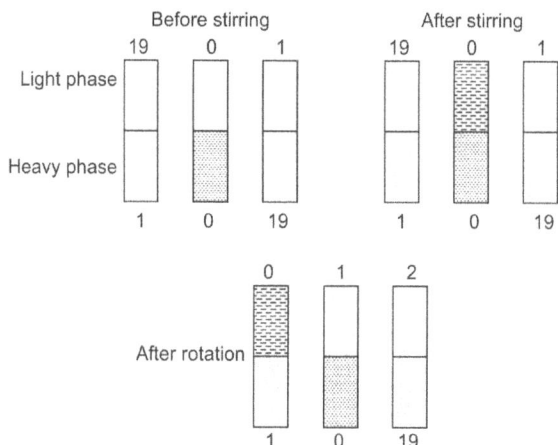

Figure 2: Qualitative Successive Repartitions During the First Operation. (Reprinted with Permission from Gwenola and Jean-Louis Burgot, Paris, Lavoisier, Tec et Doc, 2017, 93).

- Quantitatively, in order to answer the question, as usual in this kind of problem, we shall study:
  - the equilibrium condition
  - the conservation of matter

Concerning the equilibrium constant, we suppose the distribution regular, that is to say:

$$K = C_B/C_A$$

Here, B implies the *light phase*, which is sometimes called the *mobile phase*. A implies the *heavy phase* or *stationary phase*. (We keep the same symbols as in the previous chapters). The condition of conservation of the matter is:

$$Q = Q_A + Q_B$$

Q_A and Q_B are the quantities in the stationary and mobile phases. Both conditions must be always respected. Let us set up y and z, the fractions of present substance in the heavy and light phases respectively.

We first demonstrate that y and z are independent of the initial quantity of the substance to distribute into the two phases and independent of the number of the operation. We have:

$$Q = Q_A + Q_B$$

$$Q = yQ + zQ$$

We deduce:

$$y + z = 1 \tag{1}$$

Otherwise, the equilibrium condition is:

$$K = C_B/C_A$$

It can be also written as:

$$K = Q_B/V_B/Q_A/V$$

$$Q_B/Q_A = K(V_B/V_A)$$

$$zQ/yQ = K(V_B/V_A)$$

$$z/y = K(V_B/V_A) \tag{2}$$

The product $K(V_B/V_A)$ is constant for a given separation since the distribution is regular (K: constant). Let us set up:

$$K(V_B/V_A) = \alpha$$

Here, α is named the *corrected partition coefficient*. Equations (1) and (2) constitute a system of two equations in which the two unknowns are y and z. It follows that:

$$y = 1/(1 + \alpha) \quad \text{and} \quad z = \alpha/(1 + \alpha)$$

The y and z are indubitably independent of the number of the operation and from the quantity of the substance to distribute.

Hence, during the first operation, we have after agitation and before rotation:

- zQ moles in the mobile phase
- yQ moles in the stationary phase

And after rotation:

- zQ moles in column entitled 0–1
- yQ moles in column entitled 1–0

One notices that after one operation, the substance is present in two columns, which is according to the development of the binomial relation:

$$(y + z)^1 Q \quad \text{that is to say} \quad yQ + zQ$$

Study of the Second Operation

The initial state is represented in Figure 3:

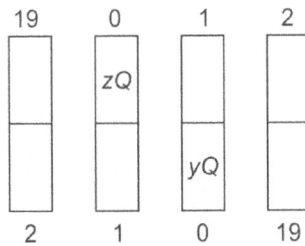

Figure 3: The Initial State of the Second Operation. (Reprinted with Permission from Gwenola and Jean-Louis Burgot, Paris, Lavoisier, Tec et Doc, 2017, 95).

- Stirring

  In the column 0–1, in order to respect the equilibrium condition and that of the conservation of the matter, the zQ moles of substance will be shared according to the constant fractions z and y, that is to say:
  - the fraction z of zQ in the mobile phase,
  - the fraction y of zQ in the stationary phase.
  - In columns 1–0, there is the partition of the fraction z of yQ, of the fraction y of yQ (Figure 4a).
  - The rotation is not followed by a stirring that belongs to the following operation. After rotation, one ends with the following distribution (Figure 4b).
  - The substance is present in three columns according to the quantities:

$$z^2Q \quad 2yzQ \quad y^2Q$$

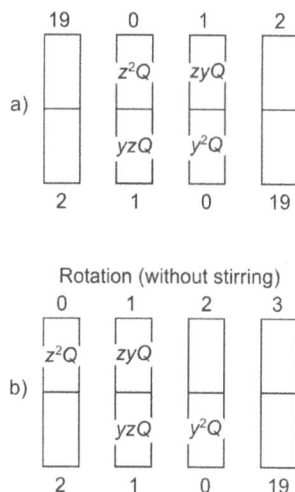

Figures 4a and b: Second Operation. (Reprinted with Permission from Gwenola and Jean-Louis Burgot, Paris, Lavoisier, Tec et Doc, 2017, 96).

In summary, after two operations, the substance is present in three columns. That is to say, in 2 + 1 columns which is according to the development of the binomial relation:

$$(y + z)^2 Q = y^2 Q + 2yzQ + z^2 Q$$

The exponent 2 is the number of operations.

Generalization

One can generalize the preceding results. If one carries out n operations, the substance will be present in $(n + 1)$ columns according to the development of the binomial relation.

$$(z + y)^n Q$$

In other words, in each column, the quantities of the substance are given by the expression;

$$C^p_n \, y^p z^{(n-p)} \, Q$$

The C^p_n is not a mysterious number. It is the number of combinations of n things that can be taken p to p. The quantity contained in the $(p + 1)^{th}$ column is given by:

$$C^p_n \, y^p z^{(n-p)} \, Q$$

Therefore, let us take again the case of two operations ($n = 2$). The quantity of substance present in the first tube ($p = 0$) is:

$$C^0_2 y^0 z^2 Q$$

Here, $C^0_2 = 1$ by convention, $y^0 = 1$; that is to say $z^2 Q$. It is what we have found. The quantity of substance present in the second tube ($p = 1$) is:

$$C^1_2 y^1 z^1 Q$$

$C^1_2 = 2/1 = 2$; that is $2yzQ$ is what we have found. The quantity of substance present in the third tube ($p = 2$) is:

$$C^2_2 y^2 z^0 Q$$

It is just the quantity found. Finally, let us take the example of 4 operations (n = 4):

- The quantity in the first tube (p = 0) is:

$$C_4^0 y^0 z^4 Q = z^4 Q$$

- The quantity in the second tube (p = 1)

$$C_4^1 y^1 z^3 Q = 4yz^3 Q$$

- The quantity in the third tube (p = 2)

$$C_4^2 y^2 z^2 Q = 6y^2 z^2 Q$$

- The quantity in the fourth tube (p = 3)

$$C_4^3 y^3 z^1 Q = 4y^3 z^1 Q$$

- The quantity in the fifth tube (p = 4)

$$C_4^4 y^4 z^0 Q = y^4 Q$$

Fraction of the Quantity of Substance Present in One Column

A graphical representation of the preceding results can be done. Let us suppose $\alpha = 3$. It follows $y = \frac{1}{4}$ and $z = \frac{3}{4}$. Let us also suppose that there are eight operations. The quantity present in the first tube (p = 0) is $(3/4)^8 Q$, where Q is the total quantity. The fraction of the total quantity in the first tube is, hence:

$$(3/4)^8 Q/Q = 0.1$$

Quite generally, the fraction of the total quantity present in the tube of rank p, that is to say in the $(p + 1)^{th}$ column is:

$$C_n^p y^p z^{n-p} Q$$

Figure 5 represents the fraction present in the tube as a function of p.

Remark: The diagram of a fraction $C_n^p y^p z^{n-p}$ of the total quantity present in a tube of rank p versus p (the rank of the tube) is a histogram.

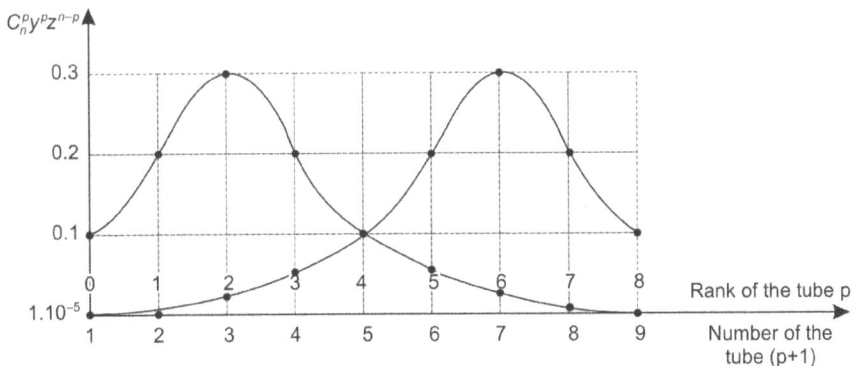

Figure 5: Fraction of the Total Quantity Present in a Tube of Rank p (Numerical Data: See Text). (Reprinted with Permission from Gwenola and Jean-Louis Burgot, Paris, Lavoisier, Tec et Doc, 2017, 98).

Since it is a binomial law, one can be certain from the mathematical standpoint that on average the elementary favorable result will be attained for the value m = ny. The favorable event is the rank of the column where located is the maximum substance. With n = 8, we find $p_{max} = 2$, that is to say, the third column.

Remark: It is important to notice that y and z are numbers inferior to unity. Once n increases, z^n and y^n possess ridiculously weak values.

The Separation of Several Compounds

Craig's method is a method of separation. Let us consider the case of two compounds. The basic hypothesis is that the two compounds distribute themselves independently into the two solvents. That means that the partitioning coefficient of one compound does not modify the value of the coefficient of the other. Let us suppose also that one of the compounds should be the precedent with $\alpha = 3$. We know that after eight operations, it is present in nine columns. Let us suppose also that the other compound shows the value $\alpha' = 1/3$ for its corrected partitioning coefficient. As a result:

$$y' = \text{¾} \quad \text{and} \quad z' = \text{¼}$$

Here again, we can trace its repartition in the different columns. The maximal fraction of the total quantity of this second substance is located in the tube:

$$p_{max} = 8 \times (3/4) = 6$$

That is to say, it is in the 7th tube. The calculations concerning the two substances show that in the third tube, there is the maximum substance 1, whereas there is almost no Product 2 in it and inversely in the 7th tube. Compounds 1 and 2 are separated.

It is evident from the mathematical standpoint that the separation is possible if only *y* and *z* have different values from one compound to the other. When the values of the corrected partitioning coefficients are nearer, the substances once separated despite this drawback are finally in elevated quantities of solvents. So, the separation can be possible when the number of columns increases.

Advantages and Drawbacks of the Method

This is a powerful method of separating substances possessing nearly identical partitioning coefficients (or distribution ratios) in a reasonable time. For example, let us consider the separation of two compounds M and N whose partitioning coefficients are respectively 0.10 and 12. Calculations show that after 10 operations, in the third tube (p = 2) when the quantities of M and N are equal to the start, there are about 30 million times more A than B in the third tube after the 10 stages. The great inconvenience of this method is the handling of great quantities of solvent.

There exists some Craig's apparatus, which contain up to 400 tubes.

Craig's Countercurrent Separation Method and Partage Chromatography

Now, we know that Craig's method is a discontinuous liquid-liquid countercurrent separation. By adopting this point of view, it appears that this method is connected to the partage chromatography that can be considered legitimate as being a *continuous liquid-liquid countercurrent separation*. This point will be discussed again later (more from Chapter 13 and onward).

CHAPTER 11

Solid Phase Extraction
and Microextraction

Solid phase extractions (SPE) are introduced to eliminate some shortcomings of liquid-liquid extraction. One of the largest criticisms is directed toward the relatively large amounts of organic solvents, which are needed. That is true for the extraction of organic products as trace levels in an aqueous medium. In addition, liquid-liquid extraction is time-consuming. It is also a tedious operation, and it presents some difficulties to be automated. Furthermore, it requires large amounts of samples that are difficult to obtain when the matrix is a biologic fluid. SPE has taken on importance in the preparation of samples. They simultaneously allow the purification of the samples and the enrichment of analytes, which is very interesting to the evaluation of trace levels.

The technologies employed for SPE include:

- Exhaustive extractions of the analyte as SPE
- Several types of non-exhaustive extractions with only a fraction of the analyte in the sample, which is extracted in the sorbent. The ultimate goal is to reach equilibrium between the phases. They are:
  - Solid Phase Microextraction (SPME)
  - Stir Bar Sorptive Extraction (SBSE)
  - Thin Film Microextraction (TFME)
  - Solid Phase Dynamic Extraction (SPDE)

SPE

Features of the SPE

SPE consists of extracting one or more analytes from a liquid sample matrix by fixing them on a natural or synthetic solid material. The extraction involves the difference between the physical and chemical properties of the solid material or sorbent and those of the analyte and the liquid sample matrix. The analyte can be nonpolar, moderately polar or polar. Hydrophobic, electrostatic and van der Waals interactions that contribute to retention of the analytes on/in solid materials are similar to those used in the elution chromatography described in Chapter 16. There is in this case one difference; the extraction is complete in the extraction processing but incomplete in the chromatography. In most cases, the retention mechanism is adsorption. However, it is also possible to use partition and ion exchange mechanisms.

The device with packing sorbent is marketed in a conventional format (cartridge—refer to Figure 1—disk and pipet tip) or in the 96-well plates, particularly suited to handling large series. In the case of the cartridge, the sorbent is immobilized between two polypropylene or sinter filters. The volume of the cartridge varies from 1 to 60 ml and is filled with 50 mg to 10 g of sorbent. A selection of cartridges is available in many sizes, shapes and types of sorbents.

The device disks have some advantages over cartridges. The diameter of the particles of the sorbent is smaller than one of the cartridges, thereby increasing the flow rate of the sample matrix solution and reducing the time of the analysis. The reduction of the volume of the bed (mass of sorbent) into the disks compared to those of cartridges limits the volume of elution solvents.

Implementation

The implementation is realized manually or automated on a liquid handler, or online with a coupling of the extraction device with liquid chromatography (LC) or gas chromatography (GC) instruments.

Manual SPE Procedure

In this case, the SPE procedure is realized before the analysis and follows generally four steps (Figure 1).

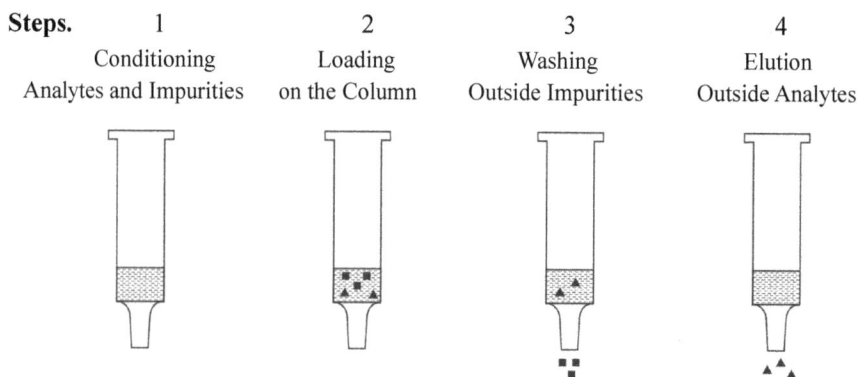

Steps.	1	2	3	4
	Conditioning	Loading	Washing	Elution
	Analytes and Impurities	on the Column	Outside Impurities	Outside Analytes

Figure 1: The Four Basic Steps of the SPE by a Sorbent Packed in a Polypropylene Cartridge. (Reprinted with Permission from Gwenola and Jean-Louis Burgot, Paris, Lavoisier, Tec et Doc, 2017, 111).

Step 1 Conditioning and Activation of the Active Sites of Sorbents. The wetting of the sorbent is necessary to ensure reproducible interactions between the sorbent and analytes. The solvent is pulled through the device by gravity or using a vacuum pump. With a hydrophobic sorbent, the wetting is realized with an organic solvent, thereafter with solvents that are more polar and more to obtain a polarity very close to that of the analyte. Generally, from five to ten clean-up operations are performed to eliminate impurities.

Step 2 Loading the Sample. The crude sample matrix is introduced into the device with a solvent of weak eluotropic strength to bring about the retention of the analytes on the sorbent. It can be possible to pull the sample matrix solution that contains the analyte and the interfering compounds or impurities through the sorbent with a vacuum-controlled workstation loading with a positive or a negative pression. There are two possible procedures. Analytes may be adsorbed by the sorbents or, on the other hand, it is possible to retain the impurities. It depends on the aim of the process.

Step 3 Washing of the Sorbent. It is necessary to eliminate some impurities adsorbed by the sorbent. The device is washed with a weak eluotropic solvent for the analyte. The washing of the sorbent is realized no less than 5 or 10 times with a given volume of solvent. The washing solvent needs to have an intermediary eluotropic strength (see Chapter 16 to learn more about eluotropic strength) weaker than the elution solvent and greater than the fixation solvent. Moreover, it is necessary that the washing solvent should be miscible at the elution solvent and the liquid sample. If this is not the case, it will be required to dry the support or the sorbent between each step.

Step 4 Eluting the Analytes. Thereafter, the analytes of interest are released from the device with a small amount of solvent with strong eluotropic strength. The final solution is enriched and purified. After that, the extract is introduced into the analysis instrument after evaporating to dryness and dissolution again in the convenient solvent. Automated processes are also used to perform multiple extractions simultaneously.

The result of the extraction depends not only on the choice of the type of sorbents or the solvents but also on the matrix of the sample. It will be better to eliminate solid particles by a first filtration and also avoid the viscous matrices that have the possibility to saturate the sorbents. To improve the selectivity in the case of biological matrices, it is necessary to realize the precipitation of proteins and the hydrolysis of glucuronides.

Process of Online Extraction by Pass Column

The SPE can be linked directly to an LC or a GC instrument. In this case, the extraction column is incorporated into the apparatus of chromatography. The apparatus is equipped with two pumps, a check valve and an extraction cartridge (5–25 mm of length x 2–3 mm inside-diameter) linked with an analytical column by an injector system of a six-port valve. The cartridge replaces the sampling loop of the conventional LC (viz., Figure 2). The process is performed in three steps:

- Firstly, the injection of a volume of the sample on the cartridge. A great part of impurities is carried along by the solvent and the analyte remains on the sorbent.

- After a sufficient time of contact, both cartridge and column are connected and the mobile phase moves through the analytes in the column.

- Finally, both cartridge and column are released to avoid introducing a too important quantity of the sample. It is provided to enhance the sensibility of the chromatography process.

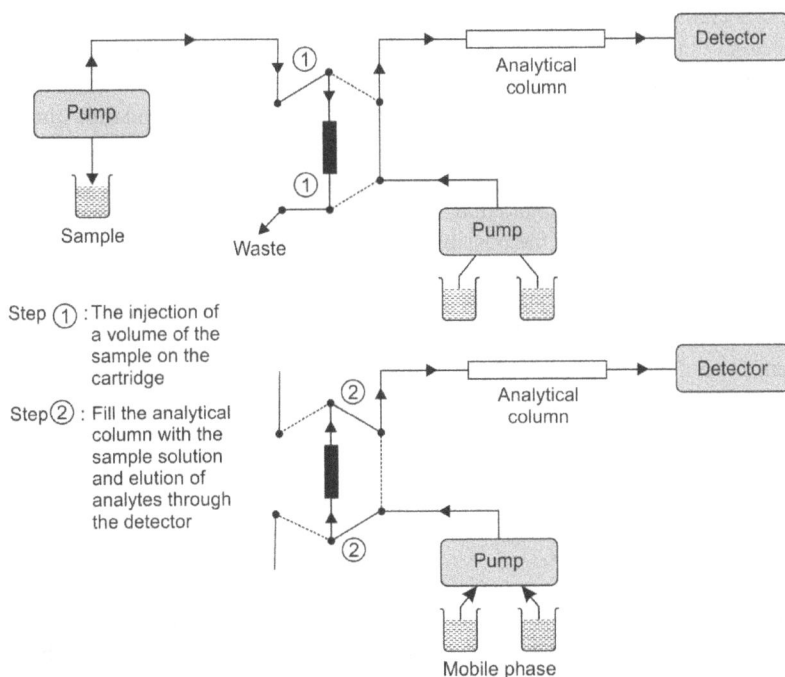

Figure 2: Online Extraction Process Coupling With a Liquid Chromatography Instrument. (Reprinted with Permission from Gwenola and Jean-Louis Burgot, Paris, Lavoisier, Tec et Doc, 2017, 113).

Parameters for the Optimization of SPE

Two parameters allow comparing the behavior of sorbents related to analytes, the *extraction efficiency* and the *factor of enrichment*. The extraction efficiency is concerned with the amount (in molarity) of the analyte that is eluted from the device and corresponds to the ratio of the amount of the analyte eluted to the amount of the analyte introduced in the device.

$$\rho = \text{amount recovered/amount percolated} = Q_r/Q_p$$

The factor of enrichment measures the concentration of analyte in the solution that is eluted from the device. It is possible to calculate it according to the formula:

$$F = C_r/C_p = \text{concentration of the analyte in the recovered solution/concentration}$$
$$\text{of the analyte in the percolated solution.}$$

$$F = C_r/C_p = (Q_r/V_r) \times (V_p/Q_p) = (Q_r/Q_p) \times (V_p/V_r) = \rho \times (V_p/V_r)$$

with:

F = factor of enrichment

C_r = concentration of the analyte in the recovered solution

Q_r = amount of the analyte recovered from the sample

V_r = elution volume

C_p = concentration of the analyte in the percolated sample solution

V_p = percolated volume

Q_p = amount of analyte percolated in number of moles

Other requirements are also used, such as the volume of the end of fixation V_f provides for the determination of the volume of the sample solution that is necessary to percolate to obtain the fixation of the analyte. It provides the volume from which analytes are eluted by the fixation solvent without being retained by sorbents. A percolation of volumes of solvents more important as V_f results in a decrease in the extraction ratio. V_f is determined by percolating some volumes of sample that become more important with an amount of the analyte that will remain constant. V_f is defined by the point where the concentration of the analyte at the outlet of the device is less than 1% of the total percolated concentration. Thus, the volume of the percolation and the volume of washing of the cartridge does not exceed the volume V_f. If this requirement is not fulfilled, the extraction ratio will be modified.

The *capacity of the sorbent* or *load capacity* provides the maximum amount, which it is possible to fix on the sorbent. It depends on the possibility of a bond between the analyte and the active sites of the sorbent. The overrun of the capacity of the sorbent decreases the extraction ratio. The capacity of the sorbent included in the device is around 1 or 5% of the mass of the sorbent. Thus, the elution volume is also optimized and does not exceed 2 or 3 times the volume of the device. A 30 ml of elution solvent for a device contains 100–500 mg of sorbent, which is generally used. If the aim of the operation is to concentrate the analyte, the volume of the sample solution introduced in the device will be important but the volume of washing solvent will be low. If, on the contrary, the aim is to purify the sample, the percolated volume is less important than the washing volume to limit the quantities of interfering species.

Type of Sorbents

The sorbents used for extraction are very similar to those used for chromatography described in Chapters 16 and 17. There exist many possibilities for choosing the appropriate sorbent. Generally, there are classified into two groups: conventional sorbents and molecular recognition sorbents.

Conventional Sorbents

Silica-Based Materials. The silica-based materials, the first to be marketed, are probably the most frequently used. They exhibit a retention mechanism linked to their framework. There are many types where the cross-linked silica with an octadecyl-derived functional group (C18) or octyl (C8) (see Chapter 17) is the very common package. There are used to get out the hydrophobic organic analytes or weakly polar from the aqueous matrices according to the partitioning reverse process (see Chapters 16 and 17). The hydrocarbons chains of the analytes and the sorbent are attracted by weak intermolecular van der Waals interactions. Moreover, it is possible to graft on the silica some aromatic groups, such as the phenyl group. The silica-base packing is barely specific and the extract also has many impurities. The silica-based materials grafted with the polar groups (diol, nitrile or primary amines, for example) provide the retention of the polar analytes from a non-polar sample matrix. The hydrogen-bonding forces or π-π, dipole-dipole, induced dipole-dipole between the analyte and the polar groups of the sorbent are responsible for the retention. Moreover, these packings can be used to adsorb and elute some compounds of similar structures so closed as isomers. The silica-based materials grafted with the ionic groups' sulfonate or ammonium with mobile counterions are used for ion-exchange of ionic species from aqueous samples matrices. In this case, the choice of the pH of the sample solution and solvents is very important. Thus, it is intended to keep the charge of the analyte as well as those of ionic groups of the adsorbent. However, silica-based materials have many drawbacks including limited stability at low or high pH and poor sensitivity due to the great adsorption power of the silanol groups.

Porous Polymers Materials. To avoid the drawbacks of silica-based sorbents, some polymeric materials are worked out. There are made of porous polymers materials such as polystyrene-divinylbenzene resins for the lipophilic part combined with N-vinylpyrrolidone polar monomers, which improve the alkaline pH stability. As for the silica materials, by the functionalization, there involve the partitioning mechanism (hydrophobic and electron-donor or acceptor interactions) and those of the adsorption/ion exchange. Consequently, they permit the extraction of hydrophobic organic analytes. The load capacity is more important than that of silica-based materials due to their high specific surface of 800 to 1,200 m^2 per gram.

Carbon-Based Materials. Carbon-based materials are interesting for sampling and concentration of air pollutants before analyzing. Carbon is an element with several allotropes (graphite, graphene, diamond, etc.). The charcoal, the older carbon form, endowed with a high specific surface (around 1,100 $m^2.g^{-1}$) provides unfinished and aleatory desorption of the analyte. It depends on chemical classes. The charcoal presents a high sensitivity to humidity. For these reasons, new types of carbon-based sorbents were marketed. There are obtained:

- By carbonization process or transformation of synthetic organic polymers in charcoal when heated. This leads to hard black marbles of carbon molecular sieves whose specific surface is often in the same value as that of charcoal.

- By physical-chemical change, such as graphitization. The carbon-graphite sorbents are more hydrophobic than silica-based materials or polymers. Moreover, they show stability more important than that of silica-based materials in a large pH range. From its low specific surface 90 to 200 m^2g^{-1}, the desorption of analytes is sometimes difficult and requires solvents with strong eluotropic strength.

- Carbon-based nanomaterials are very often investigated as sorbents. These nanomaterials include graphene (shack of planar graphitic sheets), fullerenes (polyhedral carbon cages) and carbon-nanotube. The benefits relating to their use result from a high surface area to volume ratio and also an easy functionalization procedure.

Sorbents with Mix Selectivity

There exist also mix selectivity sorbents involving both hydrophobic and electrostatic interactions that contribute to extracting acidic and basic compounds. These sorbents improve the selectivity of the extraction process. These include some Restricted Access Material (RAM) that will enable to make an extraction according to both mechanisms, such as exclusion steric and partitioning. There are made of a porous material whose surface of little pores is grafted by alkyl or ionic groups. These chemical groups are accessible exclusively to small molecules. These sorbents used in bioanalyzer also permit elimination of molecules with a high molecular weight, such as proteins.

Molecular Recognition Sorbents

The previous sorbents mainly involve too general hydrophobic interactions for only extracting the active analyte. As a result, there is a lack of selectivity. It is, above all, the case for the complex matrices that contain constituents, such as proteins, lipids and carbohydrates. Therefore, it becomes necessary to quantify the analytes by some sophisticated instruments that have high sensitivity and selectivity, such as mass spectrometry. However, this is not wholly without any problem. The presence of non-volatile impurities indeed modifies the properties of droplets produced by electrospray (see Chapter 42) during the ionization step. The result is a lowering of the signal with a decrease in the sensitivity and the precision of the method.

To prevent the low selectivity of the previous sorbents, a new concept has been developed that involves some specific biologic interactions, such as an enzyme-substrate system. There are some molecular recognition sorbents including:

- *Immunosorbents* are antibodies or immunoglobulins (Ig) that are developed against the analyte and fixed on hydrophilic solid support. Hydrophilic support is chosen to prevent non-specific interactions with the analyte. There are capable of extracting the analyte alone by spatial and electronic complementarity. These sorbents are largely used to extract proteins and peptides with very high specificity. However, their cost is important. Moreover, their temperature and pH stabilities are limited close to extreme values. Non-specific interactions are evaluated with a blank or a sorbent without antibodies.

- *Aptamers*, single-strand oligonucleotides strands of DNA or RNA, allow to specifically trap molecules. These sorbents are used in the bioassay named ELONA (Enzyme Linked Oligonucleotide Assay), which is conceived on the same principle as the assay ELISA (Enzyme Linked Immunosorbent Assay). They are more stable and less expensive than antibodies and are used in many applications.

- *Molecularly Imprinted Polymers* (MIP) are synthetic products that are obtained by polymerization of monomers around a molecule used as a fingerprint. The fingerprint molecule is taken off from the structure, leaving behind some specific holes. The holes feature a steric and functional complementarity by essentially hydrogen bonds. The MIP is interesting for the analytes whose mass is less than 2,000 Daltons. Such polymers are more stable than immunosorbents. Their major drawback is the possibility of salting-out the fingerprint molecules that are not eliminated from the holes. Consequently, this phenomenon modifies the measurement of the analyte. It is possible to avoid it by using a structural analog of the fingerprint molecule to create specific holes. Moreover, the use of analog can be interesting to save analytes, which are very often some expensive products.

Magnetic Nanoparticles SPE

For improving the speed of the extractions, magnetic nanoparticles made of magnetite Fe_3O_4 or FeO, Fe_2O_3 or Fe^{2+}, $Fe_2^{3+}O_4$ that are coated with polymers, such as divinylbenzene (DVD), n-vinylpyrrolidone and styrene, which have some adsorption properties. The balls of magnetite are introduced into the liquid samples. At the equilibria, it is easy to separate the balls with the use of magnets. The desorption of analytes is realized by elution solvent. This dispersive SPE has several advantages over conventional approaches, which mainly include easy recycling of balls of magnetite that are chemically inert. It is possible to use this process for the sample cleanup and the analyte enrichment.

Solvents or Mixture of Solvents

The solvents must comply with the objectives of each step, which are conditioning, loading, washing and eluting.

The *loading composition solvent* is chosen depending on the physical and chemical properties of the analyte and the composition of the sample matrix:

- **Physicochemical properties of analytes:**
  - Acid-base properties are the adsorption of the analyte, which depends on the value of pH.
  - Neutral compounds are extracted without pH control.

- **Physicochemical properties of the matrix:**
  - Water miscible matrix uses methanol or acetonitrile that is diluted in an aqueous phase and is the more used solvent.
  - In a non-water miscible matrix, hexane is the best choice.

The *washing solvent* has to eliminate all the impurities without modifying the bond of the analyte to the sorbent:

- For *ion-exchange extraction*, the washing step is realized by a buffer solution the ionic strength and pH values of which are controlled.
- For *reversed-phase extraction*, the washing solvent is made from 5 to 50 percent of organic solvents, such as methanol or acetonitrile diluted in an aqueous phase.
- For normal-phase extraction, the washing solvent is a non-polar solvent to remove further impurities.

  For example, the ratio solvent/mass of sorbent is 1 ml for 100 mg of sorbent.

The *eluting solvent* must avoid the interactions of the analyte with the sorbent. The most used eluting solvents are methanol, ethanol, acetonitrile or tetrahydrofuran that is mixed with water, acid or alkaline solutions or buffers. There are very often incompatible with the direct injection process on chromatography column of type reverse phase. In this case, the sample is evaporated to dryness and reconstituted with a convenient solvent.

From a practical point of view, the speed of the solvent must not be too fast or too slow to obtain the equilibria. The lower the diameter of the particles of the sorbent, the easier it is to increase the flow rate of the solvent with a reducing time of the analysis. The basicity or acidity of the solvent must be controlled for the acid-base analytes. It is easy to use a pH meter but the significance of pH in a complex media is an important problem. The viscosity of the sample matrix must be low. To avoid plugging, it is necessary to filter or centrifuge the sample matrix solution to eliminate the particles before the extraction process.

Applications

The SPE is convenient to eliminate the impurities and to concentrate to a factor of 10,000, which traces in the environmental, forensic, food and pharmaceutical analysis. SPE is largely used in the pre-columns for liquid chromatography to improve their life. We should make special mentions about some applications including:

- Pesticides and endocrine disrupters in drinking water with the C18-silica or polymeric materials.
- Lipidomic analysis as sphingolipids (silica without cross-linked).
- The pharmaceutical active ingredient of drugs, metabolites and biomarkers in biological samples (polymeric materials).
- Nitrophenolic compounds and triazine derivatives in a water sample (carbon-based materials).
- Phthalates and plasticizers are used in many sectors of industry (sorbents with mix selectivity).
- Pesticides in fruits and vegetables with a combination of liquid-liquid extraction and SPE. The name of the method is QuEChERS (Quick, Easy, Cheap, Effective, Rugged and Safe). It requires an extraction process in the presence of salts and buffers ($MgSO_4$ + NaCl + sodium citrate for example) with solvent as acetonitrile and a cleaning-up process with SPE or by dispersive SPE. In the last method, the sorbent is spread into the sample matrix solution.

In a word, the strengths of the SPE are numerous. We are citing some examples of them:

- The absence of emulsion between the two phases as is frequently seen in liquid-liquid extraction.
- The absence of adsorption of analytes on the glass as liquid-liquid extraction, even so it is necessary to be careful with hydrophobic analytes that can stick to the wall of the sampling bottle.
- The enrichment of the analyte in the sample.
- The improved stability of brittle molecules is due to the absence of the vaporization step.
- A high extraction rate is due to specific interactions.
- Automation except for the step of precipitation of proteins during a bioanalysis.
- Easy use with single-use sorbents.

While useful in many cases, SPE presents some shortcomings. Firstly, poor batch-to-batch reproducibility even if selectivity and reproducibility can be improved with the use of Molecularly Imprinted Polymers (MIP). Secondly, the silica sorbents are poorly stable at basic pH. Moreover, it is necessary to centrifuge or filter the complex sample before the SPE procedure at the risk of losing hydrophobic compounds and finally, the single-use cartridges are very expensive. One must not forget the possibly salting-out of particles that could harm the chromatographic instrument.

Non-Exhaustive Methods of Extraction on Solid Phase

Solid-Phase Microextraction

General Approach to the Solid-Phase Microextraction

At the end of the 1980s, Arthus et Pawliszyn (University of Waterloo, Canada) developed a new process of extraction of organic molecules, which were more and less polar, occasionally volatile, from an aqueous sample and even solid or gas, although they were in trace levels. It was called solid-phase microextraction (SPME). Contrary to the SPE, the compounds are partially extracted by adsorption or partitioning on a fiber made with silica covered by polymers or a suitable coating as a sorbent. The SPME is a solvent-free extraction method, which preserves the composition of the initial sample. It can be classified as a green technique with lower waste and by lowering the consumption of organic solvents. It can be automated due to its simplicity. Furthermore, the fiber could be used over one hundred times.

The fiber is introduced directly into the sample for liquid or gas or in the headspace of liquid complex matrix samples. The desorption and the analysis are performed using the thermal desorption process and gas chromatography instruments or by flow-through by the mobile phase if the analysis is carried out by the liquid chromatography system.

Theoretical Developments

The adsorption and partitioning involved do not match with a total extraction. All the aspects of the process have been studied as well as the influence of the nature and the thickness of sorbents and the extraction time. Its conditions of handling have been studied from the thermodynamic and kinetic points of view.

Distribution Study From the Thermodynamic Point of View. For example, the compounds are extracted by partitioning on a fiber recovered with a suitable coating. The analyte is distributed between the coating and the liquid sample matrix (or headspace).

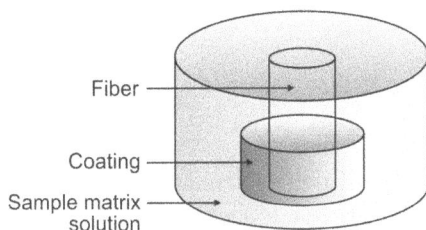

Figure 3: Distribution of a Compound S Between the Coating of Fiber and Sample Matrix. (Reprinted with the Permission from Gwenola Burgot, Solid Phase Microextraction, Techniques de l'Ingénieur, Paris, 2019, P1430V2).

Let us take a compound S in a liquid sample matrix wherein plunges the fiber as an example. The equilibrium conditions of compound S with both phases (or equality of the chemical potential) require that Relation 1 is verified:

$$K_{fl} = C_f^\infty / C_l^\infty \tag{1}$$

with

$$K_{fl}$$
$$S_l \rightleftarrows S_f$$

The equation of the conservation of the matter can be written:

$$C_o V_l = C_f^\infty V_f + C_l^\infty V_l$$

with

K_{fl} = Distribution constant of S between the liquid sample matrix and the coating
C_f^∞ = Concentration of S in the coating when the equilibrium is obtained
C_l^∞ = Concentration of S in the liquid sample matrix when the equilibrium is obtained
C_o = Initial concentration of S in the liquid sample matrix
V_l = Liquid sample matrix volume
V_f = Coating fiber volume

In order to be rigorous, K_{fl} is a ratio of the activities of compound S. In fact, it is possible to assimilate activities and concentrations in this case because the concentrations of the compound in both phases are very low. Moreover, the compounds are often ionized.

If we take, as an example, a coating made of polydimethylsiloxane (PDMS), it is possible to generalize for all the porous polymers if the concentration of the analyte is kept low. In this case, the surface area available for the adsorption process is in proportion to the volume of the coating if the porosity of the sorbent is homogeneous. For the polymers with high porosity and the sample with a high concentration of the analytes, the equations are more complicated. As a result:

$$K_{fl} = (n_f/V_f) \times (V_l/n_l) \tag{2}$$

with $n_l = n_0 - n_f$

n_0 = initial number of moles of compound S in the liquid sample matrix,
n_f = number of moles of compound S adsorbed by the coating of the fiber at the equilibrium,
n_l = number of moles of compound S in the liquid sample matrix at the equilibrium.

From the relationship (2), it is possible to extract n_f

$$n_f = K_{fl} V_f n_0 / (K_{fl} V_f + V_l) \quad \text{or more}$$

$$n_f = K_{fl} V_f C_0 V_l / (K_{fl} V_f + V_l) \tag{3}$$

If $K_{fl} V_f$ is negligible relatively to V_l the Equation (3) becomes:

$$n_f = K_{fl} V_f C_0 \tag{4}$$

This condition is, in general, met if the volume of the sample matrix V_l becomes no less than 2 ml. This relation is interesting because it shows that the quantity of the compound which is extracted is independent of the volume of the sample. Consequently, the volume of the sample is not necessarily measured with great accuracy. This fact is always interesting because it is a means to limit errors.

On the other hand, if K_{fl} is high enough that:

$$K_{fl} V_f \gg V_l$$

The Equation (3) becomes:

$$n_f = C_0 V_l = n_0,$$

The extraction of the analyte is complete. This is particularly interesting for the sample of low size, which is difficult to extract by other methods. Naturally, it is still possible to obtain a complete extraction with some compounds for which the distribution constant is not so high. For that, we realize several experiments by dipping the fiber many times in the same sample. Even so, it is time-consuming.

The relation (3) is based on several hypotheses:

• The driving of the substance S from the liquid sample matrix to the coating must only be by a diffusion process. This means that the quantity of the analyte adsorbed as a function of time could be obtained from the Fick law without any extra energy to drive the analyte.
• The analyte does not disturb the physical properties of the sorbent.

Improving the equilibrium state, it is possible to adjust a convection phenomenon by stirring regularly the sample matrix to make it more homogeneous. We note from the relations (3) and (4) that the quantity of the analyte adsorbed depends on the K_{fl} constant, V_f volume of the coating and in a word the thickness of the coating and the initial concentration of analyte in the sample. Intuitively, these results make sense.

The formula, more complicated if the extraction is realized from the headspace, is not given here and also involves the initial concentration C_0. However, its simplification is immediately obvious. The solid phase microextraction, by taking into account the differences in distribution coefficient between both phases, is frequently more selective than the other extraction methods.

Distribution Study From the Kinetics Point of View. The choice of the type of extraction (from the liquid sample or from headspace) impacts the kinetics of exchange. Indeed, when the fiber is put into the headspace above the liquid sample matrix, the compounds which are in the gas phase, ones that are more volatile, are extracted the first and the less volatile are extracted the last. The increase in the temperature accelerates the process because it affects the vapor tension. As a rule, for volatile compounds, the time to reach the equilibrium is lower if the extraction is realized from the headspace and for two reasons:

- The headspace realizes a concentration of volatile compounds.
- The diffusion speeds in a gas phase are higher than in a liquid phase.

Implementation

Description of the Apparatus.

Static Sample Matrix

- Fiber and the holder: The holder looks like a micro-syringe and presents two parts (Figure 4). These include:
  - a fiber (1 cm) coated with a polymer (7 to 100 µm thickness)
  - a stainless-steel piston

The fiber and piston can easily slide within a needle. The holder is compatible with all the fiber and can be introduced in the system injection of a chromatography instrument. To facilitate the handling, some automated apparatus can be used on the computer with pilot software to control the adsorption and desorption steps.

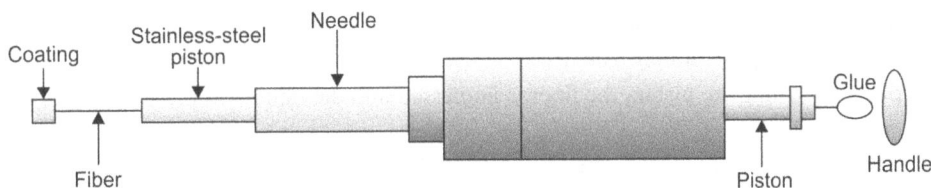

Figure 4: Holder System Syringe. (Reprinted with the Permission from Gwenola Burgot, Solid Phase Microextraction, Techniques de l'Ingénieur, Paris, 2019, P1430V2).

- Stir Bar Sorptive Extraction (SBSE): Stir bar sorptive extraction (SBSE) has been developed and is used increasingly. The principles of the extraction on the stir bar are similar to one of SPME, but the coating covers a stir bar and not a fiber that is placed into the aqueous sample matrix (SBSE). The stir bar is washed with water to eliminate salts and proteins and then is dried and incorporated into the gas chromatography instrument for thermal desorption. The desorption in the case of liquid chromatography is realized by immersing the stir bar in a solvent such as methanol, acetonitrile or a mix of both.

  The most used stir bar coating is polydimethylsiloxane (PDMS) due to its resistance in a broad temperature range and its high stability toward organic solvents. But PDMS is not used to perform extractions of polar compounds.

  SBSE has a higher extraction capacity than SPME, particularly it is possible to extract 50 to 250 times more analyte. As a result, SBSE improves the sensibility and reproducibility of the analytical instrument. The stir bar can be used more than 50 times.

- Thin Film Microextraction (TFME): In the case of TFME, the polymer is deposited on a slide. This system which could appear so close to SPE is different by no pump is used to circulate the sample. The extraction is static and the slide is introduced in the sample matrix solution. TFME permits the extraction of biological matrices as saliva *in situ* or by deposing a few microliters on the slide. TFME is also available in the 96-well plates, which are perfectly adapted with an automating system.

Capillary SPME or In-Tube Solid-Phase Microextraction IT-SPME. In capillary SPME or in-tube microextraction, the fiber is substituted by a capillary column or tube coated with the same polymers as those described for the system syringe. The sample matrix solution is no longer static. It flows through a capillary continuously or cyclically. The capillary column is introduced in a sampling loop of a chromatographic instrument. An automatic injector introduces the sample solution in the sampling loop. The analyte is adsorbed on the phase of the capillary inlet column and desorbed a second time by the mobile phase flow. The efficiency of the extraction depends on:

- the analyte equilibrium constant between the sample matrix solution and the sorbent,
- the speed of the flow in the capillary inlet column,
- the geometrical characteristics of the capillary: length, internal diameter and thickness of the coating.

The IT-SPME can be coupled with all the instruments used for trace analysis, which are LC-UV, LC-MS, CE, MALDI-MS, Cap-LC and even UHPLC (see Chapter 42). In the latter case, the instrument involves two sampling loops; one for the extraction capillary and another for injection in the analysis column due to high pressions used in the UHPLC.

Handling. The sample is introduced in a glass vial equipped with a barrel to homogenize the solution (Figure 5). The vial is closed with a plug. Consider the example of the fiber, in this case, the process evolves in three steps:

- Step 1: The fiber is inserted into a needle which is used to pierce the septum of the vial containing the analyzed sample solution. The needle is raised at 1 cm below the plug.
- Step 2: By pushing the piston, the fiber is immersed into the liquid sample or its headspace for 2 to 15 minutes.
- Step 3: By pushing the piston, the fiber retracts back into the needle. The needle is removed from the vial and introduced into the chromatography injector (GC or HPLC).

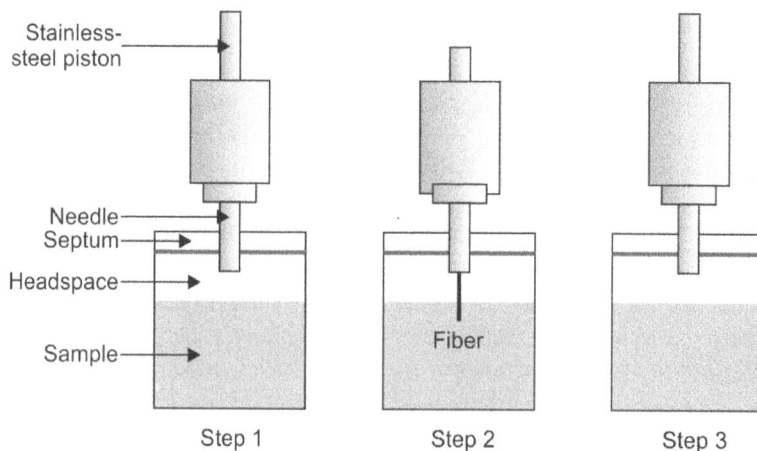

Figure 5: Adsorption on the Fiber From Solid Phase Microextraction. (Reprinted with the Permission from Gwenola Burgot, Solid Phase Microextraction, Techniques de l'Ingénieur, Paris, 2019, P1430V2).

Between two trials, the fiber is clean-up by the thermal or washing method.

The Coatings of the Fiber

The choice of the coating depends on the analyte especially its polarity, volatility and chemical properties.

There exist several groups of coatings:

- Liquid polymers
  - Polydimethylsiloxane (PDMS) is a liquid with a high viscosity. The PDMS has a very high thermal stability to 300°C and presents a high lifetime of the order of a hundred times for use. However, the sensibility of this coating is limited for the polar compounds.
  - Polyacrylate (PA), The PA is a better coating for extracting polar compounds. The partitioning mechanism is involved with these coatings for the extraction.

- Porous polymers

  Divinylbenzene (DVB) is coupled with PDMS or with polyethylene glycol (Carbowax®). The analytes are adsorbed at the surface of the coating, and the adsorption isotherm can be described by Langmuir's law but only if the concentrations of the species in both phases are low.

- Very porous materials

  This includes carboxen® carbon-based with a specific surface of 1,200 m^2/g. Its high porosity provides to the fiber to keep strongly the low size analytes and to obtain a high capacity to manage sample matrix solution. At the surface of the sorbent, the presence of numerous layers of the compound is responsible for non-linear isotherms. In this case, that means that the ratio of the concentrations is not constant.

- In the particular case of the mix PDMS-Carboxen®, the sorbent contains many low holes responsible for a supplementary screening by the size of the molecules of analytes. This coating is interesting for the small volatile compounds such as BTEX (benzene, toluene, ethylbenzene and xylene).

To increase the potential of the SPME, new coatings are developed in particular:

- Crown vinyl ether compounds arranged in formed polar cage molecules which can encapsulate many different analytes.
- Carbon-nanotube can be attached to the fiber. The main benefits are a high surface-to-volume ratio.
- Molecularly Imprinted Polymers as those described in SPE.
- Some polymers based on polypyrrole can be deposited on a platinum wire. They permit the extraction of compounds according to two processes: adsorption and electrochemical reaction. Thus, polypyrrole is oxidized with the positive potential and its positive charge permit keeping anionic compounds. The latter ones are released by the application of negative potential.

  For the first use, all the fibers involve a conditioning step between one hour to four hours before using in an injector of gas chromatography apparatus (two hours at 250°C with helium, for example) or liquid chromatography to eliminate the residues of compounding.

Optimization of the Solid Phase Microextraction

Before using SPME, it is necessary to optimize the handling parameters including:

- The choice between a direct or a headspace extraction.
- The type of polymers is almost as important as the choice of organic solvent in liquid-liquid extraction.

- The time to obtain the distribution equilibrium of the analyte between the sample solution and the sorbent.
- The size of the vials.
- The ionic strength and the pH of the sample solution.
- The time and the conditions of desorption of the compound from the polymer. For gas chromatography, the efficacity depends on the volatility of compounds, the thickness of the coating, the injector temperature, the time of exposition of the fiber and the carrier gas flow. In liquid chromatography, the flow and the composition of the mobile phase affect the results of desorption.

As regards calibration, it is necessary to link the amount of the analyte onto the fiber to its concentration in the sample matrix solution according to the most used method, such as the standard addition one. Moreover, the absence of ghost peaks must be verified with fiber without compounds before handling. Internal standards are also used and must have a structure very close to one of the analytes.

Advantages and Weaknesses

The SPME offers several advantages including:

- A short time to prepare the samples with preconcentration, extraction and clean-up procedure in one step.
- Easy automation to eliminate the manual handling of samples.
- A low sample requirement ranging 2–10 ml.
- A very few interferences with non-volatile compounds.
- No solvent use, which is good for respecting the environment. Moreover, it becomes possible to use some shorter columns in gas chromatography. The period of analysis becomes shorter and the detection limits too.
- The selectivity is interesting and linked to the choice of polymers and thus the possibility to simplify the chromatogram.
- The determination of compounds at trace levels of the order of (p.p.t) ($ng.L^{-1}$).
- Results confirm the satisfactory repeatability, reproducibility, accuracy and detection limit. For example, some results of the order ng/ml are obtained for the detection limit with good linearity, satisfactory repeatability and reproducibility. For the target analytes, recovery rates in the range of 85–115% are most commonly found.
- SPME is a better method than SPE for the long analysis process, which involves an important amount of organic solvent. The coupling of the SPME with chromatographic instruments is easy and provides quantitative use.

Despite these many advantages, SPME presents some limitations. The most frequent shortcoming is fiber breakage. The introduction of the fiber directly in the sample matrix decreases its lifetime and the reproducibility of the analysis. It is due to the presence of salts, at pH values too acidic or alkaline or of some solvents as alcohol.

Applications

SPME is a complementary tool to other methods of extraction. This has an advantage which is to achieve the sample purification and pre-concentration in one step from the extraction to the transfer in chromatographic instruments. SPME provides a large range of applications for the analysis of:

- environmental pollutants, such as pesticides (atrazine), volatile organic compounds, phenolic compounds and polycyclic aromatic hydrocarbons.
- Food (pesticides in fruits and vegetables, and wine aromas)
- Biology and toxicology (methadone, metabolites and cannabinoids)
- Pharmaceutical products (residual solvents in the bulk product)
- Forensic chemistry (explosive residue).

However, the device fragility and the low extracted quantity of analytes led to optimizing this process. A new device has been focused on: solid-phase dynamical extraction (SPDE).

Solid Phase Dynamic extraction (SPDE)

SPDE is considered to be the replacement of the fiber of the SPME by a needle whose internal surface is coated with a sorbent. The needle is immersed into the liquid sample or into its headspace for extracting the analyte. It can be operated many times before retracting the needle by pushing the piston. The needle is removed from the vial and introduced into the chromatography injector (GC or HPLC). The extracted quantity is four or six times as high as SPME. The most important applications are the extraction of pesticides in water or volatile organic compounds in food products.

CHAPTER 12

Distillation

A distillation consists in vaporizing a liquid mixture by heating it and condensing the obtained vapors with the help of suitable equipment. Hence, it is concerned with the equilibria of liquid-vapor. In general, the constituents of a liquid mixture do possess different *volatilities*. As a result, the composition of the vapor is different from that of the initial liquid mixture and the condensation of the vapor leads to a liquid; the composition of which is different from that of the starting liquid.

Distillation is a method of separation and is of great importance in the proximate analysis as it permits resolving one liquid phase in its constituents, at least in parts. However, the simple distillation does not permit obtaining the components of the initial mixture in a pure state as it is demonstrated theoretically. On the contrary, it is the case with the rectification.

The study, presented here, contains the five following points:

- Some recall of physical chemistry is necessary in order to understand the theoretical foundations of the methods. We take advantage of these recalls by already defining some parameters; the values of which condition the success of a distillation.
- The case of the simple distillation of miscible compounds with the particular behavior of azeotropes.
- The fractioned distillation (or rectification) of miscible compounds.
- Some other methodologies of distillation.
- Some applications of the distillation.

Recalls

Some Physicochemical Properties of Compounds and Distillation

Ebullition Laws

Vaporization is the change of a compound from the liquid state to the gaseous state. It can occur in different ways among which we can notice the evaporation and ebullition. Ebullition is the tumultuous vaporization by the formation of bubbles in a liquid.

First Law. Under constant pressure, the ebullition of a pure compound begins always at the same temperature, and this temperature remains invariable during the time of the ebullition. The rule of phases gives the variance by the formula as follows:

$$v = c + 2 - \varphi$$

Here, v is the variance, c is the number of independent constituents and φ the number of phases. When the two phases, i.e., liquid and vapor, are present, the value of $v = 1$. One can fix only one factor. Once the latter is fixed, all the others factors are. It is the case, for example, of the ebullition temperature, and the liquid and vapor instantaneous concentrations. In the context of the first law, the parameter which is often arbitrarily fixed and maintained constant is the pressure. As a consequence, the temperature of ebullition is constant.

Second Law. A liquid compound boils at the temperature for which its maximum vapor pressure (pressure of saturated vapor) is equal to the pressure exerted on its free surface.

Vapor Pressure

Atmospheric Pressure. The atmospheric pressure is equal to the weight of a cylindrical column of mercury of 1 cm² of base and has for height a vertical distance between the levels of mercury in the tube of Torricelli and the tank (Figure 1).

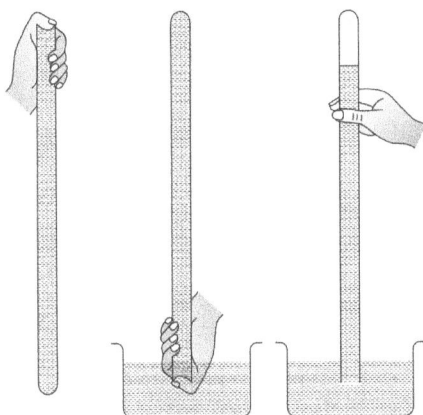

Figure 1: Atmospheric Pressure. (Reprinted with Permission from Gwenola and Jean-Louis Burgot, Paris, Lavoisier, Tec et Doc, 2017, 5).

Vapor Pressure in the Vacuum. Let us take two barometric tubes A and B of which A is the check sample. With the help of one syringe, let us introduce a few sulfuric ether drops into Tube B. (The choice of ether justifies itself by the fact that it vaporizes easily, but the experience described here is general). So long as there is very little injected ether in the tube, it is wholly vaporized and the level of the mercury decreases in Tube B (Figure 2). One says that one is facing a *dry vapor*. This dry vapor, in principle, obeys the law of perfect gases.

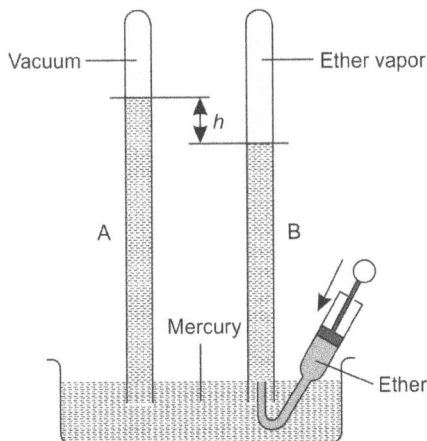

Figure 2: Production of Dry Vapor. (Reprinted with Permission from Gwenola and Jean-Louis Burgot, Paris, Lavoisier, Tec et Doc, 2017, 5).

Saturated Vapor Pressure

If the volume of injected ether is increased, one stage arrives at which the vertical distance no longer changes. At the same time, one notices the apparition of liquid ether above the mercury column. One says that one is in presence of *saturated vapor.* The height h is called saturated vapor pressure. It is only a function of the temperature and the nature of the liquid (Figure 3).

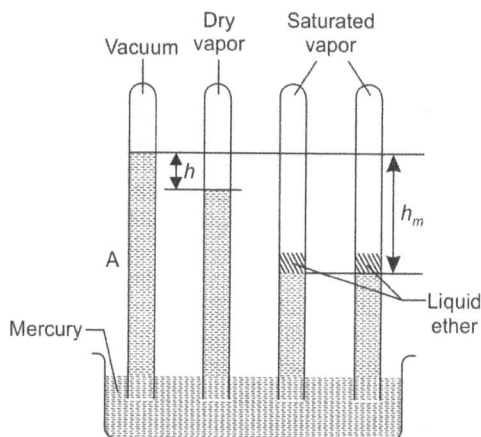

Figure 3: Saturated Vapor Pressure. (Reprinted with Permission from Gwenola and Jean-Louis Burgot, Paris, Lavoisier, Tec et Doc, 2017, 6).

Vapor Pressure in Limited Atmosphere: Dalton's Law

If in the preceding experience, air would be present in the tube in place of a vacuum, we should observe that the liquid (ether) vaporizes until its vapor pressure in the gaseous mixture (ether-air) would be equal to its saturated vapor pressure in a vacuum at the same temperature. This experimental fact is on the basis of the definition of partial pressure. More precisely, let us suppose that in the same recipient, we introduce several gases without mutual action.

One calls partial pressure p_i of each of the gases *i* when pressure it would exert if it would be alone in the total volume of the container at the same temperature. Dalton's law[7] stipulates that if the gaseous mixture has the behavior of a perfect gas, the pressure P exerted by the whole vapor is equal to the sum of partial pressures of the different constituents:

$$P = \Sigma_i \, p_i$$

Equilibrium Liquid-Vapor Diagram of a Pure Compound

The vapor pressure p of a pure liquid compound at ambient temperature and normal atmospheric pressure depends only on its nature and its temperature. If we draw the evolution of p as a function of the temperature, one obtains a curve of the type represented in Figure 4.

Such a curve is called the equilibrium liquid-vapor diagram. One sees that the vapor pressure increases with the temperature. One also notices that if the pressure exerted on the free surface is p_s, the ebullition temperature T_{eb} is fixed immediately, according to the second law of ebullition. The p_s is the saturated vapor pressure P of the compound at T_{eb} (notice that a partial pressure is symbolized by p minuscule and the saturated vapor by P majuscule).

---

[7] Dalton John, English physicist and chemist, 1766–1844.

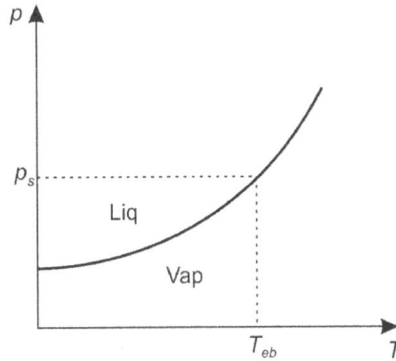

Figure 4: Change of the Vapor Pressure P of a Pure Compound as a Function of Temperature (K). (Reprinted with Permission from Gwenola and Jean-Louis Burgot, Paris, Lavoisier, Tec et Doc, 2017, 7).

The change of vapor pressure of a compound with the temperature is governed by Clapeyron's law[8] whose expression is:

$$dP/dT = \Delta_{vap}H_m/T\Delta V$$

Here, $\Delta_{vap}H_m$ is the molar (m: molar) heat of vaporization of the liquid or latent heat of vaporization. The ΔV is the change of its molar volume accompanying its transformation from its liquid state to its vapor state ($\Delta V = V_g - V_l$).

(*Remark*: It is not sufficient to be at the right temperature so that the vaporization occurs. It is also necessary that energy in the form of heat should be provided to the system. It is the latent heat of vaporization.)

If we neglect the volume of one mole of liquid with respect to one mole of gas ($V_l \ll V_g$), simplification which is often legitimate (for example, taking the case of water $V_l = 18$ cm^3 and $V_g = 22,400$ cm^3 in ordinary conditions). The vapor has the behavior of a perfect gas, one obtains:

$$V_g = RT/P$$

Clapeyron's equation becomes:

$$dP/dT = \Delta_{vap}H_m P/RT^2$$

Finally, if the latent heat changes only very little with temperature (that constitutes a legitimate hypothesis when the domain of variation of the temperature during the distillation of mixtures of liquids is weak, which is often the case), the differential equation is immediately integrable. One obtains:

$$\ln P = -\Delta_{vap}H_m/RT + Cte$$

Here, Cte is the integration constant. One can fix it if we know the ebullition temperature T_1 is corresponding to the vapor pressure P_1. For example, at the atmospheric pressure $P_1 = 1$ atmosphere,

$$\ln P_1 = 0$$

We deduce:

$$Cte = \Delta_{vap}H_m/RT_1$$

and

$$\ln P = -\Delta_{vap}H_m/RT + \Delta_{vap}H_m/RT_1$$

---

[8] Clapeyron Emile, French physicist, 1799–1864.

Behavior of a Pure Compound During its Distillation

In summary, during the distillation of a pure compound, according to what is preceding one can assert that:

- The temperature remains constant and equals the temperature of ebullition.
- The liquid phase coming from the condensation of vapors (the distillate) has the same composition as the pure compound submitted to the distillation. This is obvious since there is only one compound to distill. There is only one compound that can vaporize.

From the standpoint of separation, the distillation of a pure compound strictly affords no interest. However, the respect of the preceding characteristics by a compound constitutes a criterium of purity, but this criterium is not absolute since the azeotropes (see the section on 'The Curves of Liquid and Vapor Isobar Exhibit a Maximum and a Minimum Which Coincide') exhibit the same characteristics.

Mixture of Two Miscible Compounds: Some Properties Interesting for the Distillation

Equilibrium Liquid-Vapor

A liquid mixture of two miscible compounds presents a different behavior from that of a pure compound since its total Vapor Pressure P is a function not only of the temperature and of the nature of the compounds but also of their title in the initial mixture.

The equilibrium diagram liquid-vapor must be drawn in a system of three rectangular coordinates of the general type $P = f(T, C)$. We shall specify what parameters correspond to T and C. The variance $v = C + 2 - \varphi$ is 2 when the two phases liquid and vapor are together. As a result, for example:

- The temperature of ebullition of the mixture to distill T_{eb} and the total pressure above the system are fixed. Therefore, all other parameters, such as the vapor pressure of each compound and the instantaneous composition of the distillate, are fixed.
- The total pressure and the concentration of the mixture to distill are fixed, and consequently all the other parameters are fixed. For example, the temperature T_{eb} of the system is fixed.

(*Remark*: Some precisions are given in the study of the isothermal and isobar diagrams.)

In practice, the distillation is most often carried out at constant pressure (for example at atmospheric pressure), we shall study essentially the diagrams so-called isobar $T = f(C)$. They represent the changes in the ebullition and condensation temperatures as a function of the concentrations of liquid or gaseous mixtures for a total given pressure (Figure 5).

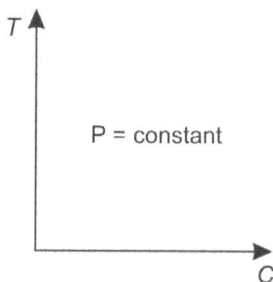

Figure 5: Isobar Diagram. (Reprinted with Permission from Gwenola and Jean-Louis Burgot, Paris, Lavoisier, Tec et Doc, 2017, 9).

Expressions of the Composition of a Binary Mixture

In order to express the composition of a binary mixture (compounds 1 and 2), for example, one can use:

- The molar concentrations
- The molar fractions

It is more convenient for the study of the distillation to use the molar fractions x. Their use permits a comparison of the values of concentrations in both phases easier to do. Let us use n_1 and n_2 as the numbers of moles of compounds 1 and 2 in the liquid phase. The molar fractions of 1 and 2 in it are:

$$x_1 = n_1/(n_1 + n_2) \quad \text{and} \quad x_2 = n_2/(n_1 + n_2)$$

Given the fact that for a binary mixture:

$$x_1 + x_2 = 1$$

Actually, one usually represents the molar fraction of the compound which is most volatile by x and that of the compound less volatile by $1 - x$ (for the binary mixtures). In the gaseous mixture, at every moment of the distillation, there exist n'_1 moles of 1 and n'_2 moles of compound 2. The molar fractions are symbolized by the letter y:

$$y_1 = n'_1/(n'_1 + n'_2) \quad \text{and} \quad y_2 = n'_2 (n'_1 + n'_2)$$

Again, in this phase, one usually represents the molar fraction of the compound which is most volatile by y and that of the compound less volatile which is $1 - y$.

Definition of the Ideal Solutions

Except in special circumstances, we only consider the case where the liquid phase is an ideal solution (Figure 6):

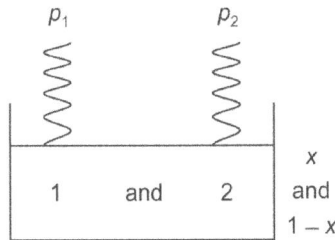

Figure 6: Definition of an Ideal Solution. (p_1 and p_2 are the vapor pressures of compounds 1 and 2 in the System). (Reprinted with Permission from Gwenola and Jean-Louis Burgot, Paris, Lavoisier, Tec et Doc, 2017, 10).

With this hypothesis that unfortunately does not correspond to the majority of the cases, the theoretical study of distillation is not too difficult.

Let us consider a binary liquid mixture of compounds 1 and 2 of compositions x and $(1 - x)$, and let p_1 and p_2 be the partial pressures of these two components above this solution. One of the possible definitions of an ideal solution is the fact that the different partial pressures p_1 and p_2 obey Raoult's law.[9] It can be written in terms of vapor pressures (in some conditions):

$$p_1 = xP_1 \quad p_2 = (1 - x)P_2$$

P_1 and P_2 are the saturated vapor pressures of compounds 1 and 2 at the same temperature. The partial pressures are proportional to their saturated vapor pressure at the same temperature. The

---

[9] Raoult, François – Xavier, French physicist and chemist, (1830–1901).

proportionality factor is their molar fraction. These relations are legitimate to define ideal solutions in the entire domain of concentrations.

(*Remarks*:

- Starting from this law, one can find another expression of Raoult's law related to the decrease of the vapor tension of the solvent, which is proportional to the molar mass of the solute.
- Raoult's law is nothing other than a particular case of Henry's law according to which:

$$p_i = k_i P_i$$

Here, $k_i \neq x_i$. Henry's law applies to non-ideal solutions and only when they are sufficiently dilute. It also concerns the solubility of gases in a liquid.)

Isobar Curves (Ideal Case)

They are the diagrams $T = f(C)$ for the total pressure P = constant. (C is the "concentration" in molar fractions x and y). They are the most interesting diagrams for the study of distillation.

Diagrams $T = f(x)$. According to Raoult and Dalton's laws, it is easy to demonstrate that:

$$x = (P - P_2)/(P_1 - P_2) \tag{1}$$

Besides, this relation is also satisfied for the case of isotherm curves. But for the study of the isobar curves, the pressure P is a constant whereas the saturated vapor pressures P_1 and P_2 change with the temperature. The variations, $P_1 = \phi_1(T)$ and $P_2 = \phi_2(T)$, are immediately accessible by application of Clapeyron's law by the relations given just above. By replacing P_1 and P_2 in the relations above, one conceives that we arrive at an expression of the kind that $T = \varphi(x)$; $\varphi(x)$ means "function of". It is the equation of a curve called *isobar of the liquid composition*.

Diagrams $T = f(y)$. One calculates y as a function of T via the relation;

$$yP = P_1 x$$

$$yP = P_1(P - P_2)/(P_1 - P_2) \tag{2}$$

Again, by using Clapeyron's equations in order to express P_1 and P_2 as a function of T, one comes to an expression of the kind; $\Psi(y)$ means "function of". The relation $T = \Psi(y)$ is the equation of the curve named *isobar of vapor composition*.

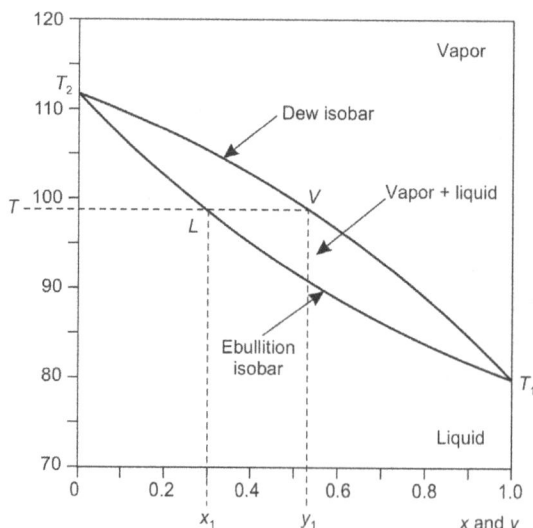

Figure 7: Isobar Curves (Ideal Case). (Reprinted with Permission from Gwenola and Jean-Louis Burgot, Paris, Lavoisier, Tec et Doc, 2017, 14).

These diagrams are built as follows. In ordinates, the temperatures are noticed. In abscissa, the compositions (in molar fractions) of the liquid and vapor phases are given attention. The compound that has the most volatile A also has the molar fractions x_1 and y_1 in phases of liquid and vapor. At the origin of the abscissa, on the right, $x_1 = 1$ (100% of A). As previously mentioned, when both phases (liquid and vapor) coexist, the system is bivariant. If the pressure is fixed (it is done by hypothesis) and if in addition, for example, the composition of the liquid mixture to distill is to be fixed, then the composition of the vapor and the temperature of ebullition are fixed (Figure 7).

The two isobars delimitate three domains. These are:

- Vapor
- Liquid
- Liquid and vapor

For example, the liquid mixture acetone (boiling point 56°C) and toluene (111°C) obeys this case. When a liquid of composition represented by x_0 boils at temperature T_0, the composition of the vapor first evolved is given by y_1 since it is richer in A $y_1 > x_1$. It is the basis of the principle of separation by distillation. The liquid that is remaining becomes relatively richer in B, and the boiling point consequently rises.

Azeotropic Non-Ideal Solutions

For non-ideal solutions, Raoult's law is no longer satisfied. In order to reason analogously (as previously seen), one must set up the following relations:

$$p_1 = x_1 P_1 \gamma_1 \quad \text{and} \quad p_2 = x_2 P_2 \gamma_2$$

Here, γ_1 and γ_2 are called the *activity coefficients* of compounds 1 and 2. Their values can be superior or inferior to unity. When they are equal to unity, one is again in the case of ideal solutions. The values of the activity coefficients indirectly measure the difference in behavior between the non-ideal system under study and the corresponding ideal one. They depend on the temperature, pressure and concentrations of the species. Thus, the activity coefficients are not constant when one describes the isobars of liquid and vapor composition. Given the two relations defining them, as mentioned above, they are determined experimentally by measurements of partial pressures.

The non-ideality of the solutions which is taken into account by the introduction of the activity coefficients has for consequence the deformation of the isobar diagrams (and also of the isothermal ones) with respect to those obtained with ideal mixtures. Two cases can be mentioned. We only consider the isobars.

The Curves of Non-Ideal Liquid and Vapor Isobars Composition Neither Exhibit Maximum Nor Minimum. They are connected to the ebullition points T_1 and T_2 of the pure constituents. If the curvatures are of opposite signs, one is not very far from the case of ideal solutions (Figure 8a). If they are of the same sign, one obtains the diagrams represented in Figures 8b and 8c.

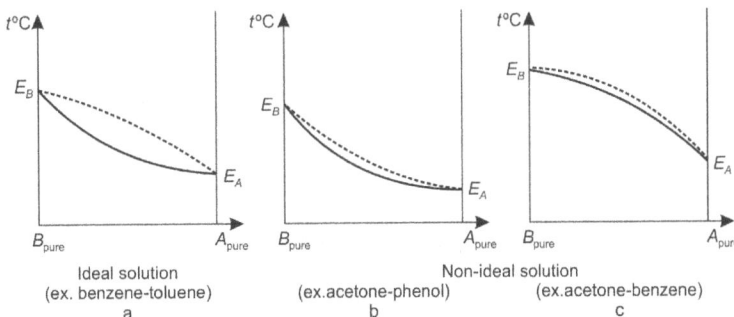

Figures 8a, 8b, 8c: Isobars: Non-Ideal Case. Curves Without Maximum Neither Minimum A, B and C. (Reprinted with Permission from Gwenola and Jean-Louis Burgot, Paris, Lavoisier, Tec et Doc, 2017, 15).

The Isobar Curves of Liquid and Vapor Exhibit a Maximum and a Minimum Which Coincides (Figures 9 and 10).

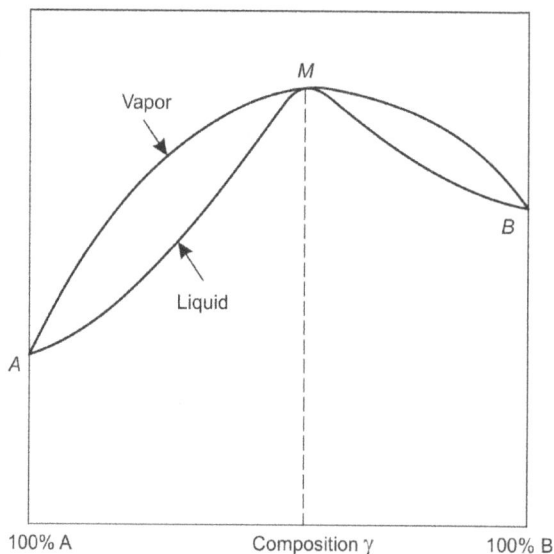

Figure 9: The System with a Maximum Boiling Point.

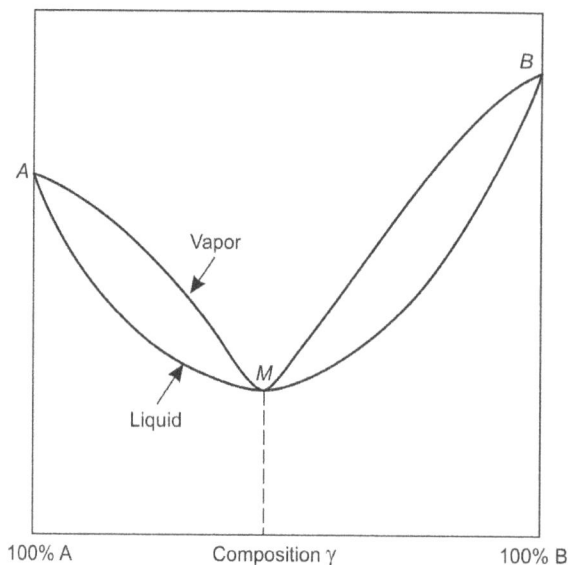

Figure 10: The System with a Minimum Boiling Point.

One notices that the particular mixtures of composition γ have the behavior of a pure compound since the vapors have the same composition as the liquids in equilibrium with them. Hence, these mixtures boil at a constant temperature. They are called *azeotropic mixtures*. The phenomenon giving rise to the formation of azeotropic mixtures is also called "azeotropism". It is called *negative* when the isobar shows a maximum. It is *positive* when it is an inverse case.

(*Remark*: The symbol γ represents only one composition. It has nothing to do with the activity coefficients γ_1 and γ_2.)

Relative Volatility

We shall now define the fundamental parameter that governs every distillation, the *relative volatility* α which is—for an ideal solution—the ratio of the saturated vapor pressures of the two pure constituents 1 and 2 at the ebullition temperature of the mixture:

$$\alpha = P_1/P_2$$

The separation is all the more difficult as the coefficient α is close unity. By convention, $\alpha \geq 1$ since one sets up the vapor pressure of the compound which is most volatile on the numerator.

The relative volatility changes with the temperature:

$$\ln \alpha = \ln P_1 - \ln P_2$$

If one expresses P_1 and P_2 by their values as a function of the temperature given by Clapeyron's equation:

$$\ln P_1 = -L_1/RT + L_1/RT_1$$

Where L_1 is the latent heat of vaporization of the compound, T_1 is its ebullition temperature for a pressure of 1 atmosphere and likewise for L_2 and T_2.

$$\ln P_2 = L_1/RT_1 - L_2/RT_2 + (L_2/RT - L_1/RT)$$

$$\ln \alpha = \text{constant} + (L_2/RT - L_1/RT)$$

Here, α is independent of the temperature when $L_2 = L_1$. This is true when the two compounds 1 and 2 are very close to the standpoint of their chemical structure. Moreover, when they are located in a narrow domain of temperature. We admit, however, this hypothesis for this study of the simple distillation and then the rectification.

Relation Between α and the Molar Fractions X and Y

Ideal Case. According to Dalton's law:

$$P = p_1 + p_2$$

If we suppose that the vapors of each compound and those of their mixture have the behavior of perfect gases and if there exist in the vapor phase n'_1 moles of compound 1 and n'_2 of 2, one can write:

$$p_1 = n'_1 RT/V \qquad P = (n'_1 + n'_2)RT/V$$

Now, by definition of partial pressure:

$$p_2 = n'_2 RT/V$$

It follows that:

$$p_1 V/PV = n'_1 RT/(n'_1 + n'_2)RT$$

$$p_1 = yP$$

And likewise:

$$p_2 = (1 - y)P$$

On another side, according to Raoult's law:

$$p_1 = xP_1 \quad \text{and} \quad p_2 = (1 - x)\,P_2$$

From the definition of the volatility coefficient, one finds by replacement:

$$\alpha = (yP/x)[(1 - x)/(1 - y)]$$

$$\alpha = y\,(1 - x)/x(1 - y) \tag{3}$$

This is the relation expressing the equilibrium liquid/vapor. It relates the molar fractions y and $(1 - y)$ in the gaseous phase to the molar fractions x and $(1 - x)$ in the liquid phase as well as to the volatility coefficient α. It is very interesting to remark that relation (3) shows a clear analogy with the general one:

$$K = C_1/C_2$$

It defines the partitioning coefficient K of a solute between two phases 1 and 2 at equilibrium, C_1 and C_2 being the concentrations in both phases at equilibrium (see Chapters 4 and 14).

Remark: When $\alpha = 1$, it comes $x = y$. The molar fractions of each compound are the same in the liquid and vapor phases and no enrichment is possible. The distillation presents no interest.

Diagram y = f(x)

It is convenient and useful (viz., under the theory of rectification according to McCabe and Thiele) to draw the values of y as a function of x by starting from the diagram isobar. The relation (2) above shows that for a given value α, one is in presence of one hyperbole of the equation:

$$y = \alpha x/[1 + (\alpha - 1)x]$$

The simple examination of the isobars shows that the hyperbole is located above the diagonal of the square $y = x$.

Non-Ideal Case. Let us introduce the real volatility coefficient α_{real} in order to consider this case.

$$\alpha_{real} = \alpha_{ideal}\,\gamma_2/\gamma_1$$

The α_{real} is defined as above and depends not only on the temperature but also on concentrations because of the presence of activity coefficients γ_i. For the azeotropes, we may see, when we consider the isobars, that the curve $y = f(x)$ intersects the diagonal $y = x$ for the concentration γ.

Simple Distillation of Miscible Compounds

Principle and Apparatus

For the simple distillation, we only investigate the batch distillation, that is to say, a discontinuous distillation. One loads the boiler once and for all with the whole quantity of the mixture. The distillation simply uses a flask, a refrigerating system and a receptacle to collect the distillate (Figure 11):

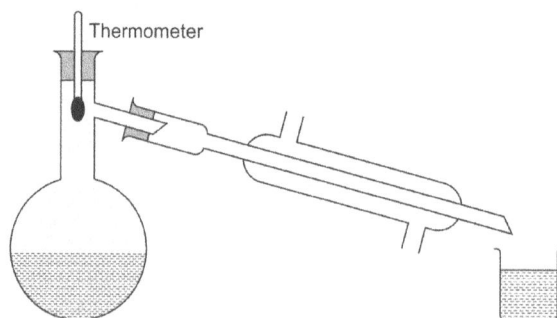

Figure 11: Apparatus for a Simple Distillation. (Reprinted with Permission from Gwenola and Jean-Louis Burgot Paris, Lavoisier, Tec et Doc, 2017, 19).

When the ebullition takes place, the infinitely small puff of vapor emitted by the liquid at a given moment and in equilibrium with it is immediately and totally condensed at the level of the cold wall constituted by the refrigerating device, without having lost any change of composition between its emission and its condensation. Then, it gives an infinitely small drop of the distillate of composition identical to its proper composition. This infinitely small drop adds to others that have already been obtained before. These conditions are purely theoretical and are never encountered in practice. The vapor indeed partially condensates itself on the wall of the flask before attaining the refrigerating system, and the exchanges take place between the formed liquid which refluxes toward the contents of the flask and the vapors subsequently produced. Therefore, the composition of the latter changes. We shall see that these conditions are those of *rectification*.

However, we shall consider that the theoretical conditions are obtained for the study of simple distillation.

Qualitative Study

Let us start by saying that the notions of the instantaneous composition of the distillate, which is that of the drop that is forming at a given moment of the distillation, must be well differentiated from the average composition, which is that of the distillate obtained at this moment since the beginning of the distillation.

Binary Mixture Giving No Azeotrope

We follow the evolution of the different phases during the distillation on an isobar diagram by starting from a liquid mixture of composition x_0 (Figure 12). Ebullition begins at the temperature T_0 and the first puff of emitted vapor; hence, the first drop of distillate has the composition y_0. Once this first puff is emitted and condensed, the composition of the liquid mixture has changed since it is enriched in the liquid the less volatile. It becomes x_1 with;

$$x_1 < x_0$$

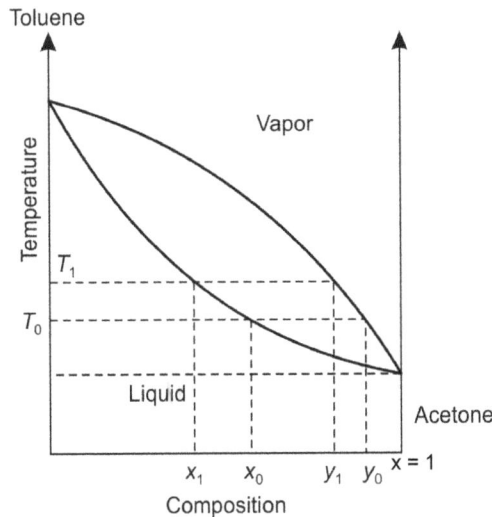

Figure 12: Composition of Phases During a Simple Distillation at Constant Pressure.

Because of this fact, the pressure remaining is fixed and the composition of the liquid phase is known (fixed now at x_1). The temperature of ebullition is also fixed according to the phases' rule. It enhances to T_1. Likewise, the instantaneous composition of the distillate is enriched with the less volatile one, which takes the composition y_1 with $y_1 > y_0$ and so forth.

Therefore, one describes the curves of ebullition and condensation until the ebullition temperature of the pure less volatile liquid. One sees that the ebullition temperature regularly increases. Concerning now the average composition of the whole distillate, it follows a different curve. Because the vapor is always richer in the most volatile compound than the liquid which has generated it (except exactly at the end of the distillation at the temperature of ebullition of the liquid the less volatile), the average distillate is always much richer in the compound which is also more volatile. When the temperature reaches that of ebullition of the compound the less volatile, all have been distilled and condensed. The distillate has the same composition as the initial liquid.

Binary Mixtures Giving One "Azeotrope"

We must have a look at Figures 13a and 13b where the forms of the diagrams are given by the negative and the positive "azeotropes".

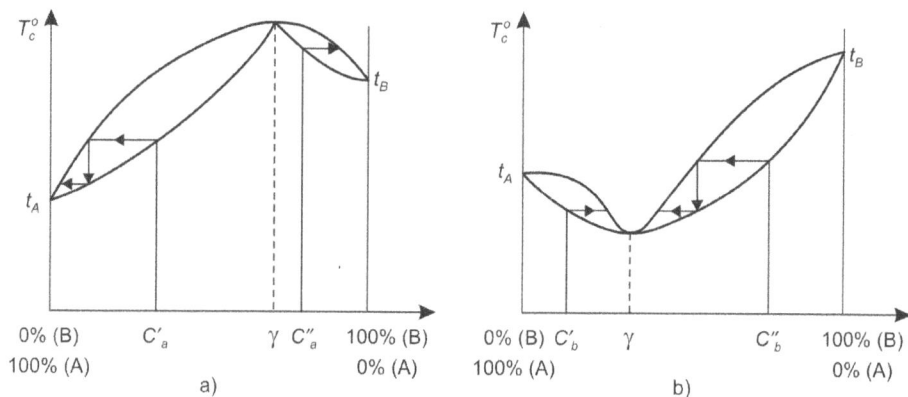

Figures 13a and 13b: a) Negative Azeotropism and b) Positive Azeotropism.

The matter is to know the possibilities of separation when one distills such mixtures. We study the case of the mixture of the two compounds A and B in which A is the most volatile. These data are simply obtained by the study of the two kinds of diagrams by following the isobars of liquid and vapor composition during the distillation. One important element of differentiation is the location of the starting points of the isobar-scanning with respect to the composition γ of the azeotrope. The results are:

- In the case of negative azeotropism, if the initial mixture possesses one composition in which A is inferior so that it has in the azeotrope then the distillation affords compound A as the heading-product and at the end of the experiment the azeotrope. If the initial title of A was superior to that it possesses in the azeotrope, the order of obtention (in time) is Compound B and the azeotrope.

- In the case of positive azeotropism, if the initial title of A is inferior to the composition γ, the distillation first affords the azeotrope and then A. If the title of A is superior to the composition γ, the distillation affords the azeotrope and then B. One never obtains compounds A or B pure in this case.

Concerning more generally the azeotropic mixtures, let us mention now that they can be solved if one changes the conditions of distillation. One has at our disposal physical and chemical means to do that. They are:

- Physical Means

These can be the reduction of the pressure. The composition of azeotrope changes with pressure. For example, in the case of the negative azeotrope water-alcohol, the title in water decreases as the pressure decreases (Table 1):

Table 1: Example of Water/Ethanol Azeotrope.

P (mmHg)	Temperature ebullition°C	Water%
760	78	4.4
70	27	0 (pure ethanol)

There is also the action of another solvent (method of Sydney Young). In the case of ethanol, the industrial process consists in carrying away water by distillation with a third constituent, i.e., benzene. If one adds anhydrous benzene to the 96° ethanol by distillation, one successively obtains a ternary azeotrope water/ethanol/benzene that distills at 64°C and then a binary azeotrope benzene/ethanol that distills at 68°C and finally the pure ethanol.

- Chemical Means

Pure ethanol can be obtained by distilling alcohol on a dehydrating agent (Mg).

Important Conclusions

- The first drop of distillate is not constituted by the compound that is most volatile and pure. One cannot hope to separate the latter from the initial mixture but only to concentrate it in the distillate, and this is possible if one does not distillate all the mixture.
- It is the same for the compound that is less volatile. This is only the last drop of liquid, that is to say, one tiny fraction of its total volume which is theoretically constituted by this pure compound (In order to obtain a distillate more concentrated in the compound that is more volatile than the initial one, one can isolate the "head portion" of the distillate; it is the most concentrated in the compound the most volatile and once more submit it to a new simple distillation by isolating again the head portion, which is still richer than the preceding one. By renewing the process several times, one ends with a distillate, the composition of which is very close to that of the compound that is the most volatile and pure, but at the expense of a quantitative yield that is weak. Here, are the conditions analogous to those of rectification but conducted to a lesser yield).

Quantitative Study (No Azeotrope Mixture Present)

We shall successively consider Rayleigh's equation,[10] the determination of the average concentration of the distillate at a given moment of the distillation, of the yield of the distillation and finally the notion of the curve of distillation.

Rayleigh's Equation

This equation connects the instantaneous composition of the distillate to the number of moles of the mixture remaining in the flask by the experimental measurement of $x_D = y$ (known most often by refractometry) through the value α (Figure 14). The x_D is the *instantaneous concentration of the drop distillate.*

---

[10] Lord Rayleigh, English physicist (1842–1919), Nobel prize in physics–1904.

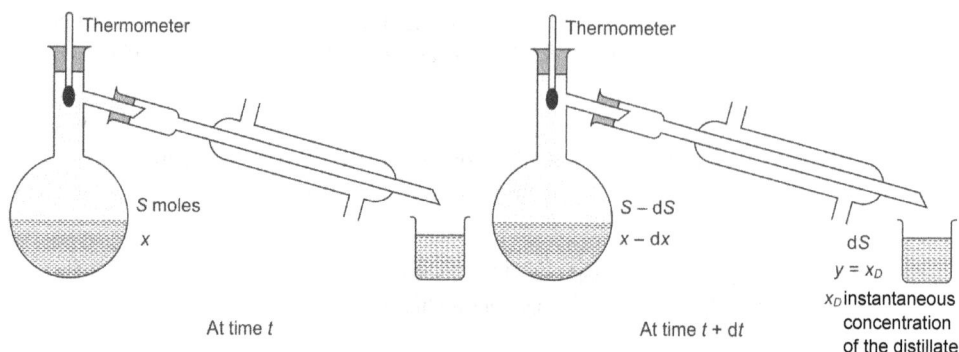

Figure 14: Symbolism Used to Set Up Rayleigh's Equation. (Reprinted with Permission from Gwenola and Jean-Louis Burgot, Paris, Lavoisier, Tec et Doc, 2017, 22).

Let us consider any moment during the distillation when the flask contains S moles of the liquid mixture of which the molar fraction is x. During an infinitely short time interval dt, a number dS moles of the mixture is volatilized to form a vapor, the molar fraction of which is y. After condensation in the refrigerating system, it gives dS moles of distillate. In the flask remain $S - dS$ moles of the mixture, the molar fraction of which changed $- dx$. The sign minus takes into account the fact that the mixture becomes poorer in the most volatile compound during the distillation.

Let us write the equation of the conservation of the matter applied to the compound the most volatile. Its quantities are:

- xS moles before the transformation
- $(x - dx)(S - dS)$ moles for the liquid remaining in the flask after the transformation
- $x_D dS$ moles for the distillate obtained after the transformation

The equation of conservation of the matter is:

$$xS = xS - xdS - Sdx + dxdS + x_D dS$$

By neglecting the infinitely small quantity of the second order dxdS

$$Sdx = - xdS + x_D dS$$

$$dS/S = dx/(x_D - x)$$

Between two moments of the distillation—for example, from the time to which the experience begins with the flask containing S_o moles of the mixture and the fraction molar of which x_o to the time t_1 for which it remains S_1 moles of the mixture in the flask and the molar fraction of which is x_1, the integration leads to:

$$\ln S_1 - \ln S_o = \int_{xo}^{x1} dx/(x_D - x)$$

Let us recall that the instantaneous composition of the distillate x_D is a function $f(x)$ of the composition x of the mixture. Therefore, the preceding relation can be written:

$$\ln S_1 - \ln S_o = \int_{xo}^{x1} dx/[f(x) - x]$$

The solution to this differential equation is more easily accessible by graphical means. Its research consists in drawing $1/(x_D - x)$ as a function of x through the relation $x_D = \alpha x/[1 + x(\alpha - 1)]$, which is simply the equilibrium relation between the instantaneous composition of the vapor and the composition of the mixture remaining to distillate ($y = x_D$). Then, one measures the area under the curve between the abscissa x_0 and x_1 (Figure 15). Let us remark that the area is negative since $x_1 < x_0$:

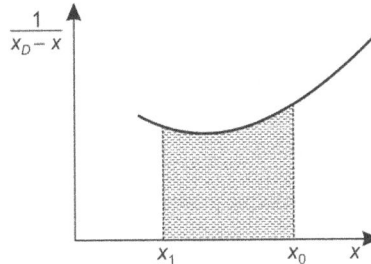

Figure 15: Graphical Solution of the Differential Equation. (Reprinted with Permission from Gwenola and Jean-Louis Burgot, Paris, Lavoisier, Tec et Doc, 2017, 24).

(*Remark*: The algebraic solution of the integral is accessible since x_D is related to x since one knows this function. It is the equilibrium curve. By supposing α constant in the domain of temperatures one is working, the integral solution is:

$$\ln(S_1/S_0) = 1/(\alpha - 1)[\ln x_1/x_0 - \alpha\ln(1 - x_1)/(1 - x_0)]$$

This relation permits to calculate S_1 if one knows $y = x_D$ through x_1 since the knowledge of x_0 commands that of x_1. The equilibrium condition indeed leads to the expression;

$$x_1 = x_0/[\alpha - x_D(\alpha - 1)]$$

If, of course, one knows α, x_0 and S_0.)

In the case of nonideal solutions, one can only integrate graphically given the fact that α varies too widely in the suitable domain of temperature. One still draws the curve $1/(x_D - x)$ as a function of x. (One does not forget that in the just previous reasonings, x_D is accessible experimentally by a measurement of the instantaneous composition of the drop of distillate for example by refractometry.)

Determination of The Average Concentration of The Distillate at a Given Moment of the Distillation

Let us consider a simple distillation by starting from S_0 moles of the mixture, the molar fraction of which is x_0 which is broken off when the instantaneous is $y = x_{D1}$. We immediately can deduce:

- On one hand the molar fraction according to the equilibrium equation
- On the other one, the number of S_i moles of mixture remaining in the flask according to Rayleigh's equation.

The distillate is then constituted by $S_0 - S_i$ moles of the mixture, the average molar fraction of which is x_{DM} which we want to determine. Let us write the equation of the conservation of the matter in the component the most volatile:

- $x_0 S_0$ moles in the liquid phase of the flask before the operation
- $x_1 S_1$ moles in the liquid phase after the operation
- $x_{DM}(S_0 - S_1)$ in the distillate after the operation

$$x_0 S_0 = x_1 S_1 + x_{DM}(S_0 - S_1)$$
$$x_{DM} = (x_0 S_0 - x_1 S_1)/(S_0 - S_1)$$

Determination of the Yield of Distillation

By definition, the yield ρ is:

ρ = number of moles of the compound present in the whole distillate/number of moles of the compound present in the whole initial mixture.

For example, for the compound the most volatile,

$$\rho = x_{DM}(S_o - S_1)/x_o S_o$$

By replacing x_{DM} with the expression previously found:

$$\rho = [(x_o S_o - x_1 S_1)/(S_o - S_1)][(S_o - S_1)/x_o S_o]$$
$$\rho = 1 - x_1 S_1/x_o S_o$$

Distillation Curves

The distillation curves permit the determination of the instantaneous composition of the distillate when a given volume of the liquid mixture has been distilled. If, initially, S_o moles of mixture and S_1 is the remaining number of moles (which has not been distilled), the percentage of a mixture having been distilled is $(S_o - S_1)100/S_o$. The distillation curves are representative of the function:

$$x_D = f[(S_o - S_1/S_o)100]$$

They change with the relative volatility. Hence, for example, for an initial composition x_o, we are faced with a bundle of curves; each one corresponding to determined relative volatility. For $x_o = 0.5$, one obtains the curve in Figure 16.

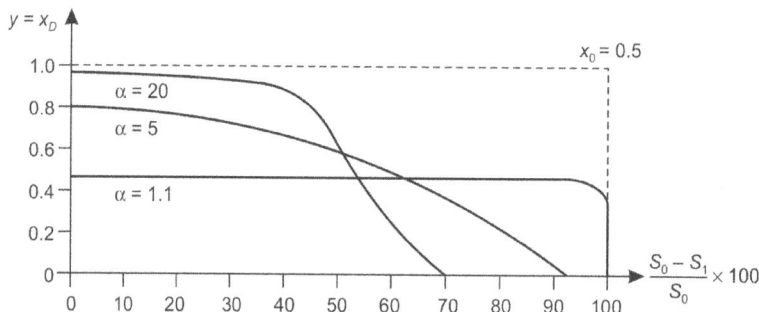

Figure 16: Distillation Curves for Some Relative Volatilities ($x_o = 0.5$). For $\alpha = 20$, One Reads 0.95 –0.90 –0.70 –0 for the First Drop, that is to Say 30%, 50% and 70%. One Can See that there is a Satisfactory Separation if $\alpha \geq 5$. (Reprinted with Permission from Gwenola and Jean-Louis Burgot, Paris, Lavoisier, Tec et Doc, 2017, 26).

(*Remark*: The point of abscissa 0 corresponds to the composition of the first drop of distillate.)

Rectification or Fractionated Distillation

Two Essential Features of the Rectification From the Standpoint of the Separation

We know that by operating by simple distillation of a binary mixture, it is not possible to separate in the pure state one or the other component of the mixture. One can only hope and that with a poor yield, for an enrichment of the distillate that is most volatile. Another drawback of the simple distillation is that during the experiment itself of distillation, the title of the distillate changes continuously. The rectification permits overcoming these drawbacks. Particularly, it permits obtaining a distillate of constant title, provided it is carried out with a stationary feeding. We shall only study the continuous rectification in detail.

Notion of Reflux

The rectification calls upon the notion of reflux. Besides, it is not the sole method of separation which calls upon the notion of reflux. Applied to rectification, reflux is the term qualifying the motion of return of one of the two currents of fluid participating in the separation. In rectification, it is the fraction of distillate that is returned to the column after separation. The origin of the increase of the efficacity of separation due to the setting up of the reflux is somewhat mysterious.

An Overview of the Rectification

The rectification is a technique that uses a repeated distillation in a column so that multiple exchanges between the rising vapor phase and the descending (refluxing) liquid phase in the column do exist. The method consists of an exchange of matter between the phases of liquid and vapor circulating at a countercurrent flow. The column is meant to put in contact with the two phases and to favor the exchanges of matter. Extraction in the system is analogous to counter-current extraction (Figure 17).

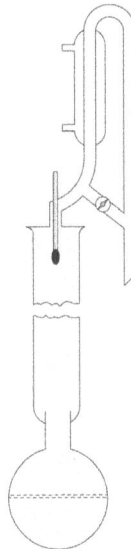

Figure 17: Rectification Apparatus. (Reprinted with Permission from Gwenola and Jean-Louis Burgot, Paris, Lavoisier, Tec et Doc, 2017, 28).

The exchanges in the column take rise in several plates (which can be fictitious) located all along the column (Figure 18).

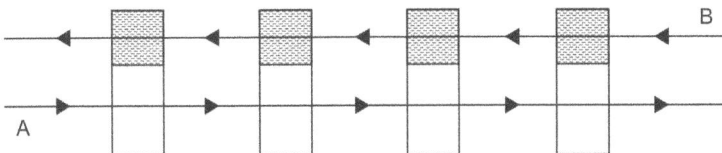

Figure 18: Analogy Device Counter-Current Extraction-Rectification. (Reprinted with Permission from Gwenola and Jean-Louis Burgot, Paris, Lavoisier, Tec et Doc, 2017, 29).

Apparatus

Actually, the column is often divided into two parts, an upper part constituted by the *rectifying column* and the bottom part constituted by the *stripping column* (Figure 19).

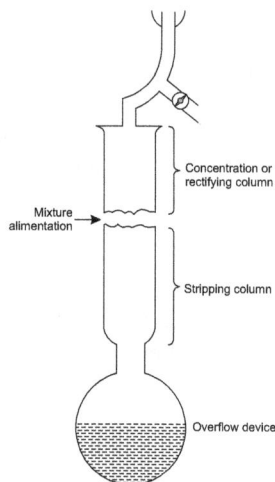

Figure 19: Rectification Column. (Reprinted with Permission from Gwenola and Jean-Louis Burgot, Paris, Lavoisier, Tec et Doc, 2017, 36).

One uses for simple distillation, a flask (a balloon being often the boiler), a refrigerating system and a receptacle in order to collect the distillate:

- A rectifying column is inserted between the balloon and the feed plate,
- The feed plate (the plate by which the feed enters) is located just between the rectifying and stripping columns. In continuous distillation columns, the feed is admitted to a plate in the central portion of the column.
- A second column is inserted between the feed plate and the flask called the stripping column. (Figure 19).

The columns are the most often columns with plates that schematically involve a series of diaphragms that are able to retain a certain volume of liquid. When the latter is reached, the liquid in excess refluxes by a device of overflow towards the inferior plate (Figure 20).

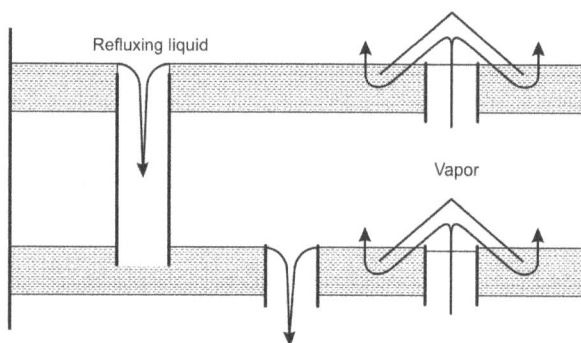

Figure 20: A Prebuilt Plate.

Some orifices permit the vapor to follow a rising trajectory and to go from one plate to another, which is upper while paddling in the liquid retained in the plates (Figures 21a, b and c).

One distinguishes two types of columns: columns with prebuilt plates and columns with no prebuilt plates. For all these columns, formerly, one defined the height equivalent to a theoretical plate (HEPT), which is the height of a fraction of a column such as the liquid and the vapor that go away it (this fraction of column) and are in equilibrium. One calculates it according to the relation:

$$\text{Height of a plate} = L/n$$

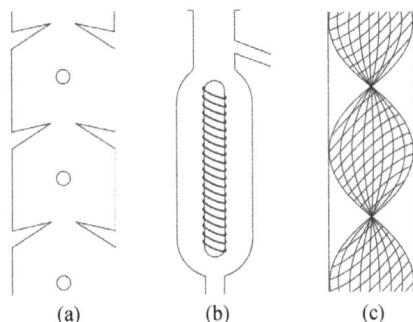

Figures 21(a, b and c): Fractionating Column: (a) Vigreux Column, (b) Rashig Rings Column and (c) Crismer Column Constituted by a Central Tube.

Here, L is the total height of the column and *n* is the number of theoretical plates.

- There are columns with prebuilt plates. There are several kinds of this type of plate. They are unlike each other because of their form that is different. These plates are conceived in order to permit efficient mixing of the liquid and the vapor (columns of Bruun and Sigwart). The equilibria are not obtained in all the plates. Hence, one defines an average efficacy E of the plates:

$$E = n/n_c$$

Here, *n* is the number of theoretical plates and n_c is the number of plates built. The efficacy can only be inferior to unity or equal to unity. The number of theoretical plates is determined as we shall see later.

- There are columns with no-built plates. In these columns, the exchange of liquid-vapor occurs between the vapor and the film of liquid condensed on the side of the column. Then, the goal is to increase their useful surface. In order to do that, one uses:

  - The Vigreux columns (Figure 21-a): Such a column of 30 cm in length corresponds to about three plates. It does permit the separation of two compounds and the difference in boiling points which is about 50°C under ordinary pressure.
  - The columns type Raschig rings are made of glass, ceramic or metal (Figure 21-b).
  - The columns of the type Crismer (Figure 21-c). They are constituted by a central tube around which rolls up a spiral of glass. Reflux takes place all along the spiral. This corresponds to a great number of plates.
  - The columns with a turning band. The filling of the column (for example, a metallic spiral) is submitted to a rotation. This warrants sustained turbulence of vapors and increases their course and their efficacy.

From another viewpoint, most columns possess auxiliary systems such as a reboiler and reflux pump, and these two devices permitting to improve the conditions of the reflux.

It is important to notice that without the reflux, no rectification would occur. It is very probable that the reflux causes differences in chemical potentials that are at the origin of the rectifying process and more precisely at the origin of the partitioning between the different phases.

Qualitative Study

When one begins to heat the flask containing the mixture, some time is necessary so that the properly so-called rectification begins. This is the period of priming. It permits to heat the whole apparatus and to do an entire rise and descent. At this time begins the period of normal running. The column becomes the seat of a double stream of opposite senses, one rising of vapor and the other descending of liquid. In each plate, the vapor and the liquid are stirring and hence can exchange matter. The successive equilibria liquid-vapor have effects that enrich the vapor phase in the compound, which is most volatile, and the liquid phase in the compound, which is less volatile. As a possible result, if the column is sufficiently lengthy, the vapor escaping from the upper plate can be composed of the most volatile compound only. At the same time, when there is a continuous feed, there is an enrichment of the boiler in the product the less volatile.

One more time, one must highlight the analogy existing between the counter-current extraction and the rectification. In the counter-current extraction, solvent A is extracted by a non-miscible solvent B. One knows that if one of the constituents of the mixture partitions preferentially in A with respect to the others, it will go away pure from the tour if the latter is sufficiently efficient. In rectification, the vapor plays the part of solvent A and the refluxing liquid that of B. As, in the countercurrent extraction, one defines a theoretical stage, in rectification one defines a theoretical plate.

Theoretical Study—Continuous Feed

This book will only deal with *continuous feed*. This means that the balloon (or the boiler) will have constant molar fractions of solutes to be separated by rectification, whatever the extent of the latter. Since the rectification separates the most volatile compound from the initial mixture and enriches the latter in a compound of lesser volatility, this means that the system must be fed continuously by the initial mixture itself with a sufficient outflow into the system. This must be carried out only by a fine adjustment of the heating of the system and the reflux device. This feed is carried out at the level of the stripping column.

The matter is to study the equilibria that occur between the liquid and vapor in the balloon and at the levels of the theoretical plates of the column. As for the extraction by a non-miscible solvent and for the simple distillation, the study is founded on the facts that by these hypotheses:

- We only study the binary miscible mixtures
- The binary mixture does not give azeotrope
- The liquid mixture has the behavior of an ideal solution
- The vapors have the behavior of a perfect gas. Let us recall that this hypothesis implicates that the laws of Raoult and Dalton are valid. The consequence is the expression of the equilibrium liquid-vapor already seen (this is the volatility expression):

$$y = \alpha x / [1 + x(\alpha - 1)]$$

It is interesting to notice that for this study, we speak of y and not of x_p, (see above):

- The study is done for the period of stationary running
- There is no loss of matter
- The quantity of matter retained in the column is negligibly related to the quantity present in the balloon
- We are in a position such as we are in ideal conditions of work, that is to say, each plate is a theoretical plate. All of them function identically and the phases that go away from them are in equilibrium.
- The heating of the flask and the setting of the reflux device (reflux pump and reboiler) are such that the flows of vapor and liquid are constant.

The followed notation is indicated in Figure 22. We suppose that the column is composed of s plates numbered from 1 to s. The inferior plate has index 1 and the superior s. Index 0 concerns the phases, which are contained in the balloon. As for the countercurrent extraction, each phase going out of a plate has a composition related to its number. For example, the vapors going out the plates

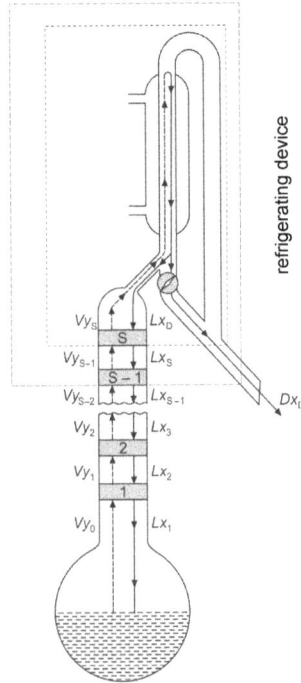

Figure 22: Symbolism for the Study of the Rectification.

1 to s will have molar fractions $y_1, y_2...y_{s-1}, y_s$. The liquid going out the plates s to 1 will have for molar fractions $x_s, x_{s-1},...x_1$.

As for the extraction by a non-miscible solvent and as for simple distillation, the study is founded:

- On the notion of equilibrium between the two phases liquid and vapor
- On the conservation of the matter.

Concerning the conservation of the matter, since the outflows of liquid and vapor are constant, one can say that during a determined time each theoretical plate:

- Receives and gives the same quantity of vapor, that is to say, V moles
- Receives and gives the same quantity of liquid, that is to say, L moles.

The quantity of V is determined by the conditions of heating of the balloon, and the quantity of L by the setting of the reflux pump at the bottom of the refrigerating device. Its stopcock permits directing all or one part of the condensed liquid toward the column at will. If D' is the number of moles collected (as distillate):

$$V = L + D'$$

One calls the *reflux ratio* θ

$$\theta = L/D'$$

$$R_D = \theta$$

The θ is also called the ratio of the *reflux to the overhead product* and is also symbolized by R_D. It is a parameter that is easily accessible experimentally. For example, it can be determined by counting the number of drops, which are going into the distillate at the level of the reflux pump. Some authors have proposed another parameter R_V. It is the ratio of the *reflux to the vapor*:

$$R_V = L/V$$

R_D changes from zero to infinity, whereas R_V changes from zero to unity. When $R_D = 0$, one has a *reflux null*. No liquid refluxes into the column. When $R_D = \infty$, one has *total reflux*. There is no distillation.

- Let us consider the following set – the superior plate s, the refrigerating system and the reflux device. Let us express the conservation of the mass of the compound, which is the most volatile:

$$Vy_{s-1} = D'x_D + Lx_S$$

The term on the left of the sign of equal represents the number of moles arriving on the plate s under the form of vapor. The term on the right is the number of moles that go away from it, a part of which $D'x_D$ is its number of moles arriving into the distillate and the other part Lx_S is refluxing.

- If, now, one considers the plate $s - 1$, one can write (by considering as previously the conservation of the matter of the most volatile solute)

$$Vy_{s-2} = D'x_D + Lx_{S-1}$$

By comparing the last two expressions, one notices that the quantity directed toward the distillate $D'x_D$ is always the same. Indeed, one must recall that the interest of the rectification versus the simple distillation is to offer the possibility of obtaining a distillate composition which has been chosen before the experiment and remains constant all along the separation.

- One can generalize the preceding relations and write:

$$Vy_{n-1} = Lx_n + D'x_D$$

One can express it as a function of θ. Therefore:

$$V = L + D'$$

Since there is no accumulation of matter on every plate. As $\theta = L/D'$, one can also write:

$$V = D'(1 + \theta)$$

$$D' = V/(\theta + 1) \quad \text{and} \quad L = V\theta/(\theta + 1)$$

Hence,

$$Vy_{n-1} = \theta V/(\theta + 1)x_n + V/(\theta + 1)x_D$$

$$y_{n-1} = \theta/(\theta + 1)x_n + x_D/(\theta + 1)$$

This is the equation of a straight line. This line represents the conservation of the matter (here of the compound that is most volatile). It is named *operational line.*

Graphic Representation of the Theoretical Plates According to McCabe and Thiele

McCabe and Thiele draw the diagram of molar fractions y/molar fractions x. (Their superior values on each of the coordinate axis are limited of course to 1.)

The diagram expressing the equilibrium liquid-vapor has the equation:

$$y = \alpha x/[1 + x(\alpha - 1)]$$

It is easily drawn (Figure 23).

- The above operational straight line expressing the conservation of the matter must also be drawn. But before doing that, we shall study its characteristics. Its slope is $\theta/(\theta + 1)$. When the reflux is null, $\theta = 0$. The slope is null and the equation is reduced at $y = x_D$. It is one horizontal. When the reflux is total, it is the equation $y = x$. The intercept is null. For $x = x_D$, $y = x_D$.

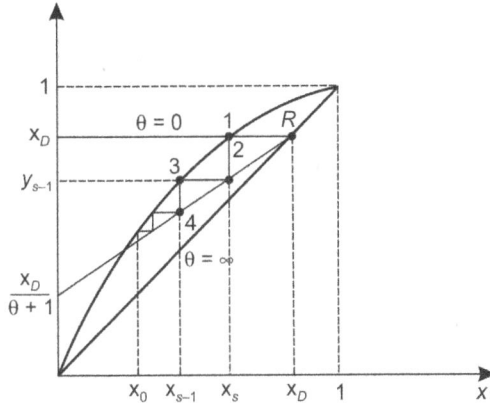

Figure 23: Graphical Representation of Operational Line and the R Point and the First Theoretical Plates. (Reprinted with Permission from Gwenola and Jean-Louis Burgot, Paris, Lavoisier, Tec et Doc, 2017, 33).

The coordinates $y = x = x_D$ defines what is named point R (Figure 23). The latter possesses a physical meaning and corresponds to the set, which is a refrigeration system plus the reflux device. As a matter of fact, if one considers what is going out of the last plate s, the conservation of the matter (in terms of the compound that is most volatile) permits writing it:

$$Vy_s = D'x_D + Lx_D$$

Concerning the operational straight line and the equilibrium curve, the liquid, condensed in the refrigerating system, has the same composition x_D as that which is collected:

$$Vy_s = x_D(L + D')$$

With $L + D' = V$

Therefore:

$$y_s = x_D$$

For $0 < \theta < \infty$, the operational line goes through the point R and has for intercept $x_D/(\theta + 1)$.

Representation of the "Theoretical" Plates on the Diagram Y = F(X)

For the sake of generalization, let us consider the last plate s. One connects it to a first point that expresses the fact that the vapor which goes out from it has the same composition as the liquid which arrives in it. It is the R point. One knows that it is located on the operational line. It is a normal point since its existence is a consequence of the conservation of matter.

The plate s is also connected to another point. It is necessary now to express the fact that the vapor that goes out of the plate is in equilibrium with the liquid that goes out from it. This is *by definition* of a theoretical plate. We have already seen that x_s and y_s are interrelated by the expression $y_s = \alpha x_s/[1 + x_s(\alpha - 1)]$.

The x_s and y_s are the coordinates of point 1 located on the equilibrium curve. The molar fraction x_s is easily determined graphically by drawing the vertical passing through point 1. The plate s is finally connected to the diagram by a third point expressing the conservation of the matter when the vapor coming from the plate $s - 1$, and the liquid is going out. It is point 2 that has the same abscissa as the precedent one corresponding to the abscissa x_s.

Consider the plate $s - 1$. As the preceding one, it is defined by three points:

- The first is point 2 which expresses the conservation of the matter between the vapor that is going away from the plate and the liquid that enters it. It is also connected to point 2 since between the two plates, there is the same phase.

- The second point expresses the equilibrium between the vapor phase and the liquid, which is going away from it. It is the point 3. It is on the equilibrium curve and has the same ordinate as the preceding.

- The third point expresses the conservation of the matter between the vapor that enters the plate and the liquid that is going away from it. It is the point 4. It has the same abscissa as the preceding. The late s – 1 is represented by the three points 2, 3 and 4 (Figure 23).

Practical Interest

Determination of the Number of Theoretical Plates of a Column. It is obviously the parameter that we seek when we are practicing rectification. Let us take the example of a mixture of toluene and benzene that we want to submit to the rectification with the reflux ratio $\theta = 2.1$. At a given moment of the experiment, the distillate and the mixture remaining in the boiler are analyzed. One finds $x_D = 0.93$ and $x_o = 0.37$. (Recall that the rectification permits having an average composition of the distillate equal to the instantaneous one). One demands the number of theoretical plates of the system knowing that $\alpha = 2.46$.

The knowledge of α and θ permits drawing the equilibrium curve and the operational straight line and also drawing the theoretical plates. Then, one traces the vertical line of abscissa $x_o = 0.37$. Then, one counts 4.5 theoretical plates (Figure 24). However, the balloon must be counted for one theoretical plate. Therefore, the column itself involves 3.5 theoretical plates.

Some points deserve further comments:

- The number of theoretical plates may be decimal
- As for the chromatography columns, the number of theoretical plates depends on the composition of mixture to rectify via the value that fixes the form of the equilibrium curve and hence the number of plates
- Although it is not evident from the above, the number of theoretical plates depends on the length of the column by fixing the remaining concentration x_o. With a different column, for the same mixture and the same chosen concentration x_D, we should have a different value of x_o
- There exist other methods of determination of the number of theoretical plates.

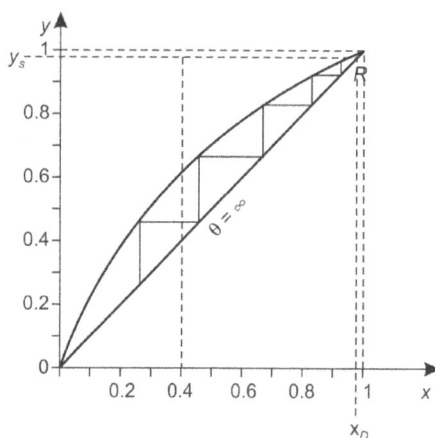

Figure 24: Representation of the Theoretical Plates. (Reprinted with Permission from Gwenola and Jean-Louis Burgot, Paris, Lavoisier, Tec et Doc, 2017, 35).

Determination of the Optimum Conditions of Rectification. Intuitively, a condition optimum for rectification involves a maximum of theoretical plates. Let us consider a mixture of benzene and toluene containing 40% of benzene. The goal is to enrich it by rectification (with a continuous feed) until a mixture at 98% of benzene. How many plates will be necessary, and what reflux ratio should be necessary so that the experiment will be satisfactory? The composition of the liquid in the balloon is $x_o = 0.40$. One gives $\alpha = 2.46$.

The problem is to place the operational line on the diagram $y = f(x)$. There is indeed no difficulty to draw the equilibrium curve, all the parameters of which are known. One can also draw the diagonal representing the infinite reflux. One knows that when the reflux is total, $x_D = 0$, and the definition of the reflux ratio $\theta = L/D$ gives $\theta = \infty$ and $y = x$. In order to place the operational line, let us draw the vertical $x = 0.4$. One immediately sees that if the operational line cuts the equilibrium curve for one abscissa stronger than 0.4 (which is where the operational curve intersects the equilibrium curve), the separation is no longer possible. If the operational line curve cuts the equilibrium curve at point M_o, the separation is possible but with a number of theoretical plates equal to the infinity. This would only be possible with a column of length infinite. This is not practicable. When the operational line intersects the equilibrium curve between M_o and $M"_o$, the separation is possible. For the point $M"_o$ itself, the separation is theoretically possible, but there is no distillation since the reflux is infinite.

Then, one notices that the minimum reflux ratio is found after the study of the operational limit straight line RM_o. It is easy to draw this line (see Figure 25). Its intercept that is measurable easily is equal to an $x_D/(\theta l + 1)$, whence θl (limit reflux ratio). In the present case $\theta l = 1.65$. The reflux ratios can be chosen or this example is located in the interval 1.65 and infinity. From a more practical viewpoint, one must realize that:

- If one chooses $\theta = 1.65$, the rectification will be quick since the reflux ratio is weak. But in these conditions, a large number of plates is necessary. This is not practical to achieve.
- If one chooses $\theta = $ infinity, a short column is sufficient with a number of plates of the order of 4.5, but the ratio reflux must be great. The rectification entails a great interval of time.
- One must seek a compromise between these two exigencies. For example, here with $\theta = 5$, a column with 6 theoretical plates is necessary.

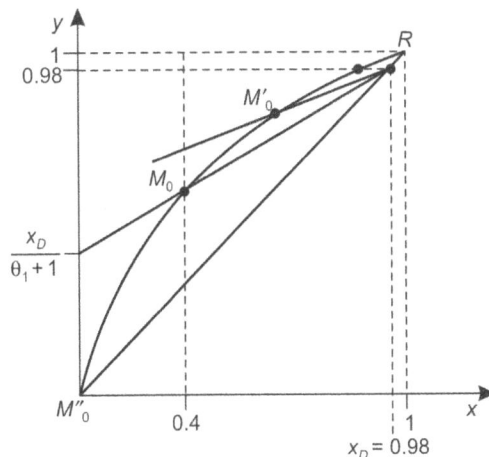

Figure 25: The Choices of the Best Practical Conditions. (Reprinted with Permission from Gwenola and Jean-Louis Burgot, Paris, Lavoisier, Tec et Doc, 2017, 35).

Practical Disposition in Order to Ensure a Continuous Feed

The mixture is generally introduced in the middle of the column and the superior part of which works as a column of concentration, according to the previous modalities. The inferior part works as a stripping column. An overflow permits evacuating the residue from the rectification composed of the mixture enriched y that is the compound the less volatile. McCabe and Thiele have equally given a graphical representation of the stripping part through notably a feed straight line. The continuous feed is often carried out in the industry that demands great volumes of mixtures. They are worked in towers.

An interesting point to mention is that the continuous feed must not disturb the thermal equilibrium of the whole system. In other words, it participates in the adjustment of the thermal equilibrium.

Other Distillations

Now, we consider briefly, the batch rectification, case of azeotropes, steam-distillation, distillation flash and molecular distillation.

Discontinuous Alimentation—Batch Rectification

In the previous study, the concentration of the mixture contained in the balloon remained constant and was equal to x_o. In the laboratory, one generally has only at disposal a small volume of mixture that is introduced in totality in the balloon at the beginning of the operation. Then, one says that we are in the conditions of batch distillation. The concentration of the compound is the most volatile of the mixture contained in the balloon changes during the distillation. If we consider a diagram of McCabe and Thiele in order to study this case (Figure 26) one notices:

- Initially, with a given column possessing some number of theoretical plates (see Figure 26) and with the chosen reflux, starting from an initial composition x_o, x_D is found and fixed immediately.

- With the same column and same reflux, when the rectification has been in progress for some time, the initial concentration has decreased and has become x'_o and the distillate x'_D with

$$x'_D < x_D.$$

In order to obtain a distillate that has always the same composition, in proportion as the rectification progresses, it is obligatory:

- Either to increase the number of theoretical plates for the same reflux ratio. This is impossible since it is impossible to change the column at each time of the distillation.

- Or to increase the reflux ratio for the same number of theoretical plates (same column). We have seen that increasing the reflux ratio is equivalent to the fact to bring closer the operational straight line and the diagonal $y = x$ (Figure 27). For the initial reflux, the compositions are x_o

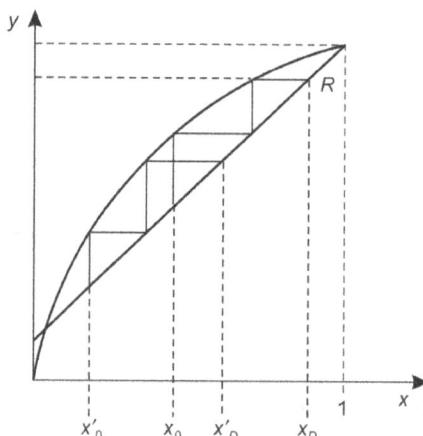

Figure 26: Evolution of Compositions x_o and x_D During a Batch Distillation. (Reprinted with Permission from Gwenola and Jean-Louis Burgot, Paris, Lavoisier, Tec et Doc, 2017, 37).

and x_D. One can see that with the second reflux, one has the same composition x_D of the distillate and the same number of theoretical plates *n* (hence, one uses the same column), although $x'_o <$ x_o. It is this solution that is adopted. It is simple indeed to turn the stop-cock of the reflux device during the progress of the distillation. In order to check the process, one watches over the reflux device as the rectification progresses, so that the temperature could remain constant.

(*Remark*: If we do not pay attention to the evolution of the rectification, the replacement of the liquid is the most volatile by a less volatile one leads to an increase in the temperature since it is necessary to increase the vapor tension of the latter until there is equality of the sum of the partial pressures with the exterior pressure. This can be done only by enhancement of temperature.)

For example, the increase in the ratio reflux can be notable. For the mixture of benzene-toluene with 40% benzene with a column of 10 theoretical plates, we find:

- At the beginning of the operation $x_o = 0.40$ and $\theta = 1.85$
- At the end $x_o = 0.10$ and $\theta = 7.10$.

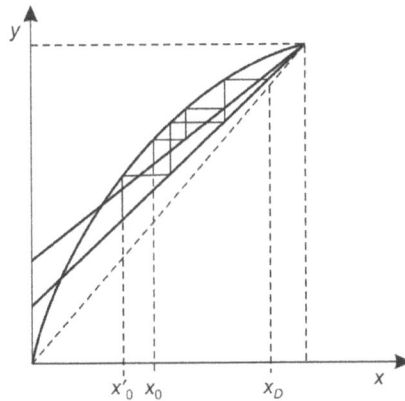

Figure 27: Increase in the Reflux Ratio to Obtain a Batch Rectification. (Reprinted with Permission from Gwenola and Jean-Louis Burgot, Paris, Lavoisier, Tec et Doc, 2017, 38).

Case of the Azeotropic Mixtures

The preceding theory applies to the case of azeotropes but:

- It is Compound A or B that is obtained in the distillate when there is a negative azeotropic mixture.
- It is the azeotrope that is obtained in the distillate when there is a positive azeotropic mixture.

By decreasing the pressure, one produces the ebullition at a temperature lower than that effective at ordinary pressure. This method is advantageous either in simple distillation or in rectification and that from different points of view which are:

- The risks of adulteration of the mixture due to the heat and the temperature and decrease under reduced pressure.
- The formation of azeotropes is often avoided.
- The efficacy of the rectification is often improved. The improvement of the efficacy results from the increase of the value of the relative volatility:

$$\alpha = P_1/P_2$$

For most binary mixtures, one approximately demonstrates that α increases when the pressure at which one works decreases. This result is objectivized by the fact that the curve of equilibrium is more incurved than it is the case under greater pressure (Figure 28).

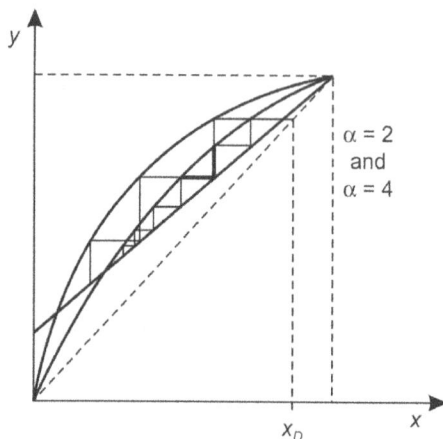

Figure 28: Rectification Under Reduced Pressure. (Reprinted with Permission from Gwenola and Jean-Louis Burgot, Paris, Lavoisier, Tec et Doc, 2017, 40).

For $\alpha = 2$, we count 5 theoretical plates. For $\alpha = 4$, we count 3 theoretical plates for the same reflux (for the same operational straight line). The number of the necessary theoretical plates decreases. Therefore, for the same column possessing the same number of plates, the efficacy is enhanced.

Steam-Distillation of a Mixture of Non-Miscible Compounds

Let us consider a mixture of the two non-miscible liquid compounds 1 and 2, the derivative 1 is the most volatile. In the liquid state, such a mixture is formed by two distinct phases (Figure 29) and each compound has the same behavior as that it has when it is sole. That is to say, it keeps its proper vapor pressure P_1 or P_2 that only depends on the temperature and its nature according to Clapeyron's law.

The second law of ebullition remains valid; the mixture enters in ebullition when its Vapor pressure P is equal to the exterior pressure. One determines the temperature of ebullition by carrying out the following construction (Figure 30).

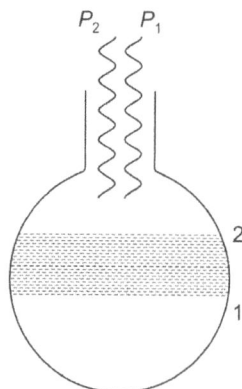

Figure 29: A Mixture of Non-Miscible Compounds. Determination of the Ebullition Temperature of the Mixture. (Reprinted with Permission from Gwenola and Jean-Louis Burgot, Paris, Lavoisier, Tec et Doc, 2017, 41).

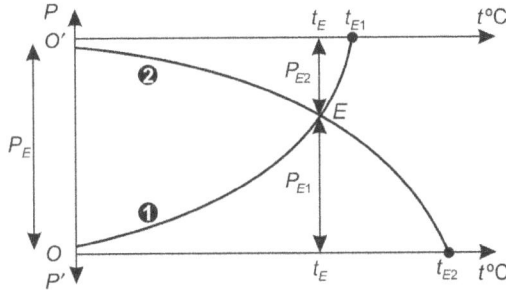

Figure 30: Determination of the Boiling Temperature of a Mixture of Non-Miscible Components. (Reprinted with Permission from Gwenola and Jean-Louis Burgot, Paris, Lavoisier, Tec et Doc, 2017, 41).

In a system of rectangular coordinates T and P, we trace curve 1 which represents the tension of vapor of compound 1 as a function of the temperature; $P_1 = f_1(t)$, which is an expression of Clapeyron's law. In the system of rectangular coordinates T' and P' of origin O', we trace the tension of vapor of compound 2 (curve 2) $P_2 = f_2(t)$. The two rectangular systems are located as follows. Both abscissa axis are parallel and are of the same direction. They are the axis of pressure P and P'. They are mutually separated by the distance OO' = P_E. The axis of ordinates (for both systems) is those of temperatures. They coincide but their directions are opposite.

Let us consider the intersection at point E of these two curves. It is evident that:

$$P_{E1} + P_{E2} = P_E$$

The sum of the vapor pressures of each constituent is equal to the external pressure. Hence, one is here at the ebullition temperature. The abscissa of Point E gives the ebullition temperature T_{eb}. At this level of reasoning, one can notice three points:

- The ebullition temperature of the mixture depends only on the nature of the two compounds and does not depend on their relative proportions. Here, is the great difference with respect to the mixture of miscible liquids for which there is ebullition when:

$$P_E = P_1 + P_2$$
$$P_E = xP_1 + (1 - x)P_2$$

In this last case, the molar fractions intervene:

- The ebullition temperature is lower than those of the two pure compounds 1 and 2 and T_{E1} and T_{E2};
- The ebullition temperature is constant during the ebullition as long as the two compounds constituting the initial mixture subsist. When one of the two has disappeared from the medium (because of its distillation is finished), the temperature goes up highly until the temperature T_{E1} or T_{E2} corresponding to the ebullition of remaining derivative.

Hence, the question is what is the composition of the vapor? One demonstrates that it only depends on the nature of the mixture and that it remains constant as long as there remain two compounds in the balloon. Contrary to the liquid, the mixture of both gases evidently only forms one phase. Thus, Dalton's law is valid:

$$P = P_1 + P_2$$

P_1 and P_2 are the partial pressures of compounds 1 and 2. It is possible to express the partial pressures as a function of the respective molar fractions y and $(1 - y)$ of compounds 1 and 2:

$$P_1 = yP \quad P_2 = (1 - y)P$$

The ratio of the partial pressures is:

$$P_1/P_2 = yP/(1 - y)P$$

$$P_1/P_2 = y/(1 - y)$$

But the partial pressures do not depend on the respective proportions of the compounds in the liquid phase. More precisely, we have:

$$p_1 = P_1 \quad p_2 = P_2$$

$$y/(1 - y) = P_1/P_2$$

P_1 and P_2 depend only on the nature of the mixture and on the temperature. Hence, for a given external pressure, the composition of the vapor depends only on the nature of the compounds and does not depend on their relative proportions in the liquid mixture.

Distillation Flash

It consists of:

- Vaporizing a well definite fraction of the liquid mixture to solve by a brutal decrease of the pressure and to wait for the vapor and the liquid to be again in equilibrium in the condenser.
- Separating the vapor from the liquid.
- And condensing the vapor.

This technique is overall used to separate the components of a mixture that boils at very different temperatures. It can be carried out in only one theoretical stage.

Molecular Distillation

Some compounds at elevated ebullition temperatures (or some others which are unstable at a given temperature) cannot be distilled without an important decomposition. Then, it may be interesting to use the process called molecular distillation. The phenomenon of condensation in distillation and fractioned distillation, particularly, results from the intermolecular collisions between the vaporized molecules that bring back to the surface of the liquid. If the collisions are to be avoided, they leave this surface and move in a straight line until they encounter a cold surface. The number of collisions is decreased when the pressure is lowered. In other words, the distance that the molecule must travel through the vapor phase in order to encounter another (*mean free path*) is enhanced. To carry out a molecular distillation, one must:

- Work at reduced pressure (value of the order of 10^{-3} at 10^{-5} mmHg).
- Fix the distribution of the liquid that must be distilled on the hot wall of the apparatus in the form of a thin film and the thickness of which is of the order of 10^{-2} and 10^{-3} mm. This is carried out by the action of a centrifugal force.
- Have a condensation surface close to the product to evaporate at our disposal.

In these conditions, the molecules that must be separated have sufficient rates in order to escape from the liquid and reach the chilled wall. It is interesting to notice that this kind of process depends no longer on the vapor pressure of the compound or in any case in the first degree.

Molecular distillation is well-adapted to the separation of organic compounds of high molecular masses, such as sterols, fatty acids, some liposoluble vitamins, etc. An apparatus is represented in Figure 31.

In summary, molecular distillation is a method of volatilization in a vacuum without ebullition.

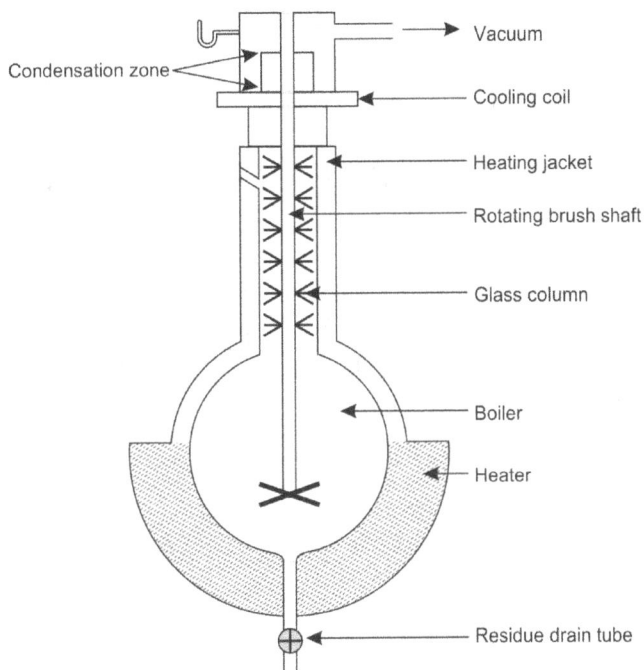

Figure 31: Apparatus of Molecular Distillation.

Applications of the Distillations

General Applications

They are innumerable. Gaseous chromatography has not diminished its importance. Actually, both methods appear to be complementary. Very often, the distillation prepares the sample that must be injected into chromatography. Often, it "pre-enriches" the sample of the second method.

Evidently, rectification is very frequently encountered in the petroleum industry in which the mixtures to solve are very complex and have more than two constituents. Several distillations and rectifications of azeotropic mixtures are carried out.

Steam distillation is used to distillate products, the ebullition temperatures of which are high. It must be mentioned that the vapor is all the richer in the compound as the vapor pressure is the highest, as is demonstrated by the above-given reasoning. This method may permit separating some substances of the neighbor constitution, such as the orthonitrophenol which is entrainable and the p-nitrophenol which is not (Figure 32).

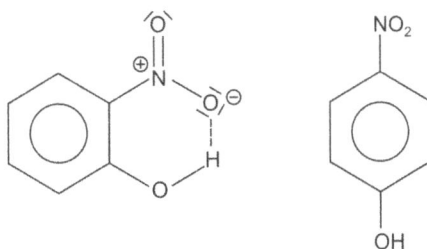

Figure 32: Structures of Orthonitrophenol and of Paranitrophenol. (Reprinted with Permission from Gwenola and Jean-Louis Burgot, Paris, Lavoisier, Tec et Doc, 2017, 43).

There are also in pharmacopoeias some general essays such as:

- The determination of the interval of distillation.
- The determination of the ebullition temperature.

Let us notice that the extraction liquid-liquid is evidently an alternative technique to the distillation for the separation of liquid homogenous mixtures.

Some Punctual Applications

Let us mention examples of punctual applications in a laboratory of chemical analysis:

- The determination of the total quantity of the volatile acids in wines (Ducloux' method);
- The determination of the indices of volatile acids soluble and insoluble in fats;
- The dosage of ammoniac and volatile bases;
- The dosage of boric acid;
- The dosage of arsenic;
- The dosage of azote according to Kjeldahl;
- The dosage of water in medicines and the aliments by driving it with solvents (Leymarie's method).

This list is very far from being exhaustive.

CHAPTER 13

Chromatographic Methods
An Overview

General Approach to Chromatographic Methods

Chromatographic methods are separation techniques based upon some different chemical and physical principles. From a historical point of view, this method appeared in 1903 when the Russian botanist Mikhail Tsvet realized the separation of various plant pigments, such as chlorophyll on a tube packed with a solid material and calcium carbonate with petroleum ether as solvent. The separation of the mixture into different colored bands on the column has given its name to the chromatographic method (from Greek *chroma* which means color). Its first separation experiment used the difference in adsorption of the pigments on the calcium carbonate; however, other mechanisms can be used. Some important steps have followed the first discovery of Tsvet. Among the highlights, there are:

- 1938 – Thin-layer or planar chromatography (N.A. Izmailov and M.S. Schraiber).
- 1939 – Ion-exchange chromatography (O. Samuelson).
- 1941 – Partition chromatography (A.J.P. Martin and R.L.M. Synge).
- 1952 – Gas chromatography on a packed column (A.T. James and A.J.P. Martin).
- 1959 – Gas chromatography on Open-Tubular (or capillary) column (M.J.E. Golay).
- 1962 – Supercritical fluid chromatography (E. Klesper).
- 1968 – High-performance liquid chromatography (J.C. Giddings and J.J. Kirkland).
- 1976–1980 – Ion-pair liquid chromatography (E. Horvath).
- 2004 – Ultra high-performance liquid chromatography (Waters company).

The origin of the mechanism of a chromatographic process is due to the fact that the components of the mixture to be separated give a very important number of continuous equilibrium with two phases:

- The stationary phase is fixed inside a column or on a solid surface.
- The mobile phase is forced to pass through the stationary phase at a given speed.

As a result, each component of the mixture is equilibrated between both phases for a very long time and run along the stationary phase at its own speed. At the end of the experiment, the component is eluted with the mobile phase.

As a result:

- If the time of migration of the mobile phase is fixed, each component covers a distance through the stationary phase so that components will be separated. If the experiment is realized with a column packed with the stationary phase, at the time t, component A is faster than component

B that will be moved through the column of a distance d_1, which is more important than those d_2 of Component B (Figure 1). Two bands appear on the column. This experiment is called *development chromatography*. It is quite possible to recover the bands that could be analyzed qualitatively and quantitatively.

- The other option is to set the length of the migration that could be the height of the column (Figure 1). As a result, each component emerges from the column at different times. This second process is called *elution chromatography*, which is the most commonly used system today.

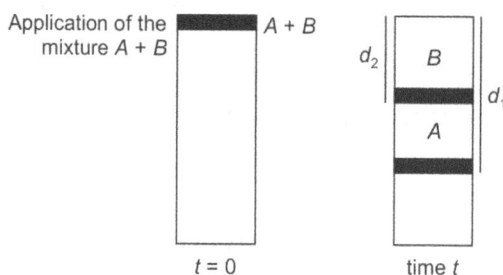

Figure 1: Separation of Two Components Through a Column Packed With a Stationary Phase. (Reprinted with the Permission from Gwenola and Jean-Louis Burgot, Paris, Lavoisier, Tec et Doc, 2017, 130)

From the Countercurrent Separation to the Partition Liquid Chromatography

Let us imagine that in Craig's method (countercurrent separation), the number of tubes (or columns) is substantially increased with keeping within them an equilibrium state and that the brewing process is made itself. Then, we have realized a *partition liquid-liquid chromatography*.

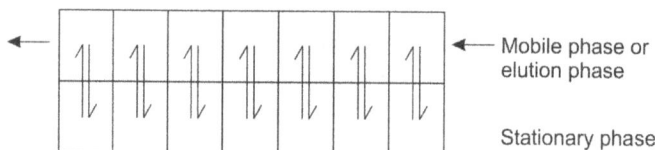

Figure 2: Different Steps of Equilibration of a Partition Liquid-Liquid Chromatography. (Reprinted with the Permission from Gwenola and Jean-Louis Burgot, Paris, Lavoisier, Tec et Doc, 2017, 131).

In Figure 2, the two vertical arrows mean that there is still an equilibrium of the analyte between both phases. Usually, the column of chromatography is presented vertically according to the Brimley and Barrett scheme (Figure 3).

In practice, the stationary phase is immobilized on inert solid particles (this means that it is not relevant to the equilibrium). The mobile phase moves through the openings between the individual beans in such a manner that it has established a sequence of material exchanges between dual phases.

The fraction of the column height which is necessary to realize the equilibrium is called by analogy with the definition of Peters in 1922 for the distillation on packed columns "height equivalent to a theoretical plate" or Height Plate H.

This is the height of the fraction of the column which is necessary to the Coefficient K is equal to C_s/C_m:

$$K = C_s/C_m \quad K = 1/\lambda$$

and λ partition coefficient of the countercurrent process.

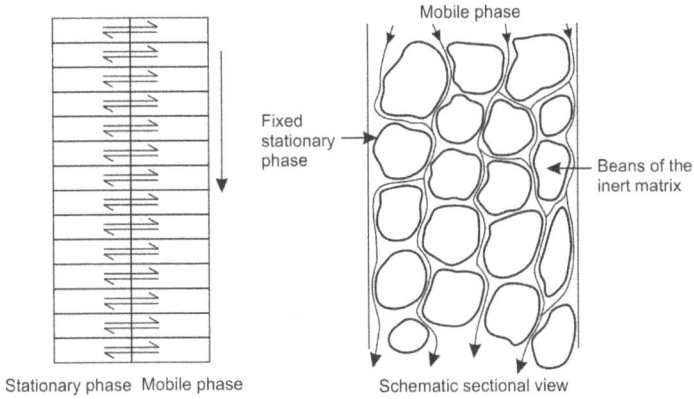

Figure 3: (a) Brimley and Barrett Scheme (b) Schematic Sectional View of a Column. (Reprinted with the Permission from Gwenola and Jean-Louis Burgot, Paris, Lavoisier, Tec et Doc, 2017, 131).

Each theoretical plate corresponds to one of the countercurrent tubes. Consequently, some of the chromatography theories are a generalization of the ones of the countercurrent process. It will be the case for the theory of plates.

The Different Steps of a Chromatography

The different steps of chromatography depend on the modes of the mobile phase migration by elution or development.

Elution Chromatography

In this case, the mixture of components is applied to the top of the column packed with the stationary phase (Step 1). The mobile phase moves through the column and the components migrate at different speeds (Step 2). Each component is recovered or eluted from the column at different times (Step 3) (Figure 4).

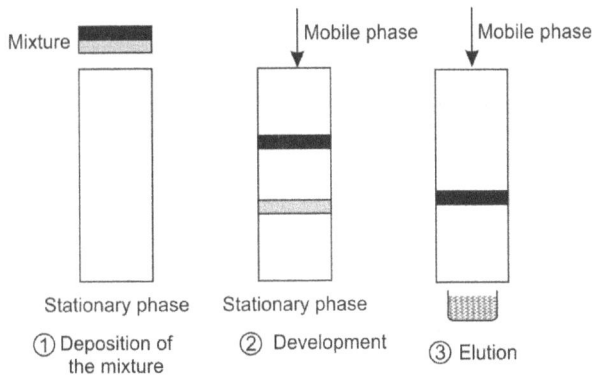

Figure 4: The Three Steps of a Chromatography Elution Process. (Reprinted with the Permission from Gwenola and Jean-Louis Burgot, Paris, Lavoisier, Tec et Doc, 2017, 132).

Development Chromatography

The components of the mixture deposited on a plate migrate under influence of the mobile phase, which runs the plate by capillarity most often. There is no elution step but it is always possible to recover the components by scraping the plate at the end of the process (Figure 5).

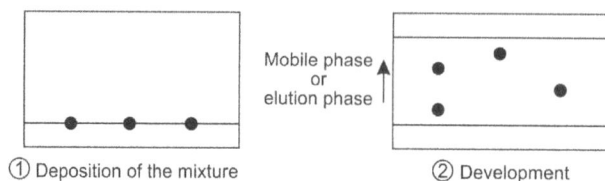

Figure 5: The Two Steps of a Chromatography Development Process. (Reprinted with the Permission from Gwenola and Jean-Louis Burgot, Paris, Lavoisier, Tec et Doc, 2017, 132).

Types of Chromatography

Chromatographic methods are classified according to a lot of possibilities. Perhaps, it means that none of them is adequate.

Classification Depending on the Type of Equilibrium

Partition

The separation process is based on two compounds not partitioned in a similar way between both phases. That means that after the equilibrium the ratio of their solubilities in both phases or partition coefficient is not equal.

$K = C_s/C_m$ (K is a temperature function as all thermodynamical constants)

C_s et C_m are concentrations in mol.L^{-1} et K a number without dimension.[11]

Adsorption

Adsorption is the accumulation of substances onto the surface of one of the phases. For example, the attachment of a substance in solution to the surface of a solid in suspension.

The adsorption is due to the forces of cohesion especially van der Waals and polar interactions. They are not specific and their exact nature is not well known.

Ion Exchange

In this case, the stationary phase has some properties of an ion exchanger. This one is a macromolecule or resin that contains some acidic or basic functional groups. They provide the exchange of some of those ions with the same charge as the analytes as shown in (Figure 6).

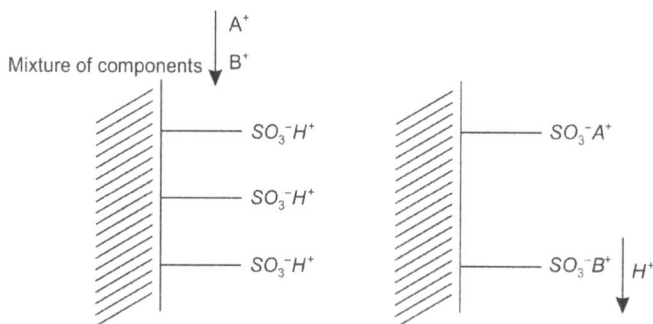

Figure 6: Ion-Exchange Chromatography. (Reprinted with the Permission from Gwenola and Jean-Louis Burgot, Paris, Lavoisier, Tec et Doc, 2017, 134).

---

[11] More precisely, to be as rigorous as possible, the partitioning coefficient is formulated in terms of activities.

The mobile phase is an aqueous solution consisting of salts, acids or alkalis. The distribution Constant K is called, in this case, the ion exchange coefficient.

Size Exclusion-Diffusion

The stationary phase is a gel or a macromolecular edifice that produces a sieving effect against the large molecules. These move faster with the flow of the mobile phase than the small molecules, which diffused freely into the gel.

Ion-Pair

The ion-pair is an entity formed by the association of two ions of opposite charge. The association is due to electrostatic interactions to which hydrophobic effects are added. These last ones are due to the rearrangement of the water molecules around the ions. Consequently, the ion-pair has the possibility to go through the organic phase.

With its type of chromatography, it becomes possible to separate the ion with reversed chromatography. It is enough to introduce in the mobile phase an ion, named counter-ion; the sign of which is opposite to one to separate. The counter-ion gives with the analyte an ion pair, which is likely to be retained by the stationary phase. This process also involves other mechanisms described in Chapter 16.

Affinity

Some authors link this type of chromatography to adsorption chromatography. It is possible to say that it is an adsorption chromatography in which the interactions between the analyte and the stationary phase are well known and specific. This process is considered as being the most selective one.

In fact, chromatographic separation is not only based on one physical-chemical process. It is not rare to have in addition to other processes, which can greatly enhance or reduce the performance of the method. For example, in partition chromatography on a silica gel, there is a superposition of a second effect as adsorption which is difficult to eliminate.

Classification Depending on the Physical State of Both Phases

That can be illustrated below:

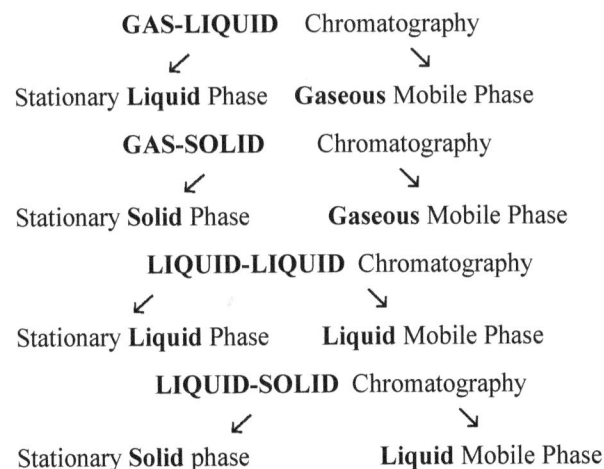

$$\begin{array}{c}
\textbf{GAS-LIQUID} \quad \text{Chromatography} \\
\swarrow \qquad\qquad \searrow \\
\text{Stationary } \textbf{Liquid } \text{Phase} \quad \textbf{Gaseous } \text{Mobile Phase}
\end{array}$$

$$\begin{array}{c}
\textbf{GAS-SOLID} \quad \text{Chromatography} \\
\swarrow \qquad\qquad \searrow \\
\text{Stationary } \textbf{Solid } \text{Phase} \quad \textbf{Gaseous } \text{Mobile Phase}
\end{array}$$

$$\begin{array}{c}
\textbf{LIQUID-LIQUID} \quad \text{Chromatography} \\
\swarrow \qquad\qquad \searrow \\
\text{Stationary } \textbf{Liquid } \text{Phase} \quad \textbf{Liquid } \text{Mobile Phase}
\end{array}$$

$$\begin{array}{c}
\textbf{LIQUID-SOLID} \quad \text{Chromatography} \\
\swarrow \qquad\qquad \searrow \\
\text{Stationary } \textbf{Solid } \text{phase} \quad \textbf{Liquid } \text{Mobile Phase}
\end{array}$$

Classification Depending on the Nature of Mobile Phase

- Liquid chromatography
- Gas chromatography
- Supercritical fluid chromatography

Classification Depending on the Physical State of Both Phases

- Planar chromatography: The stationary phase is supported by a planar surface and the mobile phase moves through the stationary phase by capillarity.
- Column chromatography: The stationary phase is trapped in a narrow column and the mobile phase moves through the column under pressure or by gravity.

Chromatographic Peaks

More often, the fluid or eluate that emerges from the end of the column is detected by some apparatus called *detectors*, which provide a signal or response detector as a function of elution time or volume of added mobile phase as shown in Figure 7. The response of the detector is proportional to the instantaneous concentration of the analyte in the mobile phase; the detector is called a differential one. The graph obtained is named a chromatogram. It is obvious that if the compounds are well separated, they appear on the chromatogram with as many peaks as compounds.

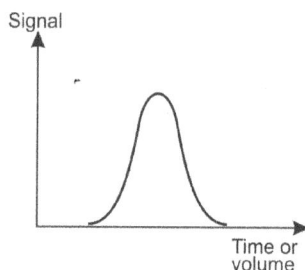

Figure 7: Chromatographic Peak.

Applications

Chromatographic methods are the most important methods of proximate analysis, and they are widely used techniques. They permit the separation of the components of a complex mixture. Moreover, in some conditions it defines, quantifies and identifies the compound as well.

CHAPTER 14

Fundamental Theoretical Aspects of Elution Chromatography

In this chapter, we give some general definitions concerning chromatography and then we mention the two principal groups of theories describing and justifying the results these methods permit to obtain. They are essentially the theory of plates and kinetic theories.

Definitions

Linear and Non-Linear Chromatography—Partitioning Isotherms

A chromatography is qualified as being *linear* when the values of ratios to which they lead C_s/C_m or Q/C_m are constant with:

- C_s is the concentration of the chromatographied substance in the stationary phase;
- Q_s is the quantity of substance in the stationary phase by unity of surface;
- C_m is the concentration of the substance in the mobile phase.

When $C_s/C_m = K$, K is called the *partitioning coefficient* of the solute. Strictly speaking, the partitioning coefficient should be defined in terms of the ratio of the activities of the solute a_s and a_m in the two phases at the partition equilibrium, that is to say:

$$K = a_s/a_m$$

Since by definition, the activities are related to the corresponding concentrations through the relations:

$$a_m = \gamma_m C_m$$
$$a_s = \gamma_s C_s$$

where the γ are the activity coefficients, one can write:

$$K = a_s/a_m \quad K = \gamma_s C_s/\gamma_m C_m$$

The diagram C_s/C_m is called the partitioning isotherm of the solute since the constancy of K (whatever the other experimental conditions are) can only occur when the temperature is fixed. In linear chromatography, the partitioning isotherm is evidently a straight line passing through the origin (Figure 1a). When the ratio C_s/C_m is not constant, the chromatography is said to be non-linear.

The nonlinearity of the isotherms can be due to the fact that the stationary and mobile phases are complex media and of strong ionic strengths that can change during a separation with a modification of the activity coefficients.

The consequence of the non-linearity of the isotherms is the deformation of the elution curve.

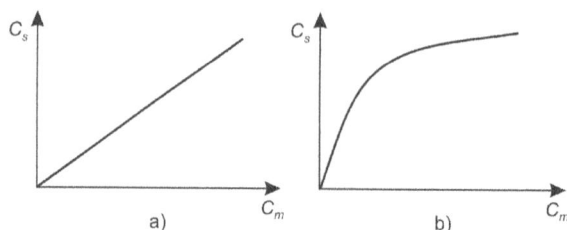

Figures 1a and b: Representation of Isotherms. (Reprinted with the Permission from Gwenola and Jean-Louis Burgot, Paris, Lavoisier, Tec et Doc, 2017, 138).

Ideal and Non-Ideal Chromatography

The conditions of ideal chromatography are subscribed when the selling out of the equilibria:

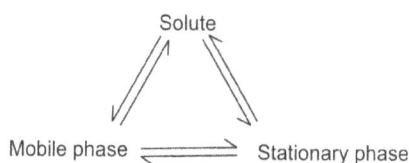

This takes place in a length of column which is infinitely short. Another manner to conceive an ideal chromatography is to imagine that the ternary partitioning equilibrium above is instantaneously reached. Practically, it is impossible to build such a column. According to some authors, it is never reached. Several theories are grounded on this hypothesis.

Brief Overview of the Theories of Chromatography

It appears, according to what is previously noticed, that the success of a chromatographic separation depends on the conjunction of two phenomena:

- The migration rates of the solutes must be sufficiently different.
- The overlapping of the chromatographic peaks must be as weak as possible. This criterion is an evident index of a good separation.

Practically, the relative importance of these both factors is wholly circumstantial. This is an experimental fact that the chromatographic peaks become larger as the chromatographic experiment progresses (Figure 2). Therefore, it is not surprising that the widening of the chromatographic peaks has been the matter of several theories, the existence of which being at the origin of the spectacular extent of these methods.

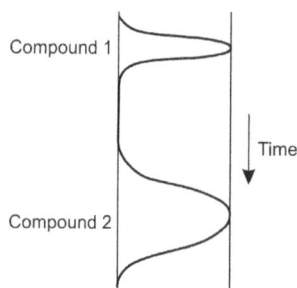

Figure 2: Widening of the Chromatographic Peak of a Solute in Proportion as the Chromatographic Experiment Goes Further. (Reprinted with the Permission from Gwenola and Jean-Louis Burgot, Paris, Lavoisier, Tec et Doc, 2017, 139).

The first theory of chromatography dates back to 1940. It is credited to Wilson. It is based on two hypotheses:

- The equilibrium between the solute and the two phases (stationary and mobile) is systematically and continuously reached.
- The diffusion of the solute in the phases is negligible.

The theory shows that for an ideal and linear chromatography, the migration on a column takes place without a widening of the peaks. It also shows that an ideal and nonlinear chromatography is accompanied by a widening and a deformation of the peak proportionally to the progress of the experiment. Wilson's theory was ameliorated three years later by de Vault and Weiss on the same bases as previously.

In 1941, Martin and Synge elaborated the theory of plates in order to explain the partition of liquid-liquid chromatography. It interprets the migration of a substance on a non-ideal column. The partitioning equilibrium is systematically reached by hypothesis, and the chromatography is linear. Finally, it neglects the phenomena of longitudinal diffusion.

It has been reexamined and developed later in order to interpret gas-liquid chromatography. N.C. Thomas was the first, from a theoretical point of view, to study the consequences of the fact that the equilibrium was not reached (1944 and 1948). In 1952, Lapidus and Amundson developed a theory that for the first time introduced some parameters, such as some kinetic coefficients of transfer of mass and longitudinal diffusion. In 1954, Glueckauf presented one equation that took into account longitudinal diffusion. From the standpoint of mathematical development, it was founded by Wilson and de Vault. These last theories have been the base of the famous theory of Van Deemter, Zuiderweg and Klinkenberg (1956). Differently, Gidding and Eyring, slightly before 1955, introduced some statistical concepts in order to describe the migration of the solutes. They served as an introduction to a stochastic theory of chromatography, which was also called the theory of aleatory walking (1958). Finally, in 1959, Gidding initiated a theoretical study called the study of generalized non-equilibrium. It is founded on the fact that the mass transfer from one phase to the other may be controlled either by a kinetic process in one or several steps or by a transfer by diffusion, which itself is a kinetic process.

We shall remain on the theory of plates, which although "unrealistic", has the advantage to introduce concepts that prove to be very fruitful. A typical example is the notion of a theoretical plate. Besides, this notion has not disappeared from the vocabulary of the other theories. Next, we shall envisage the kinetic theories of chromatography.

The Plates Theory

Generalities

Let us recall that the plates theory was elaborated by Martin and Synge in 1941 in order to interpret the partitioning liquid-liquid chromatography. It has been reconsidered and developed in order to interpret gas-liquid chromatography. It interprets the migration of a substance on a non-ideal column. In other words, plate height H possesses a finished value. It applies itself to linear chromatography and neglects the phenomena of longitudinal diffusion. If one assimilates the column to a stacking of plates of the same height H we can numerate from the top to the bottom (from 1 to N) (N being the number of plates of the column) and if in each plate, there is the same volume of stationary phase ds and the same volume of mobile phase dv (Figure 3), this means that:

- There is an exchange of matter horizontally between the two phases stationary and mobile until the condition of partitioning equilibrium would be satisfied.

$$C_s/C_m = K$$

- There is no vertical exchange (longitudinal diffusion negligible).

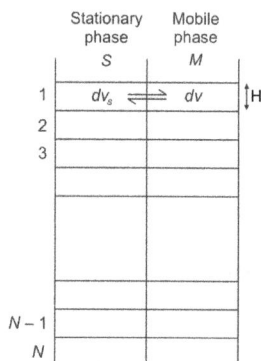

Figure 3: Representation of a Chromatographic Column. (Reprinted with the Permission from Gwenola and Jean-Louis Burgot, Paris, Lavoisier, Tec et Doc, 2017, 141).

The plates theory applies itself to an elution carried out discontinuously, that is to say by introducing the mobile eluting phase fractions by fractions of equal and very small volumes dv, that corresponds to the volume of the mobile phase of each plate.

In this study, we consider that one substance only has been introduced. The generalization to several substances is evident if the respective partition coefficients do not self-influence.

It is interesting to already notice that:

- If $K = 0$, the solute migrates at the same speed as the solvent.
- If $K = \infty$, the solute remains on the column.

These limit conditions are evidently not satisfactory.

Theory

One introduces a volume dv, noted dv_0 of mobile phase, containing a quantity Q of substance (Figure 4) into the column. It exactly fills the space devoted to the mobile phase in the first plate and one expects the substance to be shared in the two phases so that the equilibrium would be reached, that is to say:

$$C_S = KC_m$$

After this operation corresponds to the introduction of the substance into the column, the elution begins by the introduction of a first volume dv, noted dv_1 of phase mobile, which displaces the volume dv_0 from the first plate, pushing it in the second with the quantity of substance it contains. One expects that in both plates (plates 1 and 2), the substance would be in equilibrium with respect to the two phases, and then one pursues the elution by successive additions of equal volumes dv_2, dv_3, etc. One notices that when one introduces the nth volume dv_N, the volume dv_0 goes out of the column.

As in the separation at counter-current, we shall demonstrate firstly that the fraction of substance that remains in the stationary phase y and the one which goes into the mobile phase z are constant and do not depend on the total quantity to partition.

Constancy of Fractions Y and Z

Let us consider the fixation phase. If Q is the quantity of the substance to partition, one can write:

- The equilibrium condition:

$$C_S/C_m = K$$

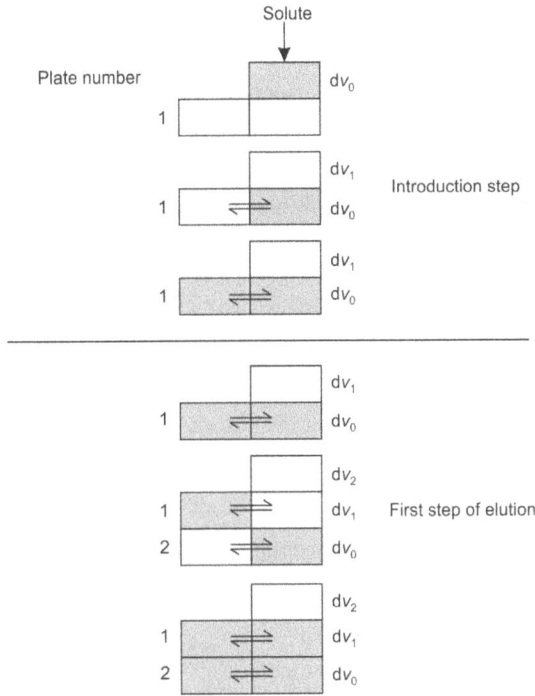

Figure 4: Introduction and First Operation of Elution of One Compound. (Reprinted with the Permission from Gwenola and Jean-Louis Burgot, Paris, Lavoisier, Tec et Doc, 2017, 142).

- The conservation of the matter

$$Q_S = yQ$$

$$Q_m = zQ$$

$$Q_S + Q_m = Q$$

Hence:

$$y + z = 1 \tag{1}$$

The condition of equilibrium for a theoretical plate is $C_S/C_m = K$. It can also be written:

$$(Q/dV_S) \cdot (dv/Q_m) = K$$

$$yQ/zQ = K(dV_S/dV)$$

Nothing precludes multiplying the numerator and the denominator of the right member by the number N of the plates of the column:

$$y/z = K \qquad K = NdV_S/NdV$$

$NdV_S = V_s$ is the total volume of the stationary phase.
$NdV = V$ is the total volume of the mobile phase.
V_s and V are constant for a given column.

Let us define by K' with :

$$K' = KV_S/V \tag{2}$$

K' is called the *corrected* (from volumes) *partition coefficient* or *retention factor of the column*. From the expressions (1) and (2), one deduces:

$$z = 1/(K' + 1) \quad \text{and} \quad y = K'/(K' + 1)$$

(*Remark*: y and z (as K) have usually their meanings inversed with respect to the corresponding quantities handled in the separation at countercurrent.)

Therefore, y and z depend only on the substance and on the characteristics of the column. The y and z are constant for a given experiment. As a result, the partition is done according to the fractions y and z, for the fixation and also for the different steps of the elution.

Some authors define the notion of efficient volume υ of a plate by the expression:

$$\upsilon = KdV_s + dV$$

By following the reasoning:

$$z = 1/(K' + 1) = 1/[K(dV_s/dV) + 1] = dV/(KdV_s + dV)$$

$$z = dV/\upsilon$$

This efficient volume υ is characteristic of the substance and the used column. With the same column, two substances with different partition coefficients each has a proper efficient volume.

Formalism of the Theory of Plates

After the phase of fixation, that is after the addition of the volume dv_0 containing all the substance to partition (and also after the addition of 0 volume of elution), the substance is present in one plate with the following repartition in the two phases:

$$yQ \quad + \quad zQ$$
$$\text{stationary} \quad \text{mobile}$$
$$\text{phase} \quad \text{phase}$$

according to the development of the Newtonian binomial relation $(y + z)^0Q$. After one step of elution, that is to say, after the addition of the volume dv_1, the substance is present in $1 + 1$ plates:

- In the first plate, it remains yQ moles.
- In the second plate, there are zQ moles.

The quantity in each plate is divided according to the development of the binomial relation:

$$(y + z)^1Q$$

It is interesting to notice that in the first plate, the yQ moles are divided according to y^2Q moles in the stationary phase and zyQ moles in the mobile phase.

After the addition of an nth volume dV_n $(n < N)$, the substance is present in $(n + 1)$ plates. The repartition in each of these plates corresponds to the different terms of the binome $(y + z)^nQ$, that is to say:

$$y^nQ \quad + \quad y^{n-1}zC_n^1Q \quad + \quad y^{(n-p)}z^pC_n^pQ \quad + ...+ \quad C_n^nz^nQ$$

quantity present in the 1st plate	quantity present in the 2nd plate	quantity present in the (p +1)th plate	quantity present in the (n + 1)th plate

Where C_n^p is the number of combinations of n things p at a time. Hence, the quantity present in the $(p + 1)^{th}$ plate after *n* operations of elution is:

$$Q_{p+1} = QC_n^p y^{n-p} z^p \text{ moles}$$

It is divided as follows:

stationary phase	and	mobile phase
$y\,QC_n^p y^{n-p} z^p$ moles		$zQC_n^p y^{n-p} z^p$ moles

(*Remark*: From the mathematical standpoint, after having added *n* volumes dv of eluting phase, the substance is present in n + 1 plates. However, from the purely physical standpoint, the quantity in some plates is no longer appreciable because of the very great value of *n* on one hand and because of the fact that *y* and *z* are obligatorily smaller than one, quantities which are at very high power. Thus, with Q = 1 g, y = 0.4 and n = 1,000, if one adds 500 volumes dv of mobile phase, the quantity of substance present in the first plate is $1 \cdot (0.4)^{500} = 10^{-199}$ g. The quantity of matter is inferior to 1 true molecule.)

Graphical Representations

The preceding results permit the studying of two important types of curves.

First Curve: Qx/Q as a Function of the Number of the Plates

This curve represents the distribution of the substance on the column at an instant of the elution when a volume of mobile phase ndv has been used (Figure 5).

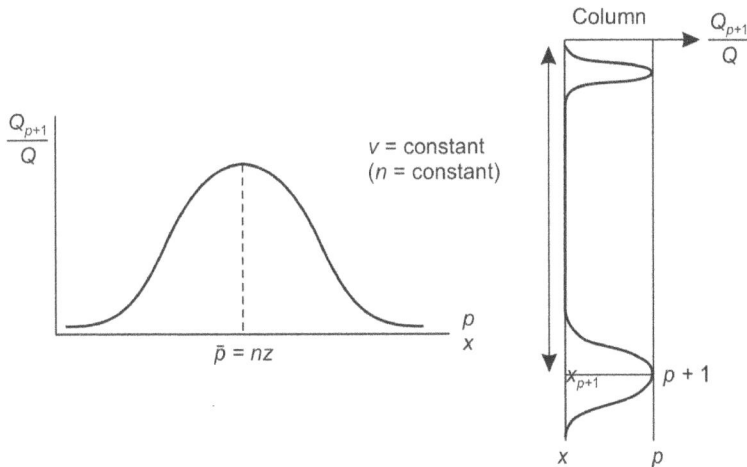

Figure 5: Repartition of the Substance as a Function of the Number of the Plates. (Reprinted with the Permission from Gwenola and Jean-Louis Burgot, Paris, Lavoisier, Tec et Doc, 2017, 145).

This curve represents the fraction:

$$Q_x/Q = f(x)_V$$

Where Q_x is the quantity of the substance present in the plate located at the distance x from the summit of the column. Since,

$$x = H(p + 1)$$

The function Qx/Q is also f(p)V. This curve represents the fraction of substance in each plate p + 1 for a constant volume v = ndv of eluting phase added. One has seen that in the (p +1)th plate, the quantity is given by the relation:

$$Q_{p+1} = QC_n^{\,p} y^{n-p} z^p$$

$$Q_{p+1}/Q = [n!/p!(n-p)!]\, y^{n-p} z^p$$

The right-hand term of this relation is the fraction of substance in the (p + 1)th plate at one instant of the elution when the volume v = ndv of phase mobile has been used. It is also the fraction of substance located at the height (p + 1)H from the summit of the column when a volume v = ndv has been added.

In partition chromatography, the plate height is generally very weak an order of $2\ 10^{-3}$ cm in chromatography liquid-liquid. This height gives 50,000 plates for a column of 1 m length. In gaseous chromatography, the plate height is of an order of 0.05 cm. This number leads to a number of about 2,000 plates per meter. One conceives that *n* and *p* take high values. One knows that when *n* and *p* are high and *y* and *z* are not too close to zero, it is possible to assimilate the binomial term:

$$C_n^{\,p} y^{n-p} z^p$$

To the expression:

$$\{1/[2\pi nyz]^{1/2}\}\exp[-(p-pbar)^2/2nyz)]$$

Thus:

$$Q_{p+1}/Q = \{1/[2\pi nyz]^{1/2}\}\exp[-(p-pbar)^2/2nyz)]$$

Here, $\bar{p}$ represents the plate where one finds the maximum substance. It is called the plate of mean rank. One must not forget that the variable is the distance *x* measured from the top of the column or equivalently *p* while *v* is constant. The curve under study is a curve of Laplace[12] – Gauss[13] centered at its maximum on the plate $\bar{p} = nz$.

For the illustration of what is above, let us consider a mixture of two substances A and B of values $z_A = 0.8$ and $z_B = 0.4$, which is treated by the addition of 500 volumes dv. This separation is successful. Substance A is present at its maximum in the plate of mean rank 500×0.8, that is to say in the 400th plate and substance B in the 200th plate $200 = 500 \times 0.4$. Each solute is present in a separated plate. These results perfectly justify the separation by chromatography.

Second Curve Qx/Qv as a Function of the Used Volume of the Eluting Phase

The second curve is that representative of the migration of the substance in a plate located at a distance *x* from the summit of the column. It expresses the variation of the fraction of substance which is in it as a function of volume *v* of the eluting mobile phase. Actually, in order to draw this curve, one places oneself at a distance *x* from the summit of the column or this is equivalent to one placing oneself in the plate *p* such that (p + 1)H = x and one notices the fraction of substance A as a function of volume *v* of mobile phase added. A particularly interesting point is such that x = 1, that is to say at the way out of the column. The curve is called the *elution curve*.

In order to draw this curve, the ratio Q_{p+1}/Q at the distance *x* (or in the plate p + 1 that one fixes *a priori*) must be expressed as a function of *v* and the volume of the mobile phase added; that is to say, equivalently, as a function of *n*.

---

[12] Pierre Simeon de Laplace, French mathematician and physicist, 1749–1827.
[13] Karl Friedrich Gauss, German mathematician, astronomer and physicist, 1777–1855.

For the $(p + 1)$th:

$$Q_{p+1}/Q = (n!/p!(n-p)!)y^{(n-p)}z^p$$

Firstly, after some rigorous mathematical calculations:

$$Q_{p+1}/Q = \{[n(n-1)----(n-p+1)]/p!\}\ y^{(n-p)}z^p$$

$$y^{(n-p)} = (1-z)^{n-p} = (1-z)^n/(1-z)p$$

$$Q_{p+1} = \{[n(n-1)----(n-p+1)]/p!\}(1-z)^n(z/1-z)^p$$

It is then possible to carry out some approximations:

$$\ln(1-z)^n = n\ln(1-z)$$

By developing in series, since:

$$\ln(1+x) = x/1 - x^2/2 + x^3/3 - x^4/4 +....$$

(Series convergent only for $-1 < x < 1$). For $x = -z$:

$$\ln(1-z) = -z - z^2/2 - z^3/3.....$$

Since $z > 1$

$$z^2/2 \ll z$$

$$\ln(1-z) \approx -z$$

$$n\ln(1-z) \approx -zn$$

On the other side, if $n \gg p$

$$n(n-1)----(n-p+1) \approx n^p$$

Example

$$100(100-1)(100-2)(100-3)(100-4) \approx 10^5$$

With this approximation, one commits a very weak error by excess. The preceding approximations lead to the following results:

$$Q_{(p+1)}/Q = (np/p!)\ e^{-nz}\ (z/1-z)^p$$

$$Q_{(p+1)}/Q = 1/p!\ e^{-nz}\ (nz/1-z)^p$$

One can suppose $1 - z \approx 1$. The fraction which is going into the mobile phase is very weak. This approximation is particularly valid and legitimate in gaseous chromatography (problem of vapor tension and molar fractions in the gaseous phase). Hence,

$$Q_{(p+1)}/Q = 1/p!\ e^{-nz}\ (nz)^p$$

We notice that the error committed here is an error in default which perceptibly compensates the error by excess committed previously.

In order to study the curve which is interesting, the remaining to do is to relate the preceding relation to the volume of mobile phase put into circulation. We have seen that $z = dv/\upsilon$ with υ being the efficient volume of a plate. The $nz = ndv/\upsilon$ is the volume v of the mobile phase, which is passed after the n operations of elution. Hence, we have $Q_{(p+1)}/Q = 1/p!\ e^{-nz}\ (nz/1-z)^p$

$$nz = v/\upsilon$$

and

$$Q_{(p+1)}/Q = 1/p!\ e^{-v/\upsilon}(v/\upsilon)^p$$

This is an equation of Poisson.[14] Let us recall that in this relation p and v are constant, the variable being v.

We shall assimilate this Poisson's curve to that of Gauss. We know that when n is very large, we can confound (by adopting a very general symbolism) the binomial expression:

$$C_n^k p^k q^{n-k}$$

With that of Gauss:

$$\{1/[2\pi npq]^{1/2}\}\exp[-(k-np)^2/2npq]$$

But also with that of Poisson:

$$e^{-np}(np/k!)^k$$

Provided the product np should be upper than 15 and inferior to 20. In the present case, it is necessary to have:

$$15 < v/\upsilon < 20 \text{ or}$$

$$15 < nz < 20$$

In chromatography, n is very great and z is very small since $1 - z \approx 1$, one supposes that this assimilation is legitimate (that is indirectly verified a posteriori). As a result, one henceforth considers that the above-found expression of Poisson is also an expression of Gauss (Figure 6):

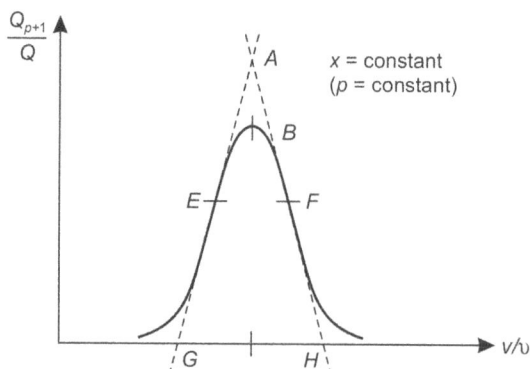

Figure 6: Representation of Q_{p+1}/Q as a Function of the Volume V. (Reprinted with Permission from Gwenola and Jean-Louis Burgot, Paris, Lavoisier, Tec et Doc, 2017, 146).

Characteristics of the Curve $Q_{p+1}/Q = f(v)_p$

As with every Gauss' curve, the studied curve exhibits the following characteristics:

- It is symmetric with respect to the vertical passing through the maximum B;
- It presents the inflection points E and F, the ordinates of which are equal to 0.6 time that of the maximum.
- The triangle is constituted by the tangents passing through the inflection points, and the corresponding portion of the axis of the abscissa is isosceles. Its summit has one ordinate equal to 1.2 time that of the maximum of the curve. Its base is so that GH = 2EF. One can assimilate the surface of this triangle to that of the curve without making an error. The curve presents the following characteristic points.

---

[14] Poisson, Denis, Simeon, French mathematician, 1781–1840.

Maximum of the Curve. The maximum and minimum are calculated by canceling the first and second derivatives. If, for the sake of simplicity, we adopt $x = v/\upsilon$ as the variable, one obtains:

$$f(x) = 1/p!x^p e^{-x}$$

By setting up:

$$1/p!x^p = u(x)$$

$$e^{-x} = t(x)$$

We obtain:

$$f'(x) = 1/p!e^{-x}x^{(p-1)}(p-x)$$

$$f'(x) = 0 \text{ for } x = \infty \text{ minima}$$

$$x = 0$$

And $x = p$ (maximum)

$$v/\upsilon = p$$

This means that at the distance x from the top of the column corresponding to the $(p+1)$th plate, there is a maximal quantity of substance when the volume of mobile phase v is passed through the column. This volume is v_{max}

$$v_{max} = p\upsilon$$

Abscissa of the Inflexion Points. One must calculate the derivative of the second order and cancel it. The first derivative is:

$$f'(x) = 1/p!e^{-x}x^{(p-1)}(p-x) \text{ or:}$$

$$f'(x) = 1/p!px^{(p-1)}e^{-x} - 1/p!x^p e^{-x}$$

It comes:

$$f''(x) = 1/p!x^{(p-2)}e^{-x}[p^2 - p - 2px + x^2]$$

$$f''(x) = 0 \quad \text{for} \quad x = 0$$

$$\text{for} \quad x = \infty$$

And:

$$x = p \pm \sqrt{p}$$

$$v/\upsilon = p \pm \sqrt{p}$$

Hence, the two points of inflection are separated by the distance $EF = 2\sqrt{p}$. As a result, the base GH is equal to $4\sqrt{p}$

$$GH = (v/\upsilon)_H - (v/\upsilon)_G = 4\sqrt{p}$$

By assimilating the Gauss' curve to an isosceles triangle, during the interval of time located between the instant where the substance appears in the $(p+1)$th plate and that it disappears from it, a volume of phase mobile complying with the following relations is passed:

$$(v_H - v_G)/\upsilon = 4\sqrt{p}$$

$$v_H - v_G = \upsilon 4\sqrt{p}$$

$$v_H - v_G = \upsilon 4p/\sqrt{p}$$

$$v_H - v_G = 4v_{max}/\sqrt{p}$$

Here, $\sqrt{p}$ is also symbolized by σ.

These results mean that when the maximum of the peak is located at the pth plate, the standard deviation is given by $\sqrt{p}$.

The Elution Curve and Consequences of the Plate Theory

We know that the fraction of substance being in a volume dv of a column possessing N plates is the one that would be in the (N +1)th hypothetical plate. In order to study the elution curve, it is sufficient to replace *p* with N in the preceding results.

- The volumes of phase mobile which have been added into the column since the beginning of the elution until the moment where the substance goes out at its maximum concentration are named total retention volume V_R. It is characteristic of the substance for a given column.

$$V_R = N\upsilon$$

Now:

$$\upsilon = dv + Kdv_s$$

$$V_R = Ndv + KNdv_s$$

$Ndv = V_m$: total volume of mobile phase in the column.

$Ndv_s = V_s$: total volume of stationary phase in the column.

$$V_R = V_m + KVs$$

This relation is important since it shows that the retention volume is characteristic of the substance for a given column (relation which is the basis of the identification of solutes in some kind of chromatography).

- The volume of the mobile phase measuring the width of the curve at its base, that is to say, the length w_V counted between the moments where the substances begin to go out of the column and finish their way out is:

$$w_V = 4Vg/\sqrt{N} = 4(Vm + KV_s)/\sqrt{N}$$

Concerning the widening of the peaks in proportion as the chromatography is developing, the expression of the width of the curve, at its basis for a plate of rank p + 1 is:

$$v_H - v_G = 4\ p\upsilon/\sqrt{p}$$

And for the curve of elution:

$$v_H - v_G = 4(V_m + KV_s)/\sqrt{N}$$

Thus, the numerator is constant for a substance and a given column. On the contrary, the denominator changes according to the number of N plates. Therefore, the greater the number of plates that a column involves, the more important the widening of the chromatographic peak at a given point of the column. At the borderline, when the number of plates tends towards infinity, the plate height tends toward zero. The column tends to be ideal.

- Reciprocally, $v_H - v_G = 4\upsilon\sqrt{N}$ and for a given point of the column:

$$v_H - v_G = 4\upsilon\sqrt{p}$$

Hence, the conclusion that for a column possessing a fixed number N of plates, the widening of the peak at a given point of the column is all the more important as the point is close to the extremity from which it is going out the mobile phase. It is maximal on the way out.

- The retention volume is related to the characteristics V_m, V_s and K of the substance which has been separated and of the used column. From another standpoint:

$$w_V = 4Vg/\sqrt{N}$$

It is possible to know V_R with an overflow meter and to measure w_V on the diagram of the chromatography by expressing it in units of volume in order to be homogeneous from this standpoint:

$$\sqrt{N} = 4V_R/w_V \quad \text{and} \quad N = 16V_R^2/w_V^2$$

From these relations, one can deduce very easily the number of plates and the plate height H:

$$H = L/N$$

L is the length of the column,
N is the expression of efficacy.

For the curve of elution, hence at the Nth plate, the standard deviation (figured in plates number) is according to what is preceding:

$$\sigma = \sqrt{N}$$

In abscissa, since the length of the column L is given by the expression:

$$L = NH$$

Where H is the height of a plate. The standard deviation figured in the same units is:

$$\sigma L = \sqrt{N}H$$

$$\sigma_L^2 = NH^2 \tag{3}$$

This relation is important. It helped Van Deemter, Zwiderweg and Klinkenberg in the obtention of their equation (read 'Kinetic Theories'). Besides, it also leads to the equation:

$$H = \sigma_L^2/L \tag{3'}$$

Since $V_R = V_m + KV_s$, N changes from one compound to another for the same column because of the coefficient K.

Critics of the Theory of the Plates

Advantages

This theory permits:

- To justify the identification of the compounds to separate with the help of parameters that themselves get into the calculations and are well defined by the theory.
- The easy determination of the number of theoretical plates of a column, that is to say, to know its efficiency.
- To justify the Gaussian form of the chromatographic peaks.
- To explain the widening of the peaks all along the unfolding.

Drawbacks

It is not permitted to identify the factors intervening on the height plate. This is due to the fact that it has been elaborated by starting from simplifying hypothesis:

- Lack of consideration of longitudinal diffusion;
- Introduction of the substance into the column with only one volume dv. This seems to be quasi-impossible practically;
- Lack of consideration of other phenomena with the help of which the stationary and mobile phase exchanges take place. These are the importance of their rates; the plates theory is a theory of equilibrium and is not a kinetic one, and it is probable that the equilibria are never reached. The theory of plates is, hence, in unreality. However, it offers great interest to introduce the concept of the plate.

Kinetic Theories

Authors have good reasons to estimate that actually the chromatographic separations are not governed by equilibria processes. The factors which are responsible for the widening of peaks are kinetic. The kinetic theories of chromatography even explained at the simple level, necessitate recalling some notions concerning the phenomenon of diffusion and some mathematical aspects which are linked to it.

Diffusion Phenomena—Fick's Laws

Diffusion is the phenomenon of displacement of a species (a molecule, ion, etc.) under the influence of a concentration gradient. It is governed by two laws named Fick's laws.

In order to simplify the task, let us only consider the displacement by diffusion along the axis of z. Let us consider two points A and B in a solution which at the instant t is not homogenous from the point of view of the concentration of the Solute C. Let C_A and C_B the concentrations in A and in B. Let also l_0 the distance which separates A and B. Figure 7 represents the profile of the concentrations along the distance l_0, the support of which is the axis of z.

Since at the instant t, $C_A > C_B$, some molecules (or ions) of solute migrate spontaneously from A to B until the moment where the chemical potentials in A and B are equal. This means that at this instant, $C'_A = C'_B$ as it is the same species in the same solution. This spontaneous migration is nothing else than a consequence of the second principle of thermodynamics.

Now, let us place at the distance l from point A along the axis of z and consider the plane normal to this axis at the distance l. The number of species crossing the surface unity of the plane by unity of time, under the influence of the diffusion, is given by the first law of Fick:

$$J = -D(\partial C/\partial z)_{z=l}$$

J is the flux of the species. $D(\partial C/\partial z)_{z=l}$ is the gradient of concentration along the z-axis for $z = l$. Using the partial derivative is necessary because the concentration can be dependent on other variables, such as time in particular. The factor of proportionality D is named the diffusion coefficient of the substance. Its unity is the $m^2 s^{-1}$. The sign minus is necessary because the flux must be obligatorily positive, the gradient is always negative in order to ensure diffusion (Figure 7).

The second law of Fick governs the profile of concentrations of one solute as a function of time and space coordinates when a concentration gradient is suddenly applied in a solution that was before equilibrium. For example, the (applied) concentration gradient may be due to a quick injection of one solute at the summit of a column.

Let us consider a parallelepiped of section unity and length dz (Figure 8). Let us make the hypothesis that the solute enters the left face and goes out by the right one. Let us suppose that

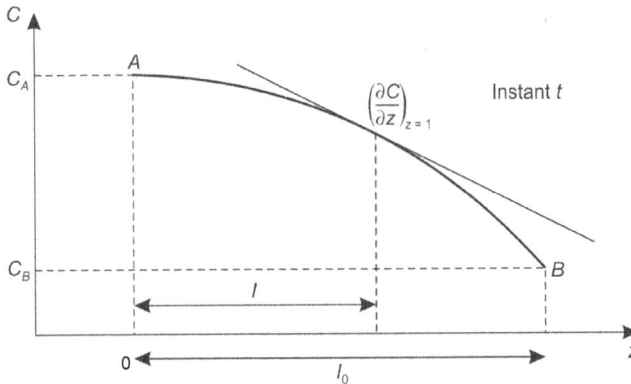

Figure 7: First Fick's Law; Profile of Concentrations. (Reprinted with the Permission from Gwenola and Jean-Louis Burgot, Paris, Lavoisier, Tec et Doc, 2017, 155).

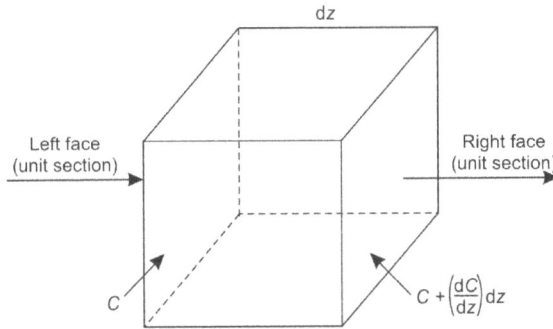

Figure 8: Fick's Second Law. (Reprinted with the Permission from Gwenola and Jean-Louis Burgot, Paris, Lavoisier, Tec et Doc, 2017, 156).

the concentration C of the solute would be a continuous function of z. If in the incoming, the concentration is C because of the difference in chemical potential, it is in the way out:

$$C + (dC/dz)dz$$

The entering flux J_L is given by the first Fick's law:

$$J_L = -D(\partial C/\partial z)$$

and likewise for the going out flux J_R:

$$J_R = -D \, \partial[C + (dC/dz)dz]/\partial z$$

The net going out flux is:

$$J_R - J_L$$

That is to say:

$$J_R - J_L = -D(d^2C/dz^2)dz$$
$$(J_R - J_L)dz = -D(d^2C/dz^2)$$

Given the dimensions of a flux (quantity of matter/time unit × surface unit), the left-hand term of the last relation has a quantity of matter/volume (surface x dz) per time unit for dimension. Thus, it is a change of concentration with time, that is to say:

$$(\partial C/\partial t) = D(\partial^2C/\partial z^2)$$

This is the second equation of Fick. It is a partial differential equation linear of the second order. One can already imagine that it is applicable in chromatography if we want to take into account the longitudinal diffusion, that is to say all along the vertical axis of the column.

(*Remark*: An equation of the same type would be also considered in order to take also into account the lateral diffusion occurring in the same horizontal section.)

Although the migration of a solute does not depend only on the longitudinal diffusion during chromatography, it is interesting to consider the integral solution of the preceding equation applied to the chromatography. This solution is obligatorily and over-simple since all the phenomena are not taken into account in the corresponding differential equation. The integral solution is:

$$n/n_{total} = \tfrac{1}{2}[1/(\pi Dt)^{1/2}]\exp(-z^2/4Dt) \tag{4}$$

Here, n is the number of moles of solute at the distance z and at the instant t, n_{total} is the total number of moles of solute "impulsed" through the horizontal section of the column of abscissa $z = 0$ at $t = 0$. Hence, it is the expression of the fraction of the number of solutes having diffused at the instant t and at the distance z. The factor $\tfrac{1}{2}$ takes into account the fact that the diffusion in the column is only operant in the sense in which z is positive.

Let us notice that the injection of the n_{total} molecules of solute at the same time corresponds to the instantaneous introduction into the column of a flux J. The name of which is "pulse" in mathematics and physics (Figure 9).

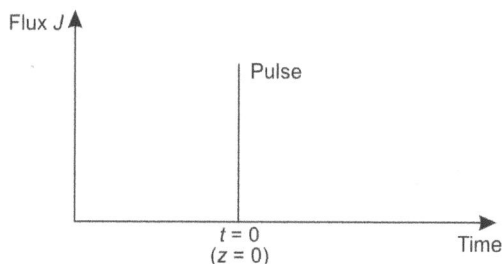

Figure 9: Instantaneous Introduction of the Solute. (Reprinted with the Permission from Gwenola and Jean-Louis Burgot, Paris, Lavoisier, Tec et Doc, 2017, 157).

Figure 10 represents the ratio n/n_{total} as a function of z and the time t. The solution equivalent to the preceding one but written in terms of concentrations is:

$$C(z,t) = [C(0,0)/2(\pi Dt)^{1/2}]\exp(-z^2/4Dt) \tag{5}$$

where $C(0,0)$ is the instantaneously injected concentration of solute at $t = 0$ and $z = 0$.

The Equations (4) and (5) are interesting for several reasons. Firstly, they are from a kinetic origin as it is just before demonstrated. Besides, the presence of the time t and of the parameter of diffusion D describes a kinetic phenomenon. On the other hand, the presence of the exponential theoretically the gaussian form of a chromatographic peak, systematically encountered experimentally, besides confirmed by the equilibrium theory of plates. Finally, the comparison of Equations (4) and (5) with the equation of the normal curve error function:

$$y = \{\exp[-(x-u)^2]/2\sigma^2\}/\sigma(2\pi)^{1/2}$$

where y is the frequency of arrival of the error, $x - u$ is the divergence with respect to the mean u and σ is the corresponding standard deviation that strongly induces the elaboration of a stochastic theory of the displacement of the solutes during a chromatographic experiment.

From a strictly mathematical viewpoint, we confine ourselves to recalling that the solution of the differential equation $y = f(x,z)$ of the second order with partial derivatives depends on

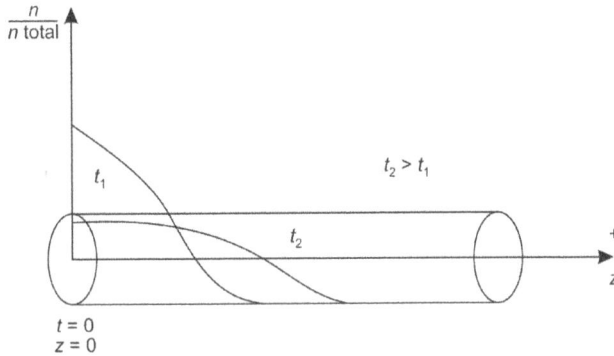

Figure 10: Fraction n/n_{total} as a Function of the Abscissa z and the Time T. (Reprinted with the Permission from Gwenola and Jean-Louis Burgot, Paris, Lavoisier, Tec et Doc, 2017, 158).

two arbitrary functions in place of two constants. The two arbitrary functions are determined in each particular case by limiting-conditions of physical nature. The borderline conditions of the chromatography are:

$$C = C_0 \quad t = 0 \quad and \quad z = 0$$

$$C = 0 \quad t > 0 \quad and \quad z \to \infty$$

These conditions at the outset exclude a solution of type $C(z,t) = Z(z)T(t)$. The integral solutions (4) and (5) result from the use of Laplace transforms, which have an interesting property to transform the equation with partial derivatives into an ordinary differential equation of second order. The Laplace transform function $F(z,s)$ entails a term of the type $\exp[-(s/Dz)^{1/2}]$, which automatically involves the presence of the term $\exp(-z^2)$ in the definitive integral solution.

Conservation of the Matter

The equations of the conservation of the matter are the outset points of the theory of Van Deemter et al. (1956) and the theory of the generalized non-equilibrium. Equations of conservation of the matter take into account other causes of migration of the solute in addition to the longitudinal diffusion. They are presented under several forms, but they are founded on the three following phenomena:

- The longitudinal diffusion takes place continually in the sense of the flux of the phase mobile since it corresponds to the migration of the zone the most concentrated towards the less concentrated (the diffusion takes place not only in the mobile phase but also in the stationary phase, including at the interface with the adsorbent when it is a chromatography of adsorption);

- The displacement under the action of the flux of mobile phase that migrates at the speed u and carries out the solute with changing rates. It is a convective phenomenon;

- The other is sorption-desorption. The word 'sorption' corresponds to the fixation of the molecule of solute in the stationary phase, whichever the physico-chemical interactions are the cause of the phenomenon. In adsorption chromatography, the sorption is the fixation by adsorption.

It is well-known that these phenomena are at the origin of the fact that the state of equilibrium is not reached and contribute to the widening of the peaks in proportion as the advancing of the solutes. The balance sheet of the matter is described, according to Giddings, by the following differential equation:

$$\partial C/\partial t = D\partial^2 C/\partial z^2 - u\partial C/\partial x + \Sigma_j(k_{ji}C_j - k_{ij}C_i)$$

C is the concentration of the solute by a unit of volume of the column. It decomposes into the two concentrations C_m and C_S in the mobile and stationary phases. The term $D\partial^2 C/\partial z^2$ corresponds to the longitudinal diffusion. The term $-u\partial C/\partial x$ represents the displacement under the influence of the flow of the phase mobile at the instantaneous u. Terms $k_{ji}C_j$ and $k_{ij}C_i$ represent the changes in masses by transfer from one phase into the other, k_{ji} and k_{ij} are apparent kinetic constants of mass changes.

Theory of Van Deemter, Zuiderweg and Klinkenberg

It also starts from the balance-sheet matter close to the preceding one that the authors write under the form of two partial differential equations, one of which being of the second order:

$$F_m \partial C_m/\partial t = F_m D \partial^2 C_m/\partial z^2 - F_m u \partial C_m/\partial z + \alpha(KC_{St} - C_m)$$

$$F_{st} \partial C_{st}/\partial t = \alpha(C_m - KC_{st})$$

Where m and st designate the phases mobile and stationary (let us notice that the term $\alpha(KC_{st} - C_m)$ is nothing different than the kinetic term of transfer of matter from the stationary phase toward the mobile phase and inversely indeed:

$$(KC_{St} - C_m) = k_{st}C_{st} - k_m C_m$$

According to the superposition of the very general theories of the equilibrium and kinetic in physical chemistry, one can write when the partition equilibrium is reached:

$$k_{st}C_{st} = k_m C_m$$

$$(C_m/C_{st})_{eq} = k_{st}/k_m = K$$

We deduce:

$$\alpha = k_m \text{ and } \alpha K = k_{st}$$

Here, α is called transfer of mass coefficient by volume unit (s^{-1}) and k_m, k is the kinetic constants of mass transfer from the mobile phase and the stationary phase towards the other. In the system of equations, Fm and Fst are the fractions of the mobile phase and of the stationary phase in the bulk phase (that is to say, in the volume of a theoretical plate). Thus,

$$F_m = dV_m/(dV_m + dV_{st}) \quad \text{and} \quad F_{st} = dV_{st}/(dV_m + dV_{st})$$

The solution of the above system of equations is:

$$C_m/C_0 = \{\beta t_0/[2\pi(\sigma_1^2 + \sigma_2^2)^{1/2}\} \exp[-(z/u - \beta t)^2/2(\sigma_1^2 + \sigma_2^2)] \tag{6}$$

Provided that the distance of migration z proceeded over by the solute should be far higher than the distance $2D/u$ and that the distance $F_m u/\alpha$, where u is the interstitial speed of the mobile phase ($m.s^{-1}$). C_0 is the initial concentration of solute introduced during the very short time t_0. The parameters σ_1^2, σ_2^2 and β are defined by the expressions:

$$\sigma_1^2 = 2Dz/u^3$$

$$\sigma_2^2 = 2\beta^2(F_{St}^2/\alpha F_m K^2)z/u$$

$$1/\beta = 1 + F_{st}/KF_m$$

The development of the sum $\sigma_1^2 + \sigma_2^2$ in the above solution leads to the expression:

$$\sigma_1^2 + \sigma_2^2 = z/u^2[2D/u + 2(uF_m/\alpha)/(1 + KF_m/F_{st})^2]$$

But, $z = nH$ where H is the height of a plate of straight section equal to unity and n is the number of existing plates until the abscissa z. It results from that:

$$\sigma_1^2 + \sigma_2^2 = nh/u^2[2D/u + 2(uF_m/\alpha)/(1 + KF_m/F_{st})^2] \tag{7}$$

Checking the equation of dimensions, the comparison with the relation (3) of the plate theory and the relation between the variance of the peak and the length of the column ($H = \sigma L^2$ relation (3')) of the theory of plates, that is to say, the relation $nH = n\sigma L^2$ suggests to set up:

$$H = 2D/u + 2 \, (uF_m/\alpha)/(1 + KF_m/F_{st})^2$$

This reasoning has been followed by Van Deemter et al. (1956). The relation introduces parameters that make the height of a plate changes with the speed u of the mobile phase, which was not possible according to the plate theory. The same authors have changed the preceding relation by giving h via introducing the notion of global resistance to the transfer of masses in the same mobile and stationary phases:

$$1/\alpha = 1/\alpha_m + K/\alpha_{st}$$

In partition chromatography gas-liquid in which the stationary phase is retained by the pores of solid support, α_{st} is given by the relation:

$$\alpha_{st} = (\tfrac{1}{4})\pi^2 D_{St} F_{St}/d^2 f$$

Where D_{St} is the diffusion coefficient in the liquid stationary phase located at the surface of the solid support and into the pores and df is the thickness of the film of the stationary phase. The authors have also expressed the hypothesis that the resistance to the transfer of masses in the bulk mobile phase is negligible, hence:

$$\alpha = \alpha_S/k$$

By reporting in the expression of H:

$$H = 2D/u + 8F_t d_f^2 KH/\pi^2(1 + KF_m/F_{St})^2 D_{St} F_{St}$$

Finally, Van Demter et al. (1956) have estimated that the longitudinal diffusion coefficient (coefficient D of the expression above) can be considered as resulting of the sum of a term figuring the molecular diffusion in the mobile phase, and another figuring the turbulent diffusion:

$$D = \gamma D_m + \lambda D_p u$$

Here, γD_m is the molecular diffusion, and γ is a number without dimension. It is called the factor of tortuosity. It takes into account the fact that to reach the distance z, the molecules of solute and mobile phase do not go through the same path among the particles of support. The expression $\lambda D_p u$ is admitted in order to figure the turbulent diffusion. The λ is called the filling coefficient, and D_p is the diameter of the particles of the solid support. After all, the relation giving H by replacing D with the preceding expression becomes:

$$H = 2\lambda dp + 2\gamma D_m(1/u) + [(\, 8/\pi^2)(KF_m d^2f)/(1 + KF_m/F_{St})^2 D_{St} F_{St}] \, u$$

By regrouping the constant terms, one can write this relation in the following way:

$$H = A + B/\overline{u} + C\overline{u}$$

where $\overline{u}$ is the average speed of the mobile phase.

(*Remark*: It is easier to handle the average speed of the mobile phase rather than its instantaneous value. The former is more accessible. Probably, their confusion leads to weak errors.)

Several points must be further highlighted in the establishment of the equation of Van Deemter:

- Firstly, from the logical and theoretical standpoints, the comparison and even the identification of the relations (3) (coming from the theory of plates) and (7) (coming from the kinetic theory) is doubtful. With the same reasoning carried further, this fact is equivalent to saying that the kinetic theory gives the same result as the equilibrium theory. This is not the starting hypothesis.

- The addition of the terms figuring the resistance to the transfer of mass is purely empirical. It is justified by the experimental results but not by some theoretical considerations as the terms in D/u and *u*.

- Nevertheless, Van Deemter's equation is in accordance with the experience.

- It presents a considerable interest in the domain of the optimization of chromatographic separation; in any case, in the choice of the speed of the mobile phase.

- Finally, through its integral solution (6), it allows us to augur a statistical interpretation of the chromatography of a solute as the parameters σ_{12} and σ_{22}. Already, these terms can be interpreted as being standard deviations that add up independently (see 'Stochastic Theory of the Chromatography').

Optimization of One Chromatography by Playing on the Speed of the Mobile Phase

Gaseous Chromatography

The curve $H = f(\overline{u})$ is one hyperbole (Figure 11). It presents the lines

$$H = A + C\overline{u} \qquad (\overline{u} \to \infty) \qquad (8)$$

and

$$H = \infty \qquad (\overline{u} = 0) \qquad (9)$$

The curve presents a minimum, that is to say, there is an optimal linear rate $\overline{u}$ for which H is minimal and the number of plates N is maximal. That means that the column has an optimal efficacy for the separation of the solute. It is such as:

$$dH/d\overline{u} = 0$$

Hence:

$$B/\overline{u}^2 + C = 0$$

$\overline{u}$, optimal $= \sqrt{B/C}$

H, minimal $= A + B\sqrt{C}/\sqrt{B} + C\sqrt{B}/\sqrt{C}$

$\qquad A + 2\sqrt{BC}$

- If $\overline{u} < \overline{u}$-optimal, $B/\overline{u}$ is preponderating
- If $\overline{u} > \overline{u}$-optimal, $B/\overline{u}$ is negligible

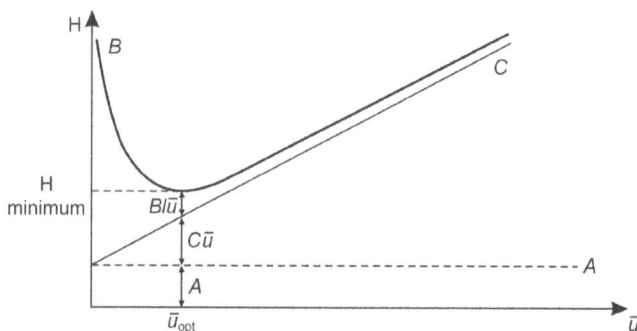

Figure 11: Representation of the Curve of Van Deemter. (Reprinted with the Permission from Gwenola and Jean-Louis Burgot, Paris, Lavoisier, Tec et Doc, 2017, 164).

(*Remark*: In the case of capillary columns of very short diameter, A = 0 since there is no filling. For equal lengths, these columns are always more effective than those which are filled.)

Liquid Chromatography

The contributions to the value of the H of each factor of enlarging of the chromatographic peaks during the migration on a column lead to an equation that is close to that of Van Demter called Knox's equation (Figure 12):

$$H = Au^{1/3} + B/u + (C_m + C_s)u$$

Here, A is a coefficient that figures the irregularity of the filling. (It is the phenomenon that leads to different speeds at different points of the same cross-section of the column). It is the check parameter of the homogeneity of the filling (usually A < 1). It is expressed in $(cm^2.s)^{1/3}$. B is the term for longitudinal diffusion in $cm^2\ s^{-1}$.

$$B = 2\varsigma\, D_m\ (2 < B < 4)$$

Here, ς is a factor of proportionality. In the liquid phase, the diffusion coefficient D_m is by far weaker than in the gaseous phase. Hence, the term B/u is negligible except if u is very weak. Cs and C_m represent the resistances to the transfer of masses into the stationary and mobile phases. They are expressed in seconds. The C_s decrease if the diameter of particles decreases.

(*Remark*: C changes from 0.02 to 0.2 with values of 1 in ion-exchange chromatography.)

Concerning gaseous chromatography, the H changes a little. For intermediary values of u, it is possible to enhance the speed of the mobile phase without diminishing too strongly the efficacy.

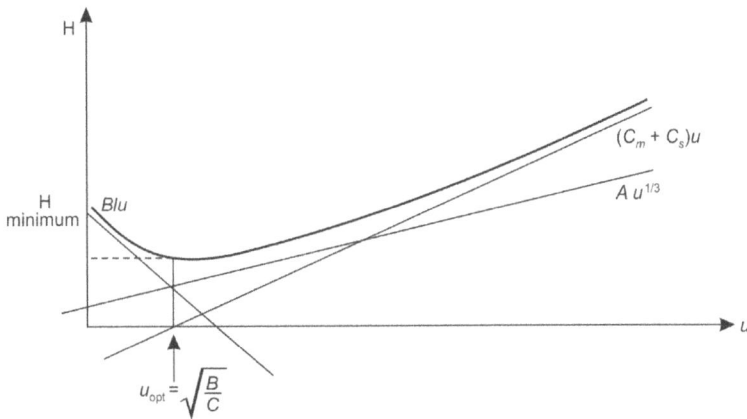

Figure 12: Representation of Knox's Equation. (Reprinted with the Permission from Gwenola and Jean-Louis Burgot, Paris, Lavoisier, Tec et Doc, 2017, 165).

Theory of the Stochastic Migration

It is essentially due to the works of Giddings and al. (1961). The solutions (4) and (5) of the differential equations (which take into account only the migration par diffusion) are normal Gauss curves of the general equation:

$$y = (1/\sigma\sqrt{2\pi})\ \exp(-x^2/2\sigma^2) \qquad (10)$$

Without taking into account the factor ½ (already mentioned). It is the same as the solution (6) of the system of differential equations of Giddings. These results suggest an aleatory migration of the solutes during the chromatographic separations. If one assimilates the expressions (4) and (5) to the above expression (10), one immediately obtains:

$$\sigma = (2Dt)^{1/2}$$

If one admits this identity, one immediately notices that the standard deviation of the covered distance by the solute increases with time. This is in agreement with the experience and also with the plate theory.

This identity, which at this instant is only a formal correspondence, is demonstrated by assimilating the motion of a solute to an aleatory motion along the z-axis. Let us consider a solute that can do successively aleatory steps of length l, starting from any given origin either in the positive sense or in the negative sense. It is easy to conceive that after N steps (N is a very great number) the total distance covered by the solute (starting from the origin) is null. The reason is the following one. The solute at each step does possess the same probability to travel in a direction as in the opposite one. Actually, the parameter which must be taken into account in this kind of motion is the average of the squares of the covered distances $<z^2>$. It is evident that according to the definition of a standard deviation:

$$<z^2> = \sigma^2$$

Hence, it remains to demonstrate that:

$$<z^2> = 2Dt$$

In order to set up the above identity (see Appendix 2).

CHAPTER **15**

Practical Consequences
of the Plates Theory
Qualitative and Quantitative Aspects
of the Chromatography

Chromatographic qualitative and quantitative analysis is based on and justified by theoretical grounds coming from the theory of plates. The same theoretical grounds permit defining criteria for quality of analyses carried out by chromatography. This paragraph is devoted to all these aspects together to optimize the chromatographic experiments. All these possibilities are given by the study of the elution curves.

Chromatographic Quantities and Factors of Identification of Solutes

Figure 1 represents an elution curve or chromatogram.

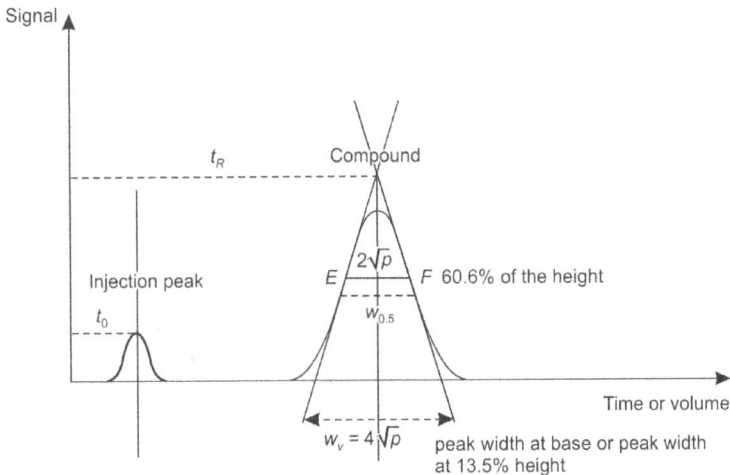

Figure 1: Elution Curve or Chromatogram. (Reprinted with the Permission from Gwenola and Jean-Louis Burgot, Paris, Lavoisier, Tec et Doc, 2017, 169).

Retention Quantities

Retention Time and Width of a Peak

In elution chromatography, one distinguishes the peak corresponding to the mobile phase followed by the peak(s) of the (different) solute(s). The symbols generally used correspond to the following definitions:

- t_m (or t_0) hold-up time or interval of time that necessitates the passing of the mobile phase through the column. With the speed u of the later in cm s^{-1}, one obtains:

$$t_m(s) = L(cm)/u(cm \ s^{-1}) \tag{1}$$

- t_R time of retention is the interval of time passed between the introduction of the compound into the column and the instant when the substance goes out of the column at its maximal instantaneous concentration.

- t'_R is the adjusted time of retention:

$$t'_R = t_R - t_m$$

- t_N is a net time of retention.
- J is the compressibility factor of the mobile phase.
- $\sqrt{p}$ or σ standard deviation is the half-width of the peak at the height of the inflection points at 60.6% height (refer Chapter 14).
- $w_{0.5}$ is the width of the peak at half height.
- w_v is the width of the peak at its basis or at 13.5% height of the peak $= 4\sqrt{p} = 4\sigma$.

Retention Volumes

The total retention volumes are the phase mobile volumes used in order to elute the solutes at their maximum concentration since their introduction into the column.

Volume of the Phase Mobile Contained in the Column

If the flow D of the mobile phase is constant, one can define the volume occupied by the phase mobile between particles of a packed column by V_0. The extra column volume, which is composed of the volume of injector, connecting lines and detector V_{ext} (or formerly named dead volume) is negligible in liquid chromatography in comparison to V_0. Also, we can consider that V_m which is equivalent to $V_0 + V_{ext}$ is equal in this case to V_0 (and t_m to t_0). The retention volume V_m is defined as:

$$V_m = t_0 \ x \ D \tag{2}$$

Where V_m is the volume of phase mobile contained in the column between the particles since the volume of the column is given by the expression Lx $\pi(d/2)^2$; here d = diameter of the column. As a result, the unoccupied volume for the mobile phase is equal to:

$$V_m = Lx \ \pi(d/2)^2 x\varepsilon_{tot} \tag{3}$$

With ε_{tot} = porosity of the column < 1 (for a good column ε_{tot} = 0.8). The porosity of the column is the ratio between the volume of the empty spaces (pores) and the total volume of the column (pores plus solid phases).

(*Remark*: As an example, for a column of 25 cm with 4.6 mm of internal diameter, the volume of the mobile phase of a good column is 3.3 mL.)

Retention Volume V_R

$V_R = t_R \times D$ is the volume necessary to elute the solutes at their maximal concentration since their introduction into the column. It is called the total volume of retention. One also uses the adjusted retention volumes V'_R and the adjusted times of retention t'_R which are related by the expression:

$$V'_R = t'_R \times D$$

But:

$V'_R = V_R - V_m$ with V'_R the adjusted retention volume. Recall (see Chapter 14) that:

$$V_R = V_m + KV_s$$

Whence:

$$V'_R = KV_S$$

One also defines the net volume de retention V_N:

$$V_N = V'_R \times J$$

The adjusted and total times and volumes of retention are characteristic for one solute for a given column. In theory, the simple measure of t_R or V_R must permit the identification of one solute. In practice, this process is uncertain since it is very difficult to obtain a perfect reproducibility of the grandeurs of retention obtained with reference chromatography and of those of the experience. This is why one prefers using the notion of the retention factor.

Retention Factor k'

The factor of retention k' does not depend on the flow of the mobile phase nor the length of the column. It proved itself more interesting than the preceding parameters.

$$k' = t'_R/t_0 = t_R - t_0/t_0 = t_R/t_0 - 1$$

$$k' = V'_R/V_m = V_R - V_m/V_m = KV_s/V_m$$

Or

$$K = C_s/C_m \text{ hence } k' = C_s V_s/C_m V_m = Q_s/Q_m$$

Here, k' is the retention factor of the column. It is very simply determined by measuring t_0 and t_R. Let us notice that the retention factor represents the ratio of the quantities of substances in the two phases.

$$KVs = V_R - V_R'$$

One has, $k'V_m = V_R - V_m$ and $V_R = V_m (1 + k')$ and

$$t_R = t_0(1 + k')$$

And

- When $k' = 0$, $V_R = V_m$, the compound is not retained by the stationary phase.
- The greater k' is, the more retained the compound is since V_R increases in proportion. Usually, k' is comprised between 1 and 10, but in practice, the best compromise seems to be between 2 and 6. A minimal value of 2 is recommended by the international conference on harmonization (ICH).[15] Too high values of k' make the peak still larger and extend the duration of the analysis.

---

[15] Organism charged with the assignment to harmonize the validity tests at the international level.

Otherwise, $L = u \cdot t_0$ with L being the length of the column and u the linear speed of the mobile phase, thus:

$$t_R = t_0(1 + k') = (L/u)(1 + k')$$

This relation permits the calculation of one of the parameters when all the others are known (for example, the calculation of *u*).

Selectivity Factor

Selectivity is a notion that is always related to two solutes. The selectivity α between two substances is given by the relation:

$$\alpha = t'_{R2}/t'_{R1} = t_{R2} - t_0/(t_{R1} - t_0)$$

If D is the constant flow of the mobile phase:

$$\alpha = (D_{tR2} - D_{t0})/(D_{tR1} - D_{t0}) = V_{R2} - V_m/(V_{R1} - V_m)$$

$$V_{R2} = V_m + K_2 V_s$$

$$V_{R1} = V_m + K_1 V_s$$

Hence, one deduces that:

$$\alpha = K_2/K_1$$

The factor of selectivity is equal to the ratio of the partition coefficients. It is also equal to the ratio of the retention factors of the column since:

$$K'_2 = K_2 V_s/V_m \quad \text{and} \quad K'_1 = K_1 V_s/V_m$$

Hence:

$$\alpha = k'_2/k'_1$$

The important point expressed by this relation is the independence of the selectivity with respect to the geometrical parameters of the column (no intervention V_s and V_m). On the other hand, the nature of the phases influences the selectivity through the intermediary of the thermodynamic constant K.

(*Remark*: A lot of the preceding relations are valid because the linear speed and the flow of the mobile phase are constant. It is not always the case, for example in gaseous chromatography for which one reason in average values of flow.)

Capacity of Separation

It represents the number of solutes *n* which can be separated in a column in some given experimental conditions. It is expressed by the following formula:

$$n = (\sqrt{N}/4R) \ln [(1 + k'_n)/(1 + k'_1)]$$

With:

  N is the plates number,
  R is resolution,
  k'_1 and k'_n resolution factors of the first and the late solute.

For example, if N = 12,000, R = 1.2, $k'_1 = 2$ and $k'_n = 12$, n = 33. Actually, the true capacity of separation is lower since some solutes do possess identical characteristics of separation.

Qualitative Aspects

The retention time is a function of numerous parameters (nature of the stationary and mobile phases, temperature, length of the column and pressure gradient) which make it impossible the characterization of one product on the basis of published results in the literature. To remedy this point, several methods have been proposed:

- The most classic method consists of comparing the times of retention of the sample and that of a standard compound.
- It is also possible to replace the determination of the grandeurs of retention with the determination of k' or by that of the selectivity α. It is sufficient to add to the sample of a supplementary solute and the selectivity of which is with respect to the solute that we want to identify.
- Let us also mention the method of overloading necessitates directly adding the check to the sample. As a result, according to the cases, there is an increase in the area of the chromatographic peaks or an apparition of supplementary peaks if there is not an identity of both substances.
- And let us also mention the use of the retention index. It is Kovats who proposed this method in 1958 and applied it for the first time to gaseous chromatography in isothermal conditions. It characterizes a solute by comparing its chromatographic behavior in gaseous chromatography to that of a series of paraffinic hydrocarbons. In isothermal conditions, the reduced retention time follows a logarithmic progression. Each n-alkane (of formula C_zH_{2z+2}) is affected by an index of retention (RI) of 100 z (Figure 2).

For:

CH_4 z = 1 index 100
C_2H_6 z = 2 index 200
C_3H_8 z = 3 index 300

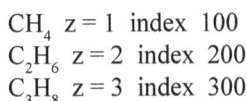

By definition, these numbers are the same for every stationary phase. An unknown substance X is characterized by a retention index, which is obtained graphically (Figure 2, z = 3, z' = 4) or by the following formula.

If x, z and z' are respectively the numbers of carbon atoms of the unknown substance and two hydrocarbons, one has:

$$RI = 100[z + (z' - z)\{\log t'_R(x) - \log t'_R(z)\}/\{\log t'_R(z') - \log t'_R(z)\}]$$

This formula is valid for isothermal conditions. The equations are more complex if there is a program of temperature. The retention indexes are more reproducible than the times of retention and are largely used in order to characterize the compounds in the complex media (essential oils, petroleum products and metabolites) by liquid or gaseous chromatography coupled with mass spectrometry.

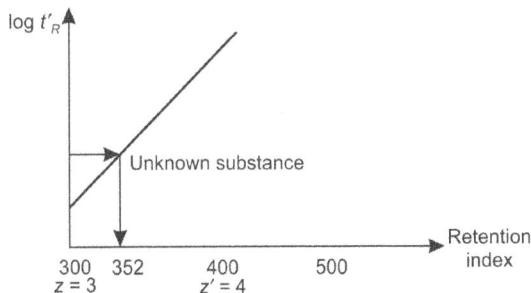

Figure 2: Characterization of One Substance According to Kovats' Method: Log t'_R = aR_I + b. (Reprinted with the Permission from Gwenola and Jean-Louis Burgot, Paris, Lavoisier, Tec et Doc, 2017, 175).

Quantitative Aspects

The chromatographic column is systematically coupled with a detector, which permits continuously sensing solutes on the way out of the column (see Figure 7 of Chapter 13). The differential detectors give a surface of peak S proportional to the total quantity of the solute submitted to the chromatography since the detector gives the instantaneous concentration C as a function of the used volume *v* of the mobile phase. Since a concentration is the ratio of a quantity of substance and a volume, one can write:

$$\int_{v1}^{v2} CdV = \text{total quantity of a chromatographied solute}$$

Where v_1 and v_2 are the volumes of mobile phase used at the beginning and the end of the elution of the compound.

It is true that in chromatographic analysis, relating the chromatographic signal to the concentration of the injected substance sets up two problems:

- The knowledge of the proportionality factor K between the surface of the peak and the injected quantity (number of moles and not a concentration).
- The knowledge of the exact volume of the solution of solute injected into the chromatographic system in order to deduce its concentration. The injectors have undergone true progress and now permit injecting some of the microliters of the sample with an increasingly low risk of error. However, the best way to follow is to use an *internal standard*.

The internal standard permits controlling the behavior of the substances along the different steps of the chromatographic determination preparation of the sample, extraction, derivatization[16] and chromatography itself. The internal standard is added to the sample at the early step of the determination. As a sample, it must be submitted to the same sources of error and particularly that linked to the variability of the injected volume. As an internal standard, it is necessary to choose molecules, the physico-chemical properties of which are close to those of the solute (pKa and solubility) in order to obtain a retention peak neighboring that of the sample. In the case of one chromatograph coupled to a mass spectrometer as a detector, the internal standard must entail at least one stable isotope of one of the atoms of the analyzed substance (see Chapter 37).

Generally, the quantitative analysis is carried out with the help of a scale of calibration. Below, we demonstrate that by adding the internal standard in the sample at the same concentration as in the check solutions, it is possible to obtain the concentration of the sample as a function of the surfaces of the chromatographic peaks without knowing the concentration of the internal standard with accuracy.

The following symbols are used in the demonstration:

- T standard solutions (calibration scales of the compound to study prepared with known concentrations).
- E for the internal standard.
- X for the sample of unknown concentration.

Internal Calibration With Only One Point of Calibration

The standard solution of known concentration and the sample solution are overcharged by the internal standard at the same concentration. Therefore, we can write as follows.

---

[16] Formation of a derivate of the analyte easier to chromatography than the analyte itself.

For the standard solution:

- $Q_T = C_T V_T = K_T S_T$
- $Q_E = C_E V_E = K_E S_E$ where S_T and S_E are the total surfaces of the peaks and K_T and K_E are the factors of proportionality
- Q_T is the quantity of substance in the standard solution
- Q_E is the quantity of the internal standard;

$$Q_T/Q_E = C_T V_T/C_E V_E = K_T S_T/K_E S_E$$

But, $V_T = V_E$ since the products are injected at the same time and in the same solution, hence:

$$C_T/C_E = K_T S_T/K_E S_E \quad \text{or} \quad K_T/K_E = C_T S_E/C_E S_T \tag{4}$$

For the sample X:

$$Qx = CxVx = KxSx$$

And:

$$Q_E = C_E V_E = K'_E S'_E$$

Here, also, for the same reasons $V_X = V_E$

This permits writing;

$$Q_X/Q_E = C_X/C_E = K_X S_X/K'_E S'_E \quad \text{or} \quad K_X/K'_E = C_X/C_E \times S'_E/S_X \tag{5}$$

But $K_X/K'_E = K_T/K_E$ since the sample and the check have a very close structure, the factors of proportionality remain in the same ratio (with a constant adjustment of the sensibility of the apparatus). We can write, according to (4) or (5);

$$C_T/C_E \times S_E/S_T = C_X/C_E \times S'_E/S$$

$$C_X = C_T \times S_E/S_T \times S_X/S_{E'}$$

C_T is known and the surfaces S_E, S_X, S_T and S'_E are measured.

(*Remark*: The concentrations of the internal standard appear no longer in the last formula. Only the areas appear.)

Internal Standardization With a Scale of Calibration

Each point of the scale corresponds to a check solution in which the internal standard is dissolved at a concentration C_E. In the sample, the internal standard is equally dissolved at the concentration C_E (Table 1).

Table 1: Composition of the Scale of Calibration and the Solution of Internal Standard.

	Sol. 1	Sol. 2	...	Sol. n	Sample
Concentration C_T	C_1	C_2	...	C_n	C_x
Concentration C_E	C_E	C_E	...	C_E	C_E

For each solution, one measures the area S_T and S_E. Then, one draws the curves S_T/S_E as a function of the ratio S_X/S_E on the straight line in order to obtain the concentration of the sample (Figure 3):

$$S_T/S_E = (K_E/K_T) \times (C_T/C_E)$$

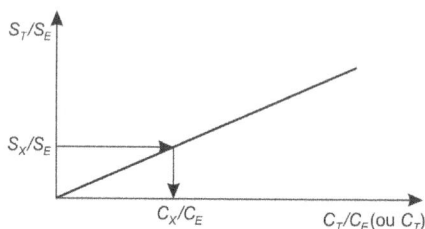

Figure 3: Graphical Representation of the Obtention of the Concentration. (Reprinted with the Permission from Gwenola and Jean-Louis Burgot, Paris, Lavoisier, Tec et Doc, 2017, 178).

Separation of Solutes and Resolution

Separation of the Chromatographic Peaks

In order to obtain a successful separation, the number of chromatographic peaks must be obligatorily equal to the number of solutes. Moreover, the peaks must be sufficiently separated.

Definition of the Resolution

The resolution (Figure 4) is the measure of the real quality of the separation of two neighbor peaks. It is defined by the relation:

$$R_S = 2(t_{R2} - t_{R1})/(w_{v2} + w_{v1}) = 1.177(t_{R2} - t_{R1})/(w_{0.5;1} + w_{0.5;2})$$

Here, $w_{0.5}$ is the width at half-height of the peak. This relation is valid for all the gaussian and symmetric peaks.

The resolution is without units. It can be calculated by measurements of times or volumes (the obligatory point being that t_R and w are expressed in the same units). The resolution is a practical datum that is easily calculated graphically. The more elevated R_S is, the better the separation. Let us suppose, for sake of simplicity, that the two peaks (the resolution of which is measured) would have the same width at their basis, $w_{v2} = w_{v1} = 4\sigma$,

- If $t_{R2} - t_{R1} = 4\sigma$, the peaks are contiguous, $R_S = 1$. The Gauss' curves are more and less wide than the isosceles triangle that overlap but only by 2%,
- If $t_{R2} - t_{R1} = 6\sigma$, $R_S = 1.5$, the overlap is negligible = –0.3%
- If $t_{R2} - t_{R1} = 3.2\sigma$, $R_S = 0.8$, the overlap is of the order of 10%.

The optimal value of R_S is 1.5 in quantitative analysis; upper values are not necessary since they unnecessarily extend the analysis.

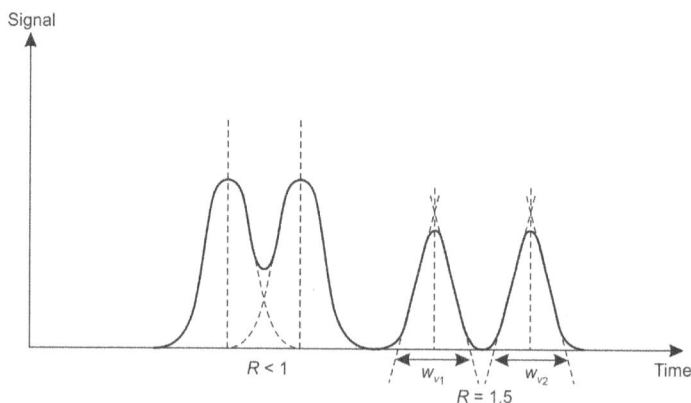

Figure 4: Resolution of a Mixture of Two Compounds. (Reprinted with the Permission from Gwenola and Jean-Louis Burgot, Paris, Lavoisier, Tec et Doc, 2017, 179).

Optimization of the Resolution

By modifying the conditions of the chromatography, it is possible to optimize the resolution, knowing that:

$$R_{S12} = 2(t_{R2} - t_{R1})/(w_{R1} + w_{R2})$$

If the peaks are very analogous, the width of the peaks at the basis is equivalent, that is to say, $w_1 = w_2 = w$ and:

$$R_{S12} = t_{R2} - t_{R1}/w$$

$$N_2 = 16(t_{R2}/w_{v2})^2 = 16(t_{R2}/w)^2$$

$$1/w = \tfrac{1}{4}(\sqrt{N_2})\, 1/t_{R2}$$

Reporting that into the expression of $R_{S1,2}$:

$$R_{S1,2} = \tfrac{1}{4}(\sqrt{N_2})(t_{R2} - t_{R1})/t_{R2}$$

$$k'_2 = (t_{R2} - t_0)/t_0$$

And:

$$t_{R2} = t_0(1 + k'_2)$$

One obtains:

$$R_{S12} = \tfrac{1}{4}(\sqrt{N_2})[\,(t_{R2} - t_{R1})/t_0\,]\,[1/(1 + k'_2)]$$

By multiplying the numerator and denominator by $t_{R2} - t_0$

$$R_{S12} = \tfrac{1}{4}(\sqrt{N_2})[\,(t_{R2} - t_{R1})/t_0\,][(t_{R2} - t_0)/(t_{R2} - t_0)]\,[1/(1 + k'_2)]$$

$$R_{S12} = \tfrac{1}{4}(\sqrt{N_2})[\,(t_{R2} - t_{R1})/(t_{R2} - t_0)\,][(t_{R2} - t_0)/t_0]\,[1/(1 + k'_2)]$$

$$R_{S12} = \tfrac{1}{4}(\sqrt{N_2})\{[(t_{R2} - t_0) - (t_{R1} - t_0)\,]/(t_{R2} - t_0)\}[\,k'_2/(1 + k'_2)]$$

$$\alpha = (t_{R2} - t_0)/(t_{R1} - t_0)$$

Hence:

$$R_{S12} = \tfrac{1}{4}(\sqrt{N_2})[(\alpha - 1)/\alpha][k'_2/(1 + k'_2)]$$

Purnell's relation

Remark: In liquid chromatography, it is possible in an isocratic flow (constant composition of the mobile phase during the separation) to have $N_1 = N_2 = N$. Then, the resolution is given by the following formula:

$$R_S = (\tfrac{1}{2})[(\alpha - 1)/(\alpha + 1)]\,[\,k'_{bar}/(1 + k'_{bar})]\sqrt{N}$$

With:

$$k'_{bar} = (k_1 + k_2)/2.$$

The formula (6) and (6') are interesting since they show that there are three parameters on which it is possible to ameliorate the resolution.

Increase in the Number of Plates N

Recall (viz., Chapter 14) that

$$N = 16V_R^2/w_v^2 \text{ or } 16t_R^2/w_v^2$$

$$N = (t_R/\sigma)^2 = 5,\,54(t_R/w_{0,5})^2 = 16(t_R/w_v)^2$$

The plate number figures the efficacy of the column. The increase of the plate number N is equivalent to a reduction of the "length" w_v. There are two manners in order to increase the efficacy of the column:

- **To lengthen the column.** In this case, if the plates' number increases, the retention volumes and the width of peaks also increase. However, the latter increases less quickly since it changes as $\sqrt{N}$. The reasoning which follows demonstrates it easily. As an example, let us consider two columns of respective lengths L_1 and L_2, such as $L_2 = yL_1$ and everything otherwise being equal:

$$V_{R1} = V_{m1} + KV_{S1}$$

$$V_{R2} = V_{m2} + KV_{S2}$$

$V_{m2} = yV_{m1}$ and $V_{S2} = yV_{S1}$ since the number of plates is proportional to the length of the column, hence:

$$V_{R2} = yV_{m1} + KyV_{S1}$$

$$V_{R2} = y\, V_{R1}$$

$$N_1 = 16V_{R1}^2/w_1^2 \text{ and } N_2 = 16V_{R2}^2/w_2^2$$

$$w^2 = 4V_{R2}/\sqrt{N_2} = 4yV_{R1}\sqrt{N_2}$$

$$L_1/N_1 = L_2/N_2 \quad \rightarrow \quad L_2/L_1 = N_2/N_1 = y$$

Thus, $N_2 = yN_1$

$$w_2 = 4yV_{R1}/\sqrt{yN_1}$$

$$w_1 = 4V_{R1}\sqrt{N_1}$$

$$w_2 = 4yV_{R1}\sqrt{y} /\sqrt{y}\sqrt{N_1}\sqrt{y} = 4V_{R1}\sqrt{y}/\sqrt{N_1}$$

$$w_2 = w_1\sqrt{y}$$

By increasing the length of the column, the maxima separate according to some ratio while the bases of the peaks have less tendency to overlap since their widths only increase in a ratio equal to the square root of the preceding. However, there is a limit to the increase of the length of a column since by staying too long in a column, the solutes may alter. This is overall the case in gaseous chromatography.

- **To decrease the plate height** by modifying:
  - The size of the particles of the stationary phase,
  - The rate of the mobile phase.

Modification of the Values k'

The behavior of the compound the most retained (Compound 2) is the only one taken into consideration. If one increases k'_2, the resolution increases but only for values such as $k'_2 < 10$. For greater values, the term $k'_2/(1 + k'_2)$ tends towards 1. Moreover, if k'_2 increases, the retention volumes also increase. k' is the function of the nature of the stationary phases and the composition of the mobile phase. One non-polar compound has its affinity for the stationary non-polar phase, which increases if the polarity of the mobile phase increases and inversely with the polar compounds.

Increase in the Selectivity

This one depends according to its definition on the values of the partition coefficients of the two solutes, that is to say on their affinities for the stationary and mobile phases. As a result, the modifications of the composition of the mobile phase or the stationary phase are efficient means in order to ameliorate the selectivity. It is also possible to modify the value of K by changing the temperature of the experience.

(*Remark*: If $\alpha = 1$, the partition coefficients are equal, the resolution is strictly impossible, whichever the efficacy of the column is. From another point of view, given the form of the mathematical function $(\alpha - 1)/\alpha$, a very weak change of α has for consequence an important modification of the ratio.)

Different methods permit optimizing the selectivity. The most known is that called the "method of the triangle of selectivity of Snyder". Snyder has quantified for each solvent the part played by:

- The acceptor-character of H-bonds of the proton
- The donor character of H-bonds of the proton
- The donor character of interactions dipole-dipole.

He has grouped these characteristics in a term named parameter of global polarity P' (Figure 5). The obtained values have permitted the classification of the solvents into eight groups. One conceives easily that replacing one of the solvents of the mobile phase with one solvent of the same group does not ameliorate the selectivity.

The optimal selectivity can be known only after chromatography of the sample with seven mobile phases the composition of which is noticed on the graphic (summits, middles of the sides of the triangle of Figure 5 and composition 0.33: 0.33: 0.33) and after comparison of the obtained selectivities.

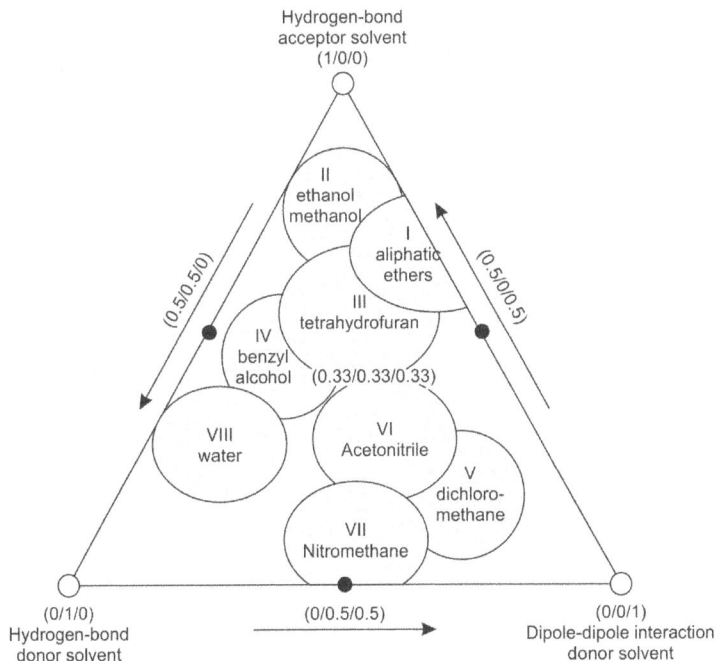

Figure 5: Classification of the Solvents or Snyder's Triangle. (Reprinted with the Permission from Gwenola and Jean-Louis Burgot, Paris, Lavoisier, Tec et Doc, 2017, 183).

As a numerical application, let us consider a separation by gaseous chromatography of two compounds A and B on a column of two meters. For the retention times and the widths of the peaks, one obtains the values:

$$t_{RA} = 330s \quad w_A = 19.7s$$

$$t_{RB} = 346s \quad w_B = 20.4s$$

$$t_0 = 30s$$

Calculate the resolution:

$$R_{AB} = 2(t_{RB} - t_{RA})/(w_B + w_A) = 2(346 - 330)/20.4 + 19.7$$

$$R_{AB} = 0.798$$

Calculate the plates number N_B:

$$R_{AB} = 1/4\sqrt{N_B}((\alpha - 1)/\alpha)[k'_B/(1 + k'_B)]$$

$$k'_B = (t_{RB} - t_0)/t_0 = 10.5$$

$$k'_A = (t_{RA} - t_0)/t_0 = 10$$

$$\alpha = k'_B/k'_A = 1.05$$

$$N_B = 5289$$

What length of the column would be necessary in order to obtain a resolution of 1.5?

$$L_2/L_1 = N_{B2}/N_{B1} = y \quad \text{and} \quad R_2/R_1 = \sqrt{N_{B2}/N_{B1}} = \sqrt{y} = 1.875$$

$$y = 3.51 \quad \text{and} \quad L_2 = 3.51 \quad L_1 = 7.02 \text{ m}$$

What height for compound B should be necessary in order to obtain a resolution of 1.5 without increasing the length of the column?

$$N_{B1} = 16(t_{RB}/w_B)^2 = 4603$$

$$N_{B2}/N_{B1} = 3.51 \text{ whence } N_{B2} = 16\ 156$$

$$N_{B2}/N_{B1} = L/h_2 \times h_1/L = h_1/h_2 = 3.51$$

$$h_2 = h_1/3.51 = (2000/4603)/3.51 = 0.124 \text{ mm}$$

A Brief Study of the Non-Ideal Behavior: Asymmetry Factor

The factor of asymmetry F_S of a chromatographic peak is a number without unity permitting to gauge the gaussian nature of the peak.

In the American, European and Japanese pharmacopeias, it is defined as the ratio of the width of the peak at the 20th of its height ($w_{0.05}$) and at two times the distance between the perpendicular lowered from the maximum of the peak and the entry of the peak at the 20th of its height (A).

$$F_S = w_{0.05}/2A$$

Instead of the preceding ratio, some other chromatographers rather use the ratio A/B measured at 10% of its height (Figure 6):

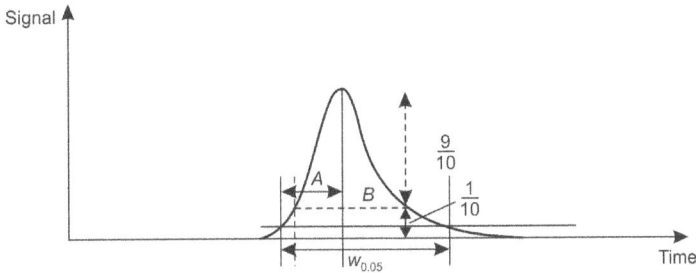

Figure 6: Study of the Symmetry of the Peak. (Reprinted with the Permission from Gwenola and Jean-Louis Burgot, Paris, Lavoisier, Tec et Doc, 2017, 184).

F_s must be located between 0.8 and 1.2 dead volumes or hold-up volumes too consequent, some adsorption phenomena or, more simply, the length of the column lead to non-linear isotherms which are objectivized by the formation of distorted peaks.

Ratio Signal Noise

If the baseline is stable, it is possible to evaluate the ratio between the signal S obtained with the sample and that N for noise emitted by the instrument during the measurement of a blank sample where there is no solute. The noise of a blank sample may have several origins. The determination of this ratio permits us to appreciate the possibilities of the chromatographic analysis, notably in terms of limits of detection.

CHAPTER 16

Basic Principles of Chromatographic Modes

We describe in this chapter some basic principles of chromatographic separations in following with the classification depending on the type of equilibrium.

Adsorption

Adsorption chromatography involves finely divided solid stationary phases named adsorbents and consequently, adsorption chromatography is referred to as liquid-solid chromatography. It can apply to the separation of compounds, including some isomeric ones. It is not suitable for the separation of very polar molecules, which are often irreversibly adsorbed on the stationary phase. The adsorption process is used for the elution and development of chromatography.

Adsorption and Desorption Processes

The adsorption corresponds with the energetic binding of gas, liquid or solid at a solid surface. It is necessary that the process be reversible in order to recover the compound. The interactions are not specific ones. The desorption of the solute can be carried out by breaking the earlier bonds with an eluent.

The retention time and the retention volume of a solute depend on the competition of molecules of solute S with molecules of the mobile phase M as both bind to the active sites of the adsorbent.

Theoretical Aspects

The process of adsorption-desorption can be presented as:

$$S_{(M.P.)} + M_{(a)} \rightleftarrows S_{(a)} + M_{(M.P.)}$$

M.P. = Mobile Phase
a = Adsorbent

Distribution Coefficient

In the case of adsorption, the stationary phase is solid; consequently, the ratio which provides the partition balance becomes the distribution coefficient K_D expressed by $K = \dfrac{q}{C}$ with:

q = amounts (masses) of the solute per unit mass of stationary phase in $mol.g^{-1}$
C = concentration of the solute in the mobile phase $mol.L^{-1}$
K is a function of temperature, the nature of the solute and that of both phases

The relation between q and C is a non-linear isotherm when the expression is given by Langmuir's law[17] (see Chapter 15).

Retention Factor k'

The retention factor k' is given by the following expression $k' = K (W_A/V_M)$ with:

W_A: mass of the adsorbent packed in the column

V_M: volume of the mobile phase in the column

The substitution of K by q/C gives $k' = (q/C) (W_A/V_M)$ or the amount of the solute in the stationary phase/quantity of the solute in the mobile phase.

Origin of Interactions in the Adsorption-Desorption Processes

These interactions between compounds and phases are various and not specific. We can mention:

Ionic

The acidic and alkali molecules could be ionized according to the value of the pH of the solution and their pKa, thus giving with the adsorbent some ionic bonds. These bonds are so difficult to break that it is therefore advisable to eliminate them.

Electrostatic Interactions

The dipolar molecules, the electrical positive and negative poles of which do not coincide are responsible for the adsorption and the elution. However, these interactions are less strong than those provided by ions but they are much more frequent.

The highly polar molecules are retained by most adsorbents themselves, i.e., polar ones. This fact could also occur for the non-polar molecules, although not possessing a dipole moment since they can be polarized at the proximity of a polar adsorbent. There is a spontaneous induction of a dipole moment.

Hydrogen Bonding

Solutes can provide according to their structures; numerous hydrogen bonds with the mobile phase which may lead to unexpected behavior. As an example, we describe the separation of the methaqualone and acetaminophen mixture (Figure 1).

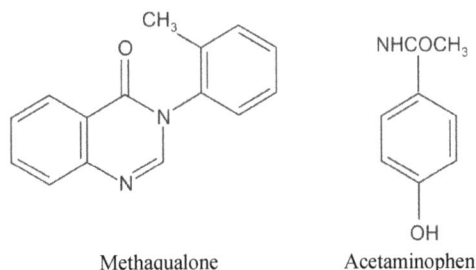

Figure 1: Chemical Formula of the Methaqualone and Acetaminophen. (Reprinted with the Permission from Gwenola and Jean-Louis Burgot, Paris, Lavoisier, Tec et Doc, 2017, 190).

---

[17] Irving Langmuir (1881–1957), American Physical Chemist, Nobel prize in Physics 1932.

It becomes obvious experimentally that the selectivity factor α between both products varies widely with the nature of alcohol used as the mobile phase. It is possible to explain that by the fact that among both products, acetaminophen is the only one that gives hydrogen bonds with the mobile phase and it is all the more important as the alcohol has a low molecular weight (methanol).

Nature of the Adsorbable Substances

The importance of adsorption depends on a number of factors:

Chemical Structure of Adsorbable Substances

The structure of the solute makes a significant contribution to the adsorption process. Some low differences in structure are responsible for the separation of isomers and of diastereomers. Polar compounds and linear carbides adsorb more easily than those which possess some branched structures. Therefore, the strength of the adsorption increases according to the order following:

Acids > Alkalis > Alcohols > Aldehydes > Ketones > Esters > Halogens > Unsaturated Carbides

Molar Mass

In the chemical similar series, the bigger the molar mass is the more important adsorption. For aromatic compounds, the number of aromatic rings is a relevant factor as shown with:

Benzene < Naphthalene < Anthracene < Naphthacene

The pH of the Mobile Phase

The adsorption of the acidic or basic compounds depends on their pK_A, the nature of the adsorbent, acid or basic, and the pH of the solvent. Thus, morphine which is not adsorbed on bentonite (hydrated aluminum silicate), around pH 8 or 9, features much better adsorption if the pH becomes more acidic or basic.

Adsorbent Substances

Adsorbents are chosen according to their ability to separate complex mixtures and their inertness to solutes and the solvent. Adsorbents must be porous, finely ground and reduced into particles of small diameter to feature an important specific activity. There exist many types of adsorbents, such as bentonite, alumina Al_2O_3, nH_2O or silica SiO_2; this one is the most used (Figure 2).

The same adsorbent involves some different properties of adsorption depending on moisture content. In the case of pure silica, for example, water molecules can be attached to the silanol groups. It decreases its adsorption capacity and as it is often claimed, the activation status is different. It follows that to have a good reproducibility of the chromatographic experiments, it is necessary to make experiments with the same activation status. It exists a parameter named β* which represents the activation status of an adsorbent. In practice, for good adsorption, the intermediary values of β* between 0 and 1 are chosen after adding some water. Some values closed 0 for activating produce irreversible adsorption.

Solvents

Overview

From the chemical standpoint, solvents must be inert with adsorbents and molecules solutes. It is rare for one solvent to provide the entire process from adsorption to elution. For each step, it is useful to choose the best solvent in function of the nature of the solutes.

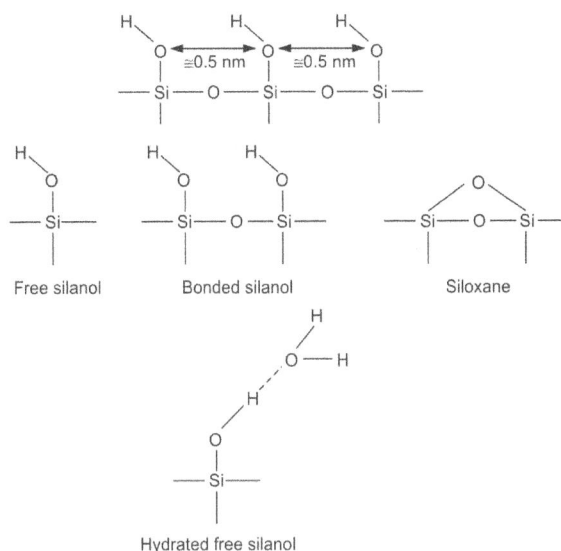

Figure 2: Different forms of Silica. (Reprinted with the Permission from Gwenola and Jean-Louis Burgot, Paris, Lavoisier, Tec et Doc, 2017, 191).

How to Choose the Solvent?

With an adsorbent polar, the solvent used to fix analytes must be as little polar as possible. On the contrary, the step of elution is started with a weakly polar solvent and then continued with solvents that are more polar.

Eluotropic Strength

In adsorption chromatography, the binding strength of the mobile phase is given by the ε_0 parameter of the solvent which has the physical signification of free standard enthalpy of adsorption in the standard conditions ($\beta^* = 1$). The higher the ε_0 parameter is, the higher it is fixed by the stationary phase and the more the solute is shifted. In Table 1, the values of ε_0 or eluotropic strength are given with the alumina as the reference adsorbent. With another adsorbent as silica, for example, the order of classification will be different. Some mixtures of solvent are frequently used to facilitate the release of the molecules that have been adsorbed.

Table 1: Eluotropic Strengths of some Solvents with the Alumina Adsorbent. (Reprinted with the Permission from Gwenola and Jean-Louis Burgot, Paris, Lavoisier, Tec et Doc, 2017, 193).

Solvent	ε_0	Polarity
Hexane	0	The less polar
Ligroin	0.01	...
Carbon Tetrachloride	0.18	...
Benzene	0.32	...
Chloroform	0.40	...
Ethyl Acetate	0.58	...
Pyridine	0.71	...
Methanol	0.95	The more polar

Example: Compounds fixed to silica with hexane ($\varepsilon_0 = 0$) are eluted with t-butyl and methyl oxide ($\varepsilon_0 = 0.48$) without separation. But if the experiment is realized with about 10% t-butyl and methyl oxide in hexane ($\varepsilon_0 = 0.29$), the separation of compounds becomes possible.

On the other hand, it is possible to introduce the delocalization notion for solvents. That means that the interaction between polar groups of the adsorbent and of the solvent does not involve a specific group, it is the case with dichloromethane or benzene. This notion is available for the solvents which possess many polar groups because each group cannot be linked to the adsorbent groups. The result is that there exist in a solvent some chemical groups that directly interact with adsorbents and others that do not.

Thus, it is not advisable to use some gradients of solvents that contain a variable composition of the mixture during the chromatography process because the time of equilibrium after each new composition should be very long.

Optimization of the Composition of the Mobile Phase

For the first time, the optimization is realized on a thin layer to find the nature and composition which are the most appropriate to separate the mixtures of compounds. The selectivity factor can be improved by choosing solvents not only with different eluotropic strengths but also with different basicity.

Partition Chromatography

Principle

Partition chromatography or liquid-liquid chromatography is a process based on the difference in partition constant of the solutes to analyze between aqueous and water-immiscible phases.

For chromatography on the column, the aqueous phase is fixed on a silica gel introduced in a glass cylindrical diameter vertical tube. Silica gel has the possibility to fix half of its weight of water without external appearance being amended. The mobile phase flows through the pores of the silica gel. It is an organic solvent. Firstly, mixtures of components are dropped on the top of the column and after the mobile phase flows through the column. The more important the solubility of the compound in the mobile phase, the faster the velocity of the compound increases. It depends on this partition coefficient, that in this case must be lower than $K = C_s/C_M$. Some experiments are also carried out in the reverse phase, in other words with a stationary organic phase and an aqueous mobile phase.

Brief History

The first experiment was achieved in 1941 when Martin and Synge[18] had made the separation of a mixture of amino acids (the separation of amino acids by adsorption chromatography was very difficult) after their preliminary acetylation. The phenylalanine and the alanine present a different partition coefficient between water and chloroform when there are acetylated:

- Acetyl phenyl alanine: $K = 4$;
- Acetyl alanine: $K = 216$;

It is a type of chromatography that has not undergone great development the first time because it has been difficult to immobilize its stationary phase on the support. Moreover, it gave some adsorption phenomena linked with these functional groups. For a long time, partition chromatography was reserved for the development process of cellulose acetate because it was easier to fix its liquid stationary phase. In the early 70s, it had been heightened by the new technologies with new linked stationary phases which had very low granulometry of materials. The detectors on the line also gave

---

[18] A.J.P. Martin and R.L.M Synge, English chemists and Nobel prize in chemistry in 1952.

a boost to the process. In these last few years, the interest in this chromatography has never stopped growing, especially with the significant innovations in this field.

Supports

The supports are made of particulate materials in a very finely divided state from which the adsorption properties are hidden. Moreover, they must be inert against stationary and mobile phases. Historically, the first products used were cellulose and kieselguhr which are fully replaced by silica gel and polymers. New supports have emerged in recent years, such as zircon or zirconium oxide but also graphite carbon particles.

The diameter of the support particles varies from 2 to 15 µm, and the size pore from 80 to 300 µm. There are traditional packed columns made of non-porous particles, which provide good kinetic performances and high mechanical stability but with specific surface area limited. On the opposite, a new system of filling a column named monolithic can improve the separation efficiency (plates/meter). Monolithic supports are made of silica or polymers filled *in situ* and consist of one piece of continuous porous networks made of macropores and mesopores:

- Macropores of 50–1,000 nm diameter between the aggregates of polymers.
- Mesopores inside the aggregates whose diameter is in order of 10–15 nm.

Consequently, the void space increases from 40% for a conventional column to 80% for a monolithic column by means of mesopores. For the same pressure, it is possible to choose a speed flow at a higher value than with a conventional column because the macropores realize the mobile phase transfer. Monolithic columns, made of porous particles, have an important surface area that gives them a great charge capacity.

Some new columns filled with core-shell nanoparticles of 1.3 at 5 µm of diameter meet with an important interest in UHPLC (ultra-high performance liquid chromatography) (see Chapter 17). In this case, a porous shell is dropped off on a solid core of silica. This technology, which provides for some homogeneous size of particles, is interesting for the analysis of the biomolecules which have some diffusion constant that is too low.

Stationary Phases

The modern stationary phases are solid phases bonded on which lipophilic groups are attached to the support surface by chemical bonds. The most used are made of aliphatic chains whose polarity and chain length of the alkyl group have different values.

Thereby, there are:

- **Normal phase chromatography:**
  - Stationary phase with hydrophilic groups bonded on silica
    - Amino or amino-propyl $-Si-(CH_2)_3-NH_2$
    - Nitro $-Si-(CH_2)_3-NO_2$ or nitrile $-Si-(CH_2)_3-CN$
    - Dialcool $-Si-(CH_2)_3-O-CH_2-CHOH-CH_2OH$
  - Mobile phase lipophilic
- **HILiC mode chromatography** (Hydrophilic Interaction Liquid Chromatography), also named an aqueous normal phase with a hydrophilic stationary phase, is used with an organic phase; for example, acetonitrile containing a small amount of water. It favors the analysis of very polar compounds (metabolites, peptides, etc.) which features a low solubility in the organic phase. This mode is often used for the analysis of the compounds by LC-MS and liquid chromatography linked to mass spectrometry. Thus, it is not possible with this method to diminish the polarity by ion-pair formation. A greater percentage of organic solvent in the mobile phase is also an advantage in LC-MS because it facilitates solvent removal (see Chapter 17).

- **Reversed phase chromatography:** The stationary phase is lipophilic in nature (alkyl chains of C4 at C30, propyl, phenyl and diphenyl) and the mobile phase is hydrophilic. If column C18, $-Si-(CH_2)_{17}-CH_3$ are the most known, column C4 is very useful to analyze peptides and proteins and column C30 to separate carotenoids or flavonoids.

- There exist also some **chiral stationary phases**, which are used to separate enantiomers of a racemic mixture. Enantiomers exhibit no difference in their chemical or physical properties in an achiral medium. However, the resolution of a racemic mixture can be achieved by:

  - An *indirect process* that is based on the interaction of a racemic mixture with a chiral selector to make diastereomers pairs of molecules that have some different properties even in an achiral medium. The diastereomers separation can be achieved on achiral chromatography. In the end, the enantiomers are regenerated by chemical reaction on each diastereomer separated. This method is limited to molecules that possess some chemical group derivatized (acid, alcohol, amine, etc.), but it is restrictive because the derivatization reaction must be quick, quantitative and the chiral compound must have a great optical purity;

  - A *direct process* by formation of *labile diastereomers* between enantiomers and a chiral stationary phase. The chiral stationary phases are classified according to the nature of the interaction between the diastereomer and the stationary phase; for example, H-bonding, dipole-dipole interactions, interactions π-π and steric effects or the function of their chemical structure. The main stationary phases are classified as follows:

    - *Synthetic phases*, the chiral selector (phenyl-glycine, dinitro-benzoyl phenylglycine, poly(meth)acrylates, poly-siloxane), is immobilized by ionic bonds on a silica support.

    - *Polysaccharides phases*, especially cellulose, amylase, chitosan and chitin consist of sugar unit chains with a helical twist of the polymer backbone. They are often brittle and do not put up with the values of pH too acid or basic and the chlorinated solvents.

    - *Cyclodextrins* phases consist of cyclic oligosaccharides chains of 6 at 12 molecules of glucose linked by a 1–4 bond. They give a truncated cone shape whose internal diameter is a function of the number of glucose molecules (Figure 3). The interior of the cavity is hydrophobic because all the free hydroxyl groups are located toward the exterior. The hydroxyl groups can be derivatized by diverse substituents to modulate their properties. Cyclodextrins can be divided into three classes; α, β and γ according to the number of glucose in the structure. They are interesting for the separation of pharmacological chiral molecules which have often one hydrophobic part and some polar groups.

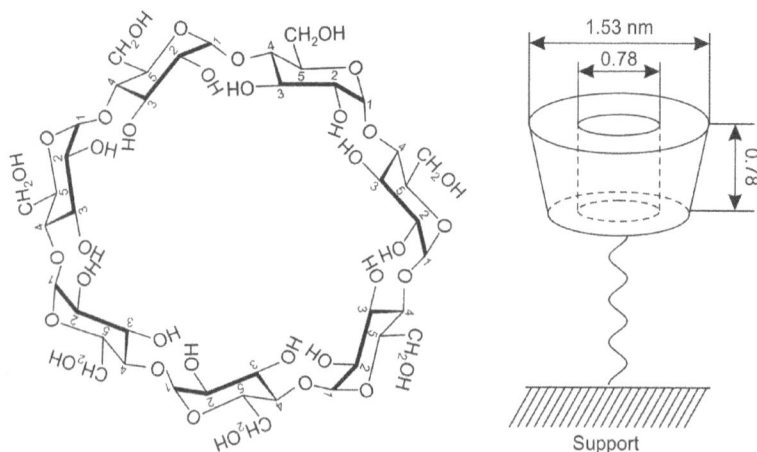

Figure 3: Cyclodextrin Molecule Linked to a Solid Support. (Reprinted with Permission from Gwenola and Jean-Louis Burgot, Paris, Lavoisier, Tec et Doc, 2017, 198).

- *Proteins* are based on chiral subunits (amino acids). One notices that the most used are alpha-1-glycoprotein, the bovine or human serum albumin, the ovomucoid or cellobiohydrolase.

- A *direct process* is by the *addition of a chiral agent* in the mobile phase but this method is rather used in preparative enantiomeric separation.

The stationary phases with cross-linked silica provide the following drawbacks:

- An instability on basic pH.

- With mobile phases very hydrophile (less than 10% of organic solvent), they collapse. It means that the alkyl chains adhere to the support and consequently, retention times are modified and the reproducibility of the analysis decreases.

These problems have led to the creation of new following phases:

- Alkyl phases with polar groups (amide, ether or carbamate) "embedded polar linked" with electrostatic repulsion which provide the analysis of polar compounds by using very hydrophile mobile phases. The analysis of the basic groups is also better because interactions with silanol groups of the support are lower.

- Some silica coated on the surface by methyl groups settled at pH 11.5. It becomes possible to use with this phase; some mobile phases are hydroxide ammonium and sodium hydroxide. These phases are interesting for the chromatographic methods coupled with mass spectrometry.

- Linked Zircons or zirconium dioxide or covered with porous carbon with a very good thermal stability and pH extreme resistance (1 to 14), which improve the efficiency of the separation. Zircons are covered with carbon for the isomers and diastereomers.

- Fluoridated silica is more hydrophilic than cross-linked silica and provides a better selectivity for the halogen compounds analysis or compounds with a carbonyl group (aldehyde, ketones and esters).

Mobile Phases

As for the absorption chromatography, the mobile phase must have:

- A strong power of solvent for compounds and high chemical resistance against the analyzed compounds.

- A strong chemical inertness.

- And must be insoluble against the stationary phase.

The mobile phases are often made of many solvents (mixes water or buffer solutions with organic solvent) and the percentage can vary along the chromatography. Some of them can be used at high temperatures (60–70°C). The increase in the temperature of the mobile phase not only decreases its polarity but also its viscosity. It means the improvement of the performance analysis. In these cases, it is better to use zircon or titanium phases more heat-proof.

Let us give some examples of organic solvents classified with less polar and more elution power in reverse phase chromatography:

- Methanol
- Acetonitrile
- Ethanol
- Propanol-1
- Isopropanol

- Dioxane
- Tetrahydrofuran

Examples of mixtures are water/methanol or water/acetonitrile.

It is possible to work according to two modes:

- *Isocratic mode* with a constant composition of the mobile phase during the chromatography.
- *Gradient mode* with a variable composition during the experiment that limits the time of analysis.

Ion-Exchange Chromatography

Principle

The ion exchange chromatography applies to a stationary phase made of macromolecules organized in a solid porous network non-soluble in the solvents and possesses the capacity to exchange its ions with those of the mobile phase. Two phases are in presence of:

- Ion exchange or network with a large surface that possesses some ions I_1 exchangeable.
- An ionic solution that possesses the ions I_2 to separate, having the same sign as the first ones which shift the ions I_1 (Figure 4).

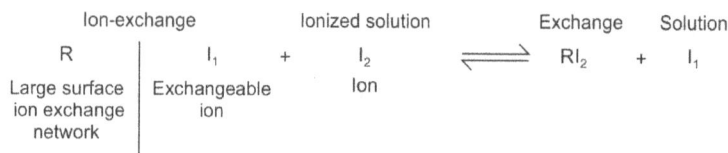

Ion-exchange			Ionized solution			Exchange	Solution
R	I_1	+	I_2	$\rightleftharpoons$		RI_2 +	I_1
Large surface ion exchange network	Exchangeable ion		Ion				

Figure 4: Ions Exchange During the Chromatography. (Reprinted with Permission from Gwenola and Jean-Louis Burgot, Paris, Lavoisier, Tec et Doc, 2017, 200).

I_1 and I_2 have obligatorily the same sign. The process of exchange is low in comparison with the other types of chromatography.

Historic Purpose

Thomson and Way (1850) studied the use mechanism of water-soluble fertilizer salt and dampened soil with a solution of potassium chloride as well as studied the composition of washing water. Thus, they found ion calcium instead of ion potassium. Actually, the revelation of the great interest which feature ion-exchangers dates from 1947. It is at this date that the works of the American atomist team had been published in the context of the Manhattan Project (building of the first atomic bomb) during World War II. Thus, they used the ion-exchange process to separate some rare elements produced by uranium fission.

In 1949, Cohn used an ion-exchange column to separate the products of the hydrolyze reaction of nucleic acids. It is interesting to know that the first works of modern molecular biology had begun in the atomist American centers.

Ion Exchangers

General Characteristics

The ion-exchangers are some solid granules more and less porous and generally gellable. They made a cross-linked of insoluble macromolecules. It is also possible to use natural products or synthesis polymers (resins).

Natural Products. There exists:

- *Zeolites* or alkali and alkaline earth metal silico-aluminate cation exchanger:
  $Na_2, Al_2O_3, 4SiO_2, 2H_2O + 2KCl$ (in solution) $\rightarrow K_2, Al_2O_3, 4SiO_2, 2H_2O + 2NaCl$ (in solution). These products are not often used because they are pH-sensitive.

- *Cellulose* which by oxidation leads to *dextran* (glucose polymer). The dextran is polymerized by glycerin in *Sephadex*® (Figure 5), a polyoside made of 1,6 linkage glucose (90%) with 10% of 1,3 linkage glucose. It is obtained by crosslinking dextran with epichlorohydrin in a basic medium and after linkage, it gives products used in ion-exchange or exclusion chromatography.

- *Crosslinked silica* possesses great mechanical properties and good efficiency. However, its major drawback is that it gets damaged if the pH of the mobile phase is superior to 8.

- *Zircons or zirconium dioxide* ZrO_2 porous framework which features a good chemical and thermal stabilities as well as a good mechanic resistance.

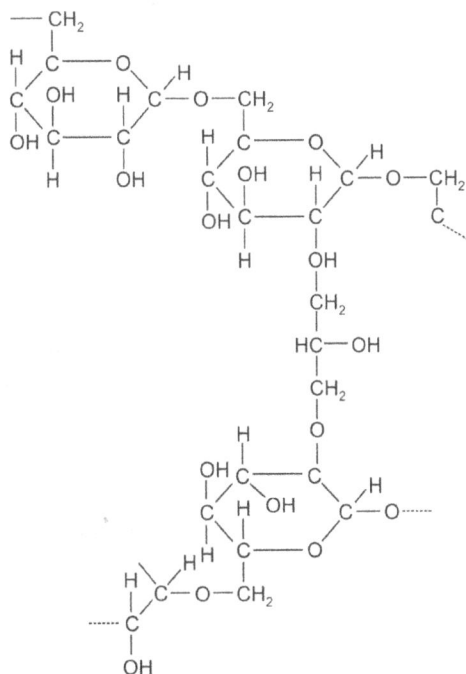

Figure 5: Framework of Sephadex® (Reprinted with the Permission from Gwenola and Jean-Louis Burgot, Paris, Lavoisier, Tec et Doc, 2017, 201).

Polymers or Resins. They are in the form of small spheres with a diameter of 0.5 mm classified according to their chemical framework and the size of the particles in MESH. The size of interval 20 to 50 MESH is equivalent to 0.85 to 0.3 mm (it is inversely proportional).

It is possible to distinguish two types of polymers:

- *Crosslinked Polymers as Polystyrene/Divinylbenzene* (Figure 6). The linkage on these polymers allows for fixing some acidic or basic groups to become anion or cation exchangers. These materials have the advantage, in contrast to silica, to be stable in a wide zone of pH. However, their exchange capacity is low. The linkage diminishes the flattening of the stationary phases in the column owing to its own weight. The styrene is responsible for a long chain, while the introduction of divinylbenzene gives some ramifications as a bridge between the chains. This results in a tridimensional polymer all

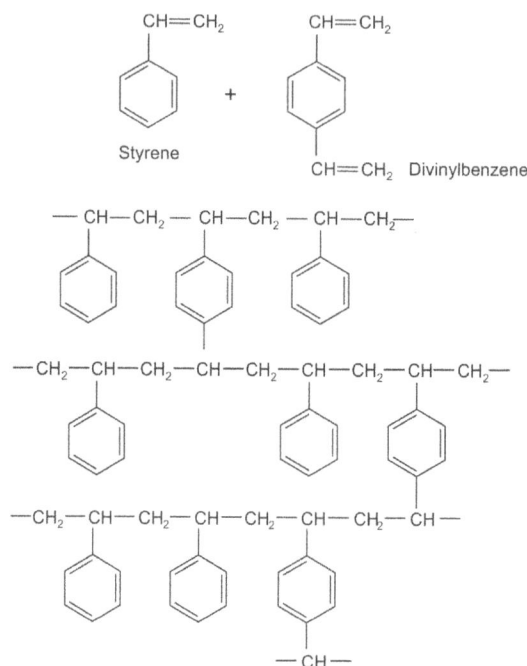

Figure 6: Framework of Styrene-Divinylbenzene Polymer. (Reprinted with Permission from Gwenola and Jean-Louis Burgot, Paris, Lavoisier, Tec et Doc, 2017, 202).

the more branched with the diameter of the very narrow pores as the percentage of divinylbenzene is high. The percentage of divinylbenzene is called the "ratio of bridging". It is referred to by the X letter followed by the value of the percentage of divinylbenzene in the initial mixture of preparation of the polymer. For example, a resin X10 contains around 10% of divinylbenzene. Ion-exchangers use X1 to X30 resins and more often X8 to X16.

• *Pellicular resins*: Minuscule spheres (0.1 μm to 0.2 μm) of polymer named latex form a continuous film of 1 to 2 μm average thickness on a support made of microspheres of silica or glass or polystyrene of 25 to 30 μm diameter. Some functional groups are fixed on the spheres of latex. The pellicular configuration is interesting because it permits concentration on a very short surface of the ionic exchange with a lower diffusion of the ions in the beans. That produces a better efficiency chromatographic and some exchange quicker. Moreover, the pellicular resin is stable in a large zone of pH.

• *Tentacule® matrix*: It exists as tentacle® technology with charged groups fixed to linear flexible chains of the polymer. This produces some exchange quicker.

Exchangers

The anionic exchangers (by reference to the group linked to the support) are cation-exchangers. It means that the ion attracted to the exchanger is a cation. Similarly, a cationic exchanger is an exchanger of the anion.

Cationic Exchanger. It is possible to use 3 types of exchangers:

• *Strong acids* with aromatic sulfonic acids (Table 1): RSO_3^- and H^+. These resins are obtained by sulfonation of the copolymer styrene-divinylbenzene. They are completely dissociated. By electrostatic interactions, protons remain in front of sulfonic rest (Figure 7).

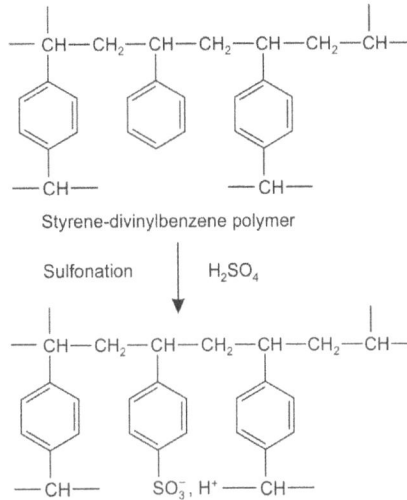

Figure 7: Sulfonation of the Polymer Styrene-Divinylbenzene. (Reprinted with Permission from Gwenola and Jean-Louis Burgot, Paris, Lavoisier, Tec et Doc, 2017, 204).

• *Weak acids* with carboxylic groups few dissociated to pH = 4 (Table 1). Synthesis is realized by copolymerization of the methacrylic acid $CH_2 = C-(CH_3)-COOH$ and of glycol bis-methacrylate as a linked agent:

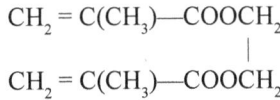

$$CH_2 = C(CH_3)—COOCH_2$$
$$CH_2 = C(CH_3)—COOCH_2$$

• *Very weak acids* with phenol group.

Table 2: Examples of Ion-Exchanger Resins. (Reprinted with Permission from Gwenola and Jean-Louis Burgot, Paris, Lavoisier, Tec et Doc, 2017, 203).

Resins	Chemical Nature	Market Name
Strong acid or cation-exchanger	Sulfonic acid RSO_3^-, H^+	Amberlite IR-120, Dowex 50w
Weak acid or cation-exchanger	Carboxylic acid, $RCOO^-$, Na^+	Amberlite IR-50, Dowex 1
Strong base or anion-exchanger	Ammonium IV R_4N^+, Cl^-	Amberlite IRA-400
Weak base or anion-exchanger	Tertiary amins (in the function of pH) R_3NH^+, Cl^-	Amberlite IR-45, Dowex 3

Anionic Exchanger. We have a large variety comprising:

• *Strong base* or ammonium IV hydroxide. There are also some styrene-divinyl benzene polymers but they had undergone positive and negative chloromethylation (Blanc's reaction) and not a sulfonation as previously (Figure 8).

• *Weak base* or poly-alkylamines whose charge is a function of the pH. These resins are commercialized either in acidic form or basic form but also as sodium salt or in the form of hydrochloride.

Exchange Capacity of the Resin. If we consider the following equilibrium:

$$R, I_1 + I_2 \rightleftarrows R, I_2 + I_1$$

It is obvious that the number of ions that have the possibility to exchange I_1 for a given amount of resin is limited. In other words, the number of fixed ions I_1 cannot have exceeded the number of active sites available on the resin. It accounts for the introduction of exchange capacity which reflects the number of ion-gram which have the possibility to be exchanged by unit mass of dry resin

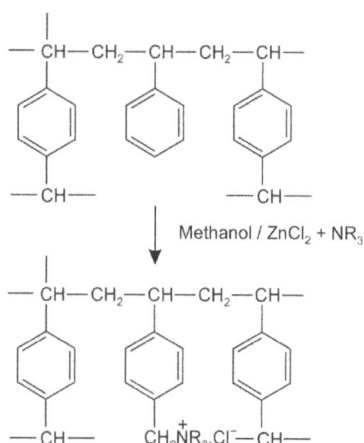

Figure 8: Chloromethylation of Styrene-Divinylbenzene Copolymer. (Reprinted with the Permission from Gwenola and Jean-Louis Burgot, Paris, Lavoisier, Tec et Doc, 2017, 204).

saturated. This parameter is expressed usually by a unit of mass of dry resin (around 5.10^{-3} eq/g for a sulfonic resin, but lower than 0.05 meq/g for some columns of very low capacity using diluted eluent). It is interesting to notice that the exchange capacity of the sulfonic and ammonium IV does not depend on pH because resins are completely ionized for all the values of pH. It is the contrary for low acid or basic resins. Thus, in a strong acid medium, a carboxylic resin has an exchange capacity of zero value. The –COOH group is not ionized. The reasoning is analogous for primary, secondary and tertiary amines in a strongly basic medium.

The measure of the exchange capacity requires to process:

- A known amount of resin in an acidic form by an excess of NaCl:

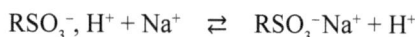

$$RSO_3^-, H^+ + Na^+ \ \rightleftarrows \ RSO_3^- Na^+ + H^+$$

Ions H^+ released are titrated by a sodium hydroxide solution.

- A known amount of resin in a basic form by an excess of NaCl:

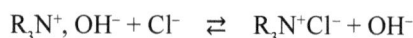

$$R_3N^+, OH^- + Cl^- \ \rightleftarrows \ R_3N^+Cl^- + OH^-$$

Ions OH^- released are titrated by a sulfuric acid solution.

These determinations provide an estimation of the amount of resin to use for the chromatographic analysis.

Retention Mechanism

Effect of an Aqueous Saline Solution on a Resin

Successively, one can notice the adsorption of ions from the solution on the resin surface, the swelling of the resin on the effect of the solution and its transformation into a gel. Then, the transfer of ionic or not solute from the aqueous solution in the pores of the resin by Donnan equilibrium and finally, the exchange reaction as such.

Exchange Reaction. If we put an ion-exchange resin in the presence of an electrolyte solution of which the free ion is different from that one of the same sign in the resin, the exchange takes place.

As an example:

$$\text{Resin R } SO_3^-, H^+ \quad \text{or} \quad R_4N^+, Cl^-$$

And

$$\text{Resin R SO}_3^-, \text{H}^+ \quad \text{or} \quad \text{R}_4\text{N}^+, \text{Cl}^-$$
$$\text{and} \qquad\qquad\qquad \text{and}$$
$$\text{Solute Na}^+\text{Cl}^- \quad \text{or} \quad \text{Na}^+\text{OH}^-$$

Exchange

$$\text{H}^+/\text{Na}^+ \quad \text{or} \quad \text{Cl}^-/\text{OH}^-$$

Equilibrium. These exchanges give equilibria characterized by thermodynamical constant as followed:

$$\text{H}_R^+ + \text{Na}_S^+ \rightleftharpoons \text{Na}_R^+ + \text{H}_S^+$$

With equilibrium constant which can be expressed by the law of mass action in terms of activities:

$$K_{H/Na}^0 = \frac{\left(\text{Na}_R^+\right)\left(\text{H}_S^+\right)}{\left(\text{H}_R^+\right)\left(\text{Na}_S^+\right)}$$

$K_{H/Na}^0$ is the thermodynamical constant. Actually, the activities of species could be evaluated only if the solution is diluted. It will be extremely rare with the ion-exchangers because the inside solution is very often very concentrated (around 5 M). That is why one reasons in terms of apparent constants available for the conditions of a given experiment. Also, in practical terms, they are the most important to provide for an exchange reaction.

Thus, it can be written:

$$K_{H/Na}^0 = (C_{Na+R}C_{H+s})/(C_{H+R}C_{Na+s})$$

The K constant is sometimes called the *separation factor* between two ions or the selectivity factor of the resin (with respect to the two ions).

We can also characterize the partitioning of one ion at equilibrium by its partitioning coefficient (apparent):

$$P = C_{resin}/C_{solution}$$

But this coefficient differs according to manipulation. It was found from this relation that for an exchange between the ions A and B:

$$K_{A/B} = P_B/P_A$$

At a general level, it can be stated that both ions are not shared in the same manner between resin and solution. In this case, we say that the resin has a greater affinity for one ion than the other. The more K moves away from 1, the higher the difference in affinities of the resin for both ions.

(*Remark*: It is necessary not to forget that the exchange between both ions of different charges is common. There is a shifting charge by charge and in this case equilibrium constant is given by an expression as:

$$2\text{Na}_R^+ + \text{Ca}_S^{2+} \rightleftharpoons 2\text{Na}_S^+ + \text{Ca}_R^{2+}$$

$$K_{Na^+/Ca^{2+}} = \frac{\left(\text{Ca}_R^{2+}\right)\left(\text{Na}_S^+\right)^2}{\left(\text{Ca}_S^{2+}\right)\left(\text{Na}_R^+\right)^2}$$

When all the ions to exchange have the same charge, the equilibrium does not depend on the dilution. It is the reason why the resin keeps a similar separation power in a very diluted solution in the case of the analysis of traces for example.

Factors Influencing the Apparent Constant. It is possible to note:

- Ionic composition of the solution.
- The concentration of the ion to separate.
- Ionic strength of the solution to analyze.
- The linkage ratio; in general, the resin becomes more and more selective when its linked ratio increases. There exists a growth of the differences of affinity for the high linked ratio.
- The capacity of exchange of the resin.
- The temperature is, just like all the equilibrium, constant.

Affinity Rules. The equilibrium constant provides the relative affinity of the ions for a given resin:

- Influence of the *charge*: The ions are all the better kept as their charge is higher. That means than in the order of increasing activities:

$$Na^+ < Ca^{2+} < Al^{3+}$$

$$Cl^- < SO_4^{2-} < Fe(CN)_6^{3-}$$

- Influence of the *nature of the ion*: The affinity of ions for the resin (carboxylic except) increases with the atomic number into a family of chemical elements:

$$F^- < Cl^- < Br^- < I^-$$

$$Be^{2+} < Mg^{2+} < Ca^{2+} < Sr^{2+} < Ba^{2+}$$

$$Li^+ < Na^+ < K^+$$

One can notice that the H^+ ion has always an affinity lower than the other ions except ion lithium.

Prediction of the Exchange on a Resin. The previous affinity rules provide a qualitative prevision. The knowledge of the equilibrium constants in some experimental given conditions allows the quantitative prevision of exchange.

Thus, the scale of affinities for a DOWEX 50* 16 (sulfonate polystyrene) is the following:

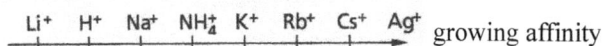

$$\underset{\text{growing affinity}}{\xrightarrow{\quad Li^+ \quad H^+ \quad Na^+ \quad NH_4^+ \quad K^+ \quad Rb^+ \quad Cs^+ \quad Ag^+ \quad}}$$

This scale is built for some equivalent proportion of ions. That means that ions Na^+ have the possibility to displace H^+ ion but not NH_4^+ ion in a great manner according to a constant K for which the logarithm is proportional to the length of segment H-Na, Na^+-NH_4^+.

It does not mean that a resin in the form K^+ cannot fix ions Na^+ whose affinity is lower. Thus, if we consider the equilibrium:

$$K_R^+ + Na_S^+ \rightleftharpoons K_S^+ + Na_R^+$$

$$K_R^+ + Na_S^+ \rightleftharpoons K_S^+ + Na_R^+$$

$$K = \frac{(K_S^+)(Na_R^+)}{(K_R^+)(Na_S^+)},$$

One can see that if the concentration of Na^+ in the solution is very high, the equilibrium is displaced to the right. Even with an unfavorable equilibrium constant, the Na^+ ions are fixed on a resin.

Shift of the Equilibria

In reality, phenomena are seldom as simple as those which are described just above. Very often, other chemical reactions are superimposed on the exchange ion reaction, not only into the resin but also in the solution. The consequence is a shift of the exchange equilibrium and a disturbance of the previous projection.

Shift of the Equilibria by Acid-Basic Exchange

• At the equilibrium in the resin, another equilibrium takes place in an aqueous solution:

Example: a solution of ion lithium Li^+ in presence of a resin cationic H^+.

$$Li_S^+ + H_R^+ \overset{K_R}{\rightleftharpoons} Li_R^+ + H_S^+$$

This equilibrium is moved onto the left in the case of a sulfonic resin because the lithium has less affinity for the resin than H^+. In some conditions, it is possible that it would not be right. As an example, for the solution which contains a basic compound, acetate ion CH_3COO^- bringing as CH_3COOLi. One has:

$$CH_3COO^- + H^+ \rightleftharpoons CH_3COOH \quad K_a$$

The global equilibrium becomes:

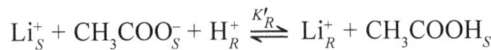

$$Li_S^+ + CH_3COO_S^- + H_R^+ \overset{K'_R}{\rightleftharpoons} Li_R^+ + CH_3COOH_S$$

With

$$K'_R = \frac{\left(Li^+\right)_R \left(CH_3COOH\right)_S}{\left(Li^+\right)_S \left(CH_3COO^-\right)_S \left(H^+\right)_R} = \frac{\left(H^+\right)_S \left(CH_3COOH\right)_S \left(Li^+\right)_R}{\left(Li^+\right)_S \left(H^+\right)_S \left(CH_3COO^-\right)_S \left(H^+\right)_R}$$

$$K'_R = \frac{K_R}{K_a} = \frac{K_R}{10^{-4.7}} = K_R \times 10^{4.7}.$$

Again, one finds the notion of equilibrium displacement. Accordingly, the ion Li^+ is completely fixed on the resin.

• The acidic-basic equilibrium can take place in the resin. In this manner, we explain the great affinity of the carboxylate resin for the ion H^+; the group COO^- is high basic. Thus, for a polyacrylate resin, the rule of affinity becomes:

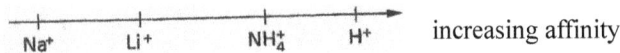

$$\xleftarrow{\hspace{4cm}}$$ Na⁺ $\quad$ Li⁺ $\quad$ NH₄⁺ $\quad$ H⁺ $\qquad$ increasing affinity

Whereas, for a sulphonate resin, the rule of affinity is modified such as:

Li⁺ $\quad$ H⁺ $\quad$ Na⁺ $\quad$ NH₄⁺ $\quad$ K⁺ $\quad$ Cs⁺ $\quad$ Ag⁺. $\qquad$ increasing affinity

Shift of the Equilibria as a Result of Complex Formation

• **Formation in the Solution**

The complex formation in solution decreases the concentration of the free ion capable of being fixed on the resin. As a result, its affinity for the resin is lowered. Thus:

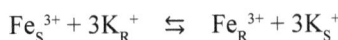

$$Fe_S^{3+} + 3K_R^+ \rightleftharpoons Fe_R^{3+} + 3K_S^+$$

As an example, with the addition of ions F^- capable of complexing ions Fe^{3+}, the concentration $(Fe^{3+})_S$ decreases. The equilibrium moves to the left. Usually, Fe^{3+} takes the place of K^+ on the resin. With F^-, it becomes possible to make a reverse reaction and to replace K^+ with Fe^{3+}.

There will, therefore, be considerable opportunities to realize very selective separations. In some cases, it will be possible to form an anionic complex; the cation cannot be fixed by the resin because it is faced with the resin as an anion brought by the complexing agent. However, it becomes possible to fix the anionic complex by an anion exchanger. This process is frequently used. For illustrating that, we describe the two following examples:

- The fixation of iron Fe^{3+} in form of chelate FeY^- with $edta^{4-}$:

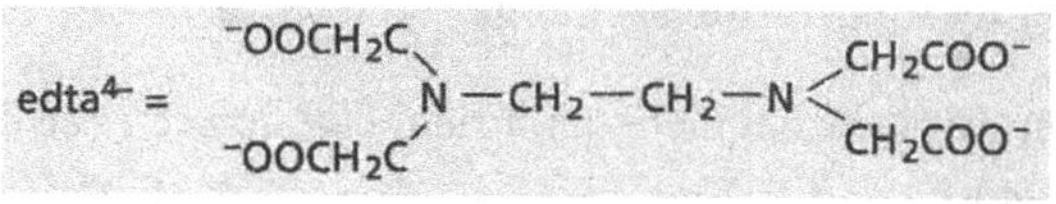

$$edta^{4-} = \quad \begin{array}{c} ^-OOCH_2C \\ \\ ^-OOCH_2C \end{array} N-CH_2-CH_2-N \begin{array}{c} CH_2COO^- \\ \\ CH_2COO^- \end{array}$$

- The separation of Ni^{2+} and Co^{2+}. The ions Co^{2+} are the only ones to form chloro-complexes and are retained on the anion exchanger resin. Ni^{2+} is not retained.

(Remark: many metal ions can form some chloro-complexes.)

In turn, the solid fixation of the complex on the resin enables the formation of complexes with usual low stability. It is the case, as an example, for $Co(SCN)_4^{2-}$ with low stability but with a good fixation on the resin, which takes a blue color (color of the complex).

- **Formation of the Resin**

The resin may itself be a complexing agent, increasing considerably its affinity for this or that ion. These resins are called chelating resins. It is another application of the formation of complexes. The affinity of a metal depends on above all the nature of the chelating group. It is interesting to know that the magnitude of energy of the electrostatic bonds is around 8–13 $KJ.mol^{-1}$ for a normal ion-exchanger, and it becomes around 60 to 105 $KJ.mol^{-1}$ for a chelating resin. The most frequent chelating group is the iminodiacetic rest, which is linked to the resin.

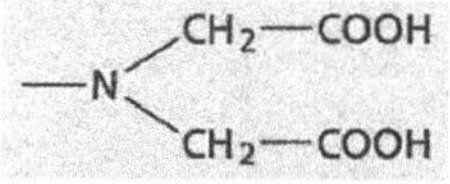

$$-N \begin{array}{c} CH_2-COOH \\ \\ CH_2-COOH \end{array}$$

One remarks that this rest is completely fixed in the edta molecule. One is linked to the resin polystyrene-divinylbenzene, according to the following scheme (Figure 9).

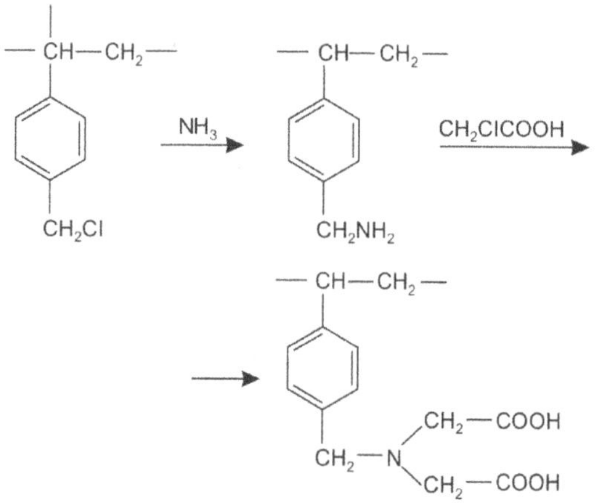

Figure 9: Linkage of Edta Molecule on the Polymer Styrene-Divinylbenzene. (Reprinted with Permission from Gwenola and Jean-Louis Burgot, Paris, Lavoisier, Tec et Doc, 2017, 211).

The iminodiacetic resin retains preferentially copper, iron and heavy metals.

Transfer by Donnan Equilibrium. We have just seen that a resin in contact with a solution can exchange its ions with others of the same sign from the solution. This exchange is only possible if ions of the solution and ones of the resin are different. If the ions are the same or even if the solute is not ionized, no exchange takes place. However, it produces some partitioning of the solute between both phases. This phenomenon is called Donnan equilibrium.

- **First Case: the Solute is an Ionic One**

Let us consider an H^+ form sulfonic resin, fully swollen in water and put in contact with hydrochloric acid. It cannot have some exchange of cations because it is the same in both phases. But it is noted that HCl goes through the inside part of the resin to obtain a partitioning equilibrium as follows:

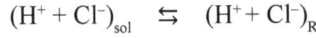

$$(H^+ + Cl^-)_{sol} \leftrightarrows (H^+ + Cl^-)_R$$

Let us apply mass action law:

$$K_{HCl} = \frac{C_{H_R^+} C_{Cl_R^-}}{C_{H_S^+} C_{Cl_S^-}}$$

Here, K_{HCl} is the apparent constant expressed in terms of the concentration of partitioning of HCl. The hydrochloric acid is completely dissociated in water. Let us say it is also in the resin (despite the high concentrations and the high value of ionic strength).

We have $C_{H_S^+} = C_{Cl_S^-} = C_{HClS}$ and $C_{Cl_S^-} \neq C_{H_R^+}$ because it is necessary to take into account the protons, which come from the resin and for which the concentration matches the exchange capacity C_E. Thus, one must write:

$$C_{H_R^+} = C_E + C_{HCl_R}$$

$$K_{HCl} = \frac{[C_E + C_{HCl_R}]C_{HCl_R}}{C_{HCl_S}^2}$$

$$C_{HCl_R}^2 + C_E C_{HCl_R} = K_{HCl} C_{HCl_S}^2$$

The interesting point for us is C_{HClR}. One notes that this concentration depends on a second-degree equation with C_E and C_{HClS}. The higher C_E the lower C_{HClR} (see Table 3).

Table 3: Concentrations of the Different Species in the Solution and the Resin.

Concentration of HCl in the Solution C_{HClS}	Concentration of HCl in the Resin C_{HClR}	% HCl into the Resin of the C_E Capacity
10^{-2} M	2.10^{-5}	0.0004%
0.5 M	5.10^{-2}	1
5	3	60

One sees that for the diluted solutions outside the resin, the Donnan equilibrium for ionic solutes in an ionic resin with a high capacity is negligible. Consequently, for aqueous diluted solutions, ions with the same sign as the functional groups of the resin are excluded from the resin.

- **The Second Case: the Solute is not Ionic**

Let us consider M the solute, its partitioning is realized according to:

$$M_S \leftrightarrows M_R$$

If we apply the mass law:

$$K_M = C_{MR}/C_{MS}$$

Even if the solution inside the resin is still very loaded with ions, a certain quantity of solute M passes through into the resin to comply with the equilibrium. In other words, the chemical potential of the solute in both phases is the same. Relatively to the previous case, according to the expression of K_M, the curve C_R in moles.kg^{-1} of dry resin in the function of C_S is in this case a right line, except for the deviation caused by problems of ionic strength. Most often shared molecules in this manner are molecular complexes, such as $HgCl_2$ and $CdCl_2$ very little ionized or some weak acids non-ionized.

- **Specification of the Donnan Equilibrium**
  Contrary to the exchangeable ions which counterbalance the ionic groups linked to the resin, ions and molecules passed into the resin by Donnan equilibrium can be eliminated by a pure water washing.

- **Case of the Mixtures of Ionic and Non-Ionic Solutes**
  In some cases, there is no sharp distinction between the ionic and non-ionic solutes. A good example is given by the weak acids or the weak bases, which are more or less ionized according to the value of pH and their pKa. For carboxylic acid RCOOH, at pH > pKa + 2 pH units, acid can be regarded as completely ionized because it is in the form RCOO$^-$.

(*Remark*: All the above considerations assume that equilibria are achieved. It means that there is no kinetic problem.)

Ion-Pair Chromatography

Definition

The formation of ion-pairs permits the separation of ions or ionic compounds. They are neutral entities, which are extractable in a low polar organic medium. This method competing with ion-exchanger chromatography is useful when partition chromatography can no longer be used. It is the case when pH adjustment does not permit by itself to eliminate the charge of the weak acids and bases. It is sometimes the case when:

- For the weak acids and weak bases, the shift of the equilibrium requires an extreme pH of < 2 or > 8 inconsistent with silica.
- For the mix of weak acids, weak basis or neutral molecules for which the ion exchange cannot be used.

Let us remember, moreover, that silica keeps adsorbent properties, and it is necessary in some cases to diminish the polarity of compounds to analyze. Thus, according to the nature of molecules, it is possible to choose the following methods:

- Strong acids or bases ⟶ ion-exchange or ion-pairs
- The mix of weak acids or a mix of weak bases ⟶ adjusting the pH to diminish ionization and to use in reversed-phase chromatography
- A mix of acids, bases or neutral compounds or amphoteric compounds ⟶ ion-pair chromatography.

Mechanisms

The ion-pair chromatography can be used according to many methods:

• *Partitioning of ion pairs between two liquid non-miscible phases*: One can impregnate the stationary phase with ions of opposite charges to those that one wants to separate. The last one gives, afterward, with the previous ions some ion pairs go into the mobile phase. The solutes separate into different compounds, and these give ion pairs that do not have the same extracting properties.

• *Ion pairs chromatography on a non-polar stationary phase*: The stationary phase is alkyl-linked silica or a copolymer of styrene-divinylbenzene. The mobile phase is a binary mix of organic solvent (methanol and acetonitrile) and an aqueous phase containing the organic counter-ion. Let us consider the different hypotheses of the mechanism of ion pair chromatography:

Model of ion exchange at the surface: The counterion with one or more hydrophobic sites is adsorbed by the alkyl chains of the stationary phase, and this one becomes equivalent to an ion-exchanger. Around 30% of the alkyl chains are coated by the counterion and the other chains are free and kept available to interact with the non-ionic compounds of the sample (Figure 10).

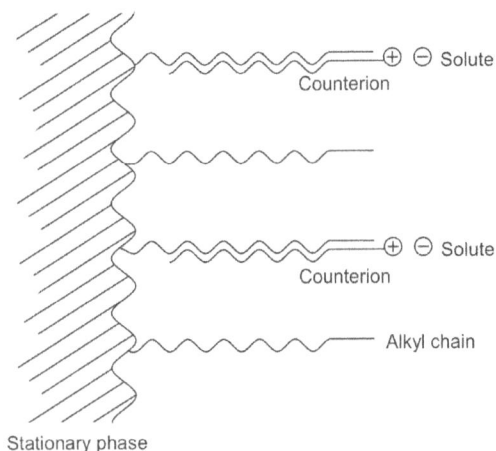

Figure 10: Model of Ion Exchange on the Surface. (Reprinted with Permission from Gwenola and Jean-Louis Burgot, Paris, Lavoisier, Tec et Doc, 2017, 216).

Model of matching in the mobile phase: In this case, there is a formation of ion-pair sufficiently stable to fix on the alkyl chains of the stationary phase. It is the case for:

- The acids at pH > 7,5: $\qquad$ $RCOO^- + A^+ \leftrightarrows RCOO^-A^+$
- The aliphatic amines at pH < 6: $\quad RNH_3^+ + B^- \leftrightarrows RNH_3^+, B^-$

Model by the formation of an electrical double layer at the interface solid-liquid: It is the model of Gouy-Chapman-Stern (see Chapter 18; electrophoresis) which is most likely. The adsorption of the counter-ion on the stationary phase leads to the appearance of potential differences between the surface of the solid and the solution. The composition of the ionic liquid layer surrounding the solid surface is, therefore, modified. This zone is called the "diffused zone". Ions with the same sign as the counter-ion are eliminated from the diffuse layer while ions with the opposite sign are pulled into it.

Factors that Affect the Formation of Ion-Pairs

Nature of the Counter-Ion

Table 4 features the major counter-ions used in ion-pair chromatography.

Table 4: Major Counter-Ions.

Compound to be Separated	Employed Counter-Ion
Strong and Weak Acids	Ammonium IV: $^+N(C_4H_9)_4$
Strong Acids	Amines III: HNR_3^+
Strong and Weak Basis	Alkyl sulphonate $CH_3(CH_2)_nSO_3^-$

Concentration of the Counter-Ion

The usual concentration is around 5.10^{-2} M except for counter-ions whose molecular mass is high (such as lauryl sulfate). In this case, the concentration is lower around 5.10^{-3} M.

Composition of the Mobile Phase

Increasing the percentage of the organic solvent (methanol, acetonitrile, etc.) of the mobile phase decreases the retention or capacity factor k' of the solute because the mobile phase is more lipophilic leads to the ion-pair.

(*Remark*: From a practical standpoint, the mobile phase will be degassed before the addition of the counter-ion to eliminate the formation of lather due to the capacity of surfactant from some counter-ions.)

Impact of the Ionic Strength

Increasing the ionic strength of the mobile phase decreases the capacity factor k' of the solute against the stationary phase by competition between the counter-ion and ions of the mobile phase such as phosphate ions used to fix the pH of the solution.

Impact of pH of the Mobile Phase

To form an ion pair, it is necessary that the solute and counter-ions be ionized. The pH is, therefore, chosen in the function of pKa of these compounds while remaining in the range of 2 to 8 in the case of linked silica.

(Remark: *It has to be noted that the column must be rinsed at the end of the experiment and be kept in presence of counter-ions.*)

Impact of Temperature

Increasing the temperature decreases the capacity factor k'. Hereafter 60°C, counter-ions such as ammonium IV show a risk of degradation (Hoffman reaction).

Size-Exclusion Chromatography

Definition and History

The size-exclusion chromatography is based on the selective retention of samples according to the size of their molecules. They are included or excluded from the porous spherical beads of the stationary phase filled by the solvent. It had been, firstly, described after observations of abnormal phenomena with ion-exchange chromatography where, in certain cases, the order of pass through the column for compounds was a reverse function of molecular weight.

It is interesting to note that contrary to other chromatography principles, the mechanism of this process is not based on direct interactions between the molecule and the stationary phase.

According to the nature of the phase mobile used, one defined:

Gel permeation chromatography if the mobile phase is an organic solvent.

Gel filtration chromatography if the mobile phase is aqueous.

Principle

The base material is a homogenous gel made of both phases:

- A *dispersed phase*, a solid called the "substrate of gel", is made of small porous beads; its size has been calibrated to obtain regular particles. Each bead arises from the link of macromolecules joined together to obtain a regular three-dimensional network.
- A *dispersant phase*, or solvent, passes through the beads generating their swelling to create some pores of a range of sizes.

The mechanism of the separation is the following (Figure 11); depending on their size, molecules are excluded or included from the pores more and less easily.

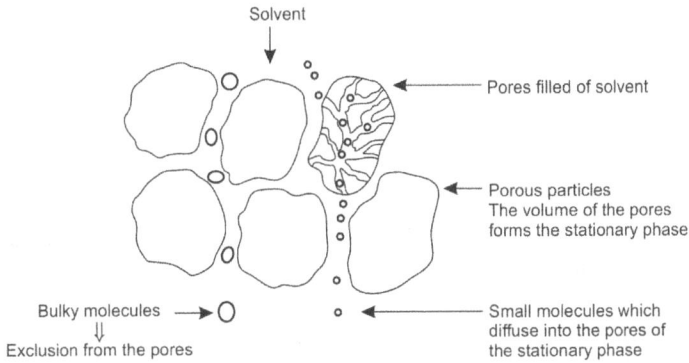

Figure 11: Separation Mechanism. (Reprinted with Permission from Gwenola and Jean-Louis Burgot, Paris, Lavoisier, Tec et Doc, 2017, 219).

Let us consider two molecules of differing sizes:

- A *bulky molecule* does not diffuse into the pores. It remains outside the stationary phase in the surrounding solution.
- A *small molecule* circulates and spreads inside and outside the pores only by the diffusion effect. Elsewhere, the molecules of solvent circulate in all the pores of the stationary phase only by diffusion effect and fill the whole volume available of the column called permeation volume. One can see that, contrary to the other chromatographic methods, the solvent is eluted the last from the column given the last peak of the chromatogram (Figure 12).

The chromatography achieves, in fact, sieving of the mix of solutes to separate.

(*Remark*: It should be noted that the partitioning, adsorption and ion-exchange phenomena can overlap.)

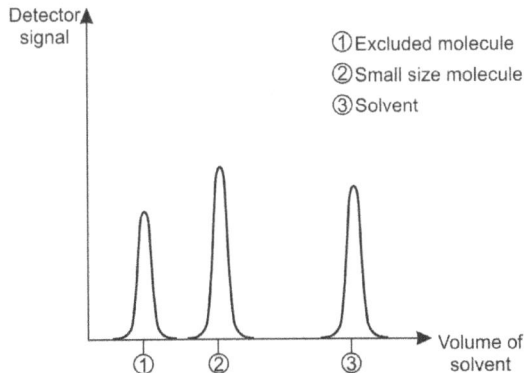

Figure 12: Size-Exclusion Chromatogram. (Reprinted with Permission from Gwenola and Jean-Louis Burgot, Paris, Lavoisier, Tec et Doc, 2017, 220).

Theories

Diffusion Coefficient

The distribution of the compound between both phases is controlled for a given gel by a constant K_d which is the diffusion coefficient:

$$K_d = C_S/C_m$$

It is the ratio of the concentration of the analyte C_S in the intragranular phase regarded as the stationary phase and the concentration of the analyte C_m in the extra-granular liquid regarded as the mobile phase. The value of this constant is in the ideal conditions (see below) between 0 and 1:

$$0 < K_d < 1$$

Molecules that are *perfectly diffuse* spread over both phases in the same manner:

$$C_S = C_m \quad \text{or} \quad K_d = 1$$

Molecules that are *fully excluded* from the stationary phase are those for:

$$C_S = 0 \quad \text{or} \quad K_d = 0$$

Between these extreme theoretical values, one finds the case of the molecules which are not fully excluded but do not have the possibility to diffuse fully through the three-dimensional network of pores. The coefficient K_d of diffusion/exclusion depends on physical properties, the size and the volume of each molecule.

(*Remark*: In some cases, K_d can be > 1. It means that the adsorption or ion-exchange phenomena overlap the exclusion/diffusion phenomena.)

Retention Volume or Elution Volume

As some compounds are not trapped in the column, the term 'elution volume' is a better option than the 'retention volume'. By analogy with the usual formula, it is possible to write:

$$V_e = V_m + K_d V_s$$

With:

- V_m: Volume of the free mobile phase between the beads or the void volume. Its determination requires high molar mass molecules that do not diffuse, such as dextran blue (PM = 2.10^6). In this case, $K_d = 0$ and $V_e = V_m$.
- V_s: Volume of stationary phase, namely volume of liquid inside the beads. It depends mainly on their porosity.
- V_s and V_m are linked to the total volume of the column V_t by the relation:

$$V_t = V_s + V_m + V_g$$

- V_g: Volume of gel substrate. Generally, this is considerably low compared to both others.

For a molecule that diffuse perfectly, $K_d = 1$ and its volume of elution is equal to:

$$V_e = V_s + V_m$$

For a molecule with no additional effect than those of size-exclusion or diffusion, the elution volume is not superior to $V_s + V_m$ and is called the volume of total permeation. This means that when such a volume flows through the column, the compounds are all eluted. It is therefore possible to restart a new chromatography with the same column.

Consequence: Retention Volume and Molar Mass

The relation $V_e = V_m + K_d V_s$ shows that there exists a linear relationship between the elution volume and K_d. As K_d is a function of the molar mass, it is possible after calibration, knowing the elution volume of a compound, to deduce with a reasonable approximation its molar mass.

Determination of K_d

Its value is obtained from the equation:

$$K_d = (V_e - V_m)/V_s$$

V_e and V_m are obtained experimentally and V_s depends on the nature of the gel but is not calculated despite the information given by the manufacturer on the porosity and the capacity of swelling. Also, K_d is substituted by a constant more easily available; K_{av} (available):

$$K_{av} = (V_e - V_m)/(V_s + V_g)$$

This constant involves not only the intragranular volume V_s but also those occupied by the substrate of the gel itself V_g. The sum $V_s + V_g$ constitutes all that is not occupied by the extragranular liquid V_m, consequently:

$$V_s + V_g = V_t - V_m$$

V_t is obtained from the diameter and the length of the column and V_e and V_m can be obtained from the elution curve.
Thus,

$$K_{av} = (V_e - V_m)/(V_t - V_m)$$

Generally, the volume V_g is negligible compared with V_s and that explain that the volume of the stationary phase must be as large as possible to improve separations. Figure 13 illustrates his point.

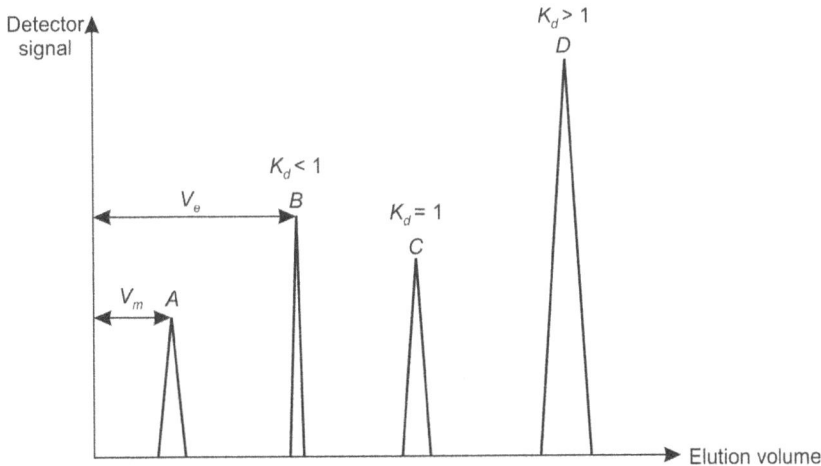

Figure 13: Evolution of the Chromatogram as a Function of the Value of the Constant K_d. (Reprinted with Permission from Gwenola and Jean-Louis Burgot, Paris, Lavoisier, Tec et Doc, 2017, 223).

The compound A is not retained by the column: $V_e = V_m$
The compound C with a constant $K_d = 1$ has an elution volume $V_e = V_s + V_m \approx V_t$
The compound D is partially adsorbed on the column and eluted later.

Plates Number

The plate number is lower than those of other chromatography but it is possible to increase it by:

• **Increasing the length of the column** (N = L/H). However, gels do not tolerate a length of the column too high because of the risk of crushing on weight. Also, it is better to use a recycling process that permits to increase fictitiously the length of the column. In this type of chromatography, all the peaks are included between the void volume V_m and the total permeation volume Ve. When the last compound is eluted, one introduces again automatically some elutes at the outlet of the detector in the system of the introduction of the mobile phase (Figure 14). It becomes possible to recycle the sample a great number of times the eluate on the same column without the risk to have overlaps between the peaks. Figure 14 depicts the improvement of the resolution with the recycling number.

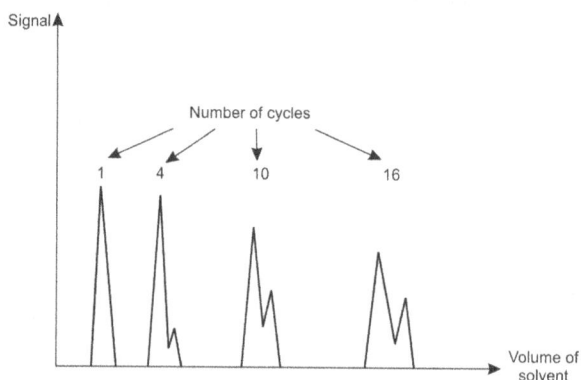

Figure 14: Recycling Chromatography: Application to Separate Two Compounds. (Reprinted with Permission from Gwenola and Jean-Louis Burgot, Paris, Lavoisier, Tec et Doc, 2017, 224).

• **Modifying the flow rate of the mobile phase** influences the height of a theoretical plate (see Chapter 14; 'Van Deemter and Knox Equation'). Thus, separations are achieved with flow rates much higher than with other chromatography. It involves that the elution times are lower.

In size-exclusion chromatography, the Van Deemter curve does not go through a low point because the term B of the relation between the plate height and the linear flow rate of the mobile phase is negligible. The lower the linear flow rate is, the better the separation of the solutes because they have time to diffuse in the available volume. Furthermore, there is no expansion of the peaks when the flow rate of the mobile phase increases because longitudinal diffusion is negligible. Thus, it is possible to separate large quantities of compounds because they do not involve interactions with active sites susceptible to becoming saturated. The limiting factor is here the viscosity of the sample which must not exceed twice that of the mobile phase.

The determination of the plate number is made from compounds fully diffusible such as:

• Methanol for the stationary phases being of type μStyragel 100 Å;
• Ortho-dichlorobenzene (in quantity < 1 μl) for the μSpherogel 500.10^6 Å.

One notes, in general, a value of 3,000 to 4,000 theoretical plates for a column of 30 cm length.

Practical Aspects

Stationary Phases

Characteristics.

• Diameter of pores: It is controlled by the degree of cross-linking beaded gel.
• Swelling Capacity: It is the amount of solvent which can be absorbed by one gram of dry gels. For example, Sephadex G10 means that the gel absorbs 1 ml of water per g dry gel; it is a highly

cross-linked gel for which the mesh is very narrow. On the contrary, Sephadex® G200 adsorbs 20 ml water per g dry and possesses a mesh that is very slack.

- Size and shape of beans of the substrate: The particles are small, and spherical with a granulometry of approximately 40 to 300 μm for traditional liquid chromatography and 20 to 80 μm for HPLC.

- Consistency: This parameter varies according to the nature and the degree of cross-linked gels classified as below:

  - Xerogels: Semi-rigid gels that do not keep their initial shape when the wetting phase is removed. Sometimes, there exist with them reproducibility problems.

  - Aerogels: Rigid gels are interesting for working with high flow rates or pression.

- Inertness: Gels must not react with the wetting or dispersant phase (in general, the salted aqueous phase is slightly acidic or basic). They deteriorate if the pH is inferior to 2 and superior to 8. There exist gels stable to organic solvents.

- Affinity of gels for analytes: It is necessary to reduce the affinity to avoid the overlap of other phenomena, such as ion exchange for ionic groups or adsorption.

Xerogels. It is possible to classify them into two groups according to the cross-linking rate:

- Soft gels are often used with aqueous solutions (except styrene/DVB which are hydrophobic gels). It is only possible to use with low pressions (< 2 bars) and with a low flow rate. The capacity or retention factor can take some value up to 3.

- Semi-rigid gels have a crosslinking rate higher and so possess different properties. The swelling is lower. That gives them some values of capacity factor between 0.8 to 1.2. They can be used with a high flow rate because they accept high pressions. Owing to their hydrophobic aromatic nature, the mobile phase is in general an organic solvent. This process is named in this case, gel permeation chromatography.

Let us only consider some examples of semi-rigid gels:

- Dextran gels or SEPHADEX® is a polyoside (see Figure 5; 'Natural Products')

These gels are not soluble in water, but they are hydrophilic compounds. They swell into an aqueous medium and in solvents, such as d.m.s.o (dimethyl-sulfoxide). They are fragile gels because they are very sensitive to oxidant compounds and acidic medium. Therefore, the primary hydroxylic groups can be oxidized in carboxylic acidic groups which are capable to give ion exchange process. By acylation or alkylation of the hydroxyl groups, it is possible to prepare some derivates which swell perfectly in an organic medium (hydroxypropyl-SEPHADEX®).

These gels allow the separation of biomolecules of molar mass up to 700 Da (G10) or of molar mass up to 30,000 Da (G50) from smaller molecules or/and to determine molecular weight (G100).

(*Remark*: Daltons (atomic mass unity). 1 Dalton = 1/12 of the mass carbon atom ^{12}C. One carbon atom has a mass of 12 u and a mole of atoms of ^{12}C has a mass of N*12 u with N, Avogadro number $(6.0221409.10^{23})$.)

- Polyacrylamide: bio-gels (Figure 15)

They are non-soluble in water but they swell properly by means of free amide groups. Bio-gel is in the form of regular beads and is not very resistant to acidic and basic mediums. They can become ion-exchangers. They allow separating molecules whose molar mass is up to 400,000 Daltons.

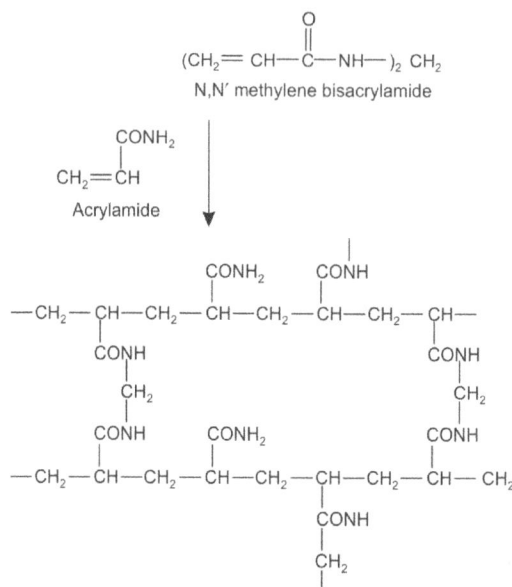

Figure 15: Polyacrylamide Gel. (Reprinted with the Permission from Gwenola and Jean-Louis Burgot, Paris, Lavoisier, Tec et Doc, 2017, 227).

- Agarose gels or sepharose

The agar-agar or gelose is a polysaccharide extracted from algae that have the property to dissolve in hot water and to give a gel by chilling. The agar-agar is constituted of two components:

- Agaropectine with carboxylic or sulfuric groups which possess ion-exchange or adsorption properties;
- Agarose non-ionic compound and linear polysaccharide, crosslinking with the D-galactose and 3-6anhydro L galactose, linked by hydrogen bonds.

- Polyvinyl gels or fractogels

Their matrix is constituted of hydrophilic vinylic polymers. The pores are made of cross-linking with numerous polymers. They confer excellent mechanical stability to the gels. They also possess chemical stability. As a result, their separator powers are high. They are used to separate some compounds whose molar mass is up to 10^7 daltons.

Aerogels. They are rigid materials, porous glass or porous silica for which the silanol groups have been deactivated by silanization to eliminate the adsorption phenomenon. There are above all used in HPLC (see Chapter 17). Aerogels do not swell in contact with the dispersant phase. The range of use is up to 10^7 daltons. They enhance the separation performance, especially the resolution. It is possible to use a mix of gels, such as acrylamide-agarose or acrylamide-dextran.

Selection Criteria for Stationary Phase

The choice is a function of the molar mass of the compounds to separate. It is necessary to calibrate each phase for drawing the curve $\log M = f(V_R)$ with M = molar mass of calibrated compounds and V_R elution volume (Figure 16).

As it is possible to see on the graph with phase A, both analytes to separate are eluted around the void volume Vm and the difference between V_1 and V_2 for each compound is very low. With phase C, both compounds are eluted with very little difference in volume. It is phase B that gives the best result because the linear part of the curve is in the range of mass of compounds to separate and the difference between the retention volume is large.

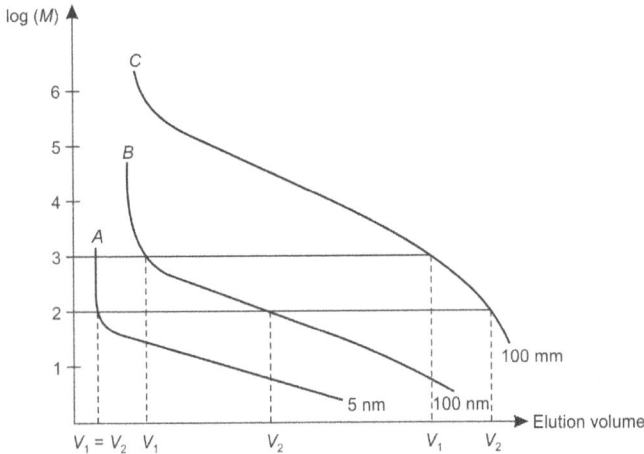

Figure 16: Choice of the Stationary Phase. (Reprinted with Permission from Gwenola and Jean-Louis Burgot, Paris, Lavoisier, Tec et Doc, 2017, 229).

If the solutes have molar mass too different, it becomes necessary to use many series columns with different exclusion limits. There also exist some columns filled with a mixture of stationary phases with different range limits. Generally speaking, a great part of separations by size-exclusion chromatography is realized on many columns of 1.5 to 2 m in length (3 to 8) putting in series.

Selection Criteria for Mobile Phase

The mobile phase has to present the following properties:

- To dissolve the sample
- To swell the gel
- To be compatible with the detection system
- Not to deteriorate the stationary phase; for example, micro-styragels must not be put in presence of water, alcohols, acetone or d.m.s.o.

Usually, the mobile phase is a pure aqueous or salted aqueous solution with different ionic strengths, but it is possible to use some organic solvent such as toluene or chloroform. Sometimes, a bacteriostatic or fungistatic product is adjusted to avoid peak deformation.

Affinity Chromatography

Affinity chromatography is the most specific in the chromatography field, at least, regarding physical or chemical phenomena as the base of the separation. That is used when the classical processes of separation are insufficient. Affinity chromatography involves specific and reversible interactions between the analyte and some specific (or ligands) in most cases of the protein nature.

Table 5: Examples of Ligand and Affinant.

Ligand	Affinant
Enzyme	Substrate or inhibitor
Antigen	Antibody
Hormone	Receptor

The analyte is specifically retained by the stationary phase, and impurities are eliminated in the mobile phase or by washing with a mobile phase of distinct nature (Figure 17).

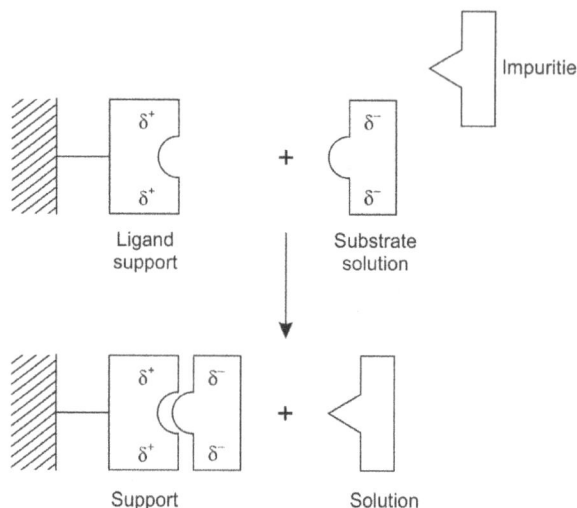

Figure 17: Interaction Between Ligand and Affinant. (Reprinted with Permission from Gwenola and Jean-Louis Burgot, Paris, Lavoisier, Tec et Doc, 2017, 234).

At the end of the experiment, it is necessary to equilibrate again the column with the immobilization solvent so that it can be used again.

Applications

The choice of the process of chromatographic separation depends on structure, molar mass, physical-chemical properties or the ionic character of the molecule that needs to be separated. Chromatographic methods are very useful to analyze and purify pharmaceutical or food products but also in environmental analysis, particularly partition chromatography. But, one can also give some examples as isomers separation by adsorption chromatography, demineralization of water and purification of biotechnology products by the ion exchange process. Another example is the determination of the molar mass of immunoglobulins IgG available as pharmaceutical products by size-exclusion chromatography.

CHAPTER 17

Instrumental
Chromatographic Methods

Among instrumental chromatographic methods which are investigated here are gas and liquid elution and thin layer development chromatographies.

Gas Chromatography

Scope and History

Gas chromatography is partition chromatography for which the mobile phase is a gas. It represents the most powerful and well-recognized technique to separate, identify and quantify the gaseous or compounds that are sufficiently volatile.

From a historical standpoint, A.J.P. Martin and R.L.M. Synge published an article in 1941 to describe liquid-liquid partition chromatography and, in a purely formal way, gas-liquid chromatography. But they had not tested this process until 1952, the date at which A.J. James and A.J.P. Martin had published the results of the separation and the analysis of methyl esters of fatty acids.

Gas chromatography is listed in the major pharmacopeia, such as the American and European ones.

Principle

One can distinguish both modes of gas chromatography depending on whether the stationary phase is a solid or a liquid:

- Partition chromatography [or gas-liquid chromatography (GLC or GC)];
- Adsorption chromatography [or gas-solid chromatography (GSC)].

The first one is far ahead of the most commonly used method. In this case, there are the partition differences of the compounds to separate between a gas phase and liquid phase that is applied. The liquid stationary phase is more often deposited on inert porous support packing a column through which an inert gas called a mobile phase or carrier gas flows. The compounds to separate and inject on the top of the chromatographic column with low concentrations (injected amounts around µg) are carried gradually in the function of some physical characteristics toward the detector by the mobile phase. The components of the sample appear to be separated successively at the output of the column where they are detected.

In gas chromatography, interactions between solute and stationary phase are the only ones responsible for the separation of the components of the sample. Thus, for a gaseous mobile phase, only vapor tension of the solute occurs for the separation; this value is independent of the nature of the gas (or mobile phase) which wraps the solute.

Instrumentation

As shown in Figure 1, the gas chromatography equipment is composed of:

- Carrier Gas Cylinder A with a Pression Regulator R to follow the entrance pressure (actual flow rate controller is measured with an electronic flow-meter);
- A sample Injector B;
- A Column C;
- An oven to control the column temperature accurately;
- A Detector D in relation to a Computer E for data acquisition and analysis.

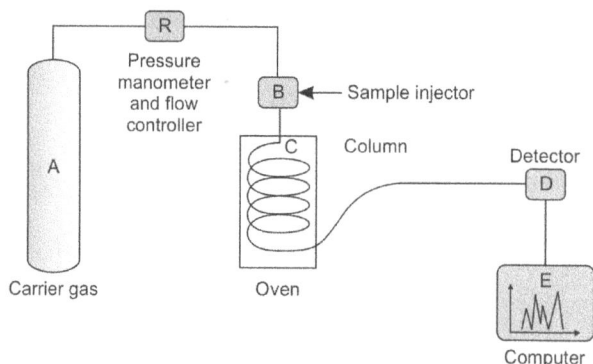

Figure 1: Major Components of Gas Chromatography Instrument. (Reprinted with Permission from Gwenola and Jean-Louis Burgot, Paris, Lavoisier, Tec et Doc, 2017, 238).

Carrier Gas

For obtaining a successful detection of the solute, the carrier gas must possess some chemical features different from those of the solute. One uses:

- Helium (90% of the separations): Inert and non-flammable gas; high separation efficiencies; expensive;
- Nitrogen: Low separation resolution; cheap;
- Hydrogen: High diffusible; good separation efficiencies but flammable gas and not completely inert.

The choice of the carrier gas is a function of the detector being used. The flow rate of the carrier gas varies from 20 to 40 ml/minute for the packed column and 0.2 at 3 ml/minute for a capillary column. The choice of the flow rate depends also on the gas carrier nature and design of the column (diameter and length).

Columns and Injectors

Guard Columns. The separation of compounds or solutes by GC is achieved at higher and lower temperatures because solutes must have a relatively high vapor pressure to be carried in the gas mobile phase. Not to disturb the temperature, the carrier gas is reheated in a guard column at the beginning of the chromatograph before the injector of the sample.

Injectors. One can distinguish many systems of injectors:

• Direct vaporization injectors (Figure 2a): Samples are introduced with a syringe through a septum (in most cases in silicone) in a stream of carrier gas in the hot chamber vaporizer. The vaporization chamber is programmed at a higher temperature than those of the column for which the sample is volatilized to be introduced as quickly as possible. This device is used with the packed columns and

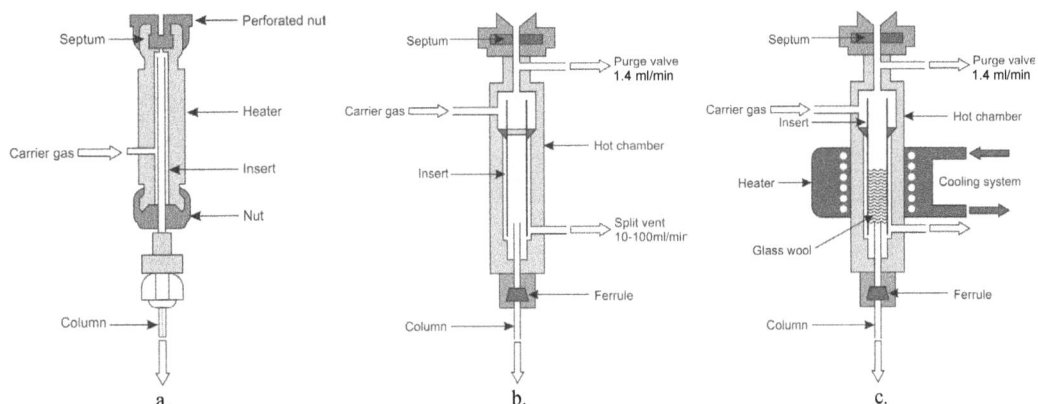

Figure 2: Sample Injectors. (Reprinted with Permission from Gwenola and Jean-Louis Burgot, Paris, Lavoisier, Tec et Doc, 2017, 239). a: Scheme of a Direct Vaporization Injector; b: Scheme of a Split-Splitless Injector; c: Scheme of an Injector-Vaporizator With Program Temperature.

the Megabore columns (see the section on 'Columns'). The injected volume varies from 1 to 10 µl and all the sample is introduced into the column.

• Split/splitness injectors (Figure 2b): The capillary column needs some lower volumes of samples from 0.1 µl to 1 µl. It is necessary to use a derivation flow system for which a part of the sample is eliminated before introducing it in the column (mode split). For trace analysis, the division system is shut. The major drawback of this mode is the selective vaporization of the most volatile components with a change in the true composition of the injected sample.

• Programmed temperature vaporizer (Figure 2c): The chamber of injection is heated gradually. It is possible with this system to inject the sample with or without split at low or hot temperatures. Moreover, it becomes possible with this type of injector to inject large volumes for very diluted samples.

• Headspace system: Its use is increasingly prevalent for the injection of very volatile compounds. This process consists of heating around 60 to 80°C liquid or solid samples placed in a small vial (1 to 3 ml) capped with a Teflon septum, which is perforated by a needle to extract a headspace gas above the sample. This process is interesting for analyzing compounds from a non-volatile matrix. When a state of equilibrium is reached, the concentration of the compound in the volume of headspace is in relation to its concentration in the matrix; it is a constant defined as the partition coefficient (see Chapter 4). Using the headspace system involves a calibration realized in the same conditions as the analysis. That means making a solution of the same nature as a sample that which is not easy for the biological matrixes.

Columns. Columns are made of glass or stainless-steel tubes enclosed in a temperature-controlled oven, which must keep the temperature constant in all the chromatographic apparatus (between 20°C and 400°C). If the temperature is not uniform, the partition coefficient will not be constant. Columns are all the more selective as they are long and narrow. One can distinguish three groups of columns:

• **Packed columns** with a finely divided solid with an internal diameter of around 2 to 4 mm, length of 0.5 to 5 m (most commonly 2 m). They produce broad peak shapes and their resolution is low. These columns are used in preparative chromatography or for gas analysis.

• **Capillary columns** or hollow Golay columns are coated with support immobilized on the tubing walls and impregnated with a stationary phase. These have an internal diameter of 0.1 mm to 0.53 mm and a length of 5 to 100 m (most commonly 30 m). Columns used in fast-gas chromatography have an internal diameter of 0.1 mm, length of 10 m and a layer of coating of 0.1 μm. Capillary columns produce sharp peaks and allow high-resolution separation on lower amounts of the samples, inferior to 1 μg. Their high efficiencies are linked to their high plates' number. Several columns of this type are available (Figure 3). One distinguishes many categories:

- WCOT (Wall-Coated Open Tubular column): The liquid stationary phase is deposited under the form of thin internal coatings on the tubing walls of a glass capillary. Columns are often substituted by FSOT (Fused Silica Open Tubular), which is more flexible and the inside walls of columns are made of fused silica covered outside by polyimide (thermal stable polymer).
- Thin layer column:
    - PLOT (Porous-Layer Open Tubular column): A solid stationary phase is deposited in a thin layer directly on the inside wall of the column.
    - SCOT (Support-coated Open Tubular column): A solid inert support is deposited in a thin layer and is impregnated with a liquid stationary phase. These columns are gradually replaced by cross-linked fused silica deposited in a thin layer.

The short capillary columns (10 to 20 m in length and low internal diameter inferior to 100 μm) provide to diminish by three-time of the analysis. Therefore, the capacity of charge is low and limits their use for analyzing traces in a complex matrix.

- **Megabore** or semi-capillary columns (internal diameter 0.53 mm): These columns have an intermediary efficiency between those of packed columns and capillary columns. They also permit the injection of a volume of solution in the same order of magnitude as those of packed columns. With regard to the latter, the megabore column has the advantage of not having problems of "leaching" of the stationary phase by gradual loss of it. Thus, the stationary phases with cross-linked silica do not move through the carrier gas.

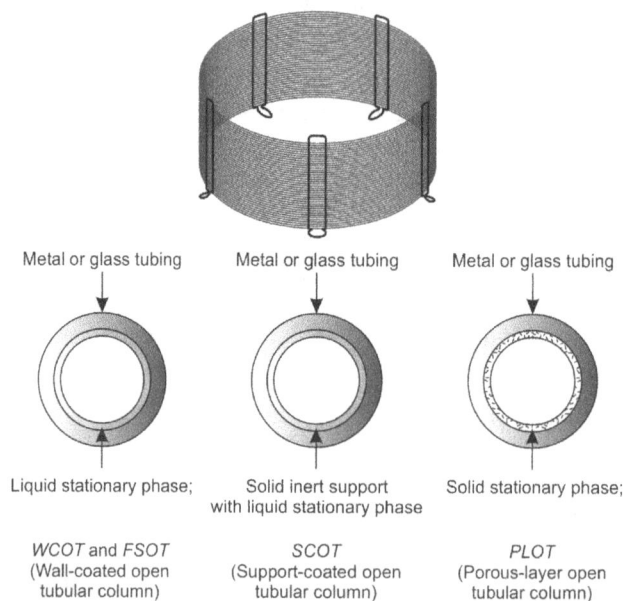

Figure 3: Capillaries Columns for GC Chromatography. (Reprinted with the Permission from Gwenola and Jean-Louis Burgot, Paris, Lavoisier, Tec et Doc, 2017, 243).

Support of the Stationary Phases. The support is made of inert solid beans on which is deposited the non-volatile liquid that plays the stationary phase. Their impregnation rate (mass of stationary phase/mass of the support) generally varies from 6% to 20%. To realize a regular impregnation, one makes a mixture of the beans of the support with a solution of the stationary phase in a solvent easy to eliminate by evaporation. It is necessary that the repartition of the liquid phase be perfectly uniform. The whole is introduced in the column with the aid of vibration apparatus to homogenize the mixture. The support must have some features:

- To have a chemical and physical inertness (for example, no catalytic properties with dihydrogen H_2);
- A large specific surface;
- Good mechanical resistance and good thermal stability;
- A large porosity to minimize the loss of charge (difference of pression between input or output of the column);
- To be as little adsorbent as possible for the solute;
- To have a homogeneous diameter of the beans.

In that respect, one notes that the smaller the beans are, the more necessary it is to use a high pression so that the gas moves through the column. We use silica, charcoal, kieselguhr and more recently, fluorocarbons for their inertness. Kieselguhr is the material most frequently used.

This diatomaceous earth is a form of hydrated silica with numerous hydroxyl groups on its surface on which solutes can be adsorbed. In partitioning liquid-gas chromatography, the phenomenon of adsorption can be reduced by chemical derivatization or silylation (see the section on 'Sample Preparation).

The diameter of the beans varies from:

- 60 to 80 Mesh (0.25 to 0.18 mm)
- 80 to 100 Mesh (0.18 to 0.15 mm)
- 100 to 120 Mesh (0.15 to 0.13 mm)

Stationary Phases. Stationary phases must be thermally stable, have a low vapor tension and do not give an irreversible association with one of the solutes. Their choice depends on the temperature of the analysis and on the polarity of components of the mixture to separate. One can distinguish:

- **Stationary solid phases:** Made of small beans too homogeneous; they fill the PLOT columns. They appear as molecular sieves such as aluminosilicate crystals or porous polymers.
- **Stationary liquid phases:** They impregnate the support or are linked to it. They are classified by the function of their polarity and the temperature that they are capable to bear.
  - **Polyether glycol:** Polar phase made of polyethylene glycol which tolerate temperatures to 225°C (CARBOWAX® 20M, DB-WAX®, HP20M®).
  - **Silicones** or poly-siloxanes (Figure 4) are used at 375°C with:
    - 50% cyanopropyl-50% phenyl-methyl-poly-siloxane (polar) (OV225® and DB225®);
    - 50% trifluoropropyl-50% methyl-poly-siloxane (intermediate polarity) (OV210® and DB210®);
    - 50% methyl-50% phenyl-polysiloxane (intermediate polarity) (HP50®, OV17® and DB17®);
    - 5% methyl-phenyl-poly-siloxane (weak polarity) (HP17® and DB5®);
    - Polydimethylsiloxane (apolar) (SE30® and HP1®).
  - **Squalanes** (saturated hydrocarbons) (apolar) are used up to 120°C;
  - **Chiral phases** based on cyclodextrin per-alkylates.

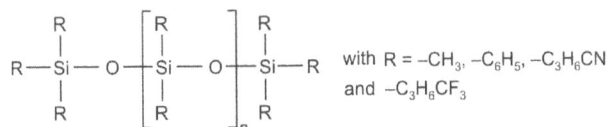

Figure 4: Poly-Siloxanes Structure. (Reprinted with Permission from Gwenola and Jean-Louis Burgot, Paris, Lavoisier, Tec et Doc, 2017, 244).

Oven. The oven must have excellent thermal stability to avoid changes in temperature that give important retention shifts. One notes that retention time can decrease up to 50% for each 15–20°C increase of column temperature.

Mostly, GC ovens are air-bath ovens equipped with high-speed fans to homogenize the temperature. The control of the initial and final temperatures of ovens is realized by a microprocessor, which permits also fixing the length of each level of temperature. Two modes are used, isothermal and temperature-programmed elution.

Types of Detectors

Detectors are immediately disposed of at the output of the column and are crossed over by the mobile phase and the solutes. They must be linear. It means that the signal provided as a function of elution time or volume of the added mobile phase is proportional to the instantaneous concentration of analytes in the mobile phase. The detector is called a differential one. One distinguishes the two groups following:

First Group: Non-Destructive Detectors. With this group of detectors, it is possible to recover components at the output of the column to characterize them by another method. They have a drawback, which is a lower sensitivity than the second group.

* **Thermal conductivity detector or Katharometer:**

Etymologically, Katharometer means apparatus which is capable of measuring purity. The principle of this detector is the following; the temperature of an electrically heated filament depends, at constant electrical energy, on the nature of the gas which wraps around it. Therefore, when a current with a magnitude I cross the resistor, the filament heats up by the Joule effect of a quantity Q equal to RI^2, which is a part according to the thermal conductivity, is passed to the walls of the katharometer (Figure 5). And thus, with all the other parameters keeping constant, its resistor becomes a function of the composition of the gas and the resistance varying linearly with the temperature according to law: $R = R0 (1 + \alpha t)$.

To put in pro-eminence the variation of the resistance, the filament is put in an arm of a Wheatstone bridge. The arm AB is wrapped around by a conductivity cell of resistance r_1. In the cell, the carrier gas flow through with the analyte. A branch AD of resistance r_2 is wrapped by another conductivity cell under a design in a double column or differential one. An arm BC with a resistance r_3 and an arm DC with a resistance r_4 round off the device. A galvanometer is placed between B and D. When the carrier gas flows through alone in both cells, the Wheatstone bridge is equilibrated with the resistance r_4. There is no current in the galvanometer. When the carrier gas and the solute flow through in the cell, the bridge is unbalanced. A voltage appears between B and D which is amplified electronically and constituted the chromatographic signal. Katharometer is a

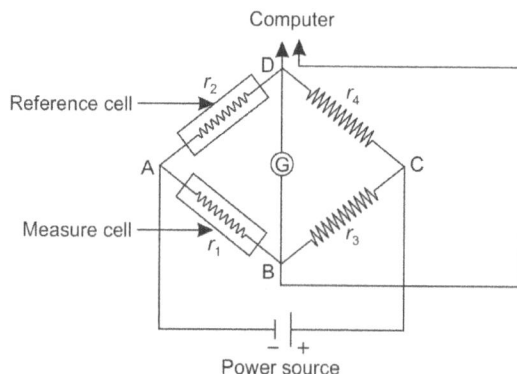

Figure 5: Katharometer. (Reprinted with Permission from Gwenola and Jean-Louis Burgot, Paris, Lavoisier, Tec et Doc, 2017, 246).

universal detector because all compounds have a thermal conductivity different from the carrier gas. This is a non-destructive detector, but it possesses a low sensitivity. It requires pressurized carrier gas as helium. The detection limit is 1 ng (10^{-9} g).

• **Electron capture detector:**

The detector is an analog of an ionization chamber constituted by an ionizing radiation source that produces only beta particles (electrons) (Figure 6). When the pure carrier gas flows across the chamber, it appears with positive ions and free electrons. If the carrier gas is nitrogen, it takes place the following ionization:

$$N \equiv N \xrightarrow[\text{only carrier gas in the column}]{\beta^-} N \equiv \overset{+}{N} + e^-$$

The probability of recombination is too low. Ions are collected by electrodes on which is applied a potential difference. That produces thus a current called basic current.

If an electro-absorbing or electrophilic compound is present in the carrier gas, there is the capture of electrons by compound:

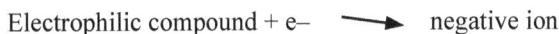

$$\text{Electrophilic compound} + e^- \longrightarrow \text{negative ion}$$

The recombination of a negative ion with ion N_2^+ is incomparably more frequent than this of N_2^+ with electron e^- (10^6 higher). As a result, the number of ions diminishes and so does the basic current. From a practical point of view, the detector is constituted by a small cell whose extremities play the role of electrodes. The ionizing radiation source is generally a thin sheet of stainless steel on which is deposited either tritium or ^{63}Ni. This detector is not universal. It is specific to electron-absorbing components, such as halogenated compounds which feature the sequence of sensitivity according to the nature of the halogen such as:

$$I < Br < Cl < F$$

It also permits the analysis of polyaromatic hydrocarbons, conjugated carbonyl compounds, nitrites, nitrates and some sulfur and phosphorus compounds. But, it is in fact non-sensitive to carbohydrates, alcohol and saturated carbides. However, it is possible to link to electrophile reagents, the compound to analyze by derivatization with a chemical reaction. The detection limit is 50 fg (10^{-15} g).

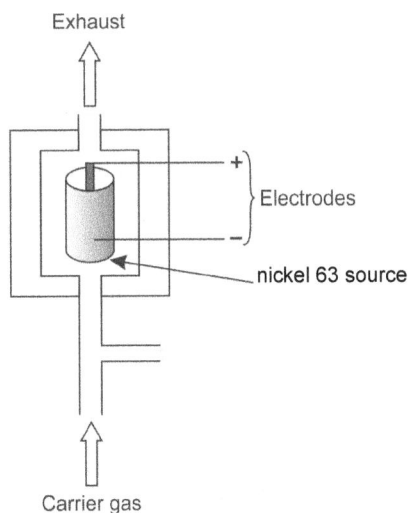

Figure 6: Electron Capture Detector. (Reprinted with Permission from Gwenola and Jean-Louis Burgot, Paris, Lavoisier, Tec et Doc, 2017, 247).

Second Group: Destructive Detectors. They destroy analytes. On the contrary, they permit a quantitative analysis more accurate than the former one.

- **Flame-ionization detector:**

The carrier gas (nitrogen) and the eluting organic molecules at the output of the column are mixed with hydrogen and are burnt in a flame where they are ionized (Figure 7). Ions are detected by a collector electrode and give an ionization current which is amplified. When the mobile phase is only constituted by the pure carrier gas, the flame produces a relatively small amount of ions and a current with a small magnitude. If an organic compound is present in the carrier gas, the magnitude of the current increases with the number of ions. This detector has a high sensitivity for all substances that contain carbon, but it is not universal. For example, it is not sensitive to H_2O, NO or N_2. The detection limit is 10–100 pg (10^{-12} g).

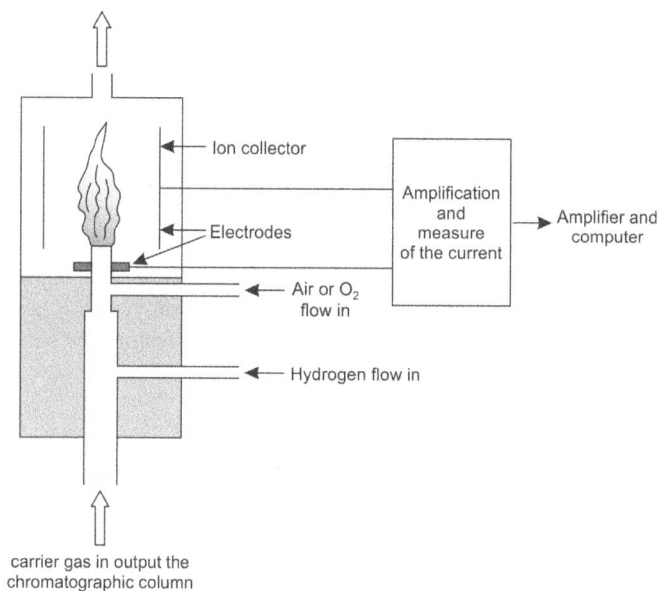

Figure 7: Flame-Ionization Detector. (Reprinted with Permission from Gwenola and Jean-Louis Burgot, Paris, Lavoisier, Tec et Doc, 2017, 248).

- **Flame Photometric detector:**

Sulfur or phosphorus-containing compounds are capable of emitting light when there are eluting in an air/hydrogen flame. Sulfur compounds emits at $\lambda = 394$ nm and phosphorus compounds at $\lambda = 526$ nm. The emitted light is collected by a photomultiplier positioned in the upper section of the flame. Some optical filters can select one or the other element (Figure 8). This detector is very selective. The detection limit is 100 pg (10^{-12} g).

- **Thermo-ionic detector:**

The principle of the thermo-ionic detector is very close to those of the flame-ionization detector but in this case, one can put a rubidium or cesium chloride bean. This bean is heated by an electric system and has a negative potential in regard to the ion collector electrode. When a compound with nitrogen or phosphorus is present with the carrier gas in the flow in the output chromatographic column, one notes that organic nitrogen and organic phosphorus compounds are adsorbed at the surface of the bean. The emission of electrons increases as well as the current. The thermo-ionic detector is extremely specific and very sensitive; indeed, 500 times more sensitive for phosphorus compounds and 50 times more for nitrogen compounds. The detection limit is 1 pg (10^{-12} g) for nitrogen compounds and 0.1 ng (10^{-9} g). for phosphorus compounds.

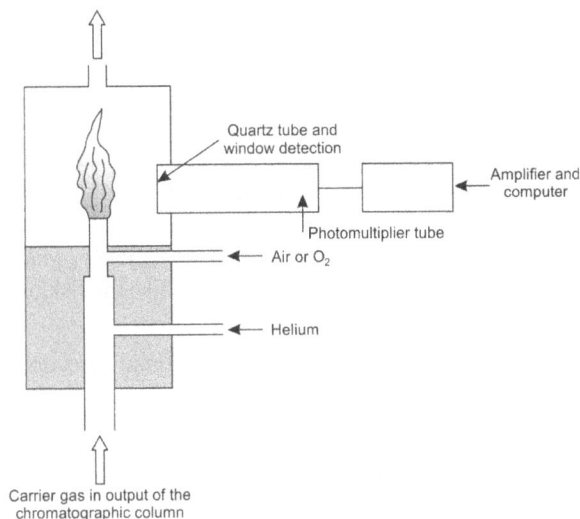

Figure 8: Flame Photometric Detector. (Reprinted with Permission from Gwenola and Jean-Louis Burgot, Paris, Lavoisier, Tec et Doc, 2017, 249).

- **Infra-red detector:**

It is rather the linkage of two processes that has become possible by the development of the Fourier-transform infra-red spectrometer. A tube with a small dead volume ensures the liaison with both apparatuses. This detector is very useful to identify some molecules from databases. It is also possible to recover the chromatogram by mathematical adjustment of the data.

- **Chemiluminescence detector:**

The chemiluminescence detector is specific to sulfur and nitrogen compounds. These compounds suffer in a pyrolysis chamber some fragmentation of the molecule and the chemical groups obtained can be converted by ozone into an excited state or radical. The radical goes back to its base state by emitting light. One measures the light with a photomultiplier tube after the selection of the wavelength of the sulfur or nitrogen group (see Chapter 22).

Mass Spectrometric Detector. The combination of mass spectrometer and gas chromatography gives multiple analytical applications. These apparatuses have undergone considerable technological evolution. The newer technology called compact GC-MS are easier to use and do not need any specific training.

The mass spectrometry has gone through great development in recent years with the designing of different apparatus for adapting to the applications, such as research or common analysis. The choice of the apparatus also depends on separation methods, i.e., gas, liquid chromatography, electrophoresis or even for inorganic compounds, the use of atomic emission spectrometry.

The principle consists in realizing a suitable method for the ionization of a molecule that generates ions. Ionized fragments are separated in the function of mass to charge ratio (m/z) and their nature and abundance are determined. Each separation method matches with a type of mass spectrometer for which the features are developed in more detail in Chapters 41, 42 and 43.

The difficulties to link GC and MS are of various kinds:

- The carrier gas and analytes in the output chromatographic column flow at the atmospheric pressure while the mass spectrometer must operate under a high vacuum to avoid collisions between the ions. The inferior pressure is for example 10^{-6} Torr for a quadripolar analyzer. Many solutions have been introduced to offset this difficulty:
  - The use of capillary columns which do not need a high flow rate of carrier gas, only 1 to 2 ml.min^{-1}. It is not necessary in this case to avoid the carrier gas.

- The removal of the carrier gas by a vacuum pump. The quality of the pump defines the choice of instrumentation of MS. For example, electron ionization is the main ionization mode used with a pump on one line.
- The detector mass spectrometry enforces the choice of carrier gas which must be easily eliminated (helium or hydrogen lighter gas).
- The stationary phases must be extremely stable even at high temperatures. Thus, the entrance in the MS detector of products of the breakdown of the stationary phase modifies the mass spectra, clogs the detector and decreases its detection limit. New stationary phases with poly-carborane chains (Si-C) instead of poly-siloxane chains (Si-O) diminish the risk of bleeding of the column and are more robust for the GC-MS separations.

Errors Attributable to GC-Detectors. To obtain an optimal detection, other criteria must be achieved such as:

- Long-term stability: In order to ensure there is no drift of the basic line (Figure 9).
- The response time: It is necessary that it is as short as possible to avoid the peak tailing.
- The linearity in a large range of concentrations.
- The reproducibility.
- High sensitivity for trace analysis.

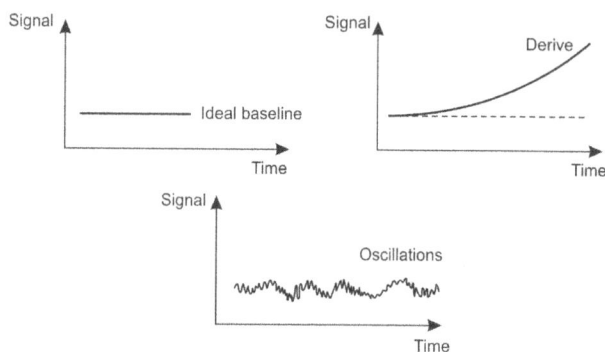

Figure 9: Derives and Oscillations of the Baseline. (Reprinted with Permission from Gwenola and Jean-Louis Burgot, Paris, Lavoisier, Tec et Doc, 2017, 254).

Fundamentals

Time and volume of retention have the meaning given in Chapter 15. Moreover, in gas chromatography one defines a net retention volume V_N:

$$V_N = J\, V'_R$$

With V'_R is the reduced retention volume (crude volume of retention of the solute minus the crude volume of retention of a non-kept compound). J is an experimental coefficient or coefficient of pressure loss which has been found by Martin and James and has been given by the relation:

$$J = \frac{3}{2}\frac{p^2 - 1}{p^3 - 1} \quad \text{avec} \quad p = \frac{P_e}{P_s}\ \frac{\text{input pressure}}{\text{output pressure}}$$

The experimental factor J takes into account the fact that, contrary to other chromatography, the volume flow rate (volume per unit time) of the mobile phase is not constant. It varies between the input and the output of the column. It increases because for crossing the column, the input pressure must be superior to the output pressure (in most cases, atmospheric pressure). Yet, as gases are very compressible, this results in a gradient of the flow rate and a gradient of the speed with a parabolic variation.

Choice of Practical Conditions

The aim is to obtain a resolution as high as possible. That means to recover on the chromatogram a number of peaks equivalent to the number of compounds in the sample. For that, it is possible to modify the efficiency, the capacity factor and the selectivity. But, before that, it must be necessary to make the analyte more volatile.

Sample Preparation

Gas chromatography is only used for volatile compounds. In some cases, it is possible to facilitate the experiment by making some derivatives easier to volatilize (derivatization). This applies, in particular, for:

- Compounds with a high boiling point;
- Compounds with a high polarity which is responsible for large and asymmetric peaks.

Three types of reaction are implemented:

- Alkylation reaction: It is the substitute of hydrogen atom more or less mobile of the solute by a radical alkyl. For example, etherification of the alcohols or esterification of the acids.
- Acylation: It consists in fixing a rest acyl R-CO on the alcohol or amine function producing some esters or amides.
- Silylation: The exchange of mobile hydrogen by a trimethylsilyl rest (CH_3Si) decreases the polarity by the removal of hydrogen bonds.

Improving the Resolution

The Purnell equation given in Chapter 15 shows that the resolution is governed by many factors, such as the square root of the plate number, the separation factor and the retention factor.

$$R_{S12} = \tfrac{1}{4}(\sqrt{N_2})[(\alpha - 1)/\alpha][k'_2/(1 + k'_2)]$$

Obtaining the Optimum Efficacity or Plate Number. The resolution is proportional to the square root of the plate number. To increase the plate number, it is necessary to decrease the plate height or increase the length of the column:

• The plate height varies with the gas flow rate or more exactly with the linear velocity of the mobile phase μ. According to the van Deemter equation, there is an optimum velocity μ which gives plate height its lowest value. To determine it, one plot experimentally, the graph plate height in the function of the linear velocity μ (see Chapter 15).

• The lengthening of the column increases the theoretical plate number; however, also the separation time and the broadening of the peaks (see Chapter 15).

Improving the Separation Factor and the Retention Factor. In both cases, it is necessary to occur the partition coefficient by modifying:

• Stationary and mobile phases: In reality, the only choice of the stationary phase is significant. Thus, the nature of carrier gas has little influence over the partition coefficient. In reality, it is not the solubility of the sample compound in the mobile phase (this notion has no physical significance because the mobile phase is a mixture of gas) which operates but its tension vapor independent of the nature of the gas;

• Temperature: The tension vapor and the solubility in the stationary phase vary with the temperature. In principle, for a compound and a stationary phase given, the partition coefficient decreases when the temperature increases. There exists an empirical relation of the type:[19]

$$\log V_R = b/T + d$$

where T is the absolute temperature in K and b and d are some constants.

These facts show clearly that working in an isotherm surrounding is important. In some cases, to improve the resolution of a mixture of compounds of very different natures, it is necessary to vary progressively the temperature to avoid the peaks too narrow and too close to each other at the beginning of the chromatography and at the end some peaks too broad for the compounds to be kept. Moreover, in this case, the time of analysis becomes so long.

To improve the separation power and the resolution, new techniques are developed as two-dimensional gas chromatography GC-GC that uses two columns with different selectivity connected in sequence; the first one is connected with a flame-ionization detector and the second is linked to the mass spectrometry. Another technique two-dimensional integral GCxGC uses two orthogonal columns which is interesting for the analysis of complex mixtures. Whole compounds eluted from the first column are transferred onto the second column which has a different polarity from the first one. This second column must be shorter to perform faster than the molecules eluted by the first column. This process improves the resolution but also sensitivity of the gas chromatography.

Improving the Speed of the Analysis

The "fast" gas chromatography allows the implementation of high throughput analysis and thus performs a larger number of samples than with conventional GC. This chromatographic technique consists in using some columns with an internal diameter between 100 μm and 250 μm and with a length inferior to 15 m and to increase the flow rate of the carrier gas and/or the temperature of the oven with a regulator of pressure and temperature. The running time of analysis is of a few minutes with "fast" chromatography to some seconds in "ultra-fast" chromatography for which the implementation is difficult. "Fast" chromatography is compatible with the major detectors used in GC and, in particular, the mass spectrometer. In "fast" GC-MS, it is appropriate to use instrumentation equipped with a time-of-flight analyzer with (see Chapters 41, 42 and 43) a high data acquisition of spectra.

Applications

Gas chromatography is a technique of:

• Qualitative analysis by the value of the parameters t_R, V_R, Kovats index or by the study of the mass spectra by comparison with mass spectra databases.

• Quantitative analysis by using the area of the peaks. Owing to low values of the injected volume, it is necessary to use an internal standard method. In GC-MS, one can adjust the analyzer to only screen some selected ions, choosing in advance in the mass spectra for their high abundance, allowing to obtain some chromatograms whose area is integrated. It is possible to use as an internal standard compound, some molecules of the compound to analyze but in the form of stable isotope-labeled molecules. This method enhances precision and accuracy.

---

[19] This relation is on the same type from that found experimentally which gives the variation of tension vapor of a solute in function of temperature.

A large number of applications are developed, in particular, in the following fields:

- **Pharmaceuticals:**
  - Active ingredient and excipient analysis: Determination of impurities, in particular, residual solvents are used during the last step of purification of the pharmaceutical products. One uses a headspace sampler, the ethylene chloride as the internal standard, and the standard addition method to identify and quantify these residual solvents.
  - Quality control of bulk drugs: Bulk drugs contain one or more Active Pharmaceutical Ingredients (API) and pharmaceutical excipients. To release a batch of bulk drugs before its distribution, it is necessary to identify and quantify the API.
  - Stability studies of pharmaceutical ingredients and bulk drugs with the research of degradation products;
  - Quality control of essential oil, aroma, and flavor compounds: For example, the use of enantioselective gas chromatography with chiral stationary phases based on cyclodextrins for controlling the risk of adulteration. Figure 10 shows the separation of both enantiomer of menthol from a reference sample of racemic menthol synthetic used for the development of the method. Consequently, it is possible to control the origin of oil essential in the function of the enantiomeric distribution of the components.

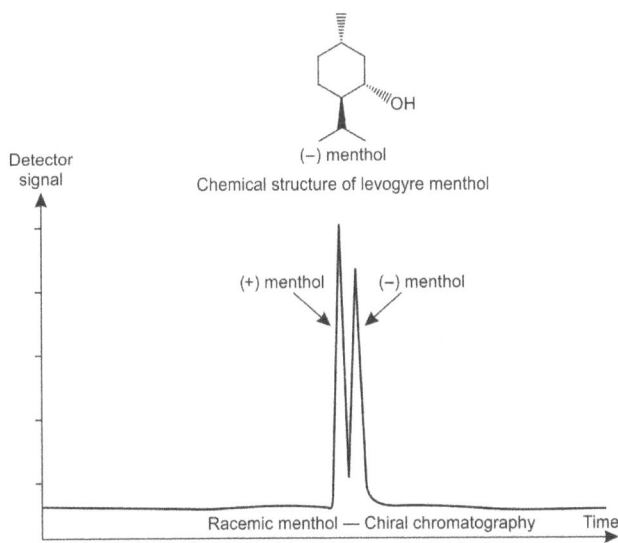

Figure 10: Separation of Two Optical Isomers of Menthol From a Sample of Menthol Synthetic. (Reprinted with Permission from C. Marin (1995)). Conditions of separation: Detector = quadripolar mass detector; Column = 50 m*250 mm inside diameter and 0.25 μm film thickness; Injector split = 250°C. Carrier gas hydrogen; Temperature program = 100°C then 2°C/min to 120°C.

- **Toxicologic studies:** Determination of ethanol and methanol for acute intoxication (Figure 11). Gas chromatography has been reported with flame-ionization detection.

- **Foods:** Determination of herbicide atrazine and their major degradation products in soil pore water sample by capillary gas chromatography with split-less injection, a 30 m*0.25 mm inside diameter *0.25 μm film thickness capillary column. The carrier gas is helium and the detector mass spectrometer.

- **Biological samples:** Determination of molecular metabolites profile related to predictive and diagnostic pathologies. An example in colorectal cancer is the one that determines some volatile organic biomarkers from routinely fluids biologics: blood, urine or saliva.

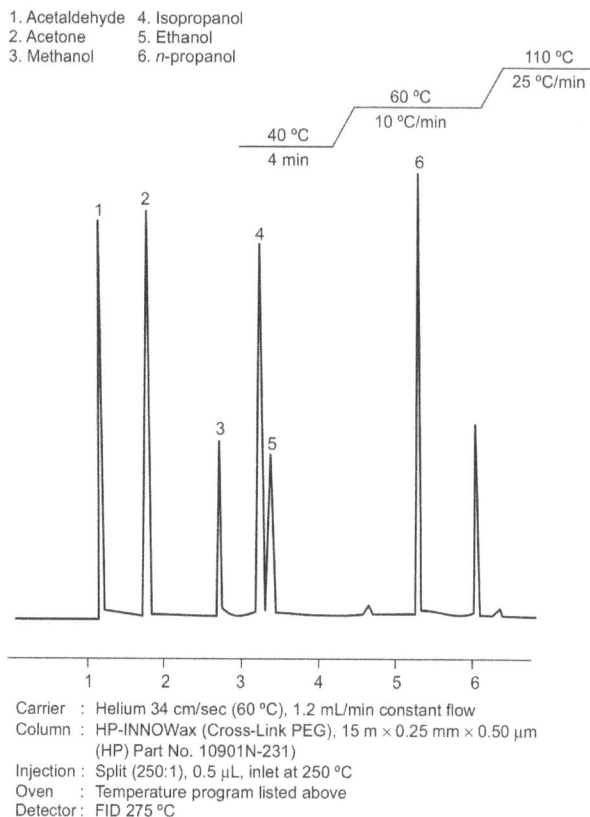

Figure 11: Separation of a Standard Mixture of Alcohols, Acetaldehyde and Acetone. (Reprinted with Permission from Agilent Technologies, 1 rue Galvani, 91745 Massy).

High-Performance Liquid Chromatography

Scope

High-performance liquid chromatography (HPLC) is a modern liquid chromatography in which the principles implemented are numerous: adsorption, partitioning, ion-exchange, etc. The first liquid chromatography methodology was too slow and did not use some online detectors. Consequently, improvements are realized such as using homogeneous and thin granulometry for the stationary phases or coupling detectors directly to the output of the column. The efficiency of the chromatographic separation increases and the experiment becomes less time-consuming (1 to 20 hours for conventional liquid chromatography and sometimes less than 5 minutes for high-performance liquid chromatography); as a matter of fact, the efficiency is inversely proportional to the square of the diameter of particles of stationary phases and the linkage of the chromatographic column to the detector permits to simplifying the experiment. HPLC is described in the pharmacopeias, such as European and American.

Comparison Between HPLC and GC

It is interesting to appreciate the differences between HPLC and GC to know better the specificities of each technique. Thus, in addition to the mobile phase and instrumentation, there are a number of differences that should be highlighted.

Number of Interactions

In gas chromatography, interactions between solute and stationary phase are the only ones responsible for the separation of the components of the sample. Thus, for a gaseous mobile phase, only vapor tension of the solute occurs for the separation; this value is independent of the nature of the gas (or mobile phase) that wraps the solute. Contrary to gas chromatography, interactions in liquid chromatography are tripartite: solute/stationary phase and the mobile phase which increases the separation possibilities (Figure 12).

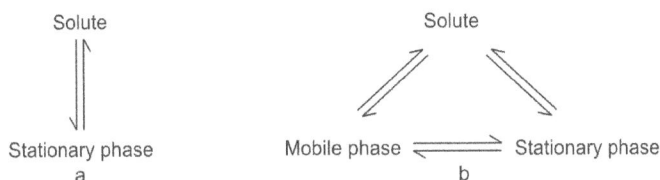

Figure 12: Comparison of Interactions Involved in GC (a) and HPLC (b). (Reprinted with Permission from Gwenola and Jean-Louis Burgot, Paris, Lavoisier, Tec et Doc, 2017, 262).

Nature of the Stationary Phase

The nature of stationary phases in HPLC is extremely varied. The drawback of thermal stability does not exist. All the mechanisms of separations described previously can be used, even ion exchange and ion-pair which are excluded in GC because ionic compounds are not volatile.

Temperature of the Separation

The temperature used in HPLC is less high in GC in compliance with the principle of GC. Yet, the intensity of molecular interactions increases when the temperature decreases. Intrinsically, the HPLC would be more efficient than the GC. In relatively recent years, high-temperature liquid chromatography (HTLC) appeared, which used mobile phases with a large ratio of water (for example, 85% water and 15% acetonitrile) brought at high temperatures (80°C to 120°C). This technique which is more environment-friendly than conventional techniques has the advantage of enlarging the choice of detectors with sometimes experiments with flame-ionization detectors (see the section on 'Applications'). It involves the use of columns on zircon or graphite which remain stable over 200°C because they are made of inorganic components.

Diffusion

The diffusion in the liquid media is lower than this in gaseous media (10^4 to 10^5 lower). This corresponds to the moving of the molecules on the action of a gradient of concentration into the medium. It is the result of the second law of thermodynamics. The particle moves to the equality of the concentrations (in reality, the activities) in the same phase. That produces a diffusive flux governed by Fick's law (sometimes called Fick's first law):

$$J_{x=xi} = -D \, (dc/dx)_{x=xi}$$

D ($cm^2.s^{-1}$) is the diffusion coefficient which is 10^4 to 10^5 times lower in liquids than in gas. Thus, if one takes the example of partition liquid-liquid chromatography, it is not interesting to eluate quickly the solute with the mobile phase if the diffusion into each phase (which ultimately orders the exchange between both phases) is low. It is this low speed of the exchange which accounts for the slowness of conventional liquid chromatography and also its lower efficiency than the HPLC. Thus, in conventional chromatography, exchanges are slower and it is necessary to have a higher height of the column to achieve equilibrium. Consequently, the height plate is higher than for HPLC. For the same length of the column, the HPLC is more efficient because it has a larger plate number.

One way to decrease the time of exchange by diffusion is to reduce the lengths of diffusion, which is explained by the use of stationary phases of very lower granulometry because the depth of the pores decreases with the size of the particles of support (the diameter of particles in HPLC varies from 2 to 10 μm).

Viscosity and Loss of Charge Pressure

The loss of charge is the difference of pressure between the input and the output of the column. The loss of charge is given by Darcy's law:

$$\Delta P = \eta LV/K°$$

ΔP reflects the loss of charge in the barye unit. The barye unit corresponds to the old CGS unit of pressure:

1 barye = 1dyn/cm^2 = 10^{-1} pascal or 10^{-6} bar
η: Viscosity of the mobile phase in SI unit, Pa. s; 1Pa. s or 1 pascal second = 10 poises = 1 Poiseuille (CGS unit)
L: Length of the column in cm
V: Linear velocity of the mobile phase in cm^{-1}
K°: Permeability constant of the column in cm^2.

In fact, K° is the function of the diameter D_p of the particles of the support and the interstitial porosity or the ratio of the volume of the column available to the mobile phase:
K° = k.D$_p^2$
Dp: Diameter of the particle in cm
k: Dimensionless factor of proportionality

We can see that the loss of charge is proportional to the viscosity of the mobile phase. Yet, the viscosity is 100 times higher for liquids than gases. That means that with an equal flow rate, it is necessary to operate with pressure 100 times higher in a liquid phase than in a gaseous phase to overcome the loss of charge. On the other hand, the loss of charge is inversely proportional to the square of the particle diameter. If the particle diameter is divided by a factor of 2 with the other factors remaining constant, the inlet pressure must be multiplied by 4 to keep the same flow rate. Yet, as it is necessary to work with very fine granulometries, it must be inevitable to work with a high pressure. The increase in the speed of separations is not linked directly to increasing the pressure but rather to the fine granulometry. That results in using high pressure which could be to 600 bars. In fact, it is possible to use more moderate pressure with shorter columns (length less than 10cm) but consequently, analysis times are longer.

Incompressibility of the Liquids

For the pressure which is used in HPLC, liquids may be considered incompressible. The flow rate and the linear velocity do not depend on the length of the column. It is not the case in GC.

Kinetic theory

Equations that link the plate height to the linear velocity of the mobile phase have some different expressions than those of GC (see Chapter 15).

Optimization of Practical Chromatography

The most interesting case is to have a high resolution with a shorter run and without a too high loss of charge. These different requirements are contradictory and the choice of experimental conditions often reflects a compromise between these different parameters. Two cases are interesting to describe:

Too Low Resolution

In the case of a too low resolution, it is possible to improve the efficiency by:

- Increasing the length of the column. But the run time increases as well as loss of charge.
- Decreasing the plate height by using fine granulometry to have a very small diameter but in this case, the loss of charge increases extensively (with the square of diameter).

Too Long Run Time

In this case, it is possible to decrease the run time by increasing the speed of the mobile phase but that occurs when there is:

- A loss of efficiency by a moderate increase of the plate height.
- An increase of loss of charge.

From a practical view, the optimization of the HPLC analysis is difficult owing to the many parameters on which it is possible to occur. For the first time, one tries to choose the best chemical conditions of the separation such as the nature and the composition of the eluent to obtain a separation factor that is not so close to 1 and a retention factor between 2 and 10.

Other possibilities have been used to improve the efficiency of liquid chromatography such as:

- Working at a higher temperature than ambient temperature to decrease the viscosity of liquids and to increase the diffusion speed of solute;
- Using the monolithic column to eliminate loss of charge (viz. Chapter 16);
- Using U-HPLC or ultra-high-performance liquid chromatography (see the section on Detectors).

Instrumentation

Figure 13 shows the key components of an apparatus of high-performance liquid chromatographic.

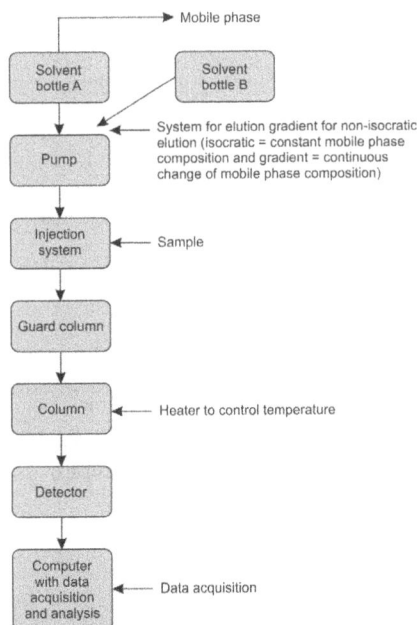

Figure 13: Scheme of Key Components of High-Performance Liquid Chromatograph. (Reprinted with the Permission from Gwenola and Jean-Louis Burgot, Paris, Lavoisier, Tec et Doc, 2017, 266).

Pumps

Pumps or solvent delivery systems must have the following features simultaneously:

- Overcoming the loss of charge higher than 350 bars.
- Giving a constant flow rate with suitable accuracy and reproducibility up to 10 ml min^{-1}. It must be possible to work with a varying flow rate to improve the resolution of a mixture.

We now use:

- Syringe pump where the mobile phase is pushed by a piston from the reservoir to the column. Syringe pumps are only used nowadays for micro or nano-chromatography.
- Reciprocating piston pump, one of the most popular, can deliver a variety of mixed or not solvents at high pressures with accurate compositional as well as for isocratic or gradient operation.

The fluid path of the pump system is composed of inert materials. The pump system is used with a mixer for gradient elution and a degasser to eliminate air bubbles that affect pump and column performance.

Injectors

The most widely used system is a loop injection mode, which ensures the injection of an accurate and reproducible volume (10 µl to 100 µl injected). This system may be automatized (Figure 14).

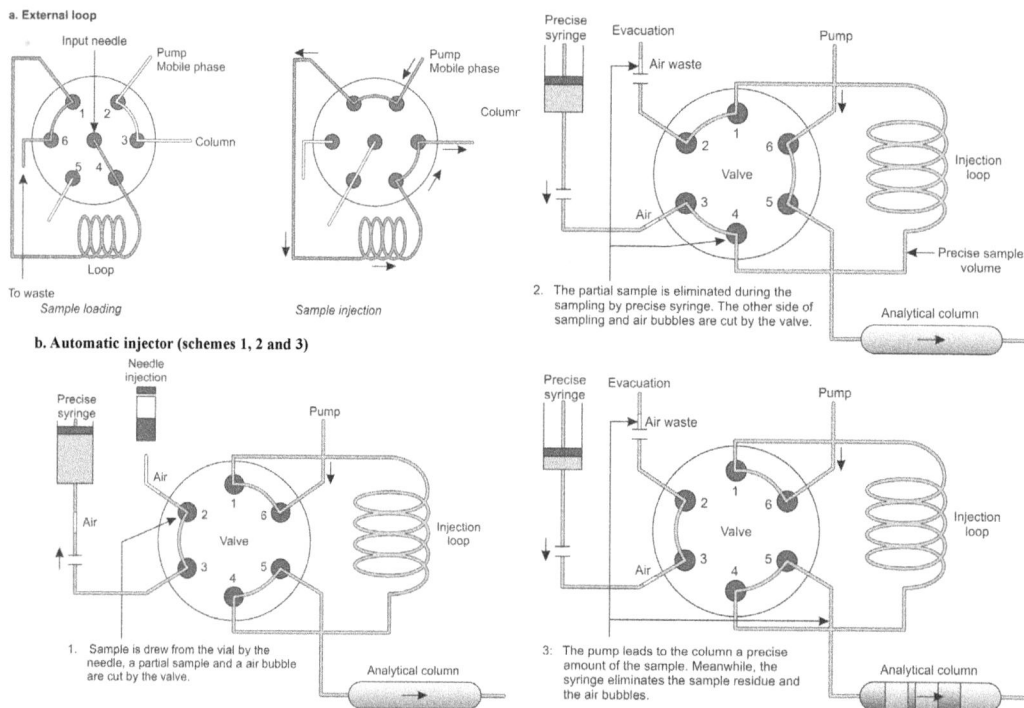

Figure 14: Scheme of Loop Injector Systems. (Reprinted with the permission from Thermoseparation products, 12 avenue des Tropiques, BP141, 91944, Les Ulis).

Columns

Columns are made in stainless steel or glass. The inner diameter of the column tubing varies from mm up to cm. Their length is around 20 cm (5 to 30 cm). Some capillary microcolumns are 3 to 25 cm in length with an inner diameter of 0.3; 0.5 or 1 mm and are commercialized to simplify the link between HPLC and mass spectrometer. They enable to obtain some very high resolution when they are used with very high pressure up to 15,000 psi (1 pound/square inch = 0.07 atm and 14 psi = 1 atm) instead of 5,000 psi with the U-HPLC. These microcolumns are also interesting for the very small samples. Their packing is difficult due to their very thin granulometry less than 2 µm. The U-HPLC undergoes great development because it permits improving the efficiency of the chromatography and reducing the run time; an experiment of 20 minutes in HPLC can be reduced by a few minutes or even seconds. The U-HPLC is very useful in the area of biomolecules analysis (glycoprotein and monoclonal antibody).

One notes the recent development of separation microsystems, such as sensors or chips, capable to ensure the preparation, separation and detection of the sample.

Detectors

The detector is a key component of a chromatograph HPLC because it permits ongoing monitoring of the separation and quantification of the solutes. As in GC, liquid detectors must be stable, linear, reproducible, and sensitive. Moreover, they should:

- Have a low time of response;
- Have a very low cell volume;
- Can be used with a gradient of elution solvents.

However, a universal detector does not exist as in gas chromatography (except mass spectrometer) but only some specific detectors for each application. The specificity is an interesting feature to quantify the traces, whereas universality is used for the study of the unknown mixture.

Ultraviolet Absorption Detectors. They are the most popular. The limits to its use are absorptions of the solvent and the solutes. The response essentially depends on the molar absorption coefficient of the solute. In the most favorable cases, the minimum amount which is possible to detect is around the ng. The response of the detector follows the Beer-Lambert law.

There exist many types of apparatus:

- At a fixed wavelength; for example, $\lambda = 254$nm;
- At a variable wavelength of 200 to 700 nm;
- Photodiode array that gives the simultaneous value of the intensity of light along the spectrum (see Chapters 19 and more to know about spectral methods). One obtains a three dimensions intensity (wavelength, time and intensity) (refer to Photo 1). This system is very useful for researching impurities in pharmaceutical ingredients.

Differential Refractometer. In this case, the difference in refractive index between the mobile phase and the eluate emerging from the column is continuously measured. It could represent a

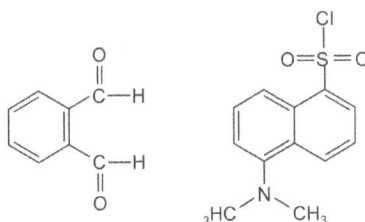

Figure 15: Chemical Structure of O-Phthaldialdehyde and Dansyl Derivatives.

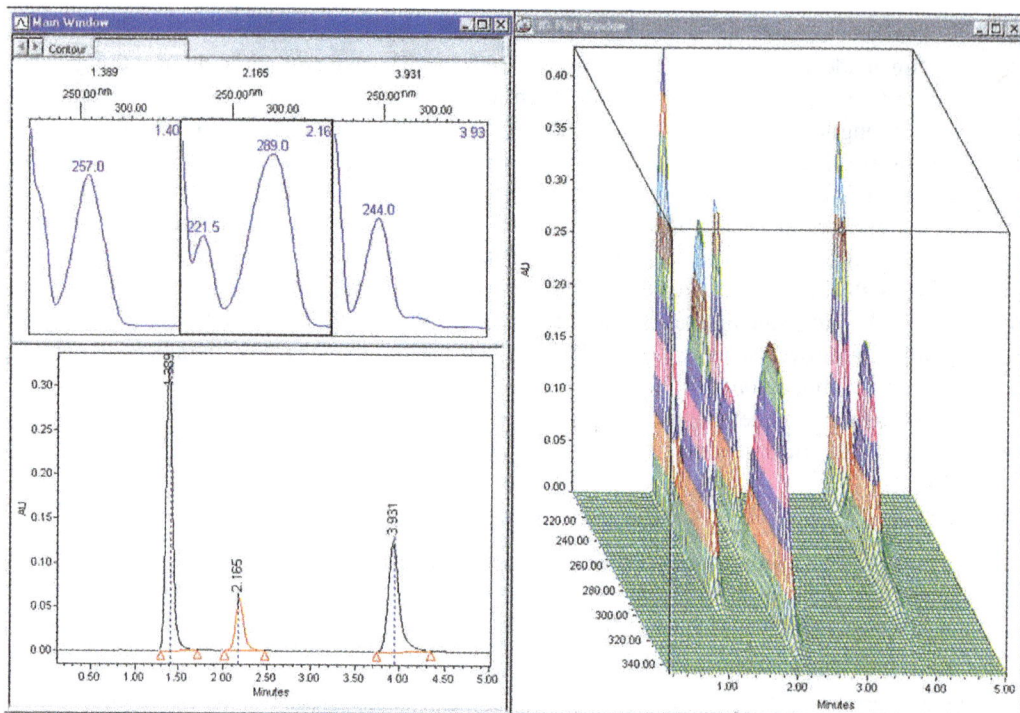

Photo 1: Three Dimensions Chromatogram Obtained with a Photo Diode Array Detector. Reprinted with the permission from Waters, BP608, 78056 Saint-Quentin-en-Yvelines.

universal detector. However, it provides a lack of sensibility and needs a controlled temperature at 0.01°C because the refractive index measurement is temperature sensitive. Moreover, its use prohibits gradients in the composition of the mobile phase.

Fluorescence Detector. A fluorescence detector possesses high sensitivity and selectivity because it is needed that the solutes are fluorescent compounds (see Chapter 22). However, it is possible to realize derivatization to generate fluorescent molecules. The reaction can take place in a pre-column or online in a post-column. The most popular derivatization agents are o-phthaldialdehyde and dansyl derivatives (Figure 15).

Electrochemical Detector. They are restricted to electroactive solutes, such as nitro derivatives, hydroquinone, aromatic derivatives, phenol, etc. The elution phase must have some electrolytes to ensure the conduction of the current:

• **Coulometric Detector:** The electroactive compound is exhaustively electrolyzed. One measures at a controlled potential to obtain the highest efficiency and the quantity of electricity (in coulombs), which is produced to transform the electroactive compound. This quantity is proportional to the amount of compound present in the detector. Moreover, one uses a potential chosen to give the limiting current or the diffusion current to obtain the highest sensibility. This system needs the use of an electrolysis electrode of a high surface (porous electrode crossed by eluate), and the time of the analysis is long because the electrolysis process is exhaustive; an coulometry detection permits noticing the end of the electrolysis.

• **Amperometric detector:** In amperometry detection, the electroactive compound is not exhaustively electrolyzed. The electrolyze process is carried out at constant potential. The measurement of current allows following the evolution of the concentration of the solute in the cell. (For a complete view, see the second volume devoted to electrochemical methods of analysis). The Amperometry detector is the most used and has high sensitivity and good selectivity with a large linear range. However, these detectors are oxygen-sensitive present in eluate because the latter is reducible.

• **Conductometric detector:** This detector is used for ionic chromatography (ion exchange or ion-pair as an example) where ions are detected continuously at the output column. Its principle is based on the measure of the resistance of the solution between both platinum electrodes functioning on alternating currents. The flow of current through an electrolytic conductor is the consequence of the ion migration of opposite signs. Unlike the metallic conductor, the conductivity of the ionic conductor depends on concentration and ions charge.

It is important to note that in ionic chromatography, the components of the apparatus (pump, injector, tubes, etc.) must not release metal products all the more as some eluent phases are very corrosive. Moreover, the apparatus must support very high pressure up to 28 MPa. For this reason, one uses specific instrumentation made of an inert and flexible material; for example poly-ether-ether-ketone (PEEK).

Mass Spectrometric Detector. Mass spectrometry is very widely used as a detector in liquid chromatography, but the link is more difficult to realize than in gas chromatography. The principle of the mass spectrometer is developed in Chapters 41 and 42. Thus, the solute in the liquid phase must have low volatility; moreover, it is necessary to eliminate the elution solvent.

To eliminate the solvent, it is possible:

• To realize selective vaporization of the solute;
• To decrease the flow rate of the mobile phase using as in GC derivation system of flux;
• And above using the micro-columns (length = 10 cm, internal diameter = 1 mm) which uses a very low flow rate (some $\mu l.min^{-1}$) and a volume of injection of 1 μl. It permits the introduction of the whole chromatographic eluate in the mass spectrometer.

To avoid the decomposition of the analyte during the step of vaporization and ionization, one uses an ionization process in the mass spectrometer at the atmospheric pressure, i.e., electrospray (ESI) and atmospheric-pressure chemical ionization (APCI). This last method is less sensitive than electrospray, but it is complementary to the first. More often, both ionization sources equip the standard mass spectrometer. Orbitrap and time of flight analyzers are also used with LC (see Chapters 41 and 42).

An interface permits the ion transfer from the source under atmospheric pressure to the mass-analyzer under a high vacuum of around 10^{-5} Torr. The most popular analyzer used with LC is the quadrupole mass analyzer and triple quadrupole (QQQ) mass spectrometer which provides high sensitivity in quantitative determination.

LC-MS is a very useful method for the analysis of traces in complex matrixes or peptides, proteins, polysaccharides analysis and also in proteomics research (studies of composition, structure and activity of cell proteins).

Nuclear Magnetic Resonance Detector. The coupling of nuclear magnetic resonance (NMR) with liquid chromatography has been delayed by the lack of sensitivity of NMR. The use of NMR instrumentation of more than 500 MHz has permitted us to consider its use for quantitative analysis. Some interesting results have been obtained; for example in the antibiotics, such as gentamycine or kanamycine. The use of HTLC (see the section on the 'Temperature of the Separation') with deuterated water allows to monitor online the NMR spectrum. Remember that the deuterium is characterized by spin quantic number $l = 1$ (in units $h/2\pi$, where h is the Planck constant), while the proton is characterized by the value $l = \frac{1}{2}$. The electromagnetic energies absorbed by the deuterium and by the proton are not in the same range of radio frequencies. Consequently, deuterium is invisible when the apparatus is configured for the proton and conversely.

Flame Ionization Detector. Initially developed for being coupled with GC, the FID can be coupled with liquid chromatography when one operates in mode HTLC.

Light-Scattering Detectors. There are very useful in size exclusion chromatography (SEC) for detecting polysaccharides and determining masses for polymers and proteins. ELSD (Evaporating light scattering detector) nebulizes the mobile phase at the output of the column. The solvent is vaporized for leaving in the gas than some particles in suspension which are submitted at a light beam that the particles will diffuse. ELSD is used for determining molar masses. MALLS (Multi-angle laser light-scattering) are useful for measuring concentration profiles and molar masses; the detection is realized, in this case, in solution (see Chapter 30).

Applications

The applications are very numerous. The process enables separating compounds of variable molecular mass (100 to 2,000 Da) from different chemical nature and even different types of isomers. HPLC is a very well complement of the GC for:

- The few volatile compounds; molecules of molar mass that are superior at 300;
- The thermolabile compounds as pharmaceuticals issued of biotechnology process or compounds of biological origin;
- The ionized compounds.

Contrary to GC, HPLC is not limited by the volatility of the sample and by its thermal stability. Moreover, the preparation of the sample before injection in the chromatograph system is easier than in GC.

HPLC is a complementary process to thin layer chromatography (TLC; see the following section to know more). It covers all the fields of applications. One can give some examples in different fields such as:

Foods

- Determination of content of carboxylic acids (lactic acid, acetic acid, malic succinic and citric acids) in the wine after derivatization.
- Analysis of the residues, additives or toxins.
- Analysis of carbohydrates (sugars, starch, etc.) In this case, one chooses the detector in the function of the concentration:
  - High: Refractometer
  - Middle: Evaporative light-scattering detection
  - Low: Electrochemical in pulse mode detector

Pharmaceuticals

- Research of impurities, identification and quantification of active pharmaceutical ingredient (API) and excipients, stability studies.
- Identification and quantification of API in the bulk product. For example:
  - Analysis of a mixture of catecholamines with and without ion-pair (Figure 16);
  - The analysis of dextrane by gel permeation (Figure 17).
- Therapeutic monitoring: Quantification of drugs with a narrow therapeutic zone is followed in biologic media (blood, plasma, etc.) to adjust their dose for each patient. These include clonazepam, amphotericin B, itraconazole, chloroquine, ranitidine and clobazam with a spectrometer UV Visible detector and levodopa with an electrochemical detector.

A: Without ion-pair

1. Norepinephrine
2. Vanillyl mandelic acid
3. Tyramine
4. Dopamine
5. Homovanillic acid

B: With ion-pair formation with sodium hexane sulfonate

Retention time, min

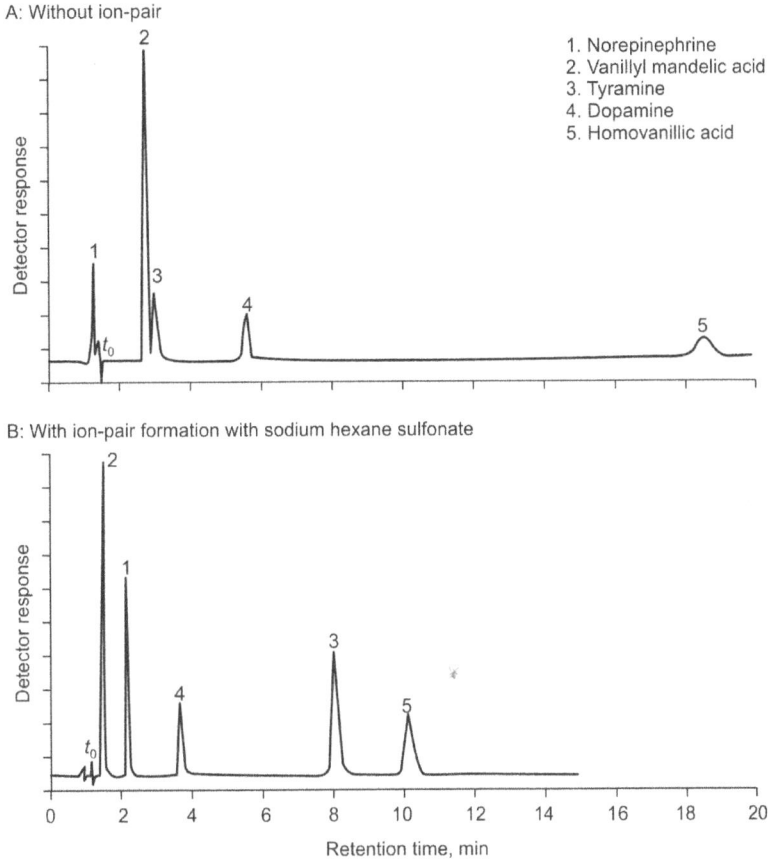

Figure 16: Liquid Chromatography of a Standard Mixture of Catecholamines. (Reprinted with Permission from Agilent Technologies, Parc Technopolis/Z.A. Courtaboeuf, 3 avenue du Canada, CS90263, 91978 Les ULIS cedex).

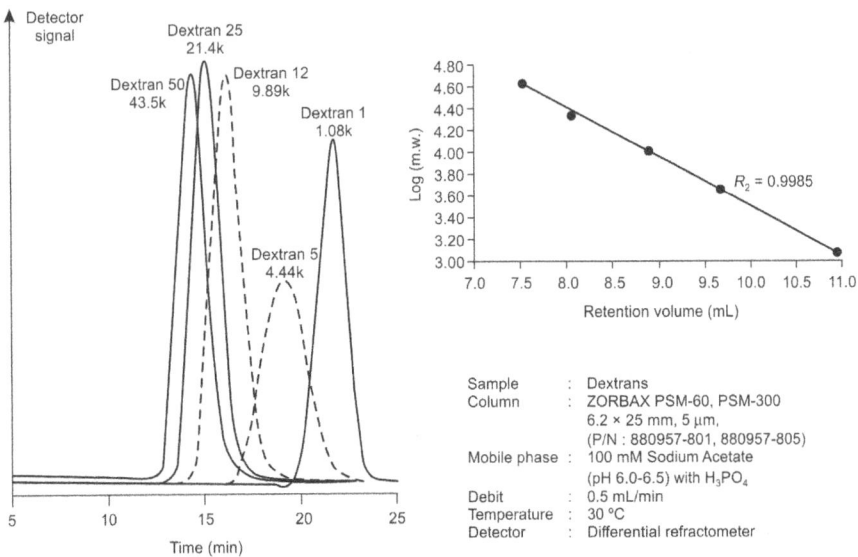

Detector signal

Dextran 50 43.5k
Dextran 25 21.4k
Dextran 12 9.89k
Dextran 1 1.08k
Dextran 5 4.44k

$R_2 = 0.9985$

Log (m.w.)

Retention volume (mL)

Time (min)

Sample	:	Dextrans
Column	:	ZORBAX PSM-60, PSM-300 6.2 × 25 mm, 5 µm, (P/N : 880957-801, 880957-805)
Mobile phase	:	100 mM Sodium Acetate (pH 6.0-6.5) with H_3PO_4
Debit	:	0.5 mL/min
Temperature	:	30 °C
Detector	:	Differential refractometer

Figure 17: Dextranes Characterization by Gel Permeation. (Reprinted with Permission from Agilent Technologies, Parc Technopolis/Z.A. Courtaboeuf, 3 avenue du Canada, CS90263, 91978 Les ULIS cedex).

Biopharmacy

Liquid chromatography is used for the analysis of the molar mass of polypeptides, proteins (Figure 18), nucleotides and proteomic analysis (identification, isolating and purification of cell proteins or biologics fluids) or metabolomics (research of biomarker).

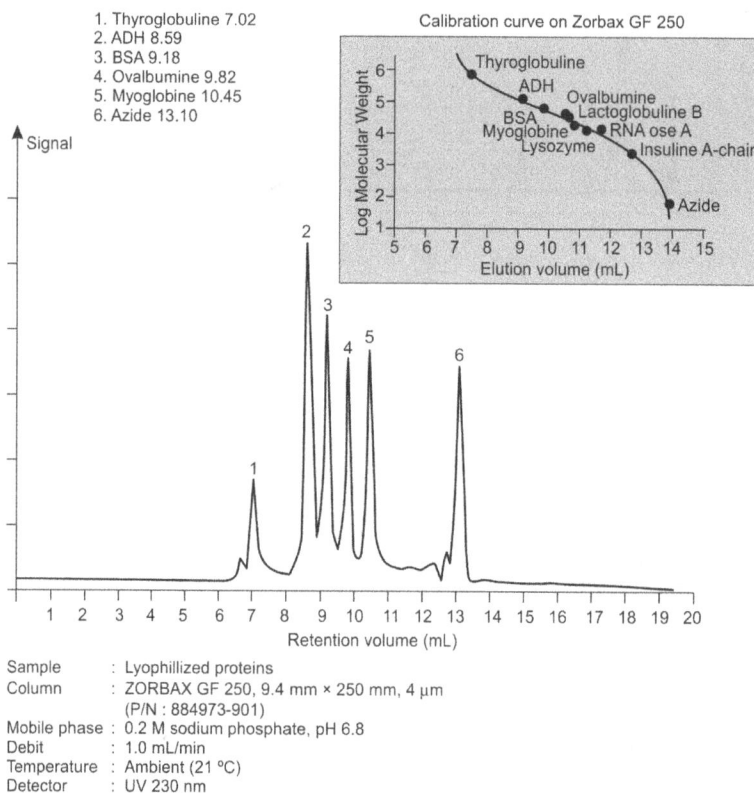

Figure 18: Standardization of Molar Mass of Proteins by Size Exclusion Chromatography. (Reprinted with Permission from Agilent Technologies, Parc Technopolis/Z.A. Courtaboeuf, 3 avenue du Canada, CS90263, 91978 Les ULIS cedex).

Planar Chromatography

The term planar chromatography involves the paper and the thin layer chromatographies.

Principle

Paper Chromatography

Paper chromatography was introduced par Consden, Gordon and Martin in 1944 for analyzing some very low amounts of a mixture. It is a partition chromatography for which a sheet of paper or filter paper plays the support for the liquid stationary phase, usually water. Thus, water molecules are held in the fibers of the paper made of cellulose molecules and make a gel which is the authentic stationary phase. The paper is crossed by the mobile phase and the partition of components of the mixture sample is realized between the gel and the solvent. In this method, one realizes only development and not an elution chromatography, the components migrating on the paper where they are indicated by a convenient process.

For some authors, the process of separation is rather an adsorption mechanism or a mixture of both types of chromatography. Furthermore, some applications based on ion exchange have been developed. It is a qualitative method described in the pharmacopeia (European, American, etc.). Each compound is identified by its retention factor R_f, which is defined as the distance run through the plate by the compound divided by the distance run through the plate by the solvent front.

Thin-Layer Chromatography (TLC)

Thin-layer chromatography has been developed in 1938 by Ismailov and Schreiber who noticed after the breakage of the chromatographic column that the separation of the components of the mixture carries in the laboratory flat on the bench. Now, it was realized on glass, aluminum or polymer plate or a quartz rod covered with a thin layer of stationary phase. It is also a development of chromatography. A drop of sample is applied on the thin layer about 1 cm from the base end of the plate. The inferior part of the plate is dipped into a sealed developing chamber which contains some milliliters of solvent. Before that, the closed chamber is saturated with the solvent for many hours.

According to the nature of stationary and mobile phases, the separation of the components is realized by different mechanisms, i.e., adsorption, partition or both phenomena. It is also possible to use ion exchange.

This mode of chromatography is suitable for qualitative and quantitative applications. TLC is described in the major pharmacopeia.

Implementation of Planar Chromatography

Instrumentation

Paper Chromatography. Papers used are specific for chromatography with a homogeneous structure, especially for quantitative applications.

Developing chambers: Enclosed space to keep saturated water or organic solvent atmosphere. Vapor tension varies with the temperature which must be kept constant.

Thin-Layer Chromatography.

- **Plates:** Many sorbents are used as a stationary phase with different mechanisms of separation:
  - Polar stationary phases:
    - Silica (adsorption or/and partition)
    - Alumina (adsorption or/and partition)
    - Cellulose (partition)
    - Diethylaminoethylcellulose (ion exchange)
    - Polyamide (partition)
  - Low polar stationary phases:
    - Modified silica gel with alkyl (C_2, C_8 and C_{18}) (partition).
  - Chiral stationary phases:
    - Modified silica gel C18 impregnated with a chiral selector (component added to give the chirality) or silica impregnated with cyclodextrin, molecules which possess some cavities with a chiral specificity.
    - Modified cellulose (same mechanism of separation that cyclodextrin impregnated silica).

The technological features of the stationary phases depend on the type of thin-layer chromatography, i.e., conventional or high-performance or HPLTC (Table 1).

Table 1: Technological Aspects of Stationary Phases.

	Conventional Thin-Layer Chromatography	High-Performance Thin-Layer Chromatography (HPTLC)
Average diameter of particles	5 to 30 μm	3 to 8 μm
Thickness of layer of sorbent	250 μm	100 to 200 μm
Homogeneity of particles	Large distribution	Narrow distribution of particles
Maximum distance of migration on the plate	12 to 15 cm	5 to 7 cm
Sample volume	1 to 5 μl	0.1 to 0.5 μl

- **Developing chambers:** Enclosed glass spaces (Figure 19).

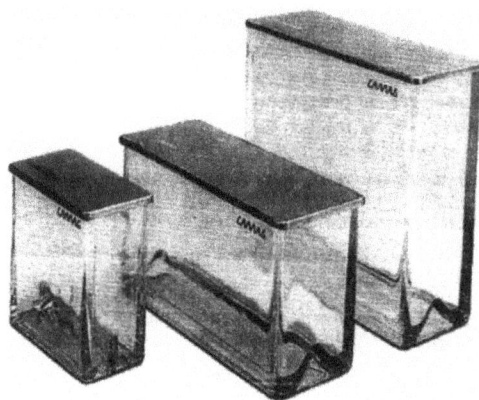

Figure 19: Developing Chambers for TLC. (Reprinted with the Permission from Camag, Sonnenmattstrasse, 11 Muttenz 1, 4132 Suisse).

Sample Applications

Samples are dissolved into the solvent, which is the less eluotropic one to avoid the beginning of the chromatography process just at the moment of the sample applications. The volume of the deposit is very low around some micro-liters.

The choice of mode sample application depends on the volume of sample which has been deposited, automation degree needed and some type of application, quantitative or semi-quantitative. One distinguishes:

- **Application by capillarity:**
  - Manually by filling some calibrated glass capillary (0.5 to 10 μl) after withdrawal of the sample by capillarity. The solution is applied by short application of the capillary to avoid too large applications. One gets the diffused and rounded spots.
  - The mechanic application uses glass capillaries calibrated from 0.5 μl to 5 μl. Spots are rounded or linear. They must be the thinner as possible. Their quality influences the efficiency of the chromatography (Figure 20).

- **Application by vaporization:**

In this case, the application is automatic by nebulization under compressed gas of the sample above the surface of the thin layer with a calibrated syringe with a needle or a stainless-steel automatic capillary. This last process is recommended for quantitative applications. The volume of deposit and the width of the strip can be controlled by a computer and software and automated with an auto-sampler (see color Photo 2).

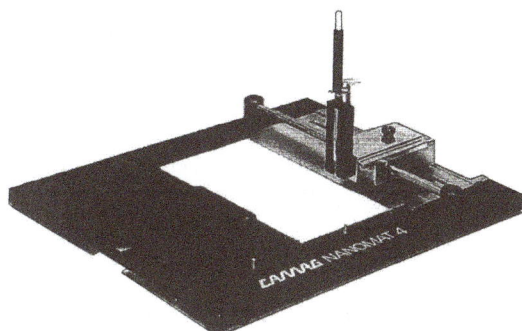

Figure 20: Mechanic Application by Capillarity With Nanomat® System for TLC. (Reprinted with Permission from Camag, Sonnenmattstrasse, 11 Muttenz 1, 4132 Suisse).

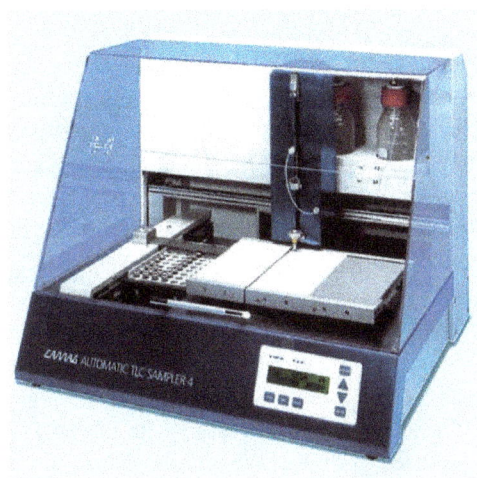

Photo 2: Device for Automated Sample Applications in Thin-Layer Chromatography. Reprinted with permission from Camag, Sonnenmattstrasse, 11 Muttenz 1, 4132 Suisse.

Mobile Phase

Paper Chromatography. The mobile phase is constituted of both phases:

- Water saturated with an organic solvent
- Organic solvent saturated with water

The well-known formula is this of Partridge:

- N-butanol = 4 parts
- Acetic acid = 1 part
- Distilled water = 5 parts

The mixture is shaken and then paused for 24 h; the inferior layer is the fixed phase and the superior layer is the mobile phase. It is possible to use a monophasic solvent; the cellulose adsorbing the water inside the solvent constitutes the stationary phase.

Thin-Layer Chromatography. One uses monophasic mixtures of many solvents of different polarities of various compositions. For example:

- Ethyl acetate for polar compounds
- Hexane for non-polar compounds

But also some mixture of solvents to adjust the polarity of the mobile phase
- Chloroform/diethylamine/acetone (50/10/40)
- Hexane/diethylamine/ethanol (78/8/16)
- Acetonitrile/methanol/tetrahydrofuran (50/25/25)

The thin layer is already impregnated with water. The rate of humidity of the thin layer must be controlled.

Development Methods

Paper Chromatography. After the application of the samples, the paper sheet is developed in vertical (ascendant or descendant) or circular mode. It is also possible to realize some bi-dimensional chromatography; in a direction with solvent system 1 and another direction with solvent system 2.

Thin-Layer Chromatography. The plate is developed by capillarity or by "forced flux":
- **By capillarity:**
  - In the classical chamber in vertical or horizontal mode (Figure 21).
  - In an automatic chamber where one determines in advance by software of the computer. The time of saturation of the chamber by the mobile phase, the distance traveled by solvent, the time and the temperature of dryness and the possibility to realize some elution gradients. This means some distances of migration are higher at each step. Here, the development is realized in many steps. This process improves the reproducibility of the R_f.
- **By "forced flux":** This method is a technological approach to thin layer chromatography. Thus, if the mechanisms of the separation are the same, the direction of the flowing of the mobile phase is different. The plate acts as an HPLC column because it is not in contact with the exterior medium:
  - The thin layer is pressurized. The solvent is pushed under a pressure (inferior to 50 bars) in the stationary phase. The process is mainly used with the HPTLC plates for which the low grain size decreases the speed of flow of the mobile phase and limits the use of the capillary process;
  - The circular and anti-circular thin-layer chromatography. The process uses centrifugal force. It is recommended to use it when the development of the chromatography in normal conditions gives a value of R_f close to zero.

It is possible, with this process to choose an isocratic or gradient elution.

Detection

Visual Detection.
- **Chemical detection:**
  - Direct for colored compounds
  - After spraying a reactant capable to react with the components of the mixture for giving colored substances. Photo 3 illustrates the chromatograms obtained for a plant extract with applications by vaporization and capillarity of the sample.
- **Physical detection** after observation under light UV/visible:
  - At 254 nm on a stationary phase impregnated with a fluorescent indicator at this wavelength. The compound absorbs the photons of incident radiation and suppresses the fluorescence emitted by the indicator.
  - At 366 nm on a stationary phase without fluorescent indicator. Molecules that are naturally fluorescent or became so to a prior derivatization reaction are detected by direct fluorometry.
- **Biological detection** after implementation of antibody-antigen reactions.

Instrumental Detection. Instrumental detection permits identification and quantification that is more precise than visual detection. One uses:

- Densitometry detector (Figure 21): Photodensitometer-scanner which provides the measure of the adsorbed or reflected light during the scanning of spots under a light beam of the range of

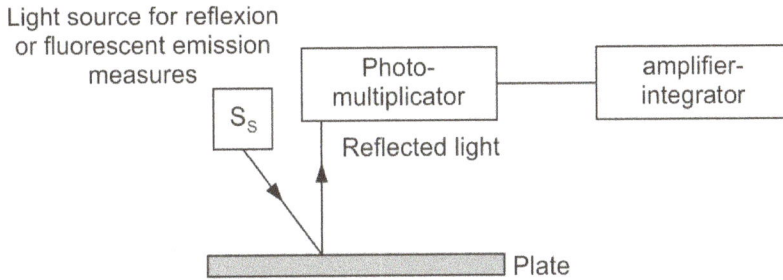

Figure 21: Scheme of a Densitometer. (Reprinted with the Permission from Gwenola and Jean-Louis Burgot, Paris, Lavoisier, Tec et Doc, 2017, 285).

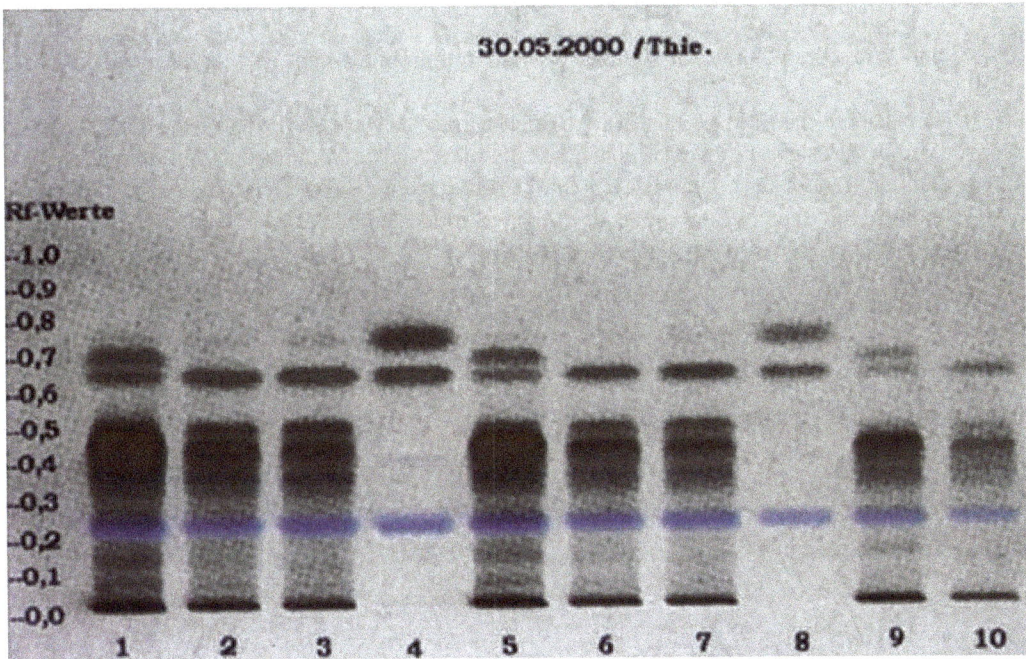

Photo 3: HPTLC Chromatogram Obtained for a Plant Extract.

wavelength of visible or UV. The video-densitometer permits the acquirement of images that are changed into a chromatogram by software. The method remains semi-quantitative by lack of reproducibility. An interesting aspect of the process is to keep the results in the computer.

- Mass spectrometers have been used to improve the possibilities to identify molecules of the mixture after putting the spot in a solvent.

Theoretical Aspects

Each of the components of the mixture migrates with a speed that depends on its partition coefficient. At the end of the operation and after revelation, each substance appears as a spot in principle rounded and symmetric. One introduces the notion of retention by the relation:

$$R_f = \frac{\text{Distance travels by the substance}}{\text{Distance travels by the solvent front}}$$

The distances are measured from the center of the original spot position and the final spot position after migration. From a general point of view (see Chapter 14):

$$K' = K\frac{dv_S}{dv} = K\frac{V_S}{V}$$

V_S and V are the volumes of stationary phase and mobile filled the thin–layer or the paper sheet.

If we calculate the ratio $Q_x/Q = f(x)v$, which represents the repartition of the substance at a distance x from the original spot position when a volume $v = ndv$ of mobile phase has been used. One knows that very few plates contain an amount of substance physically reasonable. Thus, the plate corresponding at the center of the spot is one of the substances found at the maximum concentration either the $(p + 1)$th, such as (see Chapter 14):

$$p = nz = \frac{ndv}{\upsilon}$$

Where, υ is the efficient volume of a plate and dv its mobile phase volume while the solvent front occurs at $(n + 1)$th plate. If h is the plate height:

$$R_f = \frac{(p+1)h}{(n+1)h} = \frac{p+1}{n+1}$$

And as p and n have a high value:

$$R_f = \frac{p}{n}$$

$$R_f = \frac{ndv}{\upsilon} \times \frac{1}{n}$$

v = efficient volume of a plate

$$R_f = \frac{dv}{dv + Kdv_S} = \frac{1}{1 + K\dfrac{dv_S}{dv}} = \frac{1}{1 + K'}$$

K' is the corrected partition coefficient.

Practically, one takes note that the progression of the mobile phase on the paper, the speed of this decreases and the rate dv_s/dv varies; dv tends to decrease and K' tends to increase. The measured R_f is lower than the theoretical R_f.

One can write: Theoretical $R_f = \varepsilon$ measured R_f where the ε varies from 1.1 to 1.5.

Applications

Planar chromatography is a technique whose applications are numerous in various fields, such as food, pharmaceuticals, biologics and environment. Now, its larger uses are in the field of the analysis of plants.

This chapter described the most used instrumentation of chromatography. All these processes are complementary. For example, planar chromatography is commonly used for its short time of response during the implementation of an HPLC method with the choice of solvents.

CHAPTER 18

Electro-Migration Techniques

This chapter describes some methods of separation of electrically charged compounds under the action of the electric field. The method is as well applicable to single ions as macromolecules. It is achieved for the latter in a liquid medium and in recent years on a flat porous support. Electro-migration techniques on a slab or in a tube are also grouped under the name "electrophoresis" and those achieved into a capillary are called "capillary electrophoresis". All these methods are described in the major pharmacopeia for chemical methods.

Electrophoresis was introduced by Tiselius during the 1930s from experiments of protein separation realized in a liquid medium. He noted that the proteins, positioned in electric field strength, migrated at different speeds as a function of their charges and their masses. The first methodology of Tiselius has been improved with the emergence of new techniques where the charge and the mass are not the only criteria of separation.

Overview of Theory

Electrokinetic Features

There exists at any interface a double layer of electric opposite charges similar to the plates of an electrical condenser. It is the case, for example, of the interface made of the wall of a capillary tube and the solution near the wall. (Figure 1, in this example, the wall presents fixed negative charges). When one of both charged phases moves relative to each other (here, it is an obligatory solution), important electrical phenomena occur. They are named electro-kinetic phenomena.

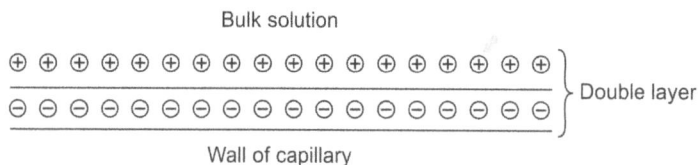

Figure 1: Electric Double Layer at the Capillary Surface. (Reprinted with Permission from Gwenola and Jean-Louis Burgot, Paris, Lavoisier, Tec et Doc, 2017, 290).

Assuming that a voltage is applied through a capillary tube filled with a solution of electrolytes (Figure 2). An electric current appears which, as intended, crosses the solution. It would also appear a less predictable phenomenon, a movement of the complete solution. That is named the "electro-osmose" phenomenon. The travel goes through the cathode because cations lead all the species, charged or not, to the cathode.

Conversely, if one applies to the initial solution a difference of pressure in place of a difference of potential with a difference of pressure responsible for a shifting of the entire solution, therefore

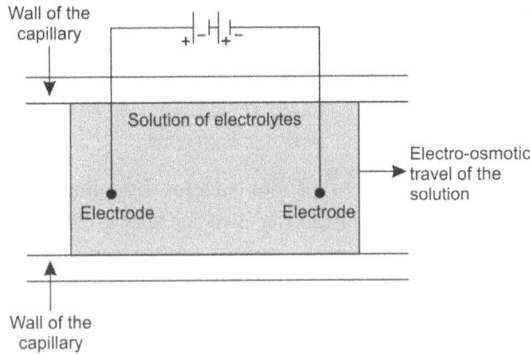

Figure 2: Electro-Osmotic Flow. (Reprinted with Permission from Gwenola and Jean-Louis Burgot, Paris, Lavoisier, Tec et Doc, 2017, 290).

it appears an electrical current between both electrodes dipping in the solution and linked by a galvanometer.

One can already underline the fact that the electro-osmotic phenomenon is from the outset of capillary electrophoresis.

Conventional Electrophoresis

Electrophoresis is the movement of a charged species on the influence of an electric field strength into a solution made of stationary electrolyte(s). In contact with the species and electrolytes, there also exists a double electrical layer. The electrophoretic phenomenon is, therefore, of the same nature as the electro-osmotic phenomenon because it is the relative shifting of both charged phases at the interface, which is important. In electro-osmosis, there is a shift of the solution against the wall of the capillary. In electrophoresis, the solution can be regarded as motionless and it is the charged species that moves. Consequently, it is the same theory that can be used for both phenomena.

Basic Theory of Electrophoresis

The differently charged species travel separately at different distances under the influence of electric field strength. One tries, here, to determine parameters that influence their travel.

Let us consider an ion under the influence of the electric field strength **E**. He is submitted to an electrical force **F** which gives it a movement towards the anode or the cathode according to its charge. The movement occurring in a liquid viscous medium, a frictional force **F'**, opposing **F**, appears and delays the particle. When **F** and **F'** are equal, the charged species have a uniform movement.

If q is the charge of the ion, the electrical force **F** expressed in the module is:

$$F = q\,E \quad \text{(in modules)}$$

The frictional force is given by the law of Stokes:

$$F' = 6\,\pi\eta r\,v$$

Where η is the viscosity coefficient of the medium, r is the radius of species supposed to be a spherical particle and v is the speed when **F** is equal to **F'**. In this case, it is possible to write:

$$q\,E = 6\,\pi\eta r\,v$$

And
$$v = q\,E/6\,\pi\eta r$$

If the field is kept for a long time enough to neglect the too small time necessary to obtain the uniform conditions, the travel d of the species at the time t is expressed as:

$$d = v\,t$$

Or
$$d = q\,E\,t/6\,\pi\eta r$$

It is usual to characterize the travel of the species by eliminating the non-experimental parameters t and E. Thus, one defines the mobility μ of the species by the expression:

$$\mu = d/E\,t \qquad (1)$$

The μ is the travel during the unit time under the influence of an electric field of one unit. Consequently, one can write:

$$\mu = q/6\,\pi\eta\,r \qquad (2)$$

If we multiply the numerator and denominator of the expression (2) by the factor ε_r, the relative permittivity of the medium, the (2) relation can be written as:

$$\mu = q\,\varepsilon_r/6\,\pi\eta r\,\varepsilon_r$$

If one remembers that the electrical potential ψ at the surface of a conducting sphere with r as a radius and q as a charge is given by the relation:

$$\Psi = q/r\,\varepsilon_r$$

The mobility μ of the species supposed as spherical can be also written:

$$\mu = \varepsilon_r\,\Psi/6\,\pi\eta \qquad (3)$$

where Ψ is the electrical potential at the surface of the species.

Inadequacies of the Elementary Theory

It is an experimental fact that the mobility of the species calculated from the experimental parameters from Equation (1) is systematically different from one calculated via Equation (2). There are many reasons for that:

- Firstly, the radius r of the species is very difficult to know above all for the large molecules such as proteins, poly-nucleotides and biopolymers which are the most studied by electrophoresis. Moreover, it appears experimentally that the radius r varies with the ionic strength of the solution.

- Secondly, it is also difficult to know exactly the charge of the species, taking into account often the presence of numerous acid and basic groups in the species in which the dissociation is a function of pH, their pKa and also the ionic strength.

- Finally, species can be delayed or accelerated by the electro-osmotic phenomenon. Moreover, in electrophoresis, the distance to travel is always superior to that measured due to a winding path to browse through the pores of the support.

A More Advanced Theory of Electrophoresis

In reality, the previous model of the spherical particle with a radius r and a charge q is not satisfactory when the radius r represents only the radius of the particle itself. A physical model more in line with the experiment than the previous one is that of Gouy, Chapman and Stern, whereby the radius of the particle must involve not only the species but also many layers of ions that screen in the solution.

The result is another expression of the electrical potential at the surface of the particle and so another expression of the mobility, as it will be seen.

According to this theory, the central ion with a radius r (the same as above) is encompassed by a layer of ions firmly bonded by electrostatic attraction to the particle. The radius of the central ion and the thickness of this layer give a radius superior to r. Moreover, beyond the layer firmly bonded, there exists an ionic cloud called diffuse layer where ions take place less bonded than the previous ones.

The total radius is symbolized by $\chi - 1$ (Figure 3).

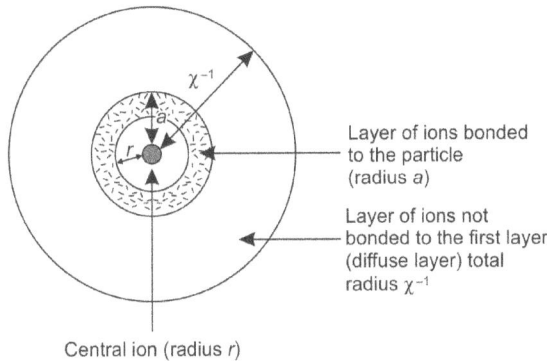

Figure 3: Scheme of the Radius a and χ^{-1} in the Theory of Gouy, Chapman and Stern. (Reprinted with Permission from Gwenola and Jean-Louis Burgot, Paris, Lavoisier, Tec et Doc, 2017, 293).

With this model, one sees with the electrostatic and thermodynamic laws (Debye and Huckel theory) that the potential ξ, which is necessary to take into account, is given by the expression:

$$\xi = q_d/\varepsilon_r \chi^{-1}$$

$$\text{with} \quad a \ll \chi^{-1}$$

Here, ξ is called electro-kinetic potential. The q_d is the charge density at the surface of the sphere equivalent to all the layers previously cited. The reverse of χ or $\chi - 1$ is called the average radius of the ionic cloud, which screens the central species. It is worth noting that this relation is more valid than for $a \ll \chi^{-1}$. With its new potential the mobility is written:

$$\mu = \varepsilon_r \, \xi/6 \, \pi\eta \tag{4}$$

This theory is more satisfactory than the previous one. In particular, theory demonstrates that χ depends on ionic strength I of the medium according to the expression:

$$\chi = k\sqrt{I}$$

The k is a proportionality constant. It legitimates, thus, the fact that the higher the ionic strength is, the lower the mobility and slower the electrophoresis.

Electro-Osmosis

Let us consider a flat surface in the bulk of the solution, which contains a density of diffuse charges q_d at a distance χ^{-1} of the wall of the capillary. This is supposed to be flat and has a charge density is $-q_d$ (Figure 4).

From the viewpoint of electrical loads, it is equivalent to a flat condenser and it is valid to consider the potential of one of the arms (this of the solution) as zero value, and the potential of the

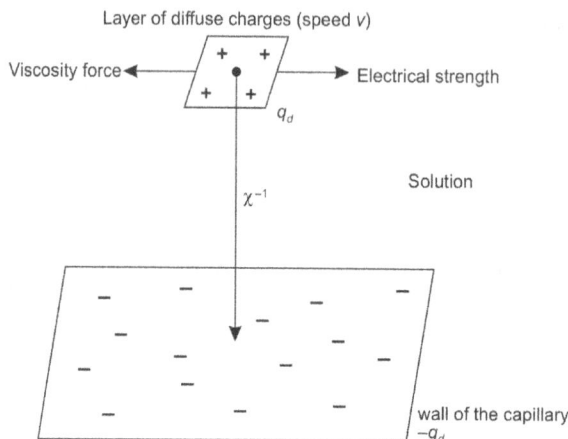

Figure 4: The Forces During Electro-Osmosis Phenomenon. (Reprinted with Permission from Gwenola and Jean-Louis Burgot, Paris, Lavoisier, Tec et Doc, 2017, 294).

other equals the potential ξ. The ξ is given by the expression of the difference of potential, which applies as a function of the distance χ^{-1} at the interior of a flat condenser (see 'Electrostatics'):

$$\xi = 4\pi q_d\, \chi^{-1}/\varepsilon_r \qquad (5)$$

The diffuse charges layer is undergone at the electrical load $q_d E$ (in the module). The layer begins to move, but it is retarded by the viscosity forces which are written as:

$$F' = \eta(dv/dx)$$

dv/dx is the gradient of speed driven by the retarding frictional force. If the speed of the fluid at the distance χ^{-1} of the wall is v, one can write with an approximation of the linear gradient:

$$dv/dx = v/\chi^{-1}$$

dx is a part of length in the perpendicular direction to both arms. One obtains:

$$F' = \eta v/\chi^{-1}$$

When the movement becomes uniform $F = F'$ and $\eta v/\chi^{-1} = q_d E$

And by introducing in this relation χ^{-1} as a function of ξ (5), one finds:

$$v/E = \varepsilon_r\, \xi/4\, \pi\eta = \mu_e \qquad (6)$$

v/E is the electroosmotic mobility μ_e or electroosmotic velocity divided by electric field strength. The μ_e depends on the potential ξ.

The electroosmotic mobility is not the same as the electrophoretic mobility (4) considering factors 4 and 6.

This is in a little contradiction with the affirmation at the beginning of the paragraph which considered electrophoretic and electroosmotic phenomena as the same. In reality, the presence of a factor of 4 or 6 at the denominator originates from purely geometric considerations and more precisely the size of the particle. The mobility for all the electrophoretic phenomena must be written:

$$\mu = \varepsilon_r\, \xi/f\, \pi\eta$$

Here, f is a numerical factor. If the particle is small, it is the spherical model which is the best ($f = 6$). If the particle is great, the double layer whose screens are closer than that one finds in electroosmotic phenomena and $f = 4$. In this case, the particle can be likened to a parallel-plate condenser than to a spherical condenser. It makes the value of the ratio a/χ^{-1} conditional upon the value of factor f.

Factors Affecting Electrophoretic Mobility

Nature of the Molecule of Interest

- Size: The small molecules migrate more easily, but their mobility also depends on other components in the solution subject to changing their solvation properties and thus decreasing their velocities.
- Charge: The charge affects the mobility of molecules. The charge is a function of pH for ionizable molecules and ionic strength and depends on the possible formation of complexes. The sign of the charge gives the direction of the migration.

Ionic Composition of Electrophoretic Buffer

- **Ionic strength**: It occurs via the value of the χ parameter. Based on the facts set out above, when χ increases, ξ decreases and the electrophoretic mobility decreases. This is precisely the case with the increase of ionic strength. But the presence of ions is necessary to obtain a flow of current, a compromise must be found with an ionic strength between 0.05 and 1 mol.L^{-1}.
- **pH**: The pH affects the ionization of weak acids and bases for which it is necessary to work in a buffered medium. For example, the modification of the electrophoretic mobility as a function of pH is described:
  Either a weak acid HA which gives in an aqueous solution:

$$HA + H_2O \leftrightarrows H_3O^+ + A^-$$

$$Ka = (|H_3O^+||A^-|)/|HA|$$

$$pH = pKa + \log |A^-|/|HA|$$

Where it is possible to assimilate activities and concentrations.

There is a shift of the ionization equilibrium all along the travel of the weak acid between both electrodes. Thus:

- A movement of ions A$^-$ toward the anode gives a shift of the equilibrium to the ionization of HA;
- On the contrary, any excess of A$^-$ ions leads to repressing or backward the ionization of HA.

The consecutive equilibria between HA and A$^-$ lead to decreasing electrophoretic mobility of A$^-$.

- **Nature of the buffer and the solution**: For constant values of pH and ionic strength, two buffers of different nature do not always give the same electrophoretic mobility. Some compounds added to the migration buffer modify also the electrophoretic behaviors. For example:
  - E.d.t.a or citric acid enables, by complexation, the separation of some mineral ions;
  - Ions issued from the complexation of carbohydrate with boric acid enable their electrophoretic separation;
  - The addition of urea modifies the electrophoretic behavior of macromolecules by breaking the hydrogen bonds between them.

Support

Some supports have some adsorbent properties and can fix the solvent or solute molecules. This results in a decreasing migration with a broadening of the band on the electropherogram. The adsorbent properties are linked to the presence on the support of some functional groups such as hydroxyl.

Electric Field Strength

The electric field strength constitutes the potential drop V by the unit length between both electrodes separated by a length l. As E = V/l, the higher the field is, the closer electrodes are. The medium between both electrodes gives a resistance R with:

$$V = R\,I$$

In considering that the higher the difference of potential V is, the higher the electrical field strength and the quicker the molecule migrates. However, the current I increases with V giving an increase in the temperature of the medium by energy dissipation in the form of heat. On these two facts, it is necessary to choose a voltage sufficient to have a quick migration but not so high as not to generate too much heat by the Joule effect. The heat release brings about the evaporation phenomena of the buffer solution. This is followed by a modification of viscosity with precipitation of slightly soluble water substances and also by a change of electrophoretic travel. Thus, if the temperature of the medium is not homogeneous, the velocities of migration become unequal generating some distortions of the band with a decrease in the resolution. Moreover, the increase in the temperature produces some convection currents. Some supporting media are more sensitive than others to the increasing temperature as is the case with the polyacrylamide gels.

Electromigration Techniques

Separation Techniques Based on Differences in Mobility

Electrophoresis in Solution

Moving-boundary or free-boundary electrophoresis is the early electrophoresis method that has been developed by Arne Tiselius (1930). The sample solution is introduced in a U-shaped glass tube and covered by a buffer solution with a lower density than that of the sample to avoid convection currents (Figure 5). Electrodes are immersed in the buffer at both ends of the tube where one applies an electric field strength. The charged species of the sample migrate freely for a time *t* and a non-completely separation of the species is obtained in each arm of the tube. This technique is little used nowadays and may help to know the electrophoresis mobility.

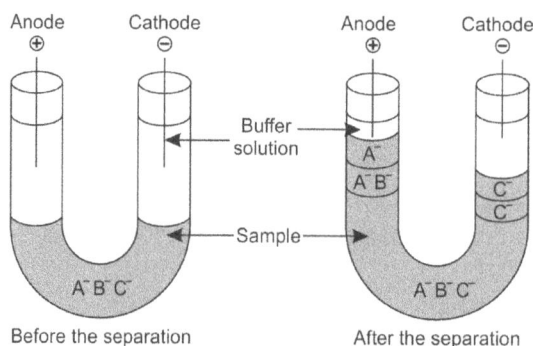

Figure 5: Electrophoresis in Free Solution. (Reprinted with Permission from Gwenola and Jean-Louis Burgot, Paris, Lavoisier, Tec et Doc, 2017, 298).

Electrophoresis on Slab

A new electrophoresis method for which the migration of molecules is realized on a porous solid supporting media impregnated with buffer or gel matrix enables the separation of molecules in a stable band or zone. The method is named "slab electrophoresis" or "electrophoresis". The mobility obtained by slab electrophoresis is lower than that of moving boundary electrophoresis for two reasons:

- Effect of supporting media;
- The electro-osmosis phenomenon is higher in the second case.

The identification of the substances is realized by the migration distances, and the quantitative analysis is possible after the coloration of zones.

The nature of supporting media is so diversified. The gels are the most used. Let us cite some examples:

- Agarose gel is mainly used for preparative electrophoresis, immune-electrophoresis and electrophoresis of DNA and proteins of molar mass superior to 300.
- Polyacrylamide gel or synthetic gel is the support that is most used owing to its high mechanical resistance, the absence of adsorption and electro-osmotic phenomena and the reproducibility of its porosity.

It is possible to realize a gradient of porosities to improve the separation of the molecules as a function of electrophoretic mobilities and sizes.

To improve the separation of molecules, it is possible:

- Working with a pulsed field. It means that the electric field strength is led alternatively during a short time in a direction and then in opposite direction. The change of direction permits the molecules with a high molar mass to travel more easily to the pores of the gel.
- Making *bi-dimensional electrophoresis* or 2D. For example, for proteomic analysis or cell proteins studies, two electrophoreses are realized sequentially to improve the resolution of the separation:
  - Firstly, electrophoresis by isoelectric focalization (see the following section on "Separation Techniques on Based on the Difference in Isoelectric Point") on a polyacrylamide gel which separates proteins as a function of charge.
  - Secondly, the gel of polyacrylamide is impregnated with sodium dodecyl sulfate which gives the proteins a negative charge. One applies an electric field that separates proteins according to the molar mass only.

Isotachophoresis

The isotachophoresis, from Greek *iso* (equal) and *tachos* (velocity), means that the molecules migrate with the same velocity v. As these molecules have not all the same electrophoretic mobility μ, it means that it is necessary to apply different electrical fields in order to have:

$$v = \mu_1 E_1 = \mu_2 E_2 = \ldots\ldots\ldots\ldots = \mu_n E_n$$

The gradient of the field is obtained by using two different buffers:

- Leading electrolyte (M) which has mobility superior to that of anions of the sample (for example, 2-morpholinoéthane sulfonate):

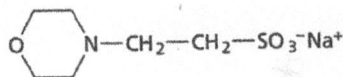

$$O \quad N\!-\!CH_2\!-\!CH_2\!-\!SO_3^-Na^+$$

- Terminating electrolyte (T) which has electrophoretic mobility inferior to that of anions of the sample (for example: 7-aminocapronate):

$$Na^+\,{}^-OOC\!-\!(CH_2)_6\!-\!NH_2$$

Both electrolytes have the same counter-ion Na^+ when one applies a voltage between both electrodes, the counter-ion migrates to the cathode and both anions must migrate with the same velocities. The result is that owing to their different electrophoretic mobilities, the development to keep the electroneutrality of electrical fields, which are different, and the establishment of a field gradient. The sample is put at the interface between leading and terminating ions. The technique permits the analysis of ions with the same sign of charge since the choice of electrolytes depends on the sign of their charge.

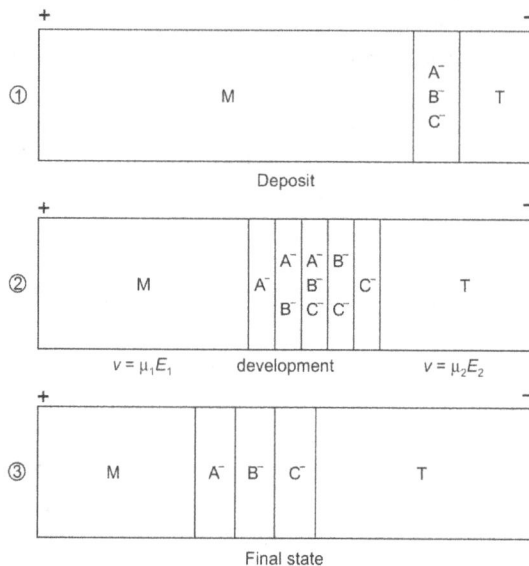

Figure 6: Isotachophoresis. (Reprinted with Permission from Gwenola and Jean-Louis Burgot, Paris, Lavoisier, Tec et Doc, 2017, 300).

Immunoelectrophoresis

This technique allows separation of the components of a sample by electrophoresis and to detect them by immunological reaction and antigen-antibody reaction whose precipitate can be seen on a gel.

Separation Techniques Based on the Difference in Isoelectric Point: Isoelectric Focusing

The amphoteric molecules move through an electrical field along a pH gradient to the zone in which the pH is equal to their isoelectric point. The isoelectric point is the value of pH for which the molecule has no net charge. One realizes a pH gradient by impregnating a gel with a mixture of amphoteric compounds substances with different isoelectric points. When one applies the voltage, the different ampholytes pull over in order of their isoelectric point and realize a pH gradient between both electrodes.

Isoelectric focusing has a high power of resolution which permits the separation of serum proteins in many bands.

Instrumentation

The instrumentation is made of (Figure 7):

- A power supply;
- A thermostatically-controlled box with solvent cells.

The highlighting of the colored bands is performed by colorimetry with Soudan black for lipoproteins and acids colorings for proteins, such as Coomassie blue (azo-naphthalene derivative) and orange acridine (di-benzo-pyridine derivative) for polynucleotides with heavy metals (silver

Figure 7: Electrophoresis Apparatus. (Reprinted with Permission from Bioblock Society, 67403 Illkirch).

salt) or fluorescent chips. The use of colorings allows the detection limits of 10 ng for proteins, 1 ng with silver salts and thus 2 ng with the fluorescent chips.

The quantification of colored bands is realized with a densitometer. The extraction of colored bands with the digestion of the proteins by an enzyme protease and introduction of proteins in a mass spectrometer. The ionization of peptides is realized by electrospray or by a technique of desorption/ionization by laser-assisted by matrix (MALDI) ionization mode which is widely used for biomolecules. This process uses a pulsed laser beam for ionizing the molecules on mono or multi-charged ions, such as type $(M + nH^+)$. The analyzer is most used with the ionization process MALDI which uses the principle of time of flight. With ionization electrospray, one uses rather the quadrupole, in particular, MS-MS (see Chapter 42).

Capillary Electrophoresis

Principle

Capillary electrophoresis is a compromise between conventional electrophoresis, described earlier, and high-performance liquid chromatography. Since the end of the 1980s, there has been significant growth. Its applications field is larger than those of conventional electrophoresis because in some conditions it is possible to separate neutral species. Thus, capillary electrophoresis is more environment-friendly than liquid chromatography because it does not use organic solvents.

The principle is based on the differential migration, of the electrical field, neutral or charged species in a narrow capillary filled with electrolytes solution.

Migration Mechanisms

The travel of substances depends on both phenomena:

Electrophoretic Migration

As for conventional electrophoresis, ions and charged particles move with a velocity that is a function of their mass, and their charge is given by:

$$V = \mu_{ep}E$$

where μ_{ep} is the electrophoretic mobility.

Electro-Osmotic Migration

Electro-osmotic migration, an artifact phenomenon in conventional electrophoresis, is responsible for separations in capillary electrophoresis. Thus, the wall of capillaries in fused silica is covered with negatively charged silanol groups (SiO^-) for pH superior to 2. Silanol groups are ionized in contact with water in SiO^-.

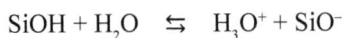

$$SiOH + H_2O \leftrightarrows H_3O^+ + SiO^-$$

The equilibrium is a function of the pH of the solution and the number of SiO⁻ increases with the pH increase. The cations of the electrolyte solution bind to the surface of the negatively charged wall by electrostatic attraction. The cationic layer moves under the electrical field toward the cathode and leads the entire solution, solvent and sample molecules. Consequently, the flow which is created gives the particles an electroosmotic mobility μ_0 adding to the electrophoretic mobility.

The intensity of the electroosmotic flow changes in the function of pH (Figure 8). The flow increases strongly between pH 3 and pH 7 until a plate is reached.

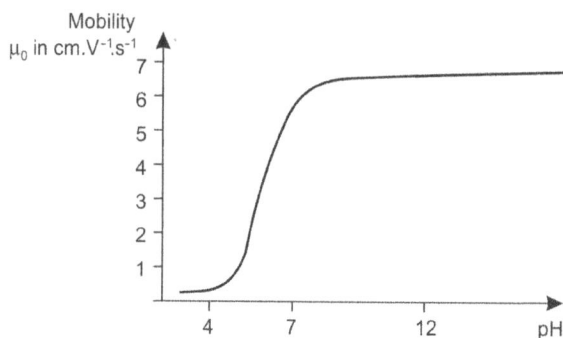

Figure 8:   Evolution of the Electroosmotic Mobility μ_0 in the Function of the pH. (Reprinted with Permission from Gwenola and Jean-Louis Burgot, Paris, Lavoisier, Tec et Doc, 2017, 303).

The travel of a solute is thus dependent on the combination of two phenomena. One defines apparent mobility as:

$$\mu = (\mu_{ep} + \mu_0)$$

$$\text{and} \quad v = \mu E = (\mu_{ep} + \mu_0)\, E$$

With μ_{ep}: electrophoretic mobility and μ_0 = electroosmotic mobility.

Thus, if an injector is positioned near the anode and a detector near the cathode:

- For charged positively species, the apparent mobility is superior to the electroosmotic mobility:

$$\mu = (\mu_{ep} + \mu_0) > \mu_0$$

- For charged negatively species, the apparent mobility is inferior to the electroosmotic mobility:

$$\mu = (\mu_{ep} + \mu_0) < \mu_0$$

- For neutral species, the apparent mobility is equal to the electroosmotic mobility:

$$\mu = (0 + \mu_0) = \mu_0$$

From a practical standpoint, the electroosmotic flow is a positive ions flow. The injector is placed closed to the anode and the detector is closed to the cathode to have a very important movement for the particles.

Instrumentation

Figure 9 gives a general scheme of a capillary electrophoresis apparatus. Let us consider the major parts of it with capillary, injector and detector.

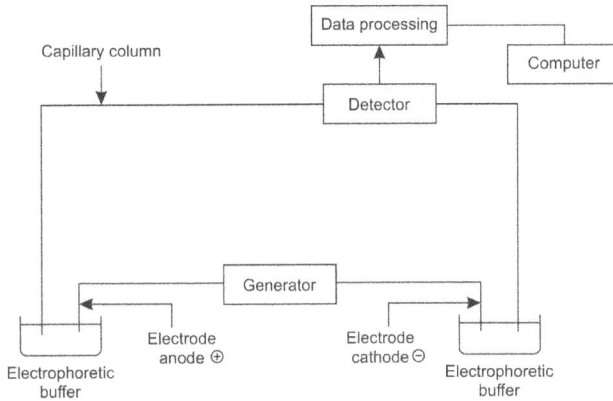

Figure 9: Scheme of an Apparatus of Capillary Electrophoresis. (Reprinted with Permission from Gwenola and Jean-Louis Burgot, Paris, Lavoisier, Tec et Doc, 2017, 304).

Capillary

Capillaries are generally in fused silica with a polyimide coating. They have an internal diameter between 10 to 100 μm and a length from 25 to 100 cm. They are filled with an electrophoretic buffer (Figure 10) and extend between two buffer cuvettes in which there are dipped electrodes. The quartz or fused silica does not absorb the UV radiation and has good conducting heat. The low diameter of the capillary gives an important ratio of surface/volume which promotes the heat dissipation produced by Joule heating. It can be thus possible to apply a high voltage of some KV. Consequently, the duration of the analysis decreases because the migration velocities become still greater.

In some applications, one uses, as in HPLC and cross-linked silica by octadecyl groups (C_{18}) to decrease the number of free silanol groups as well as modulate the intensity of the electroosmotic flow.

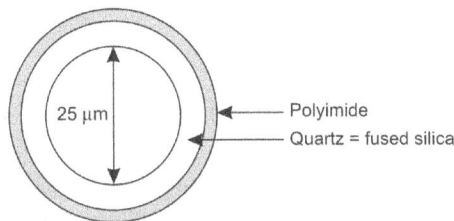

Figure 10: Scheme of a Cross-Sectional Area of Capillary Electrophoresis. (Reprinted with Permission from Gwenola and Jean-Louis Burgot, Paris, Lavoisier, Tec et Doc, 2017, 305).

Injectors

The volume of the sample introduced at one end of the capillary must not exceed 1 to 2 μl to avoid the broadening of the peaks. Two types of injectors are used:

- Hydrodynamic injection:
  - By raising height Δh, the tube of the sample in which the capillary dips during a very short time Δt. The volume (and the amount of the substance) introduced into the capillary is a function of ΔV and Δt.
  - By applying a difference of pressure between ends of the capillary during a time Δt, it is possible to introduce the sample into the capillary. With this process, the introduced volume is a function of the viscosity of the sample.

• Electrokinetic injection:

For viscous solutions, the injection is realized by the application of a voltage ΔV between both ends of the capillary during a very short time Δt for driving the sample into the capillary. The injected volume is a function of ΔV and Δt. The drawback of this technique comes from the fact that most mobile ions have concentrations that are more important than their initial concentration. The method is not used for quantitative analysis but can be interesting to concentrate and identify trace products.

Detectors

The detection can be realized through the capillary or with an interface. To install the detector on the capillary, it is necessary to eliminate the externally protected polyimide (Figure 11).

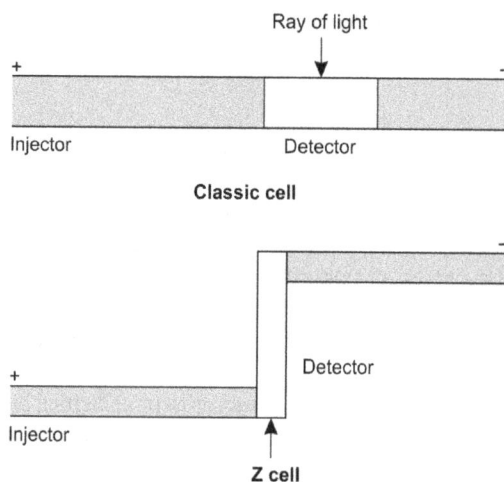

Figure 11: Scheme of Two Positioning of the detector on the Capillary (Reprinted With Permission From Gwenola and Jean-Louis Burgot, Paris, Lavoisier, Tec et Doc, 2017, 306).

The most common detectors used are of the same type as those described in HPLC such as:

• UV-Visible spectrometer, owing to the lack of organic solvent and oxygen, it is possible to use wavelengths from 190 nm. For inorganic anions and cations analysis which do not absorb most of them in UV, one adds to the electrolyte a chromophore and the most often dichromate ion for anions and imidazole and histidine for cations. It is possible to improve the resolution of the separation of cations to complexing with crown-ethers.

The optical path, equivalent to the diameter of the capillary, is very low. To improve the sensitivity of the method, it is possible to increase it with a Z-cell.

The UV lamp is sometimes substituted by an electroluminescent diode with a higher lifetime than the deuterium lamp and is capable of miniaturization.

• Fluorometer (fluorescence induced by laser) in direct mode or indirect with the addition of fluorescein or rhodamine B in the electrolyte;

• Electrochemical detectors;

• Mass spectrometer with electrospray ionization and quadrupole analyzer.

The recorded signals provide an electropherogram very close to chromatograms (Figure 12).

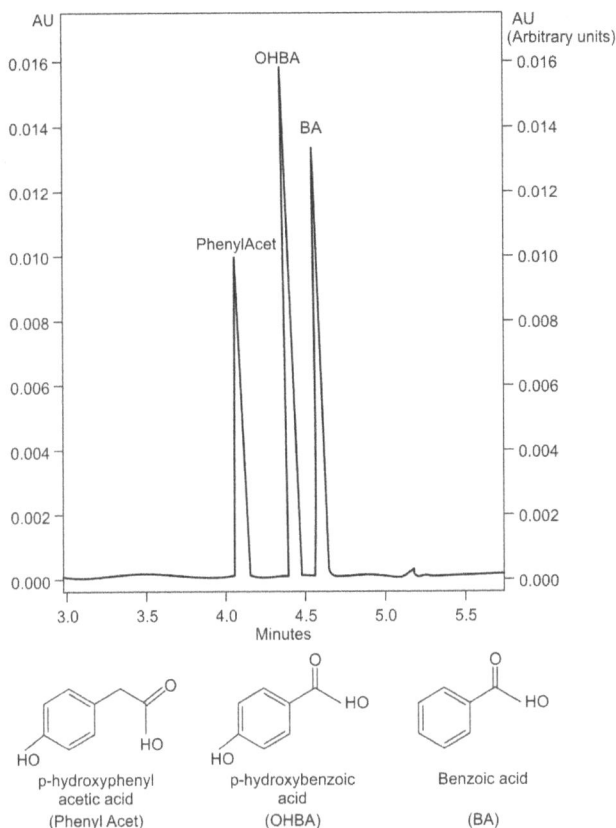

Figure 12: Electropherogram of a Mixture of P-Hydroxy-Phenylacetic Acid, P-Hydroxybenzoic Acid and Benzoic Acid. (Reprinted with Permission from Beckman Coulter France SA, ZA Paris Nord 2-33/66 BP54359, 22 avenue des Nations, 93420 Villepinte).

Modes of Separation in Capillary Electrophoresis

Capillary Electrophoresis

It is the most common method which corresponds to the process described previously. The composition of the buffer (phosphate, citrate or borate) is homogenous on the entire capillary and the electrical field is constant during the analysis. The separation of ions and neutral species is achieved with a combination of electrophoretic and electroosmotic migrations (see the section on 'Electro-Osmotic migration'). Cations are separated firstly, then are the neutral species altogether with the same velocity and finally anions (Figure 13). This technique does not permit the separation of neutral species from each other.

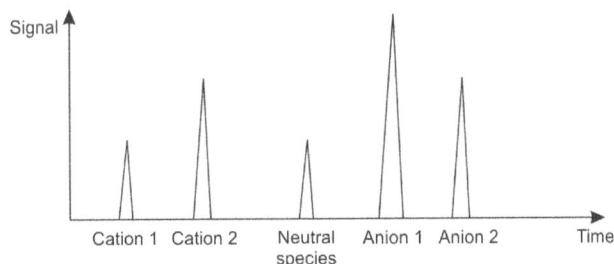

Figure 13: Separation of a Mixture by Capillary Electrophoresis. (Reprinted with Permission from Gwenola and Jean-Louis Burgot, Paris, Lavoisier, Tec et Doc, 2017, 308).

Capillary Gel Electrophoresis

Some biopolymers such as RNA or single or double-stranded DNA cannot be separated by the previous technique because they possess a ratio charge/size similar. Their separation can be achieved by a combined action of an electrophoretic separation and a molecular sieve by filling the capillary with a gel, such as a polyacrylamide gel which gives also the separation of molecules as a function of their size.

Electrokinetic Chromatography (EKC)

This process is a hybrid technique introduced in 1984 by Terabe which enables the separation of the neutral species in the same conditions as those used for the separation of cations and anions with the use of electroosmotic flow. It is also interesting for the separation of charged species that possess similar mobilities.

One adds to the buffer an ionic surfactant such as sodium dodecyl sulfate (SDS) at a concentration superior to its critical micellar concentration in order for the SDS to combine with the studied species form spherical aggregates named 'micelles'. Micelles are made of hydrophilic groups at the exterior and hydrophobic groups in the bulk of the micelles.

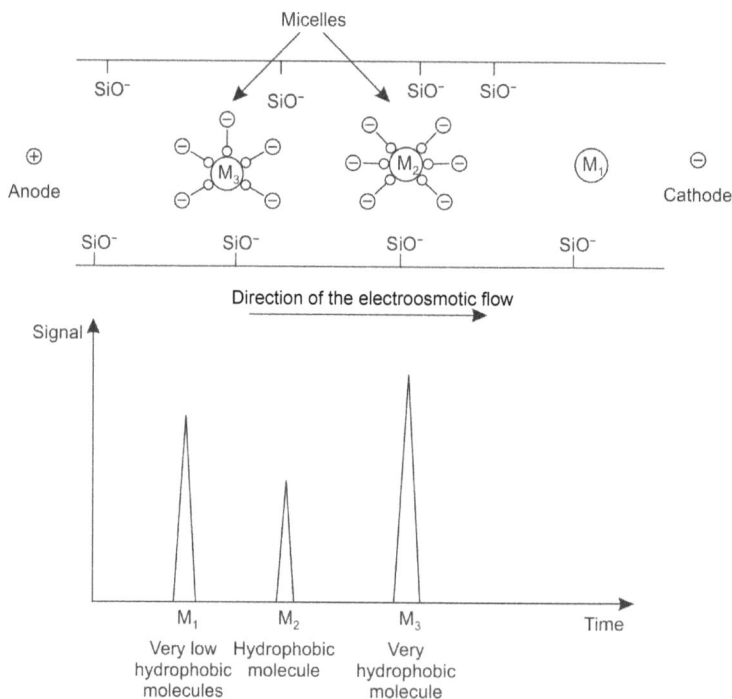

Figure 14: Micellar Electrophoresis. (Reprinted with Permission from Gwenola and Jean-Louis Burgot, Paris, Lavoisier, Tec et Doc, 2017, 309).

For the separation, species share between the micellar phase and the solution as a function of their hydrophobicity. In this technique, the separation is based on the combination of both parameters which are the charge and the hydrophobicity.

Through the solubilization power of micelles, some complex samples such as urine or blood plasma can be directly injected into the capillary.

Capillary Isoelectric Focusing (CIEF)

In this process, the capillary is filled with a mixture of ampholytes which are able to create a pH gradient. Substances migrate into the capillary until they find a similar pH to their isoelectric points (pH where their charge is null). After that, it is possible to move separated species toward the detector by hydrostatic pressure or electrolytes changing.

Capillary Isotachophoresis (CITP)

As for electrophoresis (see the section on 'Isotachophoresis'), the separation is based on an electrical gradient.

Capillary Electro-Chromatography (CEC)

Capillary electro-chromatography is a technique of separation implementing mechanisms of capillary electrophoresis and liquid chromatography. The capillary is filled with a stationary phase and the mobile phase is hydro-organic. Substances move under the effect of the electric field, and the pump system is no longer necessary to realize the shift of the liquid. The diameter of the particles can be lower, improving the resolution of mixtures. The micrometric particles of the stationary phase can be also substituted by monolithic phases prepared, for example, from methyl trimethoxy-silane.

   Neutral species are going to share as in liquid chromatography between the stationary phase and mobile phase, while the separation of charged species combines both phenomena of partitioning and electrophoretic migration.

Reversing the Direction of Electroosmotic Flow

By adding to the buffer 1,3-diaminopropane diprotonated, it is possible to modify the direction of the electroosmotic flow which becomes oriented toward the anode. That permits a better resolution for anions (Figure 15).

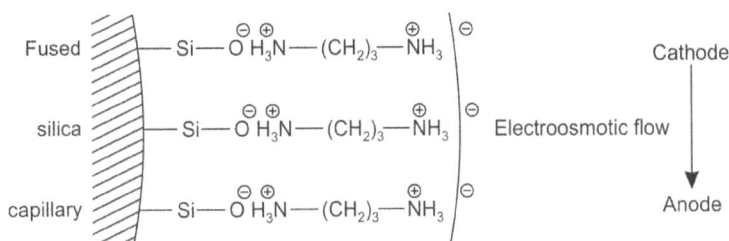

Figure 15: Reversing Electroosmotic Flow. (Reprinted with Permission from Gwenola and Jean-Louis Burgot, Paris, Lavoisier, Tec et Doc, 2017, 310).

   The 1,3-diaminopropane diprotonated gives ionized silanols some high interactions to fix it on the wall of the capillary.

Factors to Modify the Separation

Some parameters affect the separation such as:

- The pH of the buffer: It operates at two levels:
  - Firstly, on the ionization degree of molecules;
  - Secondly, on the intensity of the electroosmotic flow by the number of silanol groups that are ionized.

- Voltage: It operates directly on the value of the electrical field and consequently the mobility of the molecules.
- Geometrical characteristics of capillary: The diameter and length of the capillary have an influence on the running time of separation and the resolution. Peaks become more symmetrical when one increases the diameter of capillaries but at the same time, the resolution decreases jointly. When one wants to decrease the duration of analysis, it is possible to use capillaries shorter and increase at the same time the voltage. But the technique has some limits because the shorter capillary is, the faster the electrical resistance decreases. That gives an increase in the current intensity and consequently an increase in the heat joule.

Characteristics of the Technique

Advantages

- The analysis duration is around 5 to 20 minutes. The method is quicker than chromatographic separation which the implementation needs a certain time, especially to equilibrate the temperature and the equilibria of molecules. Thus, it appears possible to separate 15 anions in less than 10 minutes with a detection limit of around ppm.

- Resolution: The plate number for capillary electrophoresis is often superior to those obtained with the other separation techniques, especially chromatography. That can be explained by the fact that the general formula which gives the plate heigh in the function of the linear velocity of the mobile phase μ:

$$H = A + B/\mu + C\mu \quad \text{becomes: } H = B/\mu$$

In capillary electrophoresis because the only broadening source of the peaks is the length diffusion because it has no stationary phase.

- A low volume of samples injected: As with gas chromatography, the injected volume of samples is low between 1 to 10 μL and even sometimes around nL. The injected volume does not exceed 1% of the capillary volume. Consequently, with this low volume, it will be necessary to use an internal standard for each experiment.

- Separation to the room temperature: It is interesting for thermo-labile molecules. But the control of the temperature is more important in electrophoresis than in HPLC.

- Low cost of experiments: The absence of using organic solvents decreases the costs in comparison with chromatographic methods.

Drawbacks

- Sensitivity: The capillary electrophoresis is less sensitive than HPLC even if the aqueous solvent permits the use of very low UV wavelength toward 190 nm with the spectrometer detectors.

- Precision: For quantitative analysis, the precision of CE is inferior to that of chromatographic methods (HPLC and GC) owing to the low injected volume and even with an internal standard. The other difficulty is obtaining a reproducible electroosmotic flow and that contributes to the lack of precision.

Applications

They are very numerous and cover many fields such as:

- Hydrology: The analysis of mineral ions (Ca^{2+}, Mg^{2+}, Cl^-, SO_4^{2-}) and pesticides;
- Pharmaceutics: Research impurities in pharmaceutical ingredients, quantification of active pharmaceutical product (API) in bulk products (Figure 16) and the optical purity of API (Figure 17).

Buffer : 50 mM Bis-Tris Propane
: 50 mM TTAB, MeOH, pH 6.5
Column : 30/37 cm
Voltage : 18.5 kV
UV : 214 nm
Sample : 7-antidepressant drugs

1. Impurity
2. Doxepine
3. Nordoxepine
4. Imipramine

5. Desimipramine
6. Amitriptyline
7. Nortriptylin
8. Protriptyline

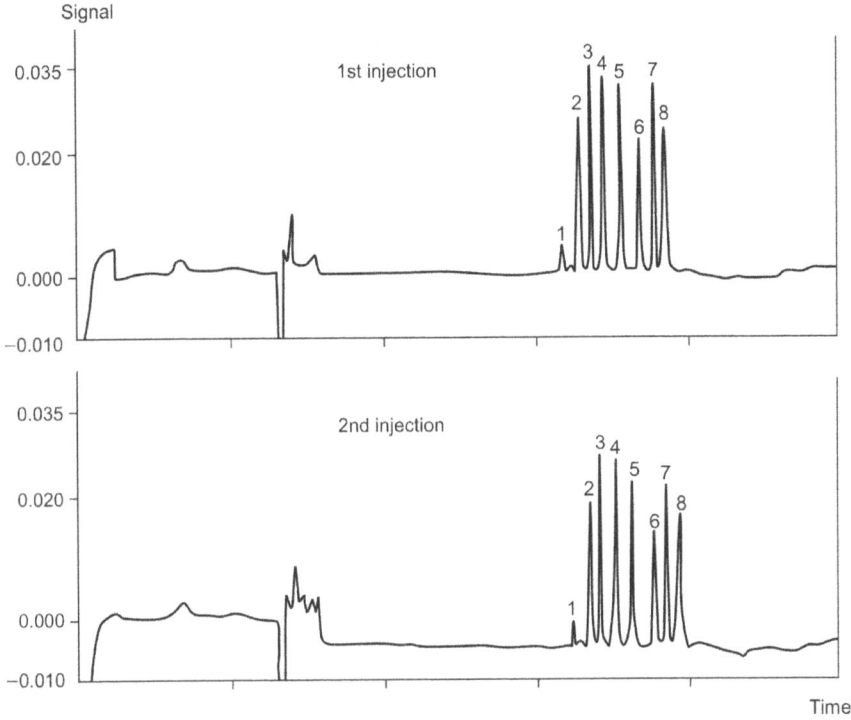

Figure 16: Separation of a Standard Mixture of Antidepressant Drugs with Reversing the Direction of Electroosmotic Flow. (Reprinted with Permission from Beckman Coulter France SA, BP54359, 22 Avenue des Nations, 93420, Villepinte).

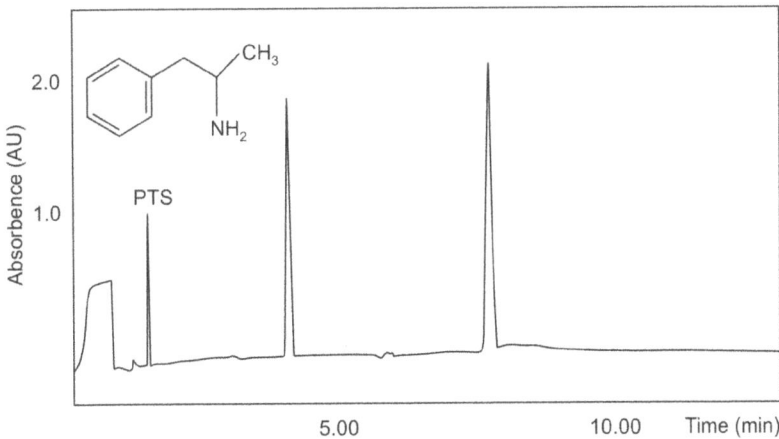

Figure 17: Separation of both Enantiomers of Amphetamine with γ-Cyclodextrin Sulfate. (Reprinted with Permission from Beckman Coulter France SA, BP54359, 22 Avenue des Nations, 93420, Villepinte).

- Biological: Serum and urine proteins (SDS PAGE and SDS give a negative charge to proteins that are separated only by size), lipoproteins (LDL and HDL markers for hypercholesterolemia), nucleic acids (DNA, RNA), therapeutic drug monitoring (phenytoin, lamotrigin, valproic acid and digoxin) and toxicological research (barbituric, paracetamol and salicylate).
- Physico-chemical constants: The pKa, kinetic constants and isoelectric point.

Part II
Spectral Methods

Introduction to the Spectral Methods

A major class of analytical methods is based on the interaction of radiant energy with matter. From a purely physical standpoint, one distinguishes indeed two kinds of grandeurs intervening in this proposition. They are:

- The matter which is made of nuclei, atoms and molecules;
- The radiations.

Both kinds of entities can exchange energy.

One knows that the functional analysis has for goal to establish the developed formula of an organic compound, once its purity and eventually its elementary analysis and its molecular mass have been determined. These two last conditions are less imperative today since some methods of functional analysis, such as molecular mass spectrometry, can provide these data.

Spectral methods, which are physical methods, are also a part of the methods of functional analysis together with some chemical methods.

But, as we shall see in the next chapters, the part played by spectral methods in analytical chemistry is larger than that devoted to the functional analysis. The great example behind this story is the multiple possibilities they offer in the domain of quantitative analysis.

Electromagnetic Radiations

Beforehand, it is interesting to recall some properties of electromagnetic radiation.

Nature of an Electromagnetic Radiation

Let us first recall that radiations propagate in space independently of any matter. In an empty space, they suffer from no loss of energy and no radiation can arise in it. Radiation cannot arise nor die without the existence of matter. The two fundamental actions, the birth of radiation and its absorption, involve an exchange of energy between radiance and matter.

Different kinds of radiant energies, which at first view are different from one another (such as radiometric waves, UV-visible, infrared, X and γ-rays); as a matter of fact, they exhibit similar properties. They are named *electromagnetic radiations*.

Electromagnetic radiation may be represented by an electric field and a magnetic field that can undergo in-phase sinusoidal radiations at right angles to each other and at a right angle to the direction of propagation (Figure 1).

The electric component of the radiation, due to the existence of the electric field, is responsible for most of the phenomena of interest for us, such as transmission, reflection, refraction, absorption, etc. However, in contrast, it is the magnetic component that is responsible for the absorption of radio frequencies in N.M.R. (nuclear magnetic resonance) and E.S.R. (electron spin resonance).

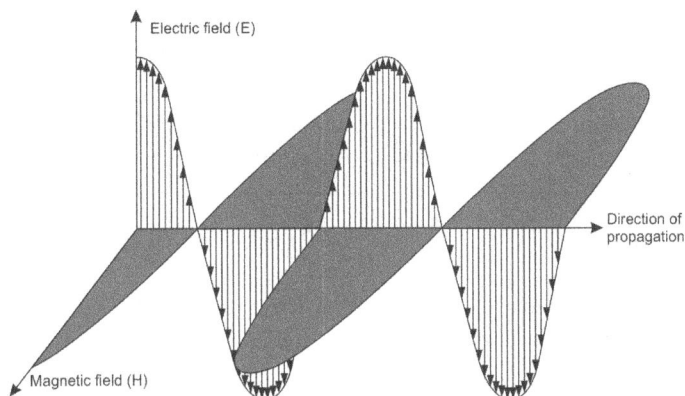

Figure 1: Nature of Electromagnetic Radiation.

Parameters of the Wave Part

Figure 2 permits recalling the usual parameters characterizing waves.

- The wavelength (symbol λ - unit: meter). It is the linear distance between any two equivalent points on successive waves. Now, one usually uses the micrometer 1 μm = 10^{-6} m and the angström unit: (symbol Å): 1 Å = 10^{-10} m.
- The frequency f or ν: Number of the oscillations of the field that occur per second. It is equal to 1/T; unit (the hertz) (1 Hz or cycle per second cps; 1 Hz = 1 cps).
- T is the period of the radiation. The usual unit of the period is the second (s).
- A is the amplitude of the sinusoidal wave. It is the length of the electric vector at the maximum of the wave.
- The velocity of the propagation v_i or C (m s^{-1}). It is equal to the multiplication of the frequency ν (in cycles per second) by the wavelength λ in meters per cycle.

$$C = \nu \cdot \lambda \qquad (1)$$

Let us recall that the frequency of radiation ν is determined by the source and is invariant. In contrast, the velocity of radiation depends on the composition of the medium through which it passes. As a result, one can prove the evidence that according to relation (1), the wavelength is also dependent upon the medium. In a vacuum, the velocity is at its maximum. It has been determined to be c = 2.99792 10^8 m/s. In the air, it differs very slightly from the value it possesses in a vacuum and, in both media, one can write safely C = 3.00 10^{10} cm/s.

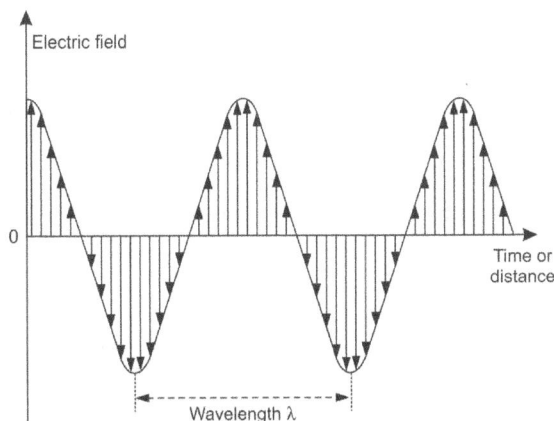

Figure 2: Planar Representation of the Electric Vector.

• The wavenumber σ in a medium is the reciprocal of the wavelength expressed in centimeter (unit: cm^{-1}; symbol $\sigma = 1/\lambda$). It is directly proportional to the frequency and thus to the energy of the radiation.

• The power P of radiation is the energy of the beam that reaches a given area per second, whereas the intensity is the power per unit solid angle (Appendix 3). Power and intensity are often used synonymously. This is not fully correct.

The Photoelectric Effect

• We now briefly take into account the photoelectric effect. The photoelectric effect is the emission of electrons from a surface irradiated by ultraviolet light. The kinetic energy of emitted electrons only depends on the frequency ν of the light received. In 1905, Einstein explained this experimental fact by admitting that the energy of radiation can be only absorbed by discrete quantities named '*quanta*'. One quantum of radiant energy of frequency ν has for value the product hν where h is Planck's constant ($h = 6.626\ 10^{-34}$ J s). Classical physics assumes that particles move along precisely defined trajectories according to Newtonian laws. It cannot explain some observed phenomena, such as the photoelectric effect.
(This fact was named *the failure of classical physics*).

• The success of the quantum theory led A. Einstein to consider that electromagnetic radiation can be also viewed as a stream of discrete particles (wave packets) of energy called *photons* where energy is proportional to the frequency of the radiation, photons propagating through space with the velocity of light.

The Failure of Classical Physics—Maxwell Equations

The behavior of electromagnetic radiation is described as being dependent on electrical and on magnetic fields $\mathbf{E}(x)$ and $\mathbf{B}(x)$. Both are interrelated by one of the Maxwell equations. They lead to the conclusion that in free space, both fields obey the equation:

$$(1/c^2\ \partial^2/\partial t^2 - \nabla^2)\ \langle\ \begin{matrix} \mathbf{E}(x) \\ \\ \mathbf{B}(x) \end{matrix}\ = 0$$

Where ∇^2 is the laplacian operator. This relation means that these fields propagate in space on the axis *x* as waves do usually with a constant speed C. $\mathbf{E}(x)$ and $\mathbf{B}(x)$ are vectors.

Bohr's Condition

Concerning the emission and adsorption spectra, there is a fundamental relation between of the energy differences and between of the fundamental and excited states of the molecule or atom and the frequency of the emitted or absorbed radiation. This is *Bohr's condition*:

$$h\nu = E_{ex} - E_f$$

Here, h is the Planck constant. E_{ex} and E_f are the energies of the absorbing species in the excited and fundamental states.

Fate of the Excited Molecules

Molecules, once excited can dissociate, either by exchanging energy (heat) with solvent molecules or by emitting radiations by falling back into the fundamental state. In this case, one speaks of *fluorescence* or *phosphorescence* phenomena, which are both called *luminescence* phenomena (see Chapter 22).

Nomenclature

There exists a certain confusion about the definitions of spectral methods of analysis, that is to say about some terms such as spectroscopy, spectrometry, spectrophotometry, spectrography, etc.

According to IUPAC, *spectroscopy* is the analysis of the electromagnetic radiations emitted, absorbed or scattered by atoms or molecules as they undergo transitions between two discrete energy states. It is, hence, a technique that permits the study of the elementary constituents of the matter based on their interaction with radiation most often electromagnetic.

Spectrometry refers to the measurement of the intensity of radiation with an electronic device. The term *spectrophotometry* applied in the past to spectroscopy by emission. It consisted of the analysis of all the metals and some other elements by comparison to another known spectrum registered on the same film, hence the name of the method. But this term is still used to qualify spectroscopies UV-visible and infrared, which, in contrast, are absorption spectroscopies.

Now, the term spectroscopy tends to be used whatever the spectral method it designates.

Spectral Regions

The principal regions of the electromagnetic spectrum are mentioned in Table 1. They are characterized by their wavelength (nm), their wavenumber σ (cm^{-1}) and their frequency (Hz).

Table 1: Principal Regions of the Electromagnetic Spectrum.

Region	Wavelength	Wavenumber	Frequency
Far-UV	10 to 200	10^6–50.10^3	30.10^3–1,500
Near-UV	200 to 380	50,000–26,300	1,500–790
Visible	380 to 780	26,300–12,800	790–385
Near-IR	780 to 3,000	12,800–3,333	385–100
Mid-IR	3,000 to 30,000	3,333–333	100–10
Far-IR	30,000 to 3,00,000	333–33.3	10–0.1
Microwaves	3,00,000 to 10^9	33.3–0.01	0.1–0.003

The range of the spectrum the most accessible is the visible region (320–780 nm) since the eye can be used in this case. Natural light is a polychromatic wave made of several monochromatic waves located in the visible region and the detail of localization of which is given in Figure 3. The extremities of the spectrum are localized:

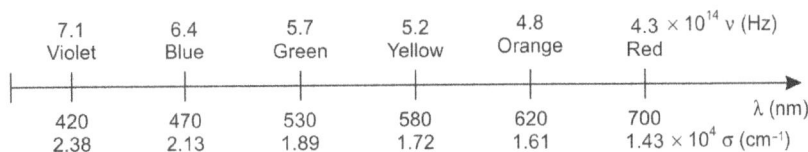

Figure 3: Electromagnetic Spectrum of the Natural Light. (Reprinted With Permission From Gwenola and Jean-Louis Burgot, Paris, Lavoisier, Tec et Doc, 2017, 318).

- The radiometric waves are "radio waves" (λ from 30 cm to 3 km).
- The X and γ rays. γ rays (λ = 10^{-12} m) are used for example in pharmacy for the sterilization of some pharmaceuticals. They are also known as being particularly dangerous for cosmonauts.

X-rays ($\lambda \approx 10^{-9}$ m) are extensively-used in the determination of crystalline structures and also in medicine for the diagnoses and treatment of some kinds of cancers.

• In other regions, let us confine ourselves to saying that microwave spectra ($\lambda = 1$ mm to 10 cm) are used in radar systems, whereas infrared radiations are used in spectroscopic chemical analysis. Natural UV radiations come from the sun but fortunately, they are absorbed by some chemicals of the stratosphere, particularly by ozone O_3. Ozone strongly absorbs near $\lambda_{max} = 255.3$ nm. Its interactions with fluorinated hydrocarbons, as a result of a decrease in O_3, are hence potentially disastrous. UV radiations are very used in the analysis, especially quantitative.

Classification of Spectra

Overall, one speaks of emission and absorption spectra. It is the first level of classification.

Emission Spectra

When atoms or molecules are submitted to very strong thermal effects or electrical discharges, they can absorb energy and become excited. By coming back to the fundamental state, they can emit radiation. This emission is the result of a transition of the atom or the molecule from the excited state to the fundamental state or, in any case, to an energy state less high than that which was excited. The energy lost during the transition is emitted as light:

• Excited atoms most often give rise to rays spectra which prove to be very useful for identifications. For example, it is in this way that the composition of some cosmic bodies has been determined (for example, by emission spectroscopy).

• Excited molecules give band spectra. They consist of a large-number of closed space lines. It is in order to account for the structure of band spectra it has been necessary to assume that the electronic, vibrational and rotational energies of molecules are superposing each other (see the section on 'Theory of Absorption Spectra').

Absorption Spectra

When a light, composed of several wavelengths goes through a compound that is at least in part transparent, a part of its radiations may be absorbed and another transmitted. The latter once resolved in its monochromatic components (with the help of a prism for example) gives a spectrum exhibiting lacuna. These correspond to the absorbed radiations and the corresponding spectrum is named the absorption spectrum. Through the absorption process, the atom or the molecule goes from a weak energy state (fundamental state) to an excited one of higher energy.

Band Spectra

Band spectra are given by molecules. They can be observed in the ultraviolet-visible and in infrared domains. Such a spectrum is constituted by a series of grooves or bands which are bright in emission and black in absorption. After a study with spectroscopes possessing a high-resolution power, each band is constituted by several rays. A band possesses a sharp limit on one of its sides. It is the head of the band. On the other side, the band grows blurred. The head can be located on one side of the band or on the other. Band spectra consist of a series of closely spaced lines that are not fully resolved. Bands arise from numerous quantized vibrational levels that are superimposed on the ground state electronic energy level.

Other Spectra

Let us mention the fact that there also exist other kinds of spectra that form without exchange of energy, such as spectra of *refringence*, which is the basis of *refractometry*, *diffraction* and *dispersion* such as those of Rayleigh and Raman. However, *Raman spectroscopy* also involves partial adsorption of the energy. Finally, one also speaks of spectra of bands for this purpose (see the section on 'Origin of the Absorption Bands').

Production of Spectra

More about spectra and their production is given in the following chapters.

A Classification of the Spectral Methods: Studied Methods

One can still class the spectral methods into two groups. The first group concerns those that are capable of providing some information on the structures of the molecules, that is to say, information which is overall used in functional analysis. There are the:

- UV-visible spectrometry
- Luminescence spectrometry
- Nuclear magnetic resonance spectroscopy
- Infra-red spectrometry
- Raman spectroscopy
- Rotatory dispersion and circular dichroism
- Molecular mass spectrometry.

However, there exist other kinds of spectrometry that have all the characteristics to be *atomic*; for example, x-ray spectrometry in which x-rays are emitted by atoms. We also shall study this method later.

Their goal is to permit the identification of elements and their concentrations. There are also:

- The atomic absorption and atomic fluorescence spectrometry
- The atomic emission spectrometry (with arc and spark spectrography and plasma sources spectrometry)
- The atomic mass spectrometry.

We shall finish this part with a chapter devoted to the criteria of purity. It involves many applications of the methodologies studied here and constitutes a kind of summary of them.

Theory of Absorption Spectra

The absorption of radiations by one substance is not uniform in the whole domain of emission and absorption of radiations. It shows maxima corresponding to the different energy levels of the absorbing molecule. These characteristic maxima have important analytical applications.

The exchanged energy during the process of absorption is present under two forms:

- The energy of the radiation itself. According to quantum theory, the exchanges of energy between radiation and matter can be only discrete. They occur only through indivisible energy quanta. The energy corresponding to one quantum is given by Bohr's relation.

  The radiation does possess high energy and all the less its wavelength is low. Therefore, X and cosmic rays are the most energetic radiations; they possess energies of the order of 10^6 electron volts (symbol eV: 1 eV = 1.60218 10^{-19} J).

- The energy of particles. The energy of particles is itself constituted by two parts:
  - The mechanical energy or potential energy. It is the part of the energy of a body that is due to its position. In other words, it is the sum of its kinetic and potential energies. They can be modified by submitting the body to external work.
  - Their internal energy depends only on their internal state as it is determined by its temperature, pressure and composition. (In certain cases, further variables such as the location of the body are necessary to define the internal energy). In thermodynamics, the internal energy is represented by the symbol U. The energy U is directly related to the first law of thermodynamics. The work done on a compound in such a way that it does not involve a change in its potential or kinetic energy is equal to the increase of its internal energy U. The internal energy is conserved in processes taking place in an isolated system. The U is a state function of the body. It is an extensive property.

More About the Internal Energy

The notion of internal energy is central within the framework of the study of spectroscopic methods of analysis. It is interesting to give some supplementary information concerning it. They come from statistical thermodynamics which itself finds part of its roots in quantum mechanics. Here, statistical thermodynamics applies to both macroscopic and microscopic systems. Solving (when it is possible) Schrödinger's equation devoted to each kind of system leads to the quantum states of the energies of macroscopic systems (phases) and microscopic ones (single molecules). In turn, knowledge of the quantum states permits acceding to functions named *partition functions* Q (for thermodynamic macroscopic systems) and f (for molecular microscopic systems). Let us now discuss the case of a perfect gas. This is by far the simplest case. If N is the number of molecules of the gas in the sample, its thermodynamic (macroscopic) partition function Q is given by the relation:

$$Q = (1/N!) \, f^N$$

Here, f is the molecular partition function of the gas, that is to say, the partition function of each molecule of gas (f has the same value for each molecule).

$$Q = (1/N!) \, (\Sigma \omega_i e^{-\varepsilon i/kT})^N$$

In this relation, the summation is over the molecular energy levels ε_0 and ε_i. T is the absolute temperature, k is the Boltzmann constant and ω_i is the degeneracy of the ith molecular energy level.

For some systems, those for which the potential energy is constant or is a particularly simple function of the coordinates of the molecules, Schrödinger's equation for each molecule of the gas can be separated into a part characteristic of its translational motion and a part characteristic of its internal motion. (This is because the energies of the internal states do not depend on the position of the molecule in space). As a result, any molecular energy ε_i may be taken as the sum of a translational part and an internal part:

$$\varepsilon^i = \varepsilon_j^{tr} + \varepsilon_i^{int}$$

This is a formidable simplification concerning the search for the solutions to Schrödinger's equation. At this point of the reasoning, Schrödinger's equation of the internal state may, indeed and in turn, be also factorized but less accurately as above. Further studies, indeed, show that ε_k^{int} may be regarded as the sum of two terms; one due to the state of electronic excitation of the molecule and the other due to the combined rotational and vibrational motion. Less accurately, finally, the latter may also be regarded as separable. If this approximation is justified, we can write:

$$\varepsilon_k^{int} = \varepsilon^{rot} + \varepsilon^{vib}$$

Moreover, there are good scientific reasons to take into account the electronic state energy of the molecule and also the internal states of nuclei ε^n together with spin variables such as:

$$\varepsilon_k^{\,int} = \varepsilon^{rot} + \varepsilon^{vib} + \varepsilon^{elec} + \varepsilon^n$$

This will not be the case in this book. On one hand, this is because there exist extremely large separations between the two lowest nuclear energy levels at ordinary temperatures. The atoms are certainly in the *nuclear* ground state. This sentence is summarized by the Born-Oppenheimer[20] principle which contends that because of the great disparity of their masses, the motions of the electrons and the motions of nuclei can be treated as independent problems with no interaction between them. This means that the vibrational and rotational motions can be considered to be quite separate from the electronic motion since the latter, which controls the forces between atoms, involves the charges and configuration of the nuclei but not their masses. Moreover, the motions of the nuclei are very slow and the intermolecular distances are parameters only very slowly changing. On the other hand, the situation is similar but not so extreme for electronic excitation. It is only for some elements that excited electronic states cannot be ignored. This is the case of some halogens at 1,000 K, such as F, Cl and possibly Br.

(*Remark*: Several monoatomic substances, as well as a few polyatomic molecules, also have multiple electronic ground states.)

For example, Figure 4 mentions the disposal of the electronic, vibrational and rotational energy levels of a diatomic molecule.

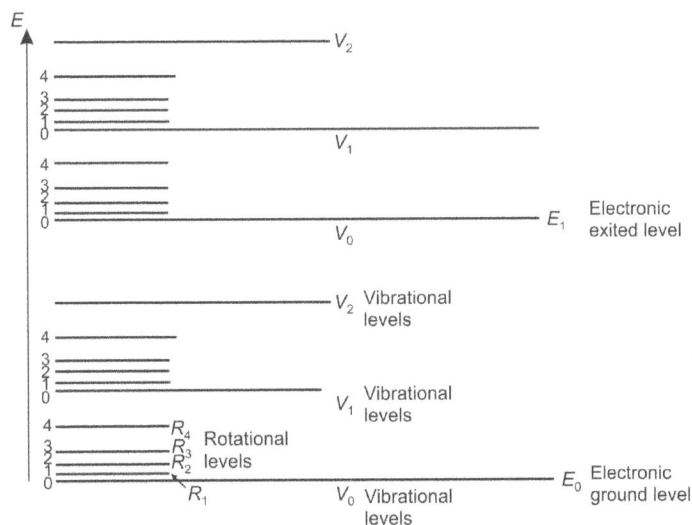

Figure 4: Electronic, Vibrational and Rotational Energy Levels of a Diatomic Molecule (Schematic).

In other words, it is possible to separate the electronic energies from the nuclear energies and separate both from the energies of rotation and vibration. With a certain approximation, it is also possible to separate the two latter ones. We shall come back to the notion of internal energy when we shall study the spectroscopy infrared (see Chapter 28).

Let us, temporarily, give some data concerning the notion of internal energy by this comment that the notion of energy is one of the most difficult questions in physics. One cannot know what

---

[20] Born, Max, German physicist, 1882–1970, Nobel prize in physics 1954.
Oppenheimer, Robert, American physicist (1904–1967), is considered as one of the fathers of A-bomb.

signifies this concept because we do not possess any image of its fundamental data. The only comment that can be certainly maintained is that there exists a quantity named energy that does not change during numerous transformations that exist in nature. For some physicists, the concept of energy is purely and simply *a strange fact*. The first law of thermodynamics is purely the result of experience.

Origin of the Absorption Bands

One part of the energy of the radiation can modify the internal energy (and not the electronic energy) of the molecule through which it is passed so that the brought energy allows the molecule to go from any one energetic state to another of upper energy (of internal energy) and the energy brought by the quantum hv must be equal to the energy difference between the two states (Bohr's condition). According to the difference of energy hv, that is to say, according to the wavelength, the internal energy change may be either that of the energy of rotation or that of vibration or both. One then obtains an infrared spectrum. If it is the electronic energy of the molecule which suffers a change, it is then a UV-visible spectrum that is obtained.

Selection Rules

All transitions between particles (molecules or atoms) and pairs of energy levels are not allowed. Their occurrence is determined by *selection rules*. They will be given as we go along this book.

CHAPTER 19

Instrumentation in Spectrophotometry

A spectrometer is an instrument that measures the intensity of an emitted or absorbed radiation as a function of wavelength of the latter. Spectrophotometers are a kind of spectrometers that permit the determination of the ratio of the powers of two beams as a function of wavelength in absorption spectroscopy.

It is known that several kinds of spectrophotometers have some parts in common. It is the case of spectrophotometers UV-visible and infrared. This is the reason why these parts are studied together in the same chapter.

Principal Components of Spectrophotometers

In this chapter, we focus essentially on UV-visible spectrophotometers but we will also mention some characteristics of other apparatus when they are devoted to other sorts of spectrophotometry.

Schematically, a UV-visible spectrophotometer is principally constituted by (Figure 1):

- A source
- A wavelength selector or monochromator
- A sample container or cell measurement
- A radiation transducer or detector
- Signal processors and readouts

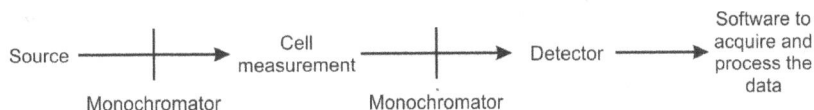

Figure 1: Schematic Representation of a UV-visible Spectrophotometer. (Reprinted with Permission from Gwenola and Jean-Louis Burgot, Paris, Lavoisier, Tec et Doc, 2017, 345).

Sources

The radiation source must be emitted in the range of frequencies in which the study of the sample is carried out. There are many possibilities for the production of light for spectroscopic investigations. The principal ones are temperature radiations and luminescence.

In temperature radiations of gases, atoms or molecules are excited to light emission by collision with other atoms or molecules. The necessary energy is derived from the kinetic energy of the colliding particles.

Luminescence includes all forms of light emission in which kinetic heat energy is not essential for the mechanism of excitation. The latter, in this case, results mostly from electron or ion collision.

The kinetic energy of ions or electrons, which are accelerated in an electric field, is given up to the atoms or molecules of the gas present and causes light emission. Another alternative is the excitation by light emission (see Chapter 22).

Lamps produce a continuous spectrum that is filtered by a monochromator. Photons are obtained by electrical discharges in the gas. Most usually, lamps that are used are:

- The tungsten lamp: Emission wavelengths of the lamp are located in the interval 350–2,500 nm. They are overall used in the visible and near-infrared IR. The lamp's tungsten-halogens (iodine lamps) show a more extended domain of use (240–2,500 nm) than the preceding ones and they possess a longer lifetime.

- The deuterium lamp is overall used in the UV domain ($160 < \lambda < 375$ nm). The xenon lamp emits in the domain 190–1,100 nm. The connection of these two lamps with a change at 350 nm permits them to cover the whole spectrum (Figure 2).

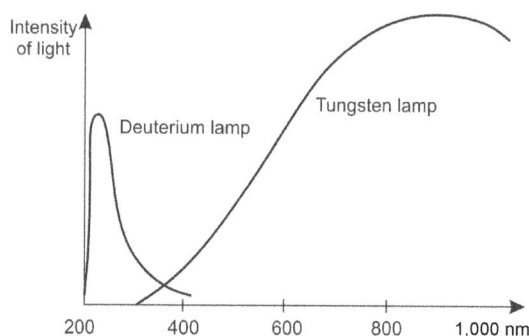

Figure 2: Emission Curves of Deuterium and Tungsten Lamps. (Reprinted with Permission from Gwenola and Jean-Louis Burgot, Paris, Lavoisier, Tec et Doc, 2017, 346).

- Laser sources: Such sources are particularly interesting in this field because they are sources of high intensities and they give narrow spectral bands. Also, their output is very coherent. A development concerning lasers is given in Appendix 4. Let us limit ourselves now to mentioning that they are used in the whole range of UV-Visible domains for the near-infrared and the beginning of the usual infrared.

Wavelengths Selectors and Monochromator

In order to obtain monochromatic radiation, it is necessary to decompose the polychromatic light provided by the lamp into its different monochromatic components. Then, a spectral band of a given wavelength must be selected. It must be as narrow as possible.

From a broad point of view, it can be said that the separation of light into its spectral components can be accomplished either by refraction or diffraction.

Diffraction is a phenomenon exhibited by all types of electromagnetic radiation. It is effective when a parallel beam of radiation is bent as it passes by a sharp barrier or through a narrow opening. Diffraction is a consequence of interference.

Refraction is an effective phenomenon when radiation passes at a certain angle through the interface between two transparent media having different densities. The refraction is the abrupt change of direction. It is a consequence of a difference in velocities of the radiation in the two media. Diffraction and refraction depend on the wavelength, but in opposite ways. The greater the latter, the greater the diffraction of light is. But the greater the wavelength, the smaller the refraction of light. For the separation of light by diffraction, gratings are used; for the separation by refraction, prisms are used (see color Photo 4). The prism method has the advantage of greater light intensity, whereas the grating method affords greater resolving power.

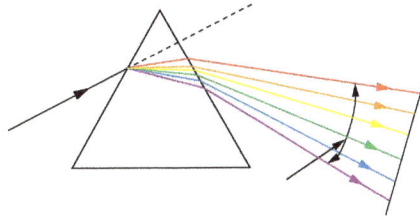

Photo 4: Dispersion of White Light by a Prism.

One notices:

- Filters which can absorb all the radiations produced by the lamp except the radiation selected. Let us mention:
  - The absorption filters are simple colored glass plates, whose spectral bands are large (about 30 nm). They are used in the visible domain. They are inexpensive.
  - The interference filters: They are used in the whole domain UV-visible. Their spectral band is of the order of 10 nm. Such a filter (Figure 3) consists of a transparent share (usually calcium fluoride or magnesium fluoride named dielectrics because the latter ones are optically transparent) surrounded by two metallic semitransparent films. The wavelength of the transmitted radiation is a function of the thickness of the dielectric.

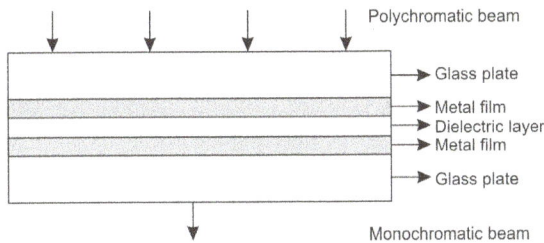

Figure 3: An Interference Filter. (Reprinted with Permission from Gwenola and Jean-Louis Burgot, Paris, Lavoisier, Tec et Doc, 2017, 347).

When light radiations perpendicularly strike the first metallic layer, one part is reflected and the other crosses through it. In turn, a part of the fraction, which has crossed it, is reflected by the second film and so forth. The successive reflections reinforce some wavelengths.

Filters are used in photometers whose working wavelengths are already chosen.

- Monochromators permit varying the wavelength of the radiation (to be absorbed) continuously in a large domain. They are composed by:
  - An entrance slit
  - A collimating lens or mirror produces a parallel beam of radiation
  - A system of dispersion of the radiation into its component wavelengths
  - An exit slit

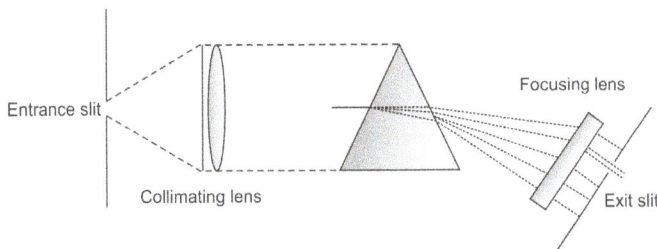

Figure 4: Representation of a Prism Monochromator. (Reprinted with Permission from Gwenola and Jean-Louis Burgot, Paris, Lavoisier, Tec et Doc, 2017, 347).

A narrow slit decreases the power of the ray and imposes a detection system provided with a strong amplification system.

Among monochromators, one differentiates prisms from gratings (Figures 5a and 5b) and (see color Photo 4).

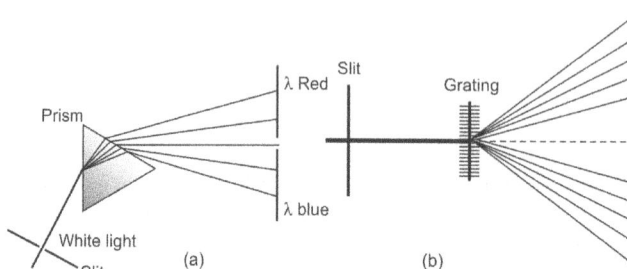

Figure 5: Dispersion of White Light 5a: By a Prism; 5b: By a Transmission Grating.

There are several kinds of gratings. They are based on the following fact. When a beam of monochromatic radiation is passing through a transparent plate, which has a large number of very fine parallel lines, its path rules it. It is split into several beams. Some of them go straight through, other beams deviate from the forward direction through angles θ which depend on the distance between lines and the wavelength of the radiation. The second type of rays is destroyed by interference in most directions; except for one of them, for one angle θ_0 for which the beams reinforce each other (Figure 6). This fact is rationalized by the notion of the "difference in the length of the path". It follows from this fact that if a beam of polychromatic radiation is passed through the grating, its decomposition in monochromatic radiations occurs.

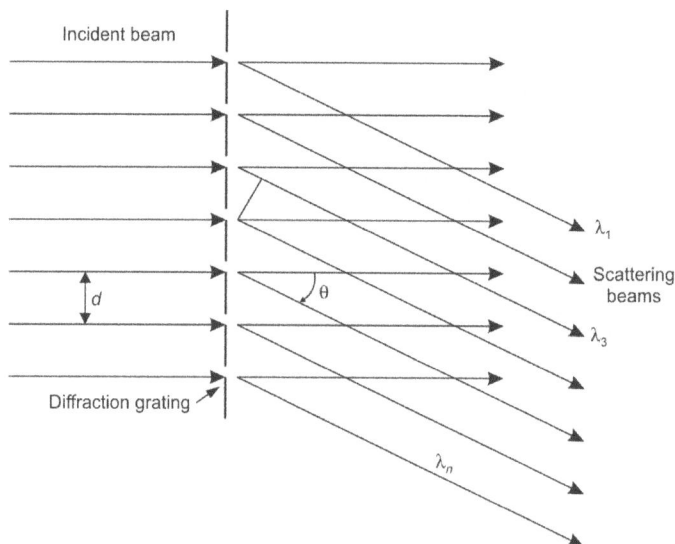

Figure 6: Diffraction at a Grating.

Let us mention the existence now of holographic gratings, the greater perfection of which with respect to the older ones is due to a better tracing of the lines.

Measurement Cells

The analysis cell is usually a parallelepiped whose base is square (see color Photo 5). Its optical path is often from 1 to 40 mm in length. It possesses two opposite polished and parallel faces. It may contain from 0.15 to 15 ml of solution. The cells that contain the samples must be made of materials that are transparent to radiations in the spectral domains of the study. One uses:

- Transparent plastic cells or ordinary glass for aqueous and non-aqueous media are devoted to analysis in the visible region. They are in quartz or fused silica for measurements in the UV region.
- Quartz microplates are provided with 96 or 384 holes, which are stable at high temperatures. They authorize their sterilization after measurement and hence analysis of biological substances often involves such temperatures which are useful.

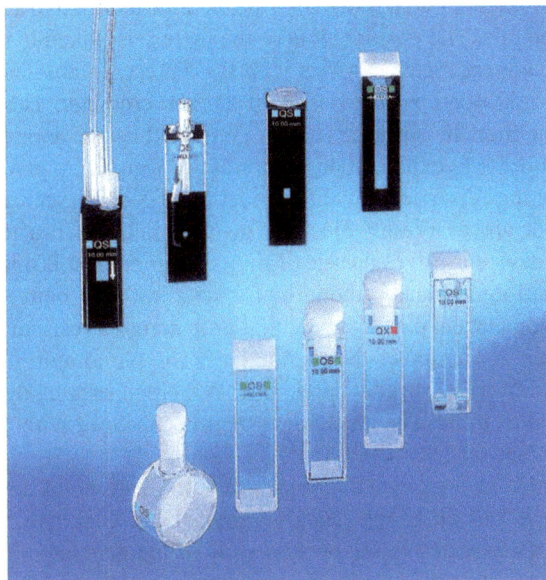

Photo 5: Measurement Cells for UV Analysis.

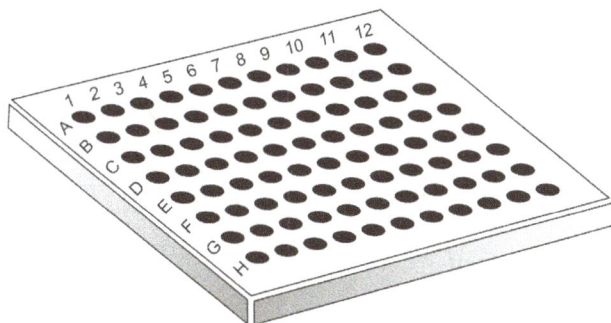

Figure 7: Microplate of 96 Holes. (Reprinted with Permission from Hellma France, 35 Rue De Meaux, 75019 Paris).

The handling of these cells is delicate and must be cautious.

Measurement of the Radiant Energy: Radiation Transducers and Photoelectric Detectors

In the early years of spectroscopy, the detectors were essentially photographic plates and the human eye. Now, radiation transducers are used. They are of two kinds, i.e., photon transducers and heat transducers.

Whatever the kind it is, a transducer must have a high sensitivity, a constant response over a large domain of wavelengths and not too a fast one. In principle, it must not give a response in case of no illumination.

• Phototransducers

We emphasize the *phototransducers* because these devices are often found in spectrophotometers. A phototransducer transforms luminous data into electrical ones, based on the photoelectric effect. Recall that the photoelectric effect is the emission of electrons by enlightened matter, in particular by metals such the alkaline ones and also by zinc negatively charged. It is obvious by the example of the use of a photoelectric cell. The latter can be an empty glass bulb. A part of its interior surface is recovered by a metal. This metallic coat is linked to the metallic terminal of a battery of accumulators, which constitutes the cathode. The positive pole of the battery is linked to a tungsten ring. It is the anode. There is a microampere meter in series with the battery. In absence of light, no current is passing but, if there is light, there is a deviation of the amperometer. The luminous energy pulls up the electrons from the metallic surface. They are attracted by the anode and circulate from the cathode to the anode. In order to explain that, light must be formed by particles called protons. An example of such functioning is provided by the production of X-rays (see Chapter 36). In summary, rays strike a photocathode and as a result, electrons go away and are attracted by the anode.

A slightly different device involves the use of a *photomultiplier*. It is often a vacuum phototube, which consists of a semi-cylindrical cathode and a wire cathode both sealed in an evacuated transparent tube. The photocurrent causes a potential drop across a resistance R which is amplified and related to a recorder. A variant of the phototube is the photomultiplier tube (PMT). It is a phototube containing additional electrodes called *dynodes*. Upon striking dynodes, there are treated as the additional ejected electrons. The resulting current is enhanced with respect to that produced with a simple phototube (see Chapter 42).

Silicon diodes can also be used as phototransducers. In this type of diode, the semi-conductor element *p* of the diode is related to the negative pole of the generator, and the element semi-conductor *n* (negative) is connected to the positive pole of the generator. This pole withdraws the elementary charges of the very mobile positive electricity (holes) toward itself of the element *p*. The holes move away from the junction p-n. Simultaneously, one observes the attraction by the positive pole of the generator of electrons, elementary negative charges. The electrons also move away from the junction. Both holes and electrons have moved away from the junction. Because of these facts, the exchanges of electricity charges are not facilitated. Only a very weak current circulates (Figure 8).

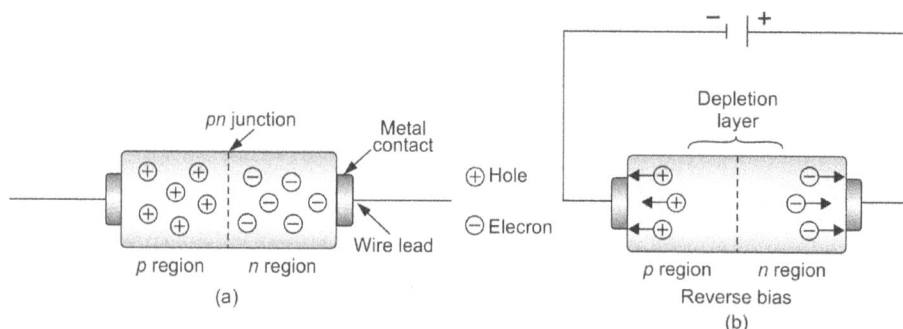

Figures 8a and 8b: Representation of a pn Junction and Direction of the Current According to the Tension. Current can Flow Under Forward Bias Conditions (a) But is Prevented From Flowing Under Reverse Bias (b). (Reprinted with Permission from D.A. Skoog et al., p. 179 (2018), Cengage Learning).

There are semiconductors, such as Si, Ge, Ga and As ones. In these compounds, the four valence electrons of the pure materials are all involved in bonds between the different atoms. The present phosphorus atom, considered as an impurity, does possess five valence electrons and as a result provides a supplementary electron. An aluminum impurity (this may also be an atom of trivalent gallium and gold) does possess one electron less than one semiconductor above. This fact creates a hole in the following manner. An electron in a normal initial location, involved in a normal covalent bond located near this impurity, may occasionally fall into this hole. The result of such a process is that the initial hole has moved from one point to another. In the piece of silicon, the holes appear to wander freely. If an electric field is applied to the piece of silicon doped in one way or by its opposite (for example, by addition of arsenic or by gallium), a current will flow which is constituted quasi-exclusively by the excess of electrons or holes. Silicon in which the current carriers are electrons is called *n*-type silicon (*n* negative), whereas that in which the predominant carriers are holes (positive) is *p*-type silicon. A diode is a pn junction (Figures 8a and b). If *n*-Si is made negative with respect to *p*-Si under the action of the external circuit, electrons flow into the nSi. At the junction, electrons and holes combine.

The result is that a current is flowing when *n*-Si is made negative with respect to *p*-Si. The diode is said to be *forward biased* (Figure 8a). If the opposite polarity is applied (Figure 8b), electrons and holes are spread out the junction, each on its proper side. As a result, there is the formation of a depletion (of electric charges) region near the pn region. There is a lack of charge carriers near it. The diode is *reversed biased*. When radiation strikes the system, holes and electrons are formed but few in the depletion layer. As a result, in these conditions, the conductance is located in the interval 10^{-8}–10^{-6} that occurring under forward bias. Under reverse bias, the current obtained with a silicon diode is of the order of 10^{-8} A. The conduction is carried out by the minority carriers (the few charges mentioned above). This reverse current is of little consequence. Silicon diodes are more sensitive than vacuum phototubes but they are less sensitive than photomultiplier tubes.

• More recently, detectors with *diode arrays* appeared. They are *multichannel transducers*. In this kind of instrument, radiation is focused upon the cell. Then, it passes into a monochromator with the help of a grating. The radiation is dispersed and falls on a photodiode array transducer. The latter consists of a linear array of several hundred photodiodes disposed on a silicon chip. The chip contains a capacitor and an electronic switch for each diode. Radiation striking any diode surface causes a discharge of its charge. The lost charge is replaced during the following cycle. The resulting charging currents which are proportional to the radiant power are amplified, digitized and stored in the computer memory. Figure 9 schematically represents a spectrophotometer with *diode*

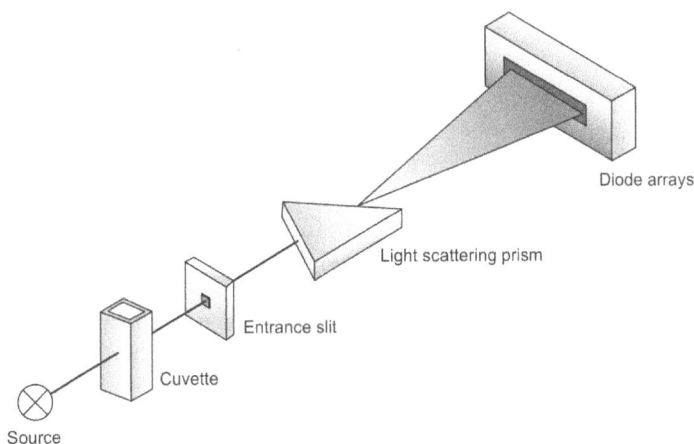

Figure 9: Scheme of a Spectrophotometer with Diode Arrays. (Reprinted with Permission from Gwenola and Jean-Louis Burgot, Paris, Lavoisier, Tec et Doc, 2017, 351).

arrays. Thus, the principle of these detectors is different from that of classical spectrophotometers. Instead of sending successively different monochromatic radiations on the sample, the latter directly receives a polychromatic light. Hence, the absorption of the entire set of radiations is carried out simultaneously. An independent detector is necessary for the measurement of the absorption of each absorption. These detectors are disposed of in arrays.

A diode array instrument is particularly useful in the qualitative and quantitative determinations of the media exiting in liquid chromatography and in electrophorese systems.

Thermal Transducers (Specifically Infrared)

The preceding phototransducers are generally not applicable in the infrared domain because photons in this region lack the energy to cause photoemission of electrons. In thermal transducers, the radiation strikes a small blackbody and is absorbed by it. The resultant temperature rise is measured. It must be noticed that the power of a typical infrared beam is very weak (10^{-7} to 10^{-9} W). As a result, the heat capacity of the absorbing element must be as small as possible. The element is designed according to this fact. The temperature change of the element can amount to a few thousand kelvins; a parasitic thermal effect of that kind which must be measured is the thermal noise of the surroundings. One way to avoid this problem is to chop the beam from the source. In these conditions, the analytic signal exhibits the frequency of the chopper after treatment. Then, it can be easily isolated electronically. Among thermal transducers, thermocouples, bolometers and pyroelectric transducers are used.

Thermocouples

Thermocouples are the most widely used infrared detectors. They are manufactured in a number of ways. They are generally equipped with a tiny bit of blackened gold foil which is the actual absorber of radiation. They are constituted of two dissimilar metals. The measured voltage by the device is nearly proportional to a difference in temperature. Further specifications are given in Appendix 5.

The Bolometer

The bolometer is another kind of thermal change detector. It is composed of two elements. One is exposed to the radiation, while the other one is wholly kept away from the latter; but it remains subject to the same other conditions. The output voltage is proportional to the ratio of the resistances of the two elements. In these conditions, the changes in ambient temperatures may be considered as being completely eliminated.

Finally, we can also mention the use of pyroelectrics crystals. The pyroelectrics crystals develop a potential difference across their opposed faces when they are heated. Among them, let us cite triglycine sulfate, barium titanate and lithium niobate; they have been used successfully as infrared detectors.

Signal Treatment

An electronic device amplifies the electronic signals of the detector which are too weak to be directly registered. A readout system is a device that converts information from an electrical domain to another one that is understandable by a human observer. The readout can be in the form:

- of a digital signal;
- of scales for example for the measurements of transmittance and absorbance;
- of a computer screen.

Configurations of Spectrophotometers

Among the spectrophotometer, one distinguishes the conventional and those with diode arrays. Whatever the kind, they can be a mono-channel or double channel.

Usual Spectrophotometers

In these instruments, the polychromatic light is dispersed in the monochromator which permits the selection of the spectral band. This radiation goes through the sample and attains the detector (Figure 10).

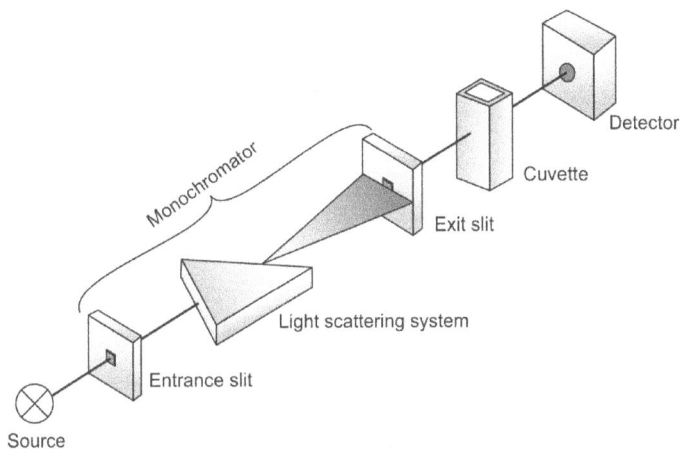

Figure 10: Scheme of a Conventional Spectrophotometer. (Reprinted with Permission from Gwenola and Jean-Louis Burgot, Paris, Lavoisier, Tec et Doc, 2017, 351).

Diode Array Instrument

(See the section on 'Measurement of the Radiant Energy: Radiation Transducers and Photoelectric Detectors').

Single-Beam Spectrophotometer

In this instrument, intensities of the reference and of the sample cells are consecutively measured during a short time of the order of a second or a minute. The time interval must not be too great because of possible changes in the intensity of the source. In this case, errors may no longer be negligible (Figure 11).

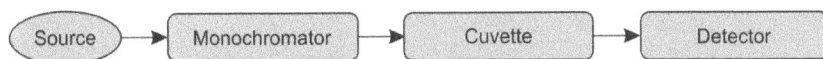

Figure 11: A Single-Beam Spectrophotometer. (Reprinted with Permission from Gwenola and Jean-Louis Burgot, Paris, Lavoisier, Tec et Doc, 2017, 352).

Double-Beam Spectrophotometer

In this instrument, the configuration of the different parts permanently permits taking into account the absorbance of the solvent. The beam successively crosses the measurement cell containing the sample and the reference cell containing the solvent and the reactants (Figure 12a). For the "true "double-beam (Model 12.b), a system of turning mirrors (chopper) permits the detector to permanently measure the transmitted intensities along one way or the other. With the photodiodes,

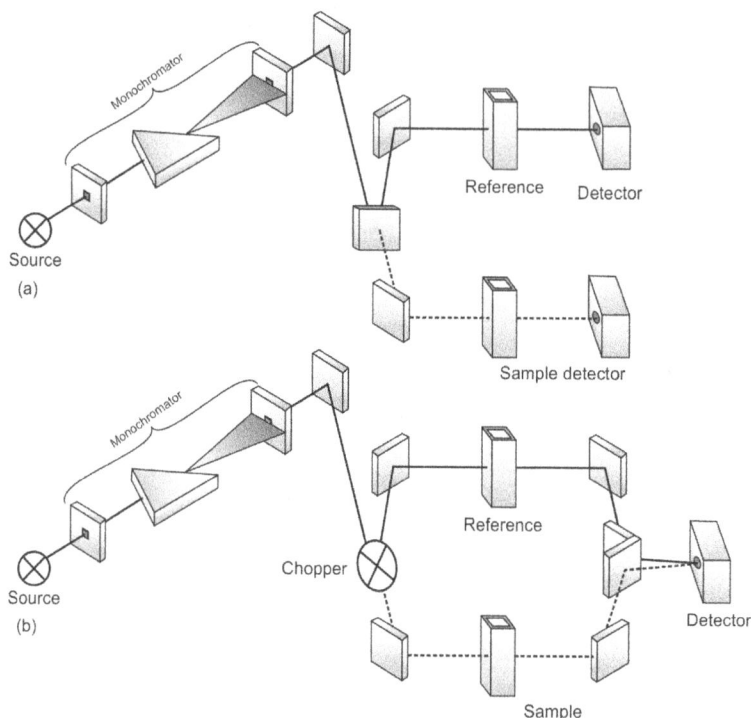

Figures 12a and b: Pseudo-Double Beam and Double-Beam Spectrophotometers. (Reprinted with Permission from Gwenola and Jean-Louis Burgot, Paris, Lavoisier, Tec et Doc, 2017, 352).

turning mirrors are replaced by fixed ones. In Model 12.a (pseudo double beam spectrophotometer), the ray is divided into two parts, each one crossing the reference and the sample cells, but the transmitted lights are not analyzed by the same detector. There exists then a supplementary origin of the error.

Finally, let us mention the fact that there exist automated systems that "work" continuously. The injection of the samples is performed by continuous-flow devices. The other various operations of one analysis (extraction, dilution and addition of reactants) are achieved with the help of the circulation of the sample in a coil through the measurement cell.

CHAPTER 20

Ultraviolet-visible Spectroscopy I
Generalities

UV-visible spectroscopy is one method of molecular absorption spectroscopy. At least in part, it permits the structural analysis of some molecules and above all, it also permits numerous quantitative determinations of different kinds of species. In addition, it reveals itself to be a very efficient tool in physical chemistry for the study of physical equilibria or kinetics in solutions.

Generalities

Species exhibiting a UV-visible spectrum are those which absorb radiations whose wavelengths are located in the domain 190–800 nm. The near UV is located in the domain of 200–380 nm, whereas the visible lies in the range 380–800 nm (Figure 1):

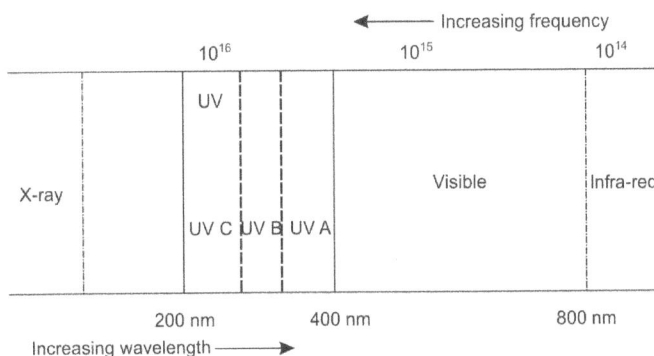

Figure 1: Spectroscopic Domain of UV-Visible.

The domain 100–190 nm is also considered as being in the domain of UV, but it is not used in the common analysis because molecular oxygen and carbon dioxide absorb in this range and this fact necessitates the use of special experimental conditions. This domain is named "the vacuum UV" or "far UV".

As it has been mentioned in the preceding chapters, when there is the absorption of a photon in the UV-visible range, there is a change in the electronic energy of the absorbing species which is accompanied by changes in vibrational and rotational energies. Each electronic and simultaneously vibrational transition corresponds to an absorption band, which is constituted of multiple transition rays of rotation. Spectra constituted by not cleared up bands are considered to exhibit a continuous spectrum. From a practical standpoint, it is nearly the case for all spectra registered in solutions.

Terminology

In literature, the following terms are often encountered. They are related to UV-visible spectroscopy. They are:

- *Absorptiometry* is the general term for chemical analysis through the measurement of absorption of radiations;
- *Colorimetry* is used in relation to the absorption in the visible spectral region;
- *Spectrophotometry* is a division of absorptiometry. It refers specifically to the use of the spectrophotometer.

Parameters of an UV-Visible Spectrum

UV-visible spectra are diagrams showing the absorption of a radiation UV-visible passes through the studied sample as a function of the wavelength of the beam. Several parameters permit the description of the absorption of radiation. They are the *absorbance*, the *transmittance*, the *molar absorption coefficient*, etc. (see Chapter 21). These terms do not have the same meaning.

For example, Figure 2 shows the structure of methylergometrine and Figure 3 its spectrum UV-visible in which its absorbance is registered as a function of the wavelength.

with R = —CONH—CH—C_2H_5
                          |
                        CH_2OH

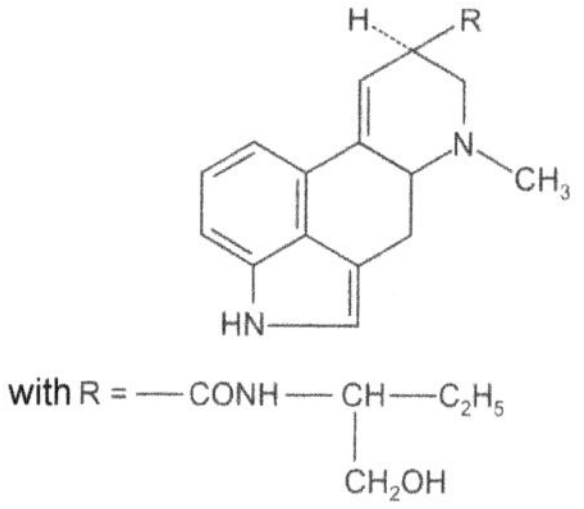

Figure 2: Structure of the Methylergometrine. (Reprinted with Permission from Gwenola and Jean-Louis Burgot, Paris, Lavoisier, Tec et Doc, 2017, 330).

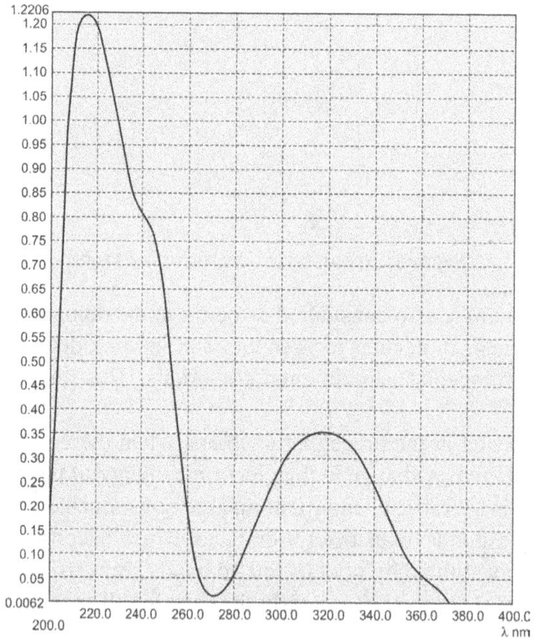

Figure 3: UV Spectrum of Methylergometrine (1.5 10^{-4} mol L^{-1}) in a Water-Ethanol Mixture (Personal Data). (Reprinted with Permission from Gwenola and Jean-Louis Burgot, Paris, Lavoisier, Tec et Doc, 2017, 331).

Two very important parameters in UV-visible spectroscopy are the wavelengths of the maxima and of the shoulders of absorption and the intensities of these maxima and shoulders measured, most often by the value of their molar absorption coefficients ε accessible through the determination of the absorbances (see Beer-Lambert law). In the case of methylergometrine, we note as maxima of absorption the values 318 nm and 215 nm with shoulders at 369 nm and 240 nm. Absorbances at 318 nm and at 215 nm permit quantifying absorptions at maxima.

In structural analysis, the whole spectrum of the studied species is registered and the two kinds of parameters already mentioned are used to identify some groups of atoms. From the qualitative standpoint, UV-visible spectroscopy is indeed essentially a method of functional analysis. Furthermore, it permits identifying the studied molecule but in some favorable cases. The identification of a species solely with the help of UV-visible spectroscopy indeed cannot be considered as being infallible and is finally unusual.

In quantitative analysis, firstly a wavelength is chosen. In preference, it is that corresponding to a maximum of absorption. The concentration of the absorbing species is secondly deduced from the value of its intensity at this wavelength but, it should not be forgotten that the intensity and the location of an absorption band depend on the solvent.

Origin of Electronic Transitions and Intensities of Bands

This paragraph only concerns transitions that are electronic, but the fact that they are accompanied by vibrational and rotational transitions in the species must be taken into account is characterized by the change of the location of an electron that jumps from an orbital of a weaker electronic energetic level to another orbital of upper energy. The electronic transitions involved are of the types:

$$\sigma^* \leftarrow \sigma ; \quad \sigma^* \leftarrow \pi ; \quad \sigma^* \leftarrow n ; \quad \pi^* \leftarrow n ; \quad \pi^* \leftarrow \pi \quad \text{(Figure 4)}$$

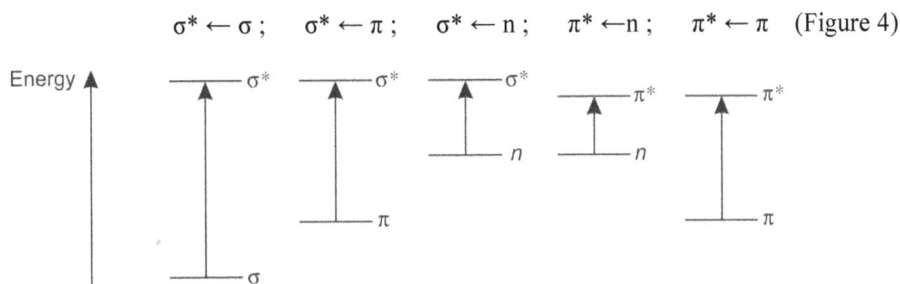

Figure 4: Schematic Representation of Some Transition Energies. (Reprinted with Permission from Gwenola and Jean-Louis Burgot, Paris, Lavoisier, Tec et Doc, 2017, 332).

σ^* and π^* are anti-bonding orbitals and n non-bonding electrons, whereas in σ and π orbitals are bonding electrons. But electrons d may also intervene. For example, the octahedral cation $[Ti(H_2O)_6]^{3+}$ absorbs in the "blue-green" and appears with the complementary color, that is to say with a pale-blue color. The necessary energy for the transition $\sigma^* \leftarrow \sigma$ is very high (Figure 4). As a result, compounds whose external electronic layer (the valence layer) is involved in a simple bond as in saturated hydrocarbons and do not exhibit absorption in usual UV. For example, propane gives a transition λ_{max} at 135 nm. Cyclopropane is an exception to this rule since it gives the corresponding transition for λ_{max} = 190 nm. This is not fully surprising. It is well-known that cyclopropane exhibits some properties of unsaturated hydrocarbons.

Thus, a UV-visible transition is invariably characterized by the presence of multiple bonds in the structure of the absorbent molecule. They are at the origin of transitions involving antibonding orbitals and also non-bonding orbitals. The latter ones are brought by some atoms such as oxygen, nitrogen, sulfur and halogens. Such electrons give rise to transitions of the types $\sigma^* \leftarrow n$ and $\pi^* \leftarrow n$.

The interpretation of electronic transitions requires not only the knowledge of electronic configurations including those of spin but also that of the symmetry elements of the intervening orbitals, as it is noticed now.

The Dihydrogen Molecule

According to the theory of molecular orbitals, the electronic configuration of dihydrogen in the fundamental state is $(1\sigma)^2$. The molecule of dihydrogen results from the coupling of the orbitals 1s of each of the constitutive hydrogen atoms H_A and H_B. A σ bond is formed when the molecule of dihydrogen is formed. It is filled by the two electrons coming each one from one of the hydrogen atoms, hence the symbol $(1\sigma)^2$. Figure 5 shows the energy levels 1s and 2s and 2p of the hydrogen atom. Figure 6 shows the bonding molecular orbital $(1\sigma)^2$ and the antibonding orbital $1\sigma^*$ (unoccupied in its lower electronic state) of the dihydrogen molecule and its filiation from orbitals 1s.

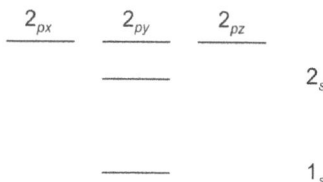

Figure 5: Energy Level 1s of the Hydrogen Atom.

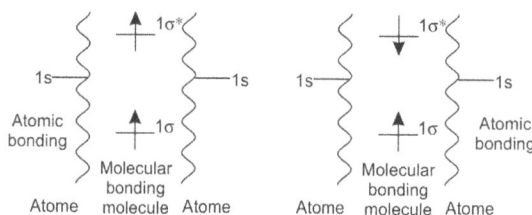

Figure 6: Bonding Molecular Orbital $(1\sigma)^2$ and its Antibonding Counterpart $1\sigma^*$ in Dihydrogen Molecule.

One or both "initial" electrons can be located in the anti-bonding orbital $(1\sigma^*)$ in an excited state (Figure 7).

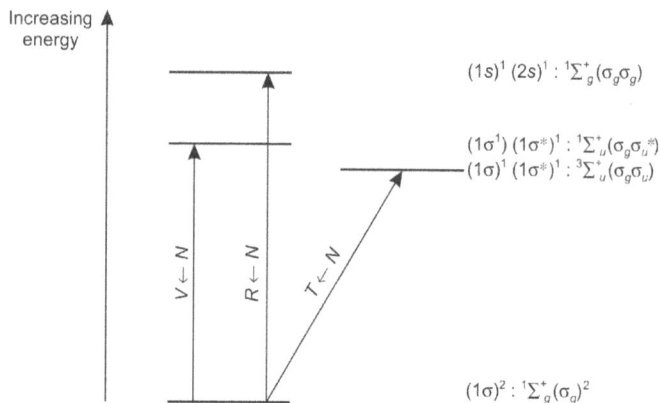

Figure 7: Some transitions of dihydrogen (Reprinted with Permission from Gwenola and Jean-Louis Burgot, Paris, Lavoisier, Tec et Doc, 2017, 334) .

A bonding molecular orbital σ has axial symmetry. It is said g (*Gerade* in the German language means even). In contrast, an anti-bonding orbital is named ungerade: u (odd). Figure 7 shows the aspects of the orbitals 1σ and $1\sigma^*$. (By combining atomic orbitals to form molecular orbitals the importance of the symmetry of the different orbitals is considerable).

Thus, by absorption of one photon, several possibilities of evolution are possible:

• One electron is raised to orbital $1\sigma^*$ with inversion of spin. The new electronic configuration of the molecule is $(1\sigma)^1(1\sigma^*)^1$. The multiplicity of the new state is given by the expression $2S + 1$

where S is the total spin. Since there is an inversion of spin, they become parallel in the excited state. The total Spin S is equal to 1:

$$S = +1/2 + 1/2 = 1$$

Hence, the multiplicity is 3. This transition $1\sigma^* \leftarrow 1\sigma$ is often named transition $T \leftarrow N$ (N for normal and T for triplet). It is by far less likely than the following one. It never occurs in the usual UV domain since it is not located in the domain of usual UV.

• One electron is raised to orbital $1\sigma^*$ but without inversion of spin. It is still a transition $\sigma^* \leftarrow \sigma$ but the excited state is now a singlet state (S = 0). As a general rule, the energy of the singlet state is higher than that of the corresponding triplet state. This transition is often called transition $V \leftarrow N$ (V for valence). It is by far more likely than the preceding. It never occurs in the usual UV domain. It is intense and predominant in the spectrum of dihydrogen.

• It is also possible that an electron can be raised to the level 2s (transition $R \leftarrow N$: R for Rydberg). It occurs in a domain still farther than the preceding ones.

Concerning now the raising to the 2s level of the two electrons of the initial level 1σ, it is not out of the question. It is only unlikely. Figure 7 summarizes these considerations:

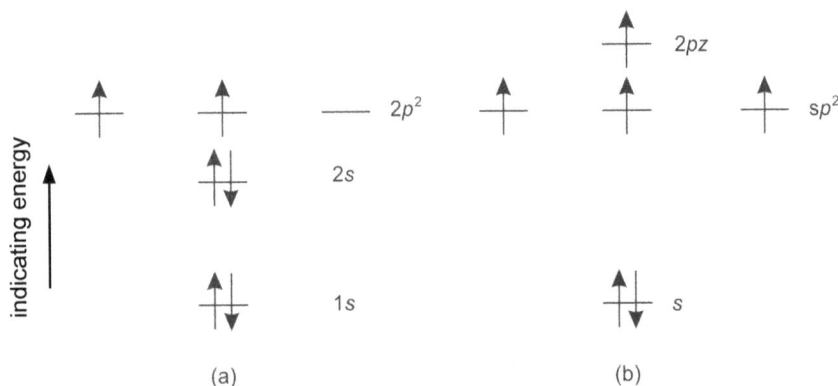

Figure 8: Energy Level in Ethylene Molecule (a) Fundamental State and (b) Hybridization sp^2 State.

To the fundamental state (N state) corresponds the spectroscopic state $^1\Sigma^+_g(\sigma)^2$. $(\sigma)^2$ means that there are two electrons in the bonding electrode σ. The symbol σ itself means that this (molecular) orbital is symmetric relative to the symmetry center of the molecule. The symbol g means that the state of the molecule is symmetric relative to the elements of symmetry of the group of the molecule and that the molecule is linear. Exponent 1 indicates the multiplicity of the state. The sign + is related to the sign of the wave function.

The Ethylene Molecule

For the absorption in the usual UV-visible, only electrons π and π^* of ethylene are concerned. We represent the atomic orbitals of carbon in its fundamental state and under its hybridization sp^2 state in Figures 8a and 8b. There is no doubt that in ethylene, both carbon atoms are hybridized sp^2.

In Figure 9, we represent the electronic configuration of ethylene (both carbon atoms being hybridized sp^2). We can consider that any carbon atom is only bonded by three σ bonds, one to each hydrogen and one to the other carbon. In those three bonds once formed, each carbon has, therefore, another electron in the remaining (nonhybridized) atomic 2p orbital (that is to say, $2p_z$ orbital) which is perpendicular to the plane of sp^2 orbitals. This fact is a consequence of the kind of hybridization sp^2.

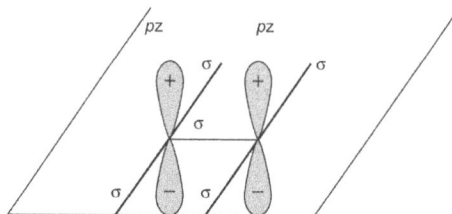

Figure 9: Overlapping Orbitals in the Ethylene Molecule.

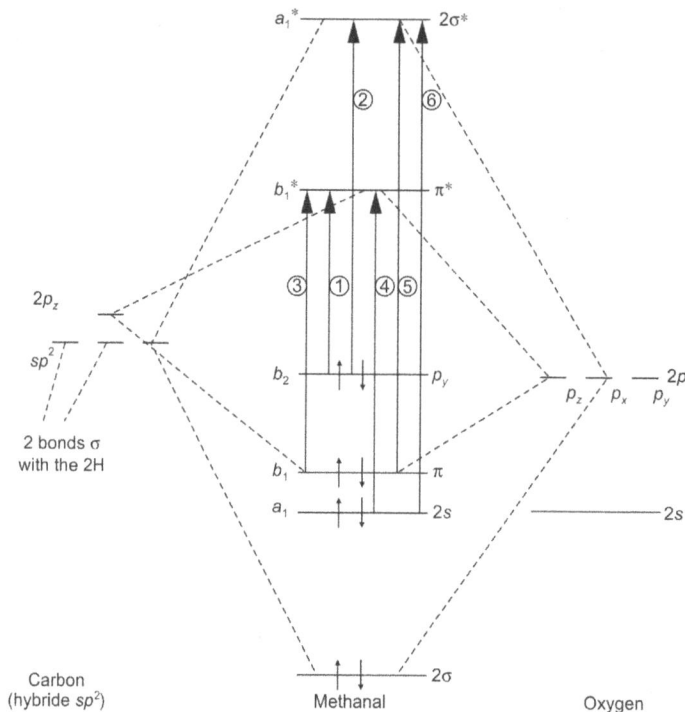

Figure 10: Electronic Energetic Levels of Methanal and Electronic Transitions (Reprinted with Permission from Gwenola and Jean-Louis Burgot, Paris, Lavoisier, Tec et Doc, 2017, 335).

The two parallel orbitals $2p_z$ overlap sideway and generate two new orbitals a bonding and an antibonding one. They are, respectively, named π and π^*. The π orbital has the shape of ellipsoids, one above the plane of sp^2 orbitals and the other under (Figure 9).

In ethylene, the energy of orbitals which are of interest for the transition UV are (by increasing order of energy values):

$$\sigma , \pi , \pi^* , \sigma^*$$

In the fundamental state, according to the diagram of molecular orbitals, four electrons may be considered in the transitions UV-visible which are obligatorily of the type $\pi^* \leftarrow \pi$. They are, in the fundamental state, the two electrons of the bonding σ orbital and the two electrons of the bonding π orbital. In the excited state, one of the two electrons of the bonding π orbital is raised into one of the anti-bonding π orbital.

The Formaldehyde

According to the theory of molecular orbitals (whether the followed reasonings are taken into account or not, hybridization or not, occurrence or not of delocalized orbitals σ), the electronic energetic levels are given in Figure 10.

$$CH_2 = CH - CH = CH_2$$
$$(\lambda_{max} = 210 \text{ nm})$$

$$CH_3 - (CH = CH)_6 - CH_3$$
$$(\lambda_{max} = 390 \text{ nm})$$

$$(\lambda_{max} = 235 \text{ nm})$$

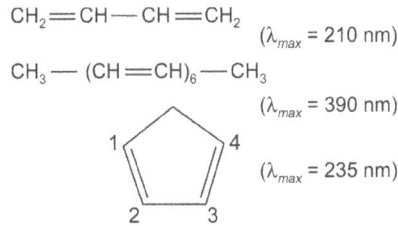

Figure 11: Structure of 1,3 Butadiene, Tetradecahexa-2,4,6,8,10,12 ene and Cyclopentadiene 1,3-Diene.

Symbols a and b are in relation to the symmetry elements of the orbitals. The orbital a_1 is the orbital 2s of the oxygen atom. It contains two non-bonding electrons, named electrons n. The orbital π, symbolized by b_1, comes from the sideway coupling of the orbitals $2p_z$ (the only one which is not hybridized) of the carbon atom and oxygen. The orbital b_2 is the initial doublet $2p_y$ of oxygen. It corresponds to still non-bonding electrons n. b_1^* is the orbital π^*. σ_1^* is the lowest orbital σ^*.

Six electrons are interesting for the study of the transitions in the UV domain (2 electrons 2s of the oxygen atom, 1 unpaired electron 2p of non-hybridized carbon and 2 electrons 2p of oxygen). The fundamental state is written $a_1^2 b_1^2 b_2^2$. Each state possesses an abbreviated spectroscopic symbol. Its abbreviated spectroscopic state is $^1A_1$. The 1 in the pre-raised location expresses the multiplicity of the state and A is the symmetry of the whole electronic configuration in relation to the elements of the symmetry group of the molecule. The index 1 is simply the number of the term. Once again, it must be noted that the capital letters A or B are related to the whole configuration, whereas the normal letters are related to the orbitals.

The excited levels are symbolized by:

$a_1^2 b_1^2 b_2 b_1^*$	spectroscopic state $^1A_2$
$a_1^2 b_1^2 b_2 a_1^*$	" $^1B_2$
$a_1^2 b_1 b_2^2 b_1^*$	" $^1A_1$
$a_1 b_1^2 b_2^2 b_1^*$	" $^1B_1$
$a_1^2 b_1 b_2^2 a_1^*$	" $^1B_1$
$a_1 b_1^2 b_1^2 a_1^*$	" $^1A_1$

The permitted transitions (for the symmetry reason) are:

$a_1^2 b_1^2 b_2 a_1^*$	$^1B_2 \leftarrow {}^1A_1$	transition 2
$a_1^2 b_1 b_2^2 b_1^*$	$^1A_1 \leftarrow {}^1A_1$	transition 3
$a_1 b_1^2 b_2^2 a_1^*$	$^1A_1 \leftarrow {}^1A_1$	transition 6
$a_1 b_1^2 b_2^2 b_1^*$	$^1B_2 \leftarrow {}^1A_1$	transition 4
$a_1^2 b_1 b_2^* a_1^*$	$^1B_1 \leftarrow {}^1A_1$	transition 5

The transition $^1A_2 \leftarrow {}^1A_1$ ($a_1^2 b_1^2 b_2 b_1^* \leftarrow a_1^2 b_1^2 b_2^2$) is not allowed for reason of symmetry. Because of the occurrence of reactional interactions, it takes place. However, its intensity is very weak. For $\lambda = 270$ nm, we measure $\varepsilon = 15$. It is characteristic of the function carbonyl. It is one of the bands Q $\leftarrow$ N. Transition 3 is a typical transition $\pi^* \leftarrow \pi$ (transition V $\leftarrow$ N). It is very intense. It is located at about 185 nm for all the carbonyl compounds. Transition 5 is also a usual one ($\sigma^* \leftarrow \pi$). Other transitions involve non-bonding electrons. They are of the kind ($\pi^* \leftarrow$ n) and ($\sigma^* \leftarrow$ n). These reasonings do not take into account multiplicities that are over 1.

Concerning the intensities of rays, a band is all the more intense as the transition which is at his origin is more probable. Some transitions are said to be forbidden since they are very poorly probable owing to the symmetry elements of the species both in its fundamental and excited states.

It is the case of transitions ($\pi^* \leftarrow n$) which are, as a result, not very intense. An intense transition is characterized by a very weak duration of the excited state (10^{-9} s). Contrarily, a weakly intense transition is associated with a duration of the excited state of the order of 10^{-6} s.

UV-visible Spectra and Molecular Structures

Chromophore Groups

The existence of bands is related to the occurrence of some functional groups absorbing in the UV-visible domain in the molecular structure of the absorbing species. These groups are named *chromophores*. They are characterized by the systematic presence in their structure of multiple bonds. Table 1 mentions some chromophores, their λ_{max}, their intensity and the solvent used to measure these parameters.

Table 1: Major Chromophore Groups.

Chromophore	Example	λ_{max} in nm	ε l.mol^{-1}.cm^{-1}	Solvent
C=O aldehyde	Acetaldehyde	290	17	Hexane
C=O ketone	Acetone	279	15	Hexane
COOH	Acetic acid	208	32	Ethanol
CO$_2$R	Ethyl acetate	211	57	Ethanol
CONH$_2$	Acetamide	178	9,500	Hexane
		220	63	Water
Ethylene	Oct-1-ene	177	12,600	Heptane
Alkyne	Oct-2-yne	178	10,000	Heptane
Aromatic	Benzene	255	215	Hexane

Conjugation

Table 1 shows that the presence of only one chromophore does not obligatorily lead to the appearance of absorption bands in the visible, even neither in the near UV. When, in the same molecule, coexist two such isolated chromophores, we notice that the absorption characteristics are the result of only one additivity effect.

In contrast, the fact that two chromophores are neighboring brings together their electronic energy levels. It is said that there is a conjugation of the orbitals of the two chromophores. Thus, the 1,3-butadiene absorbs at about 210 nm, whereas ethylene does absorb at about 190 nm (Figure 12).

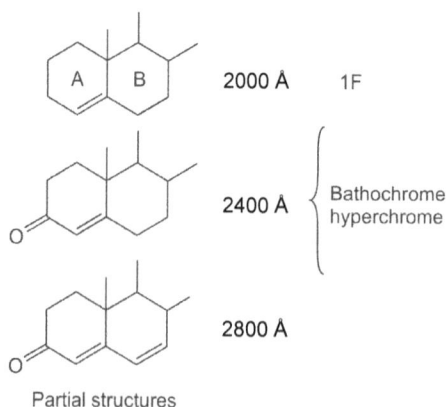

Partial structures

Figure 12: Partial Structures of Cholestene, Cholestenone and Cholestadienone.

The tetradecahexa-2,4,6,8,10,12-ene whose structure possesses 6 conjugated double bonds does absorb in the visible domain (λ_{max} > 390 nm). It must be remarked that conjugation not only displaces the absorption band toward the visible (bathochromic shift and the reverse; hypsochromic shift) but also increases the value of the molar coefficient (hyperchromic effect–reverse hypochromic effect). Concerning cyclopentadiene 1,3-diene, it is less conjugated than benzene although it is conjugated as is 1,3-butadiene.

Conjugation may occur not only between double bonds but also between multiple bonds of every origin. Another example of conjugation which can be evidenced by spectrophotometry UV-visible is that provided by the ethylenic and ketonic groups. Let us compare the absorbances of cholestene, cholestenone and cholestadienone (Figure 12).

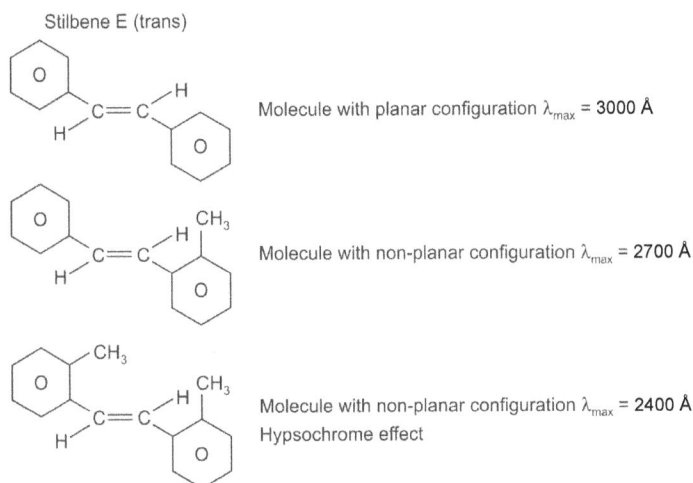

Figure 13: Structure of Derivatives of Stilbene.

The cholestene only possesses one double bond. It normally absorbs at 200 nm. The cholestenone which possesses a keto group conjugated with the preceding double bond absorbs at 240 nm. There is for this molecule simultaneously a bathochromic effect and a hyperchromic effect. The latter corresponds to an increase in the coefficient ε. This double effect is still more marked with the cholestadienone since it absorbs at 280 nm.

Conjugation is closely linked to the flatness of the molecule. The coupling of both chromophores becomes null when the planes containing each of them form a dihedral angle higher than 15 degrees. For example, the Stilbene E possesses an absorption maximum near 320 nm (Figure 13). Its α-monomethylated derivative is already less planar than the preceding one and does possess its absorption maximum near 290 nm. Its derivative α,α' – dimethyl is still less coplanar. It absorbs at 280 nm. The steric impeachment is accompanied by a hypsochromic shift generally accompanied by a hypochromic effect.

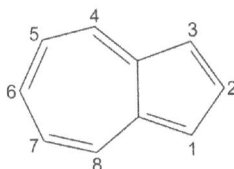

Figure 14: Structure of Azulene.

Despite the conjugation, of course, the sensed color remains complementary to that absorbed. For example, the azulene which absorbs in the orange-red (λ = 690 nm) appears to be dark-blue (Figure 14). The complementary colors are mentioned in Table 2.

Table 2: Absorbances and Colors.

Wavelengths in nm	Absorbed Color	Complementary Color
650–780	Red	Blue-green
595–650	Orange	Turquoise blue
560–595	Yellow-green	Purple
500–560	Green	Red purple
490–500	Blue-green	Red
480–490	Turquoise	Orange
435–480	Blue	Yellow
380–435	Violet	Yellow-green

Two colors are complementary when their superposition gives white color. The color of a solution is in relation to its capacity for absorption or reflection. The human eye sees the complementary color of the color which is absorbed. For example, if one substance in solution absorbs at $\lambda = 540$ nm, this means that the only absorbed color is green and the eye sees the purple red.

Auxochrome Groups

With chromophores, one also distinguishes some atoms like those of halogens or groups of atoms as the rest amino, hydroxyl which contain fewer free electrons than chromophores do possess. They are generally named *auxochromes*. They do not absorb per se in UV. Their effect is to increase or decrease the absorption of chromophores. Their effect depends on their location with respect to that of the chromophore. They work by developing electrostatic induction or electronic resonance effects.

Empirical Classification of UV Bands

In addition to the nomenclature V ← N and R ← N, already encountered, there exists another empirical classification of bands. It is based on the fact that the spectra of organic molecules permit the observation of four kinds of absorption bands, which is seldom simultaneously present. They are the bands R, A, E and B.

- R Band: They correspond to the presence of isolated chromophores. Their molar absorption is weak $\varepsilon < 100$. They are transitions of the kind $\pi^* \leftarrow n$;
- A Band: They appear when the molecule is a system of conjugated double bonds (for example, butadiene). They are transitions $\pi^* \leftarrow \pi$. Their molar absorption is high ($\varepsilon > 10,000$).
- E and B Bands: They characterize the spectra of aromatic molecules. E bands are characteristic of systems formed by 3 cyclic conjugated double bonds. They are also $\pi^* \leftarrow \pi$ transitions. This classification is rudimentary.

Part Played by the Solvent

Most often, the studied substances are in solution. Schematically, one distinguishes two kinds of solvents: the polar and the non-polar. Polar solvents, particularly those which are able to share hydrogen bonds, take part with the observed derivatives. As a result of this fact, there are modifications of the position and of the intensity of the bands which can be of some help for their identification.

For the transitions $\pi^* \leftarrow n$, there is a shift toward the weak wavelengths (toward blue) of the absorption of the band when one changes a non-polar medium for a polar one. This is due to a lowering of the energetic level of the fundamental state, which must be in relation to the association of the polar solvent with the free n electrons. For the transition $\pi^* \leftarrow \pi$, one notices a change towards high wavelengths (toward red) because, in this case, the molecules responsible for these transitions show a transition state more polar than their fundamental state, hence more solvated. As a result, there is a lowering of the excited state.

These considerations permit the identification of the transitions simply; for each spectral band, one notices and compares the location in polar and non-polar solvents. The bands characteristic of the transition $\pi^* \leftarrow n$ are displaced toward blue in polar media and the transitions $\pi^* \leftarrow \pi$ toward red.

In conclusion, we can say that some rules have been developed which correlate the known absorption data of substituted dienes with alkyl substituents (Rule of Woodward and Fieser). They have been extended to other basic structures with success, whatever the solvent. One example is provided by the works of Woodward, Gillam and Evans, Fieser and Fieser, concerning the $\alpha\beta$ unsaturated ketones which exhibit UV-visible spectra characteristics. According to these authors, the absorption maximum of $\alpha\beta$ unsaturated ketone depends on the number of substituents that are present in the $\alpha\beta$ position. Another factor is the fact that the substituent in position β still intervenes when the double bond is juxtanuclear. These rules have even resulted in many instances of a reassignation of some structures which are wrong.

CHAPTER 21

Ultraviolet-visible Spectroscopy II
Applications Analytiques

This chapter is essentially devoted to the quantitative analysis via UV-visible spectroscopy.

UV-visible Spectroscopy and Quantitative Measurements—
Beer-Lambert's Law

In 1729, P. Bouguer[21] have found that equal thicknesses of matter absorb proportional (unequal) quantities of light. In 1760, the German physicist and astronomer J. Lambert, who knew Bouguer's works, gave one relation more mathematical than the preceding one. In 1852, A. Beer, professor of mathematics in Bonn, found the accurate relation. Moreover, he noticed that his relation equally applied to solutions.

Let us consider a cuvette of length 1 containing the solution of the absorbing compound at concentration C. Let us also consider one thickness of solution dx (Figure 1). The radiative power (radiative energy by time unity) (unit W: Watt) brought by monochromatic radiation which enters into this thickness is ϕ and the power going out is $\phi + d\phi$. According to the works of the preceding authors, the absorbed power $d\phi$ is proportional to (see Figure 1):

- The thickness dx;
- The entering radiative power;
- The concentration C.

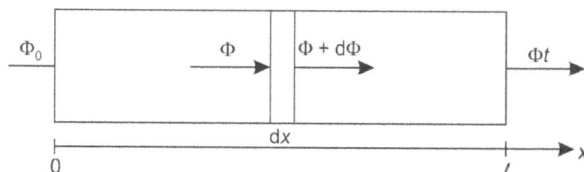

Figure 1: Setting Up Beer-Lambert's Law. (Reprinted with Permission from Gwenola and Jean-Louis Burgot, Paris, Lavoisier, Tec et Doc, 2017, 340).

As a result:

$$d\phi = -k\phi\, C dx$$

Here, k is a proportionality factor, hence:

$$d\phi/\phi = -KC\, dx$$

---

[21] Professor of hydrography of the King of France (1698–1758).

If ϕ_0 and ϕ_1 are the powers at the entrance $(x = 0)$ and at the going out $(x = l)$ of the cell, the integration between these limits gives:

$$\int_{\phi_0}^{\phi_1} d\phi/\phi = - KC \int_0^l dx \quad \text{and} \quad \ln(\phi_1/\phi_0) = - kCl$$

The ratio ϕ_1/ϕ_0 is named the *transmittance* T of the solution $(0 < T < 1)$. One prefers to handle the *absorbance* A (also called optical density[22]) defined by:

$$A = \log(\phi_0/\phi_1)$$

According to what is preceding:

$$A = k'/C \quad \text{with} \quad k' = 2,3k$$

The proportionality factors k and k' depend on the nature of the solute, solvent, temperature and the wavelength of the used radiation. According to the expressions of the concentration, the factor k' may be symbolized by:

- The ε is if C is expressed in mol L^{-1}. In these conditions, ε has for unit $l.mol^{-1}.cm^{-1}$ with l in centimeters. The ε is called the *molar absorption coefficient of the solute*;
- $A_{1cm}^{1\%}$ is if C is expressed in g per 100 ml of solution. The ε quantifies the intrinsic absorption of the solute. This is the value that is retained together with that of the corresponding wavelength for the structural analysis of the solute. The value of ε is all the higher as the probability of transition is itself high. It can change in a vast domain of values from some tens up to tens of thousands.

Beer-Lambert's law:

$$A = \varepsilon\, l\, C$$

is remarkably convenient to handle since the absorbance, directly given by the spectrophotometers for a given wavelength, is in linear relation with the concentration C, whereas the transmittance T is related to the absorbance by the equation:

$$A = -\log T$$

Remark: The law of Beer-Lambert may be obtained by considering the radiative powers per unit of surface, that is to say ϕ/A where A is the area of the portion of the solution of thickness dx. The reasoning is identical to the preceding one. The ratio ϕ/A is named luminous intensity. The unit is the watt per m^2 and the symbol I. Then, the Beer-Lambert law becomes:

$$\log I_0/I = \varepsilon\, l\, C$$

Where I_0 and I are the initial and transmitted intensities. Since each portion has the same area A, we can write:

$$\log I_0/I = \log \phi_0/\phi_1$$

The Beer-Lambert law is often expressed in terms of luminous intensities in literature.[23]

---

[22] This term is no longer recommended by IUPAC.

[23] Notice that the luminous intensity has the dimensions of a flux; it is the number of photons that cross an unit of surface by unit of time.

Properties and Limits of Beer-Lambert's Law

• The law is additive. For example, in the presence of two compounds A and B of respective concentrations C_A and C_B, the absorbance A of the solution is given by:

$$A = A_A + A_B$$

$$A = \varepsilon_A l C_A + \varepsilon_B l C_B$$

for a given wavelength and with an optical path length l. The additivity permits the dosage of mixtures and the determination of equilibrium and kinetic constants (Figure 2).

• For the measurements in the domain of physical chemistry, we must first recall that absorbance. The law function of concentrations and not of activities and that we often observe divergences with respect to linearity between the experimental and theoretical absorbances. These divergences are due to different parameters which are studied hereafter.

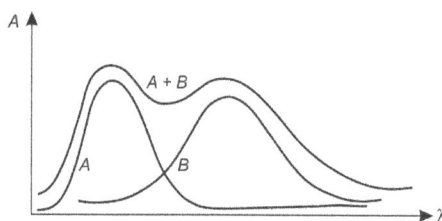

Figure 2: Spectrum of a Mixture of Two Constituents. (Reprinted with Permission from Gwenola and Jean-Louis Burgot, Paris, Lavoisier, Tec et Doc, 2017, 342).

Nature of the Incident Radiation

The incident radiation must be monochromatic. A passing band of incident radiation too large, (that is to say insufficiently monochromatic) leads to a lack of precision on the value of the absorption coefficient and the consequence of which is a divergence from the linearity of the absorbance (Figure 3).

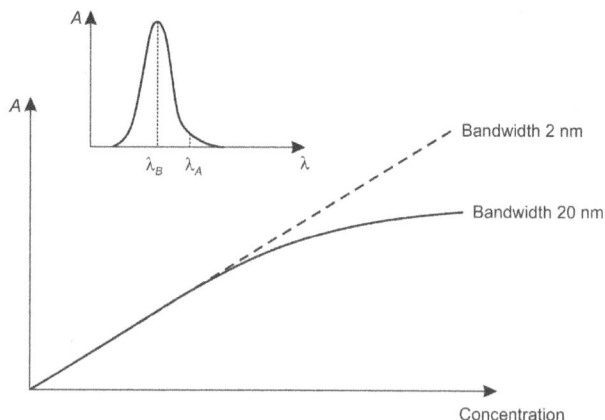

Figure 3: Influence of the width of the Band-passant on the Linearity. (Reprinted with Permission from Gwenola and Jean-Louis Burgot, Paris, Lavoisier, Tec et Doc, 2017, 342).

In quantitative analysis indeed one tries to exploit the wavelength, which gives the greatest absorbance (here λ_B). When this is the case, with a large band-width, the changes in absorption coefficients ε are all the higher as we stay closer to the maximum λ_B than to λ_A (see Figure 3).

Then, some divergences with respect to linearity may be noticed when there exist small quantities of parasitic radiations in relation to diffusion phenomena of light on some different optical parts. This light may reach the detector without having to cross over the measurement cell.

Let us recall that diffusion is a mechanism of transport of species under the influence of a gradient of concentration (see 'electrochemistry' book).

Nature of the Medium

Limpidity

Using spectrophotometry UV-visible can be only reserved for limpid solutions. (The study of suspensions is performed either by turbidimetry or by nephelometry).

Participation of the Solute to Chemical Equilibria

The solutes involved in acid-base, redox or complexation equilibria possibly exhibit spectra, which evolve during the establishment of the equilibrium. For example, the phenol may ionize according to the pH by giving a phenate ion (Figure 4):

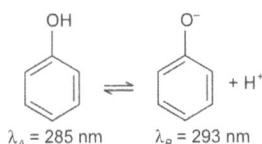

λ_A = 285 nm λ_B = 293 nm

Figure 4: Equilibrium Phenol/Phenate and Influence of the Wavelength Maximum of Absorption. (Reprinted with Permission from Gwenola and Jean-Louis Burgot, Paris, Lavoisier, Tec et Doc, 2017, 343).

More precisely, during the establishment of one acid-base equilibrium, measurements are performed in buffered solutions. For the determination of the acidity constant Ka, the pH values at which measurements of absorbances must be carried out with accuracy. The determination is grounded on the expression of the total absorption:

$$A = \varepsilon_{HA} C_{HA} l + \varepsilon_{A^-} C_{A^-} l$$

From the definition of the constant of acidity K_a:

$$[HA] = [H^+] \, C / (K_a + [H^+]) \quad \text{and} \quad [A^-] = K_a \, C / (K_a + [H^+])$$

Here, K_a is the dissociation constant of the acid and C is the total (analytic) concentration of the acid. K_a is immediately accessible by considering these two equations. It is sufficient to measure the coefficients ε_{HA} and ε_{A^-} at pH values of very acid and very basic and to measure the total absorbance A at the working pH values (see the section on 'Determination of Thermodynamic Constants').

Measurements can be performed at the isobestic point, where both molar absorption curves (those of the acid and the conjugated base, for example) cross each other when such point(s) exist(s). This is the case when the conjugated acid and base exhibit the same value of their molar absorption curve at a given wavelength(s) (Figure 5). The occurrence of isobestic points is interesting because it permits detection if there are only two or more species participating in the phenomenon.

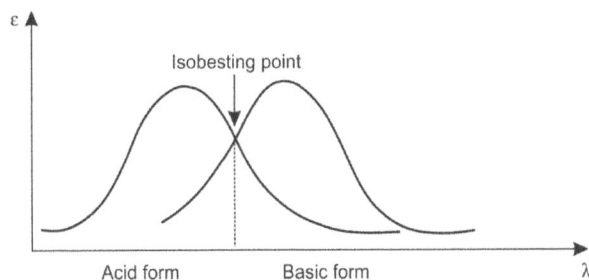

Figure 5: Definition of an Isobestic Point. (Reprinted with Permission from Gwenola and Jean-Louis Burgot, Paris, Lavoisier, Tec et Doc, 2017, 345).

Concentration of the Sample

At high concentrations, molecules react with each other from the strictly sole physicochemical point of view. This fact can for example modify their absorption of light. Some can form dimers which can be responsible for modifications of the molecular absorption coefficient. For example, methylene blue has a coefficient that increases by 88/100 at 436 nm when its concentration passes from 10^{-5} mol L^{-1} to 10^{-2} mol L^{-1}. Therefore, they must be prepared in media whose ionic strengths are weak.

Analytical Conditions of the Measurement of Absorbances

Different Quantitative Protocols

Several methodologies of measurement of absorbances can be adopted. They consist of:

- The measurement of absorbances at a given wavelength at different concentrations;
- The recording of the absorbances at a given wavelength. (This is a stability test);
- Titrations: The cell in which the titration is performed is located in the optical path.

A spectrometric measurement can play the part of an indicator of the endpoint. In this case, the molar absorption coefficient ε_T of the titrant must be very different from that of the titrated substance $\varepsilon_{substrate}$. One also can follow the appearance of the formed product if a suitable wavelength permitting the determination of the absorbance of the formed product of titration can be highlighted (Figure 6). It is a matter of absorptiometric titrations.

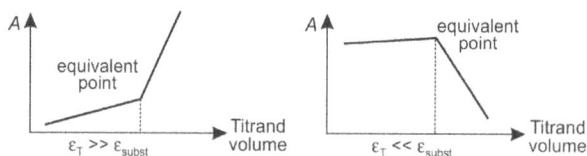

Figure 6: Examples of Evolution of the Absorbance During an Absorptiometric Titration. (In the Absorptiometric Titrations, the Evolutions of the Absorbances of the Titrant or the Titrand may also be followed if the values of their extinction molecular coefficients permit that). (Reprinted With Permission From Gwenola and Jean-Louis Burgot, Paris, Lavoisier, Tec et Doc, 2017, 354).

Optimization of the Measurements

Whichever the used methodology, the greater the absorbance is the less the light intensity striking the detector is and the risk of a lack of precision is great. There are three principal kinds of errors. They can add to each other. They are:

- The fundamental noise of the light source. It is considered as being constant. Recall that it arises from the physical properties of the electrical system. It cannot be eliminated.
- The fundamental noise of the photomultiplier devices (especially of the dynode).
- The light reflections and diffusions along the optical path.

These errors are cumulative and lead to a nonlinear change in the relative error on the measurement of the concentrations with the help of absorbances (Figure 7).

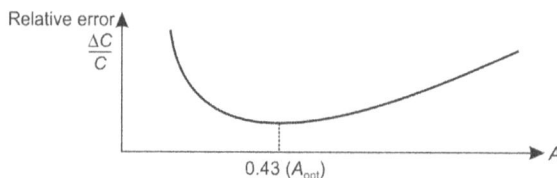

Figure 7: Distribution of the Relative Errors as a Function of the Absorbance. (Reprinted With Permission From Gwenola and Jean-Louis Burgot, Paris, Lavoisier, Tec et Doc, 2017, 354).

There is a particular absorbance for which the ratio of the relative error about the concentration is minimal together with the relative photometric error. Since a relationship between the absorbance and the concentration does exist, this means that a concentration exists such as the relative error due to the reading on the galvanometer producing a minimal error relative to the concentration. It is the concentration that gives the value absorbance A or transmittance T which follows:

$$A = 0.4343 \quad \text{and} \quad T = 0.37$$

These values are found as follows:

$$A = \log(I_0/I) = \varepsilon \, lC \quad \text{whence} \quad I/I_0 = e^{-2.303 l \varepsilon C}$$

$$dI/dC = -2.303 \varepsilon l \, e^{-2.303 l \varepsilon C}$$

Since $\varepsilon l = A/C$

$$dI/dC = I_0 \, e^{-2.303 l \varepsilon C} [-2.303 \, A/C]$$

$$dI/I_0 = -2.303 \, e^{-2.303 l \varepsilon C} [-2.303 \, AdC/C]$$

$$dC/C/dI/I_0 = [-0.4343 \, e^{2.303 l \varepsilon C}]/A = E$$

E is the ratio of the relative error about the concentration and of the photometric relative error. In order to find the optimal absorbance, we must cancel the derivative dE/dC:

$$dE/dC = d/dC \, \langle -0.4343 \, e^{2.303 l \varepsilon C]} \, \langle /\varepsilon \, l \, C$$

$$dE/dC = e^{2.303 l \varepsilon C} /C[0.4343/\varepsilon l C - 1]$$

$$dE/dC = e^{2.303 l \varepsilon C} /C[0.4343/A - 1]$$

$$dE/dC = 0 \quad \text{for} \quad A = 0.4343$$

That is to say

$$T = 0.37$$

The ratio of the relative errors remains weak for the values of transmittances located between 20 and 60%. The optimal domain of measurement is located in absorbances between 0.4 and 0.8 for the quantitative ones without forgetting the fact that exploiting absorbances stronger than 1.5 (or 2 according to the apparatus) and there no longer exists linearity of the answer of the detector.

Some Kinds of Applications

Spectrophotometry is a method easy to use. It is the easiest means of detection in liquid chromatography. Numerous applications have been developed, especially in the field of pharmacy.

Qualitative and Functional Analysis—Identification of Substances

The method detects chromophore functional groups but it does not permit the identification of molecules with certainty. It must be always completed by other methods giving physical information from spectral origins (infrared, RMN and mass) or physicochemical ones, such as the melting points, ebullition points and the obtained colorations after reactions with judicious reactants.

In contrast, we shall see that infrared spectra constituted by sufficiently sharp bands, located in another spectral domain, permit the identification with the quasi-certitude of the studied compound.

We have already mentioned that the spectroscopy UV-visible permits revealing the presence of conjugated systems and those with multiple conjugations. Let us also mention the facts that the method can permit to reveal the presence of some aromatic compounds. It is the case of aromatic derivatives possessing π non-linking electrons on some substituents such as $-OH$, $-NH_2$, etc. We have also mentioned that spectrophotometry UV-visible can show that there is some hindrance to the rotation of a suitable group in a molecule. Hence, the method may be of some help in stereochemistry.

Quantitative Analysis

This is a prime choice method for quantitative analysis. This is due to its performance in the domain. Its limits of detection are located between 10^{-4} and 10^{-5} mol. L^{-1}. They change with the values of the molar absorption coefficients ε. The accuracy and precision change from 1 to 4%.

Problems of Standardization

The spectrophotometry UV-visible as other kinds of spectrometry is a relative method. It involves the realization of a calibration curve A = F (C) with the help of solutions of known concentrations. When the analysis is performed upon a complex matrix, such as pharmaceutical forms, alimentary or about the environmental field, it is necessary prior to the analysis to check preliminary the absence of interferences due to the medium. This may be done by comparing the slopes of the calibration lines obtained by dissolving standards in a solvent and in the medium with the matrix. From another standpoint, if the fundamental noise is too high, the method named "the standard addition method" can no longer be used. A means to obviate this problem is to use mathematical methods to process the data.

Analysis of Mixtures

Direct Methods. First, a basic algebraic method may be used. It is grounded on the additivity of Beer-Lambert's law. The absorbance of the mixture is determined at n wavelengths if the mixture contains n compounds. For example, let us consider the analysis of three substances A, B and C at three wavelengths λ_1, λ_2 and λ_3. The absorbances of the medium are given by the equations:

$$\text{At } \lambda_1 \quad A_1 = \varepsilon_{1A} l\, C_A + \varepsilon_{1B} l C_B + \varepsilon_{1C} l\, C_C$$
$$\text{At } \lambda_2 \quad A_2 = \varepsilon_{2A} l\, C_A + \varepsilon_{2B} l C_B + \varepsilon_{2C} l\, C_C$$
$$\text{At } \lambda_3 \quad A_3 = \varepsilon_{3A} l\, C_A + \varepsilon_{3B} l C_B + \varepsilon_{3C} l\, C_C$$

This methodology is only suitable for compounds whose spectra are sufficiently different from each other. A matricial calculation is recommended for solving this or analogous mathematical system.

Using Derivative Spectrophotometry. The derivative spectrophotometry method consists in transforming the absorption spectrum (zero order) into a spectrum of primary derivatives (or secondary or even superior orders). With this methodology, differences even imperceptible on the zero-order spectrum can be observed (Figure 8).

The first-order spectrum shows the slope of the zero-order spectrum at each of its points. That is to say, it is a diagram $dA/d\lambda$ as a function of λ. The derivative $dA/d\lambda$ remains proportional to the concentration as it is shown by the simple derivation of Beer-Lambert's law.

The derivative spectrophotometry exhibits the following advantages:

- It leads to a better resolution of the spectra than that obtained with the zero-order because the wavelength is obligatorily maximal:
- It permits better possibilities for analysis of trouble media and mixtures.

Let us take the example of the stability of the mixture pentoxifylline–nicergoline in solutions of glucose at 5% for perfusion (documents come from our laboratory) (Figure 9).

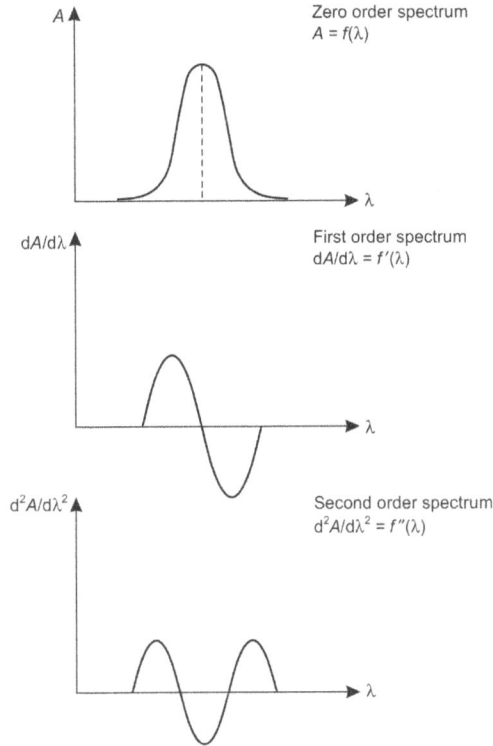

Figure 8: Derivative Spectra. (Reprinted With Permission From Gwenola and Jean-Louis Burgot, Paris, Lavoisier, Tec et Doc, 2017, 358).

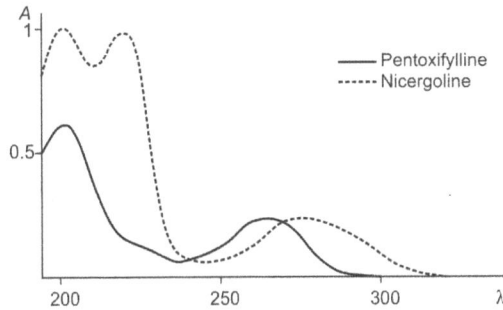

Figure 9: Spectra of the Nicergoline and Pentoxifylline. (Reprinted With Permission From Gwenola and Jean-Louis Burgot, Paris, Lavoisier, Tec et Doc, 2017, 358).

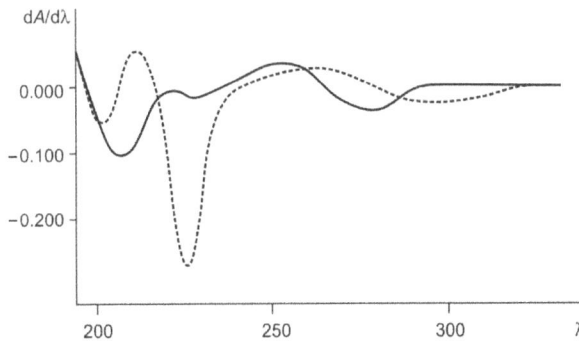

Figure 10: First Order Derivative Spectra of Pentoxifylline and Nicergoline in Mixture. (Reprinted With Permission From Gwenola and Jean-Louis Burgot, Paris, Lavoisier, Tec et Doc, 2017, 359).

The analytical exploitation of the first-order spectrum uses the methodologies of the points of cancellation. It consists of the measurement of the absorbance in first-order spectrophotometry at the maximal absorption of the pentoxifylline. It permits the cancellation of the signal due to this molecule $(dA/d\lambda) = 0$ and quantifies the nicergoline; inversely, by placing ourselves at the maximal wavelength of absorption of nicergoline, the pentoxifylline only participates in the absorbance.

Indirect Methods—Mathematical Processing of Data. Numerous software of data processing permit finding of the composition of mixtures. They are least squares methods, regressions and methods in the principal component.

Determination of Non-Absorbent Substances

Determination of Some Inorganic and Even Organic Substances that Do Not Absorb in the UV-Visible Domain. It can be, however, performed after the formation of colored and stable complexes with organic chelating agents. Let us mention, for example:

- Copper in the presence of diethyldithiocarbamate;
- Lead in the presence of diphenyl-thiocarbazone;
- Nickel in the presence of dimethylglyoxime;
- Iron in presence of o-phenanthroline in an acidic medium;
- The free chlorine with the diethyl p-phenylene diamine;
- The nitrates with the sulfanilamide and the N-(1-naphtyl)-ethylenediamine after their reduction in nitrites.

Measurements of Enzymatic Activities. They are largely grounded on the differences in absorbances between the reduced and oxidized forms of the cofactors NAD and NADP. One example is provided by the determination of the glutathione peroxidase according to the method of Paglia and Valentine:

glutathione peroxidase

$$2GSH + ROOH \leftrightarrows GSSG + ROH + H_2O$$

where ROOH is the cumene hydroperoxide:

$$GSSG + NADPH + H^+ \leftrightarrows 2GSH + NADP^+$$

The glutathione peroxidase activity is computed by the decrease of the absorption at 340 nm of NADPH which is oxidized in $NADP^+$ (Figure 11).

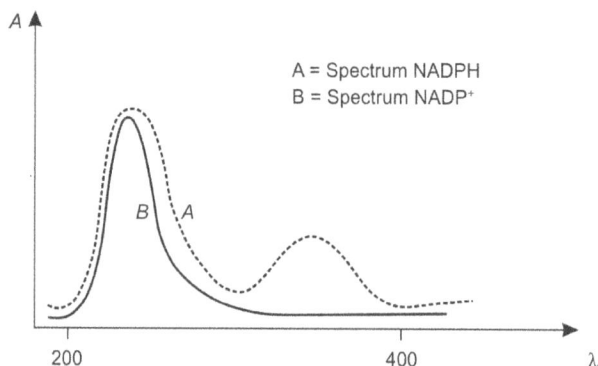

Figure 11: Spectra of NADPH (A) and $NADP^+$ (B). (Reprinted With Permission From Gwenola and Jean-Louis Burgot, Paris, Lavoisier, Tec et Doc, 2017, 360).

Quality and Purity Control of the Raw Pharmaceutical Materials

The presence of impurities may modify the UV-visible spectrum of the system. This can be perceptible by changes in the positions of the maxima and by the arrival of shoulders or supplementary peaks due to the envelopment of the various peaks. All the pharmacopeia provides several protocols in order to evidence these substances, whichever they are synthesis intermediaries or degradation products.

Determination of Thermodynamic Constants

Determination of the Equilibria Constants of Formation of Complexes

A modern methodology permitting to determine the thermodynamic constants in solution, notably in calculating by spectroscopy UV-visible is to proceed by simulation with information technology. In the occurrence, it consists in calculating the absorbances of a solution in which one or several successive complexes occur in different experimental conditions and comparing them to the values determined in the same experimental conditions. The calculated absorbances require knowing the concentrations of all the species, including those of the complexes; that is to say, finally, knowing the sought equilibrium constants. The comparison of the calculated absorbances with the tried constants and the experimental absorbances is realized by calculating a cost function that is minimized by proceeding with rational changes of the tried constants. This strategy is probably the best one. The difficulty is often a mathematical one. It is the research of the search of constant which minimizes the cost function.

Determination of Other Equilibrium Constants

Other equilibria or kinetic constants may be obtained by following the same strategy. It is the case for the determination of acid-base constants. The bases of these determinations have been already given (see the section on 'Properties and Limits of Beer-Lambert's Law'). Other constants such as kinetic constants can be determined, according to the same principle. We take the example of the determination of the pK_a value of the couple HX/X^-, which is a weak acid.

The measurement of the total absorbance A at different pH permits to determine pK_a of a couple of acid-base. Henderson's relation indeed:

$$pH = pK_a + \log [X^-]/[HX]$$

involves the concentrations (activities) of the acid and basic forms which are accessible by spectrophotometry UV-visible. The conditions of success of the whole experiment are the following ones;

- The buffer must not absorb the radiations;
- The spectra of the pure acid and basic forms must be obtained previously;
- The choice of the wavelength must correspond to the greatest difference between the absorbances of both forms. It is the better one from the standpoint of precision,
- The study of the changes of the absorbances in the spectra of the couple HX/X^- at different pH values must be carried out at this wavelength.

The principle is that at a given pH value, the total absorbance A of the solution is:

$$A = (\varepsilon_i C_i l + \varepsilon_M C_M l)$$

Where ε_i and ε_M are the molar absorbances coefficients of the ionized X^- and molecular HX forms of the couple and l is the length of the cell measurement and C_i and C_M are the concentrations of the ionized and molar forms. We have:

$$C_i + C_M = C$$

Where C is the total concentration or (analytic concentration) of the weak acid.

One can write:

$$A = C_i\varepsilon_i l + (C - C_i)\varepsilon_M l$$

Once and for all, it is not difficult to demonstrate that;

$$[X^-] = K_a C/([H^+] + Ka) \quad [HX] = [H^+]C/([H^+] + K_a)$$

$$[X^-] = C_i \quad [HX] = C_M$$

$$C_i = (A - C\varepsilon_M l)/(\varepsilon_i - \varepsilon_M)l \quad C - C_i = (C_i\varepsilon_i l - A)/(\varepsilon_i - \varepsilon_M)l$$

As a result:

$$pK_a = pH - \log (A - C\varepsilon_M l)/(C\varepsilon_i l - A)$$

$C\varepsilon_M l$ and $C\varepsilon_i l$ are the absorbances of the compound at the (analytic) concentration C, respectively, in acidic and basic media. They are experimentally measured by working at pH sufficiently acidic and sufficiently basic; that is to say by respectively working at pH < pKa –2 and pH ≈ 11–12.

By multiplying the last expression by $[H^+] + Ka$, by developing the forming terms and by regrouping those in Ka and H^+, we arrive at the expressions:

$$Ka = [H^+](\varepsilon_M C - A)/(A - \varepsilon_i C)$$

$$pKa = pH + \log(A - \varepsilon_i C)/(\varepsilon_M C - A)$$

$\varepsilon_i C$ is the absorbance of the solution (of the studied substance) at the same concentration as previously in the same cell the optical path of which l at the same wavelength, but at a pH value so that it is fully ionized. $\varepsilon_M lC$ is the absorbance of the solution, the studied solution at the same concentration as previously, in the same cell of optical path l at the same wavelength. Yet, it is at a pH so that it is fully under the molecular form. Therefore, for an acid HA, we find the expression:

$$pKa = pH + \log(A - Ai)/(A_M - A)$$

We would find for acid of the form BH^+:

$$pKa = pH = \log(A - A_M)/(Ai - A)$$

Since in the preceding example, the concentration C is weak (about 10^{-4} mol^{-1}), activities and concentrations would be, as a rule, assimilated and the obtained ionization K_a be the thermodynamic constant K_a°. But the presence of a buffer, especially if it is multivalent, leads to problems with ionic strengths. This is the reason why the above obtained constant is the conditional one. Nevertheless, the spectroscopy UV-visible can also help to carry out calculations taking into account the problem of activities.

Colorimetry

The term colorimetry comes from the fact that the first determinations consist of the visual comparison of the colors of the sample and solutions of standards of known concentrations. Colorimetry is the direct comparison of the transmitted luminous flows. Colorimetric measurements use "white and polychromatic lights" in the visible domain.

The domain where it can be applied is not only limited to the determination of derivatives possessing chromophore groups in the visible. It is possible indeed to transform a substance that does not possess any chromophore into another, which can absorb light by complexation. On the other side, in the case of a mixture of two absorbing derivatives, the chemical transformation of one of them into another in contrast with the preceding can absorb light and makes it easier for the analysis of the initial mixture.

• **Colorimetry by a Visual Method:**

The simplest colorimeters are used for routine determinations and are visual devices that permit comparisons. They are presented in the form of tubes or polyacrylate discs the mass of which brings the range of measurements as printing on them (see color Photo 6). The disc is submitted to rotations until the colors of the sample and the disc are matched, that is to say until the identities of colors.

Numerous tests are presented in the form of strips. They can be compared with the colors of the scale printed on the device. In this case, that is rather one analysis of the reflected light. This is one application of reflectometry. There is reflection when radiation crosses an interface between media that differ in refractive index.

Colorimetry has given rise to very numerous applications in various domains such as:

• Environment, in particular, the pollution of waters;
• Pharmacy with the control of different waters, in particular the water for hemodialysis, raw materials, medical disposals and so forth;
• Public health with for example the control of the proportion of oxidants;
• The food;
• The clinical biology.

• **Fiber Optics:**

Relatively to UV-visible spectroscopy, it seems interesting to put forward the device entitled "the fiber optics". They can be considered a new component of a UV-visible spectrophotometer. They

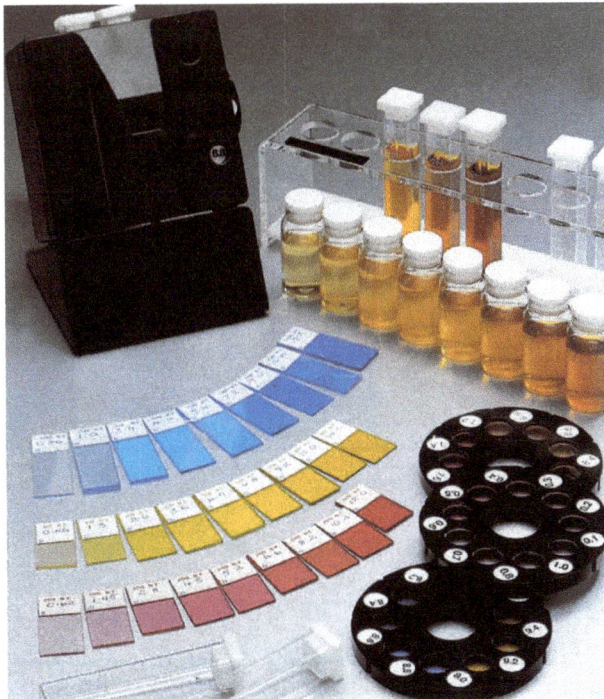

Photo 6: Colorimetry by a Visual method.

can also be found as a component of other optical instruments. According to materials constituting the fibers, they can transmit ultraviolet-visible and infra-red radiations at a distant place. It seems that it is in the first domain that they are the most used.

Optical fibers are fine strands of glass or plastic. They can transmit radiation along distances of the order of one hundred meters. The diameter of fibers is in the range of 0.05 to 0.6 cm. Light transmission in an optical fiber is carried through total internal reflections. For that, it is necessary that the fiber should be coated with a special material. Also, both materials, the fiber and the coating material do possess refractive indexes of well-definite values. Probably, they are the most used in the medical domain. It is because of their flexibility that permits the transmission of images in particularly difficult conditions. A device of the kind "fiber optics" is also used in infrared spectroscopy (see Chapter 29).

On the Apparatus

Numerous checks permit qualifying the apparatus as early as the beginning of its use or for later. The checks consist of:

- The research of the fundamental noise. It is based on the measurements of zero and 100 per 100.
- The study of the deviation of the basic line from the horizontal straight line.
- The verification of the accuracy of the measured absorbances. This may be done with the help of a solution of potassium dichromate. The values of the absorbances at given wavelengths are checked.
- The research of the parasitic light or the test of the diffused light.
- The verification of the resolution power. This check is performed by a study of the obtained spectrum with a mixture of toluene and hexane.
- The verification of the accuracy and the precision of the wavelength values. The check of the scale of wavelengths is done by the study of some compounds such as holmium oxide and also by the position of some rays.

CHAPTER 22

Molecular Luminescence Spectrometry

Fluorescence, Phosphorescence and Chemiluminescence

This chapter is principally devoted to the spectroscopy of molecular fluorescence. But the spectroscopy of phosphorescence is also somewhat investigated and this is also the case of chemiluminescence as well. These three types of methods are based on the phenomenon of luminescence because they are brought about by the absorption of photons.

Molar spectrofluorometry is essentially a method of quantitative analysis. It is grounded on the measurement of the intensity of the emitted light by fluorescence by a substance after excitation by photons coming from the domains of visible and near UV. In some conditions, this intensity is a linear function of the concentration of the responsible substance. Spectrofluorometry is recorded in various pharmacopeia.

Origin of the Phenomenon of Fluorescence and Phosphorescence

The absorption of UV-visible lights by various molecules engenders a population of these species which lie in electronic excited states (Figure 1). We know that each electronic state does possess numerous levels of vibrational energy and the excited molecules split up in these different levels (Arrow 1). Most often, this state is singlet. This means that there was no turning over of the electron undergoing the transition during the absorption. Therefore, after the latter, all the electrons have their spin paired. The total spin S of the molecule is S = 0. The relation 2S + 1 immediately gives its spectral multiplicity[24] of the considered energy level, which is 1.

---

[24] It corresponds to the number of ways in which the quantum number S (here S) can be oriented in space.

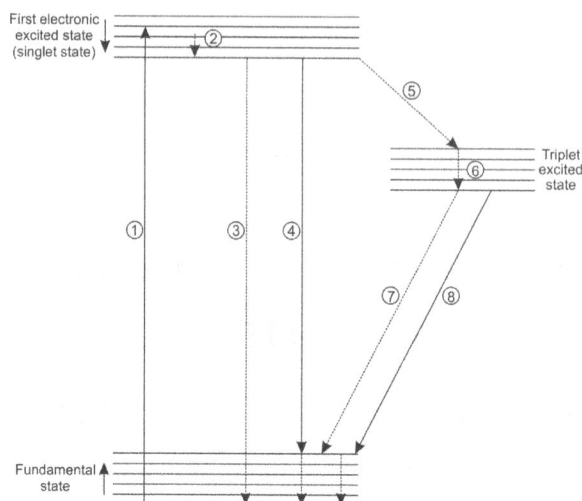

Figure 1: Jablonski's[25] Diagram (Simplified Representation). Directions of the Arrows are Those of the Transitions and Correspond to a Non-Change of Sign). (Reprinted with Permission from Gwenola and Jean-Louis Burgot, Paris, Lavoisier, Tec et Doc, 2017, 370).

Then, the excited molecules lose their vibrational energy by falling into the lowest vibrational energy of the excited state. This process is named vibrational relaxation. It is not radiative. It is represented by Arrow 2. The energy that is lost during the vibrational relaxation goes to the profit of the solvent molecules. After the vibrational relaxation, the excited molecule can then return to the fundamental electronic state either by emitting photons (Process 2) or by a non-radiative process (Arrow 4). This case is named internal conversion. In the first one, it is the phenomenon of *fluorescence* (Arrow 3). Fluorescence is a phenomenon of *luminescence* by which a molecule undergoes a transition from an excited singlet state to the fundamental electronic singlet state. The lifetime in the excited state is weaker than 10^{-8} s. Given the phenomenon of vibrational relaxation in the excited state and given the fact that the molecule goes again in the fundamental electronic state into a vibrational level higher than that it had before excitation, the radiation emitted by fluorescence is of lower energy (Arrow 3). Hence, it is located at a higher wavelength than that absorbed (Process 1). The fluorescence phenomenon is also followed by a vibrational relaxation in the fundamental state.

The return to the fundamental state by a non-radiative phenomenon is named *internal conversion* (Process 4). In this case, the energy is also lost to the profit of solvent molecules and there is heat production. The competition between fluorescence and internal conversion phenomena is quantified by the *quantum yield of fluorescence* Φ, which is the ratio of the number of emitted quanta and the number of absorbed quanta (see the section on 'Factors influencing the intensity of fluorescence').

With some other compounds, another phenomenon can occur. It is named *intersystem crossing*. In this case, a molecule situated in the lowest vibrational level of the excited state (a single state) goes upper in a triplet state. This state possesses a level of energy intermediary between those of the fundamental and excited electronic states. The intersystem crossing is carried out with a turning over of spin (Process 5). Then, the molecule falls into the lowest vibrational state of the triplet state by vibrational relaxation (Process 6). The lifetime of the triplet (10^{-4} to 10 s) is much higher than that in the preceding singlet. After the intersystem crossing, the molecule returns then into the fundamental state by emission of radiation (Process 8). This kind of luminescence is called *phosphorescence*. The phosphorescence can be defined as being one emission of the radiation corresponding to the transition from a triplet state to a fundamental singlet state. This transition can also occur without

---

[25] Alexandre Jablonski, Polish physicist (1898–1980).

the emission of radiation (Process 7). The lifetime of the excited state is much longer than that of the singlet because the emission of phosphorescence is due to a forbidden transition as is indicated by quantum mechanics.

(*Remarks*: The excitation in the singlet state may be immediately followed by the emission of fluorescence without the occurrence of vibrational relaxation in the excited state. The energy of emission is, therefore, the same as that of absorption and the wavelengths of the emitted and absorbed radiations are the same. One speaks then of fluorescence by resonance. This is a process of weak probability.)

The lifetime in the excited state is precisely stated by the relation giving the fluorescence intensity as a function of time:

$$F = F_0 \, e^{-kt}$$

Where F is the fluorescence intensity at time t and F_0 at that time is t_0. The k (s^{-1}) is the spontaneous emission coefficient, and e is the basis of the Naperian logarithms ($e = 2.718...$).

The mean lifetime of the excited state τ is defined by:

$$\tau = 1/k$$

It is of the order of some picoseconds (10^{-12} s) to some nanoseconds (10^{-9} s). It corresponds to the time necessary for the fluorescence to be equal to:

$$F = 0.36 \, F_0$$

An increase in temperature favors the intermediary intermolecular bumps between species and the dissipation of energy developed as heat. This point as a result reduces the intensity, quantum yield and lifetime of fluorescence.

Two kinds of spectra are interesting in fluorometry. The *excitation* spectrum is recorded by scanning the wavelength of the primary radiation while monitoring the emitted luminescent one. The obtained curve is nearly identical to the absorption spectrum. This cannot be unexpected when one considers theoretically Figure 1, which explains why the maxima and minima of the excitation and absorption UV-visible spectra should be superimposable. It is not exactly the case for instrumental reasons. The chosen wavelength of excitation must correspond to a strong excitation. The second type of spectrum is the *emission* spectrum. It is obtained by locking the excitation wavelength at a well-defined value and by measuring the fluorescence intensity as a function of the wavelength of the emitted light.

The transitions of excitation involve more energy than transitions of emission except that of resonance which is of common value. The maximal wavelength of emission is, therefore, greater than that of absorption which was at its origin. The corresponding shift $\Delta\lambda$ is called the Stokes'[26] shift.

(*Remark*: Some molecules, although being in the fundamental state, can be in a vibrational energy level higher than the lowest one. The absorption transition is, hence, less energetic and spectra of emission and excitation exhibit a partial recovery. Figure 2 illustrates this point.)

If all the molecules are originally in the ground state, the weakest energy they can absorb during the Process 1 is equal to the greatest energy transition possible in fluorescence.

The optimal wavelength of excitation which corresponds to a maximal intensity of fluorescence does not obligatorily correspond to the absorption maximum of the compound. Previously to an experiment, it is necessary to measure the fluorescence intensity for each value of excitation wavelength in order to find the one which optimizes the phenomenon (Figure 2).

---

[26] George Gabriel Stokes, Britannic mathematician and physicist (1819–1903).

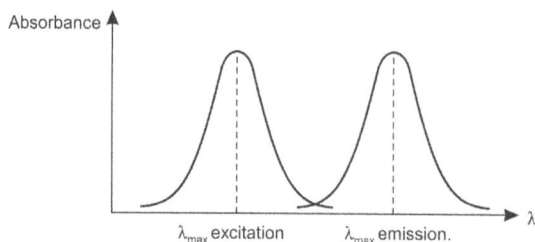

Figure 2: Overlapping of Excitation and Fluorescence Spectra. (Reprinted with Permission from Gwenola and Jean-Louis Burgot, Paris, Lavoisier, Tec et Doc, 2017, 373).

Among the possible electronic transitions, only the transitions $\pi^* \leftarrow n$ and $\pi^* \leftarrow \pi$ give rise to fluorescence phenomena. The spectrum of emission is identical, whichever the wavelength chosen for the excitation is. But the intensity of the fluorescence is maximal at the maximal excitation wavelength. The excitation spectrum is, as a matter of principle, identical to the absorption spectrum of the studied compound. One difference between these spectra can be attributed to the occurrence of impurities in the sample. The occurrence of impurities modifies more easily the excitation spectrum rather than the absorption spectrum.

(*Remark*: These phenomena have applications that go further in the field of analysis. We mention the cases of lamps, paints, markers, etc. Notably, it is the case of fluorescent lights, the working principle of which is the following one: a filament located in a tube emits electrons by heating. They are accelerated by a potential difference and collide with the mercury vapors contained in the tube; the mercury atoms which are excited go back to the fundamental state by emitting visible light. If the emitted light contains all of the visible spectra, one obtains "a white radiation" also called "cold radiation".)

Quantitative Aspects of Fluorescence

Let us consider a substance S in a cuvette, submitted to a luminous monochromatic beam of intensity I_0. Let us call the intensity emitted by fluorescence as F (Figure 3):

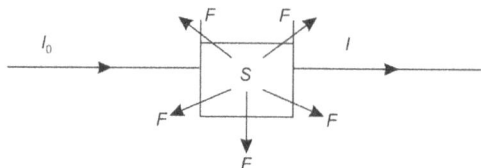

Figure 3: Irradiation of a Fluorescent Substance by a Monochromatic Beam. (Reprinted with Permission from Gwenola and Jean-Louis Burgot, Paris, Lavoisier, Tec et Doc, 2017, 373).

One measures the intensity of the radiation emitted under an angle of $90°$ with respect to the incident beam after it has crossed over a filter which stops the latter. However, it gives way to the fluorescent beam.

The light absorbed by the substance S follows Beer-Lambert's law, that is to say:

$$I = I_0 e^{-\varepsilon Cl}$$

Here, I is the intensity of transmitted monochromatic incident light and I_0 is that of the incident monochromatic light. The ε is the molar absorption coefficient, S is the concentration of substance S (mol L^{-1}) and l is the length of the cuvette.

Beer-Lambert's law applies to the intensity transmitted by the solution, but it does not apply to the intensity of fluorescent light. However, it is possible to express the intensity of fluorescence F as a function of the intensity of the incident radiation I_0. Since it is impossible that there exists emission by fluorescence without absorption of light, it is logical to assert that the intensity of fluorescence F is proportional to the absorbed luminous flood given by the difference $(I_0 - I)$. Moreover, as in solution, the intensity of fluorescence is different from point to point of the cell. The measured intensity of fluorescence is the resultant of the local intensities of fluorescence. Hence, we can write:

$$F = \varphi \, (I_0 - I)$$

Where φ is a proportionality factor and it is called the *quantum yield* or the *yield of the fluorescence*. Hence, we can write:

$$F = \varphi \, (I_0 - I_0 \, e^{-\varepsilon Cl})$$

$$F = \varphi \, I_0 \, (1 - e^{-\varepsilon Cl})$$

Applying McLaurin's relation gives:

$$e^{-x} = 1 - x + x^2/2! - x^3/3! + \ldots..$$

When $x \ll 1$, the terms $x^2/2! - x^3/3! + \ldots$ become negligible with respect to x and $e^{-x} = 1 - x$. Generally, $\varepsilon Cl < 0.02$ and $e^{-\varepsilon Cl} \approx 1 - \varepsilon Cl$. As a result, we find for the expression of F:

$$F = \varphi I_0 (1 - 1 + \varepsilon Cl)$$

$$F = \varphi I_0 \varepsilon Cl$$

The intensity of the radiation of fluorescence is directly proportional to the concentration of the substance S in the cuvette. This relation suffers the same limits of applications as Beer-Lambert's law on which it is grounded.

The quantum yield is defined by the ratio:

$$\varphi = \text{number of emitted photons/number of absorbed photons}$$

Since

$$E = h\nu \quad \text{or} \quad E = hC/\lambda$$

$$\varphi = E_F/E_A \times \lambda_F/\lambda_A$$

Relation in which E_F is the energy of the fluorescent radiation, E_A is that of the absorbed radiation, λ_F is the wavelength of the fluorescent radiation and λ_A is that of the absorbed one. The value of φ changes from 0 and 1. It is characteristic of the studied substance and is defined for a given temperature and a given solvent. (0 corresponds to the absence of fluorescence and one to a maximal fluorescence). The φ may considerably change from one solvent to another; about 10 to 100. It is independent of the value of the excitation wavelength.

Let us consider the following example of the calculation of a quantum yield. It is the example of the photolysis of gaseous hydroiodic acid HI by radiation, the wavelength of which is $\lambda = 2{,}537$ Å. It was found that the absorbed energy was $3.07 \, 10^2$ Joules and decomposed 0.0013 moles of HI. The reaction of decomposition is:

$$2HI \leftrightarrows H_2 + I_2$$

The energy of the quantum at 2,537 Å is given by $E = h\nu$;

$$E = 6.63 \, 10^{-34} \times (C/\lambda)$$

Since $C/\lambda = \nu$. As a result:

$$E = 6.63 \, 10^{-34} \times (3 \times 10^8)/(2{,}537 \times 10^{-10}) = 7.83 \, 10^{-19} \text{ J}$$

The yield of the fluorescence process is measured by the value of the quantum yield. This is an important parameter from the theoretical point of view since it permits a better grasp of the rates of the radiative and non-radiative processes together with the values of corresponding energy transfers. Quantum yield, indeed, is related to the kinetic constants k_r and k_{nr}, which are the constants of deactivation by the radiative and non-radiative ways according to the expression:

$$\varphi = k_r/(k_r + k_{nr})$$

The measurement of the absolute quantum yield is difficult and involves using a special apparatus. In contrast, the measurement of the relative quantum yield is easy. One compares the emission spectra obtained, in the same conditions, from the studied product X and a standard one S; the quantum yield of which is known. Knowing the quantum yield of the standard product φ_S, one immediately computes the unknown one φ_X. The used relation is:

$$\varphi_X = (A_S/A_X)(F_X/F_S)(n_X/n_S)^2\varphi_S$$

Here, A_S and A_X are the absorbances of products S and X at the excitation wavelength, F_S and F_X are the areas under the emission curves expressed in numbers of photons and n_X and n_S are the refraction indices of the solvents into which the products are dissolved. In this relation, the absorbances play the part of a measurement of the absorbed photons and area F is a measurement of the emitted photons.

Factors Influencing the Intensity of Fluorescence

Increase in the Fluorescence

Impurities

The presence of impurities may be responsible for an increase in fluorescence. One must be watchful about the quality of the used solvents. The detergent residues and plasticizers can provoke an increase in fluorescence.

Complexation

The formation of a complex between the fluorescent species and a chelating substance may increase the rigidity of the molecular system and because of this fact, it may increase the fluorescence. For example, let us mention the complex given by the 8-hydroxyquinoline and Zn^{2+} (Figure 4). Let us also cite the tetracyclines which give a complex with Al^{3+}.

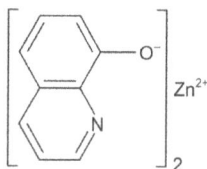

Figure 4: Complex of 8-Hydroxyquinoleine With Ion Zn^{2+}. (Reprinted with Permission from Gwenola and Jean-Louis Burgot, Paris, Lavoisier, Tec et Doc, 2017, 376).

Decrease of the Fluorescence or Quenching

Quenching From Chemical Origin

Quenching of Collision. The molecule of the quenching agent reacts with an excited molecule, which then becomes inapt to fluorescence. Molecules bringing halogenated and oxygenated groups are quenching agents of this kind.

Quenching by a Change of Structure. Acido-basic substances may have their intensity of fluorescence increasing or decreasing according to the pH of the solution. For example, aniline is fluorescent in the interval of pH 6 < pH < 14. If the pH-value decreases, an anilinium ion is formed. It does not fluoresce. It is also possible to meet intermediary cases in which the acid and base forms exhibit different emission spectra and/or quantum yields.

From another standpoint, the reaction of transfer of the proton is very fast and can occur during the lifetime of the excited state. The pK_a value of a molecule in an excited state is different from its value in its fundamental state. It is true that in some cases, some acids are up to 10^6 times stronger in their excited state. This may be explained by the fact that the energy necessary to produce the ionization is weaker in the excited state than in the normal state.

Quenching of Physical Origin

It is due to the absorption of the radiation emitted by the substance itself. This is called *self-quenching*. It decreases the quantum yield. The example type is phenol. It is possible to decrease the phenomenon by diluting the sample.

Influence of the Solvent

The solvent may modify the fluorescence of a substance by the influence of its properties (refraction index and viscosity). But, it also can:

- Absorb the incident ray and, as a result, decrease the intensity I_0;
- Absorb the radiation of fluorescence (effect of "internal filter");
- Diffuse the incident light. The irradiation of a solvent by a monochromatic exciting light provokes the occurrence of interactions between photons and the electrons of the solvent molecules. They appear under the form of an emission of a light of diffusion, polychromatic, which is perpendicular to the incident optical path. It is composed of (Figure 5):
  - Of an intense ray of the same wavelength as the one of the excitation radiation. This means that the photons diffused by the electrons of the molecules of solvent have the same energy as the photons of the incident radiation. One speaks then of Rayleigh diffusion[27] (One says then that the phenomenon is due to an elastic impact).
  - Of a series of less importance than the preceding ones (100 to 1,000 times weaker than the Rayleigh ray), the frequencies of which correspond to the difference Δv ($v_i = v_{ex} \pm \Delta v$). Δv will be the interval of frequencies between the rays. (See chapter devoted to the Raman[28] spectroscopy).

The determination of the wavelengths by emission by Rayleigh and Raman is to check the accuracy of fluorometry. For example in water, if the excitation is located at 254 nm, Raman's peak must be at 278 nm.

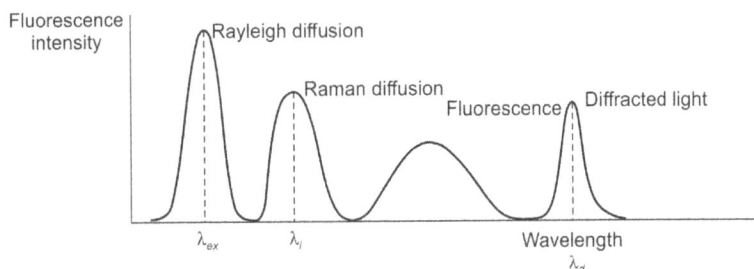

Figure 5: A Spectrum of Fluorescence, Diffusion Raman and Diffracted Light. (Reprinted with Permission from Gwenola and Jean-Louis Burgot, Paris, Lavoisier, Tec et Doc, 2017, 378).

---

[27] John William Strutt Rayleigh, English physicist (1842–1919), Nobel prize in physics (1904).
[28] Chandrashekhara Venkata Raman, Indian physicist (1888–1970) Nobel prize in physics (1930).

Influence of the Temperature

The increase of temperature decreases the quantum yield and hence the intensity of fluorescence by favoring the impacts between the molecules and the dissipation of the energy gained by the exciting forms under thermal energy.

Apparatus

There exist two kinds of apparatus:

- Spectrofluorometers permit a record of the spectrum (Figure 6). They work with two monochromators. In some systems, the detector is not located at a 90° angle to the incident beam. It may be located at a small angle or even in a straight line. The last configurations may be advantageous in some cases.
- Fluorimeters, are working most often with two filters, a primary for the absorption and the second for the transmission of fluorescent radiation. Some show ratios of fluorescence which are used for the reading of microarrays of 96 holes or even more and of very weak volumes (0.5 ml). The system simultaneously measures the fluorescence emitted by the sample and by the standard solutions and a standard of fluorescence (quinine sulfate or rhodamine B). This process is extremely used in combinatorial chemistry to suppress the fluctuations of intensity of the lamp.

The detector is located at a right angle of the incident beam in order to avoid the interference of the transmitted light.

Source

One uses two kinds of lamps:

- Xenon lamps with an electric arc. The discharge between the two electrodes is at the origin of the excitation of xenon, which emits a spectrum of rays superimposed on a continuous spectrum (between 230 and 700 nm) but the intensity of these lamps is weak in the far UV domain. Moreover, it shows a peak at about 460 nm.
- Lamps at mercury vapors emit a discontinuous spectrum (254 and 366 nM).

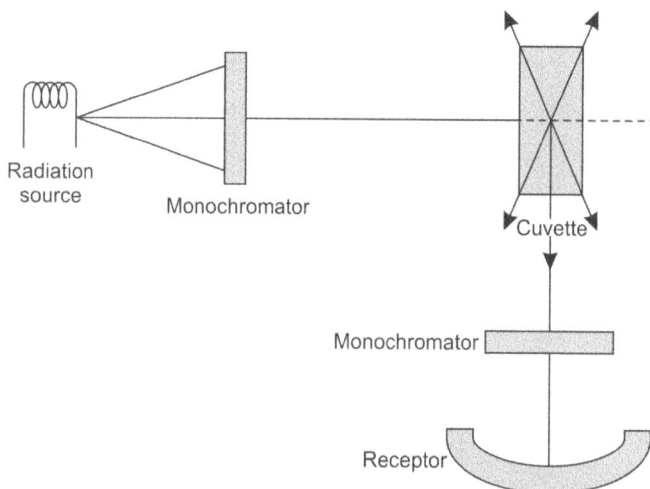

Figure 6: Scheme of a Spectrofluorometer. (Reprinted with Permission from Gwenola and Jean-Louis Burgot, Paris, Lavoisier, Tec et Doc, 2017, 379).

Excitation Monochromator

It permits the selection of the wavelength of the excitation light corresponding to the strongest absorption. It is constituted by a grating bordered by two slits located on each side. The width of the slits is expressed in bandwidths (1, 2, 5 and 10 nm). A larger slit gives rise to a stronger intensity but to a weaker resolution.

Cells

One uses cells in quartz of cylindrical or square sections. They do not show a flat surface. Numerous applications are developed notably with the use of arrays manufactured in synthetic black quartz.

Emission Monochromator

Usually, they are of the same type as those of excitation. One uses a secondary filter in order to avoid taking into account the reflection and the diffusion of excitation radiation and one selects the ray which is endowed with the greatest intensity.

Detector and Treatment of information

The detector or photomultiplier converts the emitted light into an electric current, which is amplified and recorded on a digital or numerical screen.

Fluorescence and Structures of Fluorescent Molecules

Fluorescence is a radiative process. It is often in competition with processes that are not radiative. For the phenomenon of fluorescence to be operating, several conditions must be subscribed:
- The excitation wavelength must be higher than 250 nm. Radiations of weaker wavelengths are too energetic and may induce the breaking of bonds.
- The electron which is submitted to the transition must not be in a bonding orbital.

The structures of molecules which are the most fluorescent do possess:
- One or several aromatic nuclei (such as naphthalene and anthracene) and some heterocycles (quinoleines, coumarine, indole, etc.):
- Electro-donator groups, such as $-NR > -NH > -OR > -OH$, activate fluorescence. Let us notice, as a consequence, that electro-withdrawing groups, such as $-COOH$, $-COOR$, $-CHO$, $-COR$, NO_2 and NO, decrease the intensity of fluorescence. The molecules bringing groups in which the effects $-I$ and $-M$ are in conjugation for the withdrawing electrons are not fluorescent. For example, phenol is fluorescent, whereas benzoic acid is not.
- In order to be fluorescent, molecules must be *rigid*. Hence, the quantum yield goes from 0.2 for the biphenyl to 1 for the fluorene since "the methylene bridge" located between both aromatic cycles of the latter enhances the rigidity. This favors the emission of photons by fluorescence (Figure 7).

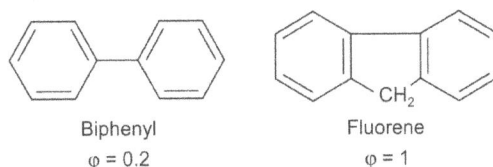

Biphenyl
$\varphi = 0.2$

Fluorene
$\varphi = 1$

Figure 7: Compared Quantum Yields of Biphenyl and Fluorene. (Reprinted with Permission from Gwenola and Jean-Louis Burgot, Paris, Lavoisier, Tec et Doc, 2017, 380).

Some examples of fluorescent molecules are given in Figure 8.

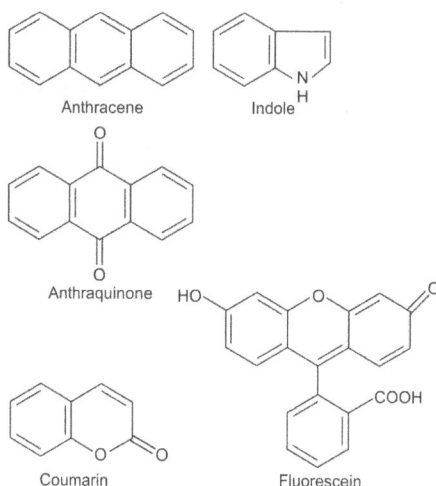

Figure 8: Some Examples of Fluorescent Molecules. (Reprinted with Permission from Gwenola and Jean-Louis Burgot, Paris, Lavoisier, Tec et Doc, 2017, 381).

Applications of Fluorescence

Qualitative Applications

Using the two processes of absorption and emission makes the spectrofluorometry more specific as spectrophotometry is UV-visible. The identification of species in mixtures by this method remains possible due to multidimensional techniques. They consist in recording the fluorescence intensity as functions of multiple parameters, wavelengths of excitation and emission and the lifetime of the excited state.

The study of spectra is devoted to:

- Wavelengths of the intensity maxima of the excitation;
- The fluorescence quantum yields;
- The displacements of these maxima in relation to the molecular modifications.

Quantitative Applications

Spectrofluorometry is a fast method, the detection limit of which is very weak (until about 10^{-9} mol L^{-1} for the most favorable cases). It is more selective than the spectrophotometry UV-visible since all the substances absorbing in this domain are not obligatorily fluorescent. This is an advantage when impurities are present. Moreover, the direct measurement of the emitted light makes the need for amplification of the signal less necessary in absorption. For these reasons, fluorometry constitutes a detector that is used for many applications in liquid chromatography.

Among the numerous molecules of pharmaceutical interest which can be determined by fluorometry. Let us mention, some alkaloids such as Ergot de Seigle, quinidine, quinine, morphine, codeine, corticosteroids products and cardiotonic glucosides (digoxin and digitoxin). Among the biochemical applications, one can get measure in the urine of homovanillic acid that is a catecholamine metabolite (Figure 9) for screening tumors. It reacts with potassium ferricyanide in an alkaline medium giving, hence, a strongly fluorescent solution.

The method is also used for the analysis of vegetal substances, such as chlorophylls and oils. Some of them are pharmaceutical primary matters.

The analysis of non-fluorescent molecules by fluorometry may require their coupling to fluorescence markers. The most used product in this context is the o-phtalaldehyde (OPA) which gives fluorescent complexes with numerous molecules. Some aromatic amino acids (tryptophane, tyrosine or phenylalanine) may be used as fluorescence markers. This is the case in *bioanalytical chemistry* and more precisely in the domain of the methodologies called *molecular recognition*. Let us cite, in the domain of molecular recognition, the use of fluorescent probes and the applications of immunofluorescence.

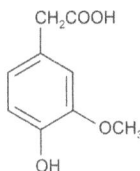

Figure 9: Homovanillic Acid. (Reprinted with Permission from Gwenola and Jean-Louis Burgot, Paris, Lavoisier, Tec et Doc, 2017, 382).

Fluorescent Probes

Fluorescent probes can detect biochemical changes in single living cells. For example, by a non-radiative *fluorescence resonance energy transfer (FRET)*. The cAMP can be measured in living cells. We recall simply for this purpose that a fluorescent molecule excited can dispose of the energy of an absorbed photon in each of two ways, i.e., by fluorescence or by a non-radiative process. That is to say, a FRET. The fluorescent probe may be a protein. This is the case of the naturally occurring *green fluorescent protein (GFP)*. It is reactive by a part of its molecule constituting its chromophore. Its chromophore would result from a rearrangement of three amino-acids serine, tyrosine and glycine.

The fluorescence of some amino acids may be sometimes inhibited in proteins but a conformational change of the tertiary structure of the latter ones may liberate it.

The classical organic markers are, sometimes, limited by their weak *luminance*. The luminance is the characteristic of a color which permits it to be classed by equivalence with a term belonging to a series of grey colors, going from white to black. They are also marked by their fast loss of the fluorescent signal. For some biological applications or some applications in "bio-imaging", new inorganic markers are used, such as semi-conductor crystals of nanometric height (1–10 nm) named "quantum dots" (QDS). They permit obtaining one emission of fluorescence more stable and more important than the preceding ones.

As a result, there is an enhancement of the "sensitivity" (detection limit) of the method. The most used QDs in biology are the chalcogenides[29] of cadmium. Nanoparticles of cadmium tellurate are promising for medical imaging. Due to a suitable choice of the conditions for synthesis and growing of the particles, it is possible to displace their emission of fluorescence of 600 nm (near infra-red) which is the optimal zone for the "bio-imaging". These nanoparticles can be linked in a covalent manner to biomolecules directly involved in cellular metabolism. Therefore, they can be coupled with some substrates such as enzymes, antibodies, AND, proteins, viruses, active pharmaceutical products and redox probes.

Applications of Immunofluorescence

Immunofluorescence is also largely used in molecular recognition. Let us recall that molecular recognition can be defined as the ability of a biomolecule to interact with one other particular type of biomolecule like a key fitting a lock.

---

[29] Chemical species possessing one element of the group VI of the periodic chart in its structure, such as sulfur, selenium or tellurium under the form of anions.

Combining an immunologic reaction of recognition antigen/antibody with fluorometry is widely used in clinical biology or the determination of plasmatic concentrations of medicines in the framework of therapeutical monitoring. The fluorescent marker is easier to handle than radioactive markers, but they lead to less sensitivity (to a higher limit of detection) than the second ones. Moreover, the later determinations may be disturbed by the autofluorescence of proteins.

Among the numerous developed methodologies; let us mention the polarization of fluorescence and the methodology based on lifetime measurements. Both seem to be most used.

Polarization of Fluorescence. The principle of this method is as follows. By interposing a polarizer between the incident beam and the cell, one obtains a polarized light, the characteristic of which being that it vibrates in only one direction with a defined wavelength (Figure 10).

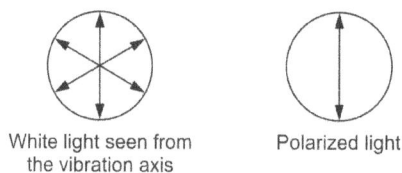

White light seen from   Polarized light
the vibration axis

Figure 10: Polarization of the Incident Light. (Reprinted with Permission from Gwenola and Jean-Louis Burgot, Paris, Lavoisier, Tec et Doc, 2017, 384).

Fluorescent molecules are excited by the polarized light and the intensities of the fluorescent radiations I// and Ip, determined in parallel and in perpendicular directions to that of the incident light, are measured. One can, then, define the polarization ratio P:

$$P = (I// - Ip)/(I// + Ip)$$

The equation, so-called Perrin's equation shows that the polarization ratio is directly related to the volume V of the molecule responsible for the fluorescence at the temperature T to the viscosity η of the medium and to the lifetime of the excited state τ, according to the relation:

$$1/P = 1/P_0 + (1/P_0 - 1/3)\, RT\tau/\eta V$$

Here, P_0 is the initial polarization.

The volume of molecules influences their rotation speed. Therefore, a small molecule such as fluoresceine spins very quickly, whereas an antibody which is a protein of high molecular weight, is animated only by a slow spin.

When a fluorescent substance is irradiated by polarized light, the polarization plane of the incident line is preserved if the molecule stays motionless (or quasi-motionless) as is the case for molecules of high molecular weights. As a consequence, small molecules which spin quickly "reemit" a polarized light in different polarization planes. As a result, there is a depolarization of the incident light with a decrease in the polarization ratio.

This methodology is applied to the case of the dosage in the plasma of some medicines. In the case of digoxin,[30] for example, the dosage benefits of a reaction of competition towards an antibody between the medicine (antigen: Ag) present in plasma and a reactant containing the same substance associated with a marker of variable nature (enzymes, radioactive atoms and fluorescent substances AgF). In this example, it is a fluorescent substance (fluoresceine, 7 – hydroxy-coumarin, umbelliferone, etc.).

One adds a known quantity of antibody anti-digoxin and another known quantity of digoxin coupled with (labeled by) umbelliferone. Both antigens compete to fix oneself on the sites of the antibodies (Figure 11).

---

[30] Cardiotonic molecule with a narrow therapeutic zone which necessitates a therapeutic monitoring.

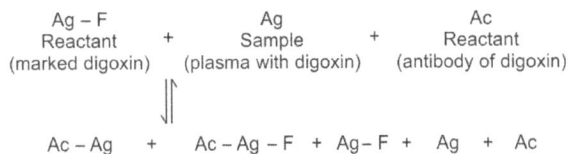

Figure 11: Competition Free-Antigen and Labeled-Antigen Toward the Antibody. (Reprinted with Permission from Gwenola and Jean-Louis Burgot, Paris, Lavoisier, Tec et Doc, 2017, 385).

When the concentration of digoxin in the plasma is low, the fraction of the complex $Ac - Ag - F$ in the final mixture is high. The result is that the major part of the fluorescent substance is in the form of a complex of high molecular weight. This compound is little mobile in solution and as a result, the light depolarization remains weak (Figure 12).

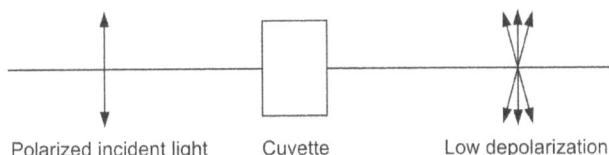

Figure 12: Change of the Polarization of the Light for a Low Concentration of the Antigen. (Reprinted with Permission from Gwenola and Jean-Louis Burgot, Paris, Lavoisier, Tec et Doc, 2017, 385).

In contrast, when the concentration of digoxin in plasma is high, the fraction Ac-Ag-F is low with respect to that of AgF and the depolarization of the emitted light is important since the labeled molecule AgF is spinning quickly (Figure 13).

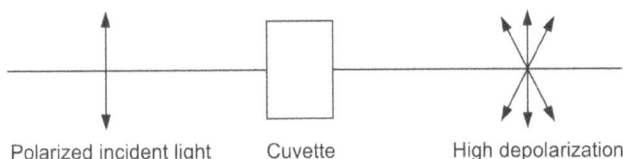

Figure 13: Change of the Polarization of the Light for a High Concentration of the Antigen. (Reprinted with Permission from Gwenola and Jean-Louis Burgot, Paris, Lavoisier, Tec et Doc, 2017, 386).

From a practical standpoint, one studies the evolution of the polarization ratio as a function of the concentration in antigen (Figure 14).

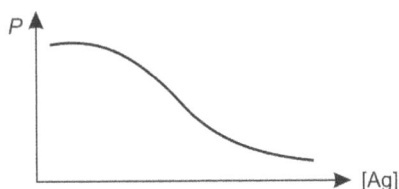

Figure 14: Evolution of the Polarization Ratio as a Function of the Concentration of the Antigen in the Medium. (Reprinted with Permission from Gwenola and Jean-Louis Burgot, Paris, Lavoisier, Tec et Doc, 2017, 386).

Fluorescence With Lifetime Measurements. Fluorescence with lifetime measurements is a particular methodology of fluorometry. It permits avoiding the interferences encountered with the classical technique of fluorometry such as the *fundamental* noise due to the autofluorescence of the proteins and due to light *scattering*. With fluorescence with lifetime measurements, antigens and antibodies are marked by europium. An example is provided by the dosage of thyroid-stimulating factor (TSH) for the exploration of thyroid function. The methodology uses the properties of luminescence of chelates of ions of *lanthanides* as ion europium Eu^{3+}. Their fluorescence is characterized by an emission peak very narrow, the wavelength of which is markedly different from that of the radiation of excitation. It is also marked by a decrease in emissions whose duration is long. This particularity comes from the fact that some excited states of the ion Eu^{3+} are of less energy than those of their ligands. (These differences in energy are weak).

Fluorometry for Detection in Liquid Chromatography and Electrophoresis

Fluorometry constitutes an important method for detecting the way out of the different components of a sample as they appear at the end of chromatography and also of capillary electrophoresis. These points are studied in Chapters 17 and 18.

Phosphorimetry

Phosphorescence differs from fluorescence from two marked points. Firstly, it possesses a relatively long decay. The second is that when one compares phosphorescence and fluorescence, there is a shift towards longer wavelengths for the former. Few organic compounds exhibit phosphorescence in solution in particular at room temperature. In contrast, many more are phosphorescent at a much lower temperature, such as liquid nitrogen. The lowering of temperature seems to enhance the probability of transition from the excited electronic state to the triplet state. It also diminishes the occurrence of the non-radiative phenomenon of return into the electronic ground state.

From the standpoint of applications, let us mention the fact that although having received few applications in particular at room temperature, several organic compounds with conjugated ring systems can be dosed by phosphorimetry, even in the form of traces. The best solvents to use in phosphorimetry are those which solidify at liquid nitrogen temperature.

Instruments for analysis by phosphorimetry are somewhat similar to fluorometers and spectrofluorometers. But there exists a noticeable difference between the two types of apparatus. It takes its origin in the fact that when a compound phosphoresces, it also fluoresces. The apparatus must be able to distinguish the two luminescences. The phosphorimeter is equipped with a device that alternately irradiates the sample and after a delay measures the phosphorescence.

Usually, phosphorimeters are equipped with Dewar flasks with quartz windows since a great number of analyses are performed at liquid nitrogen temperature.

Chemiluminescence

The phenomenon of chemiluminescence is produced when a chemical reaction gives rise to an electronically excited species which emits light as it returns to its ground state. Clearly; it is a phenomenon of luminescence. The phenomenon may be summarized by the scheme:

$$A + B \rightarrow C^* + D$$

C^* is the excited state of C.

$$C^* \rightarrow C + h\nu$$

But there is a difference in principles between fluorescence and chemiluminescence phenomena. The excited species may transfer its energy to another species which then produces emission.

Chemiluminescence reactions are encountered in a number of biological systems. Then the phenomenon is called *bioluminescence*. Several organic compounds are also capable of chemiluminescence.

The apparatus for chemiluminescence is very simple. There are essentially photometers. In this case of luminescence, no selection of wavelength is necessary.

Here, we mention some applications:

Analysis of Gases

For example, the nitrogen monoxide NO is determined according to the reactions:

$$NO + O_3 \rightarrow NO_2^* + O_2$$

$$NO_2^* \rightarrow NO_2 + h\nu$$

We see that ozone (produced by an ozonizer) oxidizes the nitrogen monoxide brought by the atmosphere. The luminescence radiation ($\lambda = 600$ to $2{,}800$ nm) is followed by a photometer.

Other interesting examples are provided by the determination of atmospheric sulfur, compounds such as sulfur dioxide. This gas is burnt in a hydrogen flame and gives a sulfur dimer:

$$4H_2 + 2SO_2 \rightarrow S_2^* + 4H_2O$$

The radiation occurs at peaks at 384–394 nm. It comes from the reaction:

$$S_2^* \rightarrow S_2 + h\nu$$

Analysis of liquids

Compounds that are analyzed by chemiluminescence in the liquid phase do possess in their structure the function hydrazide –C(=O) –NH –NH$_2$. These derivatives react with strong oxidizing agents, such as dioxygen and hydrogen peroxide. They give products of oxidation with chemiluminescence. For example, the luminol is transformed into the dianion3-amino o-phthalate, according to the following scheme (Figure 15) with chemiluminescence at about 425 nm:

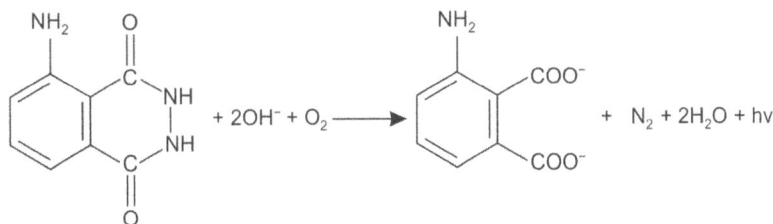

Figure 15: Oxidation of Luminol with Chemiluminescence.

Analysis of Organic Species with Occurrence of Chemiluminescence

As it has been said already, it is possible to extend chemiluminescence to derivatives that are not involved in this phenomenon. Simple organic compounds are capable of exhibiting chemiluminescence. The extension of the number of derivatives capable to give rise to the phenomenon can occur. The excited species may transfer their energy to another species which then may produce emissions.

As other examples, a number of biological organisms (many marine organisms of the kind "firefly") emit light with a range of frequencies which can spread from UV up to the red end of the visible spectrum.

Here, we follow the developments of Whittaker and Nelson and Cox.

The light emitted by a firefly is the result of visible emission from an exciting product derived from luciferin. It is one of the efficient chemiluminescent systems. The enzyme luciferase catalyzes the oxidation of luciferin to an intermediate, which loses CO$_2$ to form the exciting product that is the source of light emission (Figure 16).

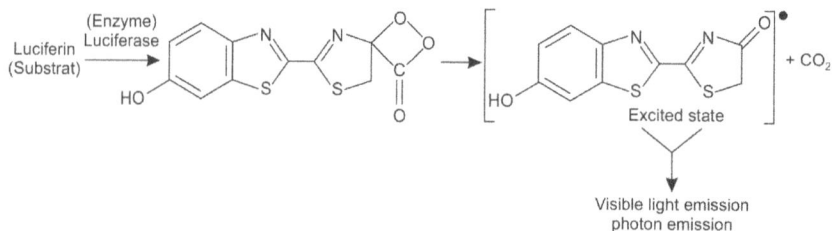

Figure 16: Successive Reactions Leading to the Bioluminescence of Fireflies.

Bioluminescence requires considerable amounts of energy. In the firefly, ATP is used in a set of reactions that converts chemical energy into light energy. The generation of a light flash requires activation of luciferin by an enzymatic reaction involving pyrophosphate cleavage of ATP to form luciferyl adenylate. In the presence of molecular oxygen and luciferase, the luciferin undergoes its process accompanied by the emission of light. Luciferin is regenerated from oxyluciferin in a subsequent series of reactions.

The number of chemical reactions that produce chemiluminescence is relatively small. However, the selectivity, simplicity and extreme sensitivity of the method are such that it is a method of the future.

CHAPTER 23

Turbidimetry and Nephelometry

Turbidimetry and nephelometry are analytical methods based on the scattering of light by a solution containing dispersed solid particles.

Introduction and Definitions

The term scattering covers a variety of phenomena. When it concerns the interaction of radiant energy with matter, it implies a change in the direction of propagation (Figure 1).

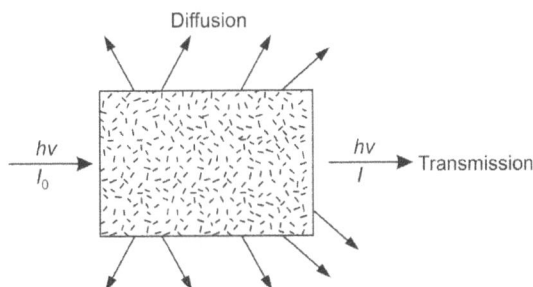

Figure 1: Diffusion and Transmission of Light in a Suspension. (Reprinted with Permission from Gwenola and Jean-Louis Burgot, Paris, Lavoisier, Tec et Doc, 2017, 387).

These methods permit defining the turbidity of a medium which some authors consider as the reduction of transparency of a liquid in presence of solid particles. When radiation crosses over a medium in which the solid particles are dispersed, one fraction of them is diffused in all directions giving the impression of turbidity to the "mixture" and the other fraction is transmitted without encountering any dispersed particle.

Turbidimetry is the measurement of the intensity of the transmitted radiation by the medium. It is determined along the axis of the incident beam. Light has been lost by absorption, reflection and diffusion. Turbidimetry is adapted to media of strong turbidity in which solid particles are big.

Nephelometry (or nephelemetry[31]) is based on the measurement of the intensity of diffused light and measurement is performed at the right angle of the direction of the initial ray. It is used for the analysis of media of weak turbidity.

Turbidimetry and nephelometry are in some instances quantitative methods of analysis.

---

[31] From the Greek *nephelé* which means cloud.

Theories of Nephelometry and Turbidimetry

In these methods, one considers that the diffusion due to the particles is "elastic". This means that there is no change of energy but only a change of direction of light when the latter is diffused by a particle.

The intensity of the diffused radiation depends on the number of particles in the solution, their length, their form, the relative value of their refraction index, the medium and finally the wavelength of the initial radiation. More precisely, the intensity and the direction of the diffused radiation depend on the length of the particle with respect to the value of the wavelength:

• If the length of the particle is weaker as the ratio $\lambda/20$, the diffused light is symmetrical with respect to its center and is in practice equally distributed according to all the directions. One speaks of *Rayleigh's diffusion* since Rayleigh's theory initially developed for the gas can be applied to liquids (Figure 2).

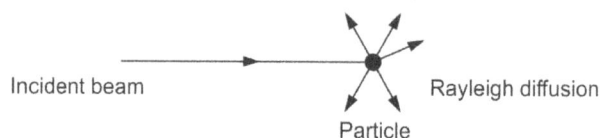

Figure 2: Rayleigh's Diffusion. (Reprinted with Permission from Gwenola and Jean-Louis Burgot, Paris, Lavoisier, Tec et Doc, 2017, 388).

If the diameter of particles is in its order of value equivalent to that of λ, the diffusion is asymmetric. The intensity is no longer the same in all directions. One speaks then of the diffusion of *Rayleigh-Debye*[32] (Figure 3).

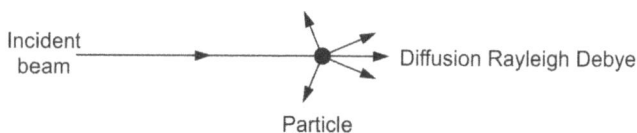

Figure 3: Rayleigh-Debye's Diffusion. (Reprinted with Permission from Gwenola and Jean-Louis Burgot, Paris, Lavoisier, Tec et Doc, 2017, 388).

Finally, if the diameter of particles is higher as λ, one notes Mie's diffusion[33] which is characterized by a diffusion angle very weak (Figure 4).

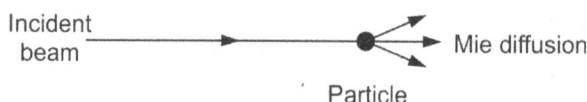

Figure 4: Mie's Diffusion. (Reprinted with Permission from Gwenola and Jean-Louis Burgot, Paris, Lavoisier, Tec et Doc, 2017, 389).

Therefore, proteins such as immunoglobulins G (IgG) with a diameter of the order of 200 A° give rise to Rayleigh's diffusion, whereas the behavior of immunoglobulins IgM (diameter; 350 A°) rather enters into the group giving the diffusion of Rayleigh-Debye.

From the quantitative standpoint, for both methods, all the parameters which have an influence on the diameter of particles are important; that is to say, the concentrations of reactants, the order of their introduction, the rate of mixing, the time of reaction, temperature and the ionic strength.

---

[32] Peter Joseph Wilhelm Debye, Dutchman physicist and chemist (1884–1966). Nobel prize in physics (1936).
[33] Gustav MIE German physicist and chemist (1868–1957).

Turbidimetry

The decrease of the intensity of an incident beam of radiation following through a "dilute suspension" is given by a relation of the type as that of Beer-Lambert if the diameter and the concentration of particles are weak (Figure 5).

$$I = I_0 e^{-\tau l c}$$

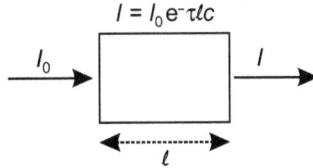

Figure 5: Relation Between the Incident and Transmitted Intensities in Turbidimetry. (Reprinted with Permission from Gwenola and Jean-Louis Burgot, Paris, Lavoisier, Tec et Doc, 2017, 389).

I is the intensity of the transmitted ray in the initial direction, I_0 is the initial intensity, l is the optical pathway and τ is the turbidity coefficient (the factor of proportionality):

$$\log (I_0/I) = klC$$

Here, k = 2.303 τ. In the first approximation, this relation is used in the turbidimetric analysis in the same manner as Beer-Lambert's law. A calibration curve log (I_0/I) as a function of C is drawn. The sole solvent is used as a reference for I_0. As with the treatment of Beer-Lambert's law, the measurement is compared to those obtained with the calibration curve drawn for the same experimental conditions as the measurement.

The precision of the method is better with solutions whose turbidity is high, for which one notices a significant decrease in the transmitted light.

Nephelometry

If the number of particles in the suspension is weak, if their diameter is lower than the wavelength λ of the incident beam and if their refraction index *n* is not too different from that of the medium n_0, the intensity of the diffused ray is given by Rayleigh's law:

$$I_D = KI_0[(n^2 - n_0^2)/(n^2 + 2n_0^2)] NV^2/\lambda^4$$

Here, I_0 is the intensity of the incident beam and V is the mean volume of the particles supposed spherical. N is the number of particles and K is a constant taking into account the observation angle.

We can see that I_D depends on the number N, all other things being equal. Then, for particles that sensibly have the same dimension, the measurement of I_D permits to have access to their "concentration". From the experimental standpoint, very often, one notices a linear relation I_D as a function of the concentration of the scattered particles. Again, the measurement of the unknown solution is compared to those of a calibration curve. The studied media must be stable. A trend towards too quick a sedimentation induces important changes in the number of particles per volume unit.

Nephelometry is more precise and more "sensitive" than turbidimetry.

Apparatus

Sources

Usually, in both techniques, the sources of radiation provide beams, the wavelengths of which go from 340 nm to 650 nm. Hence, one uses:

- The mercury arc;
- The xenon arc;
- The tungsten lamp with a monochromator (band passant 400 to 550 nm);
- The lamp laser helium-neon which emits at $\lambda = 632.8$ nm;
- The lamp laser helium-cadmium ($\lambda = 441.6$ nm).

The laser lights being a little scattered, it is not necessary to use a collimator as is the case with the tungsten lamp.

It must be recalled that the fraction of the diffused light remains weak. Therefore, a sensitive detection system must be used in nephelometry.

Configurations of Apparatus

Turbidimetry

The apparatus is constituted by energetic sources, a wavelength selector, a measurement cell (containing the sample) the faces of which are parallel, a device permitting to eliminate the diffused light and a detector connected to a galvanometer and a recorder (Figure 6).

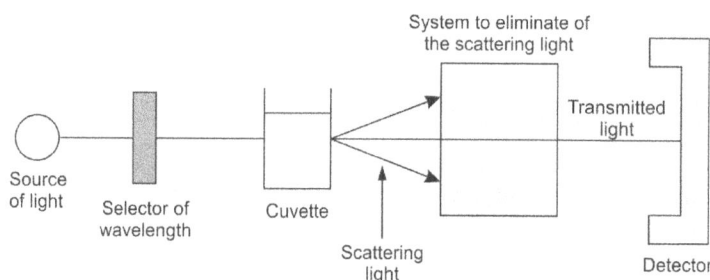

Figure 6: Scheme of a Turbidimeter. (Reprinted with Permission from Gwenola and Jean-Louis Burgot, Paris, Lavoisier, Tec et Doc, 2017, 391).

After having gone through the measurement cell, the transmitted light in the axis of the incident beam contains the transmitted light itself whose radiating axis remains parallel to the axis of the incident beam and a part of the diffused light which is deviated and is located in a cone; the head of which is located at the coming out of the emerging beam.

It can be eliminated:

- By enlarging the interval between the cell and the detector at the maximum possible, but in this situation there occur great energy losses.
- By putting a system of slits (or a metallic cylinder with grooved conduits) between the cell and the receptor. Only the rays which are staying in the initial direction can go across the system (Figure 7).
- By interposing a convergent lens and a diaphragm between the cell and the detector. The lens only concentrates the parallel rays of the transmitted light on the aperture of the diaphragm. This eliminates the diffused radiations which are not parallel (Figure 8).

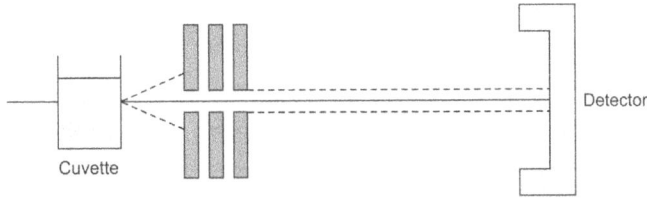

Figure 7: Elimination of the Diffused Light in Turbidimetry by a System of Slits. (Reprinted with Permission from Gwenola and Jean-Louis Burgot, Paris, Lavoisier, Tec et Doc, 2017, 392).

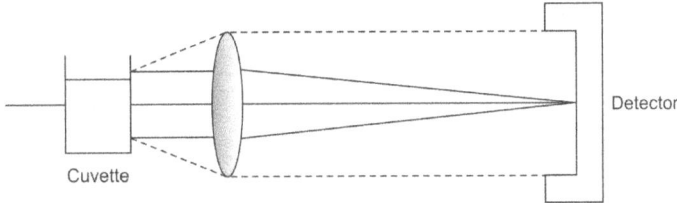

Figure 8: Elimination of the Diffused Light in Turbidimetry by Using a Convergent Lens. (Reprinted with Permission from Gwenola and Jean-Louis Burgot, Paris, Lavoisier, Tec et Doc, 2017, 392).

More recent apparatus permits measuring the transmitted light and the diffused one. Their use is suitable for samples of great turbidity.

A photometer or a classical spectrophotometer can be used as a turbidimeter if it possesses a system permitting to eliminate the diffused light.

Nephelometry

A luminous ray is transformed into a monochromatic ray by interposing a filter. The wavelength is chosen so that the particles do not absorb. This is imperative, given the principle of the method. The ray is reflected at 90° by a mirror which sends it on the cell measurement. The light which is diffused in all directions is reflected several times by a spherical mirror in all direction which sends the quasi-totality of the diffused light toward the detector (a photoelectric cell connected to a milliammeter and a recorder) (Figure 9).

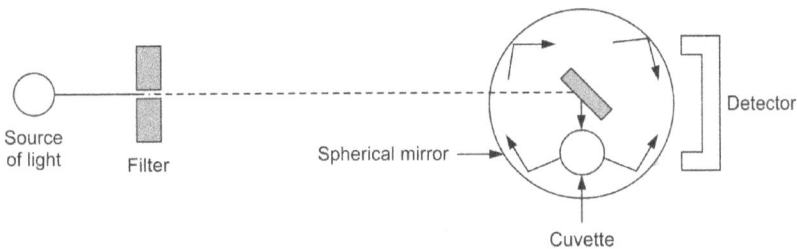

Figure 9: Scheme of a System of Nephelometer. (Reprinted with Permission from Gwenola and Jean-Louis Burgot, Paris, Lavoisier, Tec et Doc, 2017, 392).

Applications

Let us mention:

- The analysis of industrial wastewaters and river or lake waters.
- Determination of numerous ions.

The most important is without any doubt the determination of the ion SO_4^{2-}. Sometimes, the turbidimetric methods are capable to attain mass fractions of the order of some parts per million with precision from 1 to 5%.

Table 1: Some Applications of Turbidimetry and Nephelometry.

Species to Titrate	Method	Suspensions Obtained	Reactants
Ag^+	T, N	AgCl	NaCl
Au^+	T	Au	$SnCl_2$
Ca^{2+}	T	CaC_2O_4	$H_2C_2O_4$
Cl^-	T, N	AgCl	$AgNO_3$
K^+	T	$K_2Na\,Co\,(NO_2)_6$	$Na_3Co\,(NO_2)_6$
Na^+	T, N	$Na\,(UO_2)_3[Zn\,(H_2O)_6]\,(OAc)_9$	$Zn\,(OAc)_2$ and $UO_2(OAc)_2$
SO_4^{2-}	T, N	$BaSO_4$	$BaCl_2$

Turbidimetric Titrations

Turbidimetric measurements permit localizing the end of the titration reaction when it involves one precipitation reaction. According to the principle of turbidimetry, there is a change in the absorbance provided the end point of the titration reaction is not reached. Afterward, the absorbance remains constant. Experimentally, one notices, however, that it is slowly decreasing. This may be due to phenomena of sedimentation and dilution.

Precipitation in Liquid Media Through a Reaction Antigen/Antibody

In this case, one takes advantage of the competition antigen of the sample/antigen marked by latex particles or by particles marked by apoferritin versus sites of specialized antibodies.

The formation of insoluble complexes antigen-antibody enhances the diffusion of light. If the antigen is an important quantity in the sample, the formation of the complex marked antigen/antibody decreases and the diffusion of the light decreases as well (Figure 10).

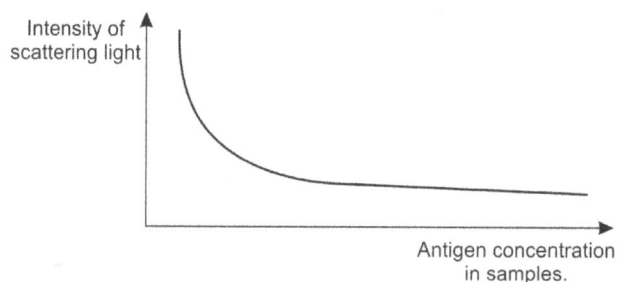

Figure 10: Change in the Intensity of the Diffusing Light During an Immunochemical Determination. (Reprinted with Permission from Gwenola and Jean-Louis BURGOT, Paris, Lavoisier, Tec et Doc, 2017, 394)

As examples of this kind, let us mention the determination of some antiepileptics in serum.

Clinical Biology

A large range of applications benefit from these methodologies for the dosage of some proteins in the serum and urinary ones, such as protein C-reactive, albumin, transferrin, α-foetoprotein, α1-antitrypsine, immunoglobulins IgM, IgG, IgA, IgD or IgE.

Food Industry

The dosage of the IgG in milk and colostrum appeals to the use of these methodologies. To conclude, it can be said that nephelometry exhibits better performances than turbidimetry if the suspension is dilute. This fact limits the problems due to the auto-absorption and reflection. Today, however, turbidimetry is more developed than the preceding methodology since it can be used with a classical spectrophotometer. The detection limit of turbidimetry is lower than that of nephelometry. Moreover, it is easily automated.

CHAPTER 24

Nuclear Magnetic Resonance Spectroscopy with Continuum Waves

General Principles

In this chapter and the following ones, we are interested in the magnetic character of electromagnetic radiations. This was not the case until now. In this first chapter, we confine ourselves to the study of continuum waves N.M.R. (C.W. N.M.R.) (sometimes called conventional N.M.R.). In the following chapter, we shall proceed to a brief study of the spectra and we shall mention some analytical applications of the method. We shall also study the electron paramagnetic resonance (E.P.R.), which is also called electron spin resonance (E.S.R.). Fourier transform N.M.R. (F.T. N.M.R.) (also named pulsed N.M.R.) will be studied in the next one. In this study, we are essentially interested in liquid systems or solutions. In both cases, narrow signals (peaks) are obtained. *High resolution N.M.R.* applies to these samples. *Wide-line* or *large bands of N.M.R.* applies to solids. It will not be studied here.

N.M.R. in Brief

N.M.R. is founded on the fact that the nuclei of some chemical elements exhibit the behavior of tiny magnets and hence do possess a magnetic moment. In absence of a magnetic field, the energy of one magnet does not depend on its orientation. It is the same for the energy of the nuclei. But when a magnetic field applies in a given direction to a sample containing these elements, the nuclei gain some quantity of energy depending on the intensity of the field. The N.M.R. methodology studies the transitions between the energy levels of the atomic nuclei in a magnetic field. This energy is quantified.

In an N.M.R. experiment, the transitions between the different permitted levels are induced by electromagnetic radiations, the frequencies ν of which are related to Bohr's rule:

$$\Delta E = h\nu$$

ΔE is the energy difference involved in the transition, ν is the frequency of the radiation (Hz or s^{-1}), which induces the transition, and h is Planck's constant (J.s). Transitions are located in the domain of radio frequencies (roughly in the interval 4–900 MHz). The signals resulting from these transitions constitute the N.M.R. spectrum. In slightly different terms, one can say that N.M.R. spectroscopy is grounded on the measurement of the absorption of electromagnetic radiations. The analyte must be placed in a magnetic field in order that the energy states (between which transitions can occur) can be materialized.

For their works about N.M.R., Bloch[34] and Purcell[35] have jointly been awarded the Nobel prize in physics in 1952.

Principle

Angular Momentum of a Mobile

According to quantum mechanics, the angular momentum of an isolated mobile **L** can take only discrete values. It is quantified. Its measure L (sometimes symbolized by J) is expressed as its function of the quantum number l of the particle by the expression:

$$L = \bar{h}[l\,(l+1)]^{\frac{1}{2}} \tag{1}$$

l is capable to take only entire or demi-entire values. In this expression:

$$\bar{h} = h/2\pi$$

Here, *h* is Planck's constant.

$$\bar{h} = 1.05457 \ 10^{-34} \ \text{J.s.}$$

The quantum number l introduces itself during the resolution of Schrödinger's equation describing the particle in its rotation. The component of the angular momentum in the direction z'z and L_z is also quantified (Figure 1). Its measure is expressed as a function of m_l:

$$L_z = \bar{h}\,m_l \tag{2}$$

m_l is a new quantum number. It can only take the following 2l + 1 values which are l, l – 1 ,..., –l + 1, –l. A vectorial model takes into account these considerations. Let us consider Figures 1a and 1b and call θ the angle formed by the angular momentum **L** and its component L_z (Figure 1):

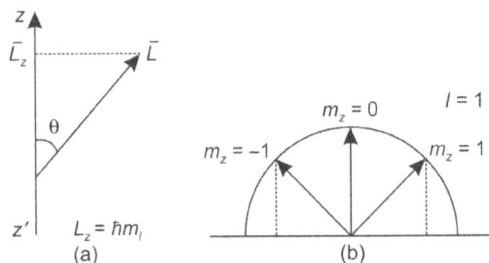

Figures 1a and 1b: Vectorial Model of the Angular Momentum. (Reprinted with Permission from Gwenola and Jean-Louis Burgot, Paris, Lavoisier, Tec et Doc, 2017, 411).

According to Relations (1) and (2), one finds:

$$\cos \theta = m_z/[l\,(l+1)]^{\frac{1}{2}} \tag{3}$$

Let us suppose l = 1. The m_z can take only the three values, i.e., 1, 0 and –1. The angular moment **L** can place itself only for the values π/4, π/2 and 3π/4 (Figure 1b).

---

[34] Felix Bloch. American physicist (1905–1983).
[35] Edward M. Purcell. American physicist (1912–1997).

Nuclear Spin

The existence of the nuclear spin was suggested by Pauli[36] in 1924 in order to explain the hyperfine structure of atomic spectra. Proton has a quantum number of spin $I = \frac{1}{2}$. It is the same as for the neutron. The result is the existence of a global spin for the whole nucleus.

N.M.R. depends on the existence of a nuclear spin. Nuclei for which $I = 0$ do not give rise to a direct effect. The numbers of spin I (numbers characterizing the value of the spin) can be entire, half-entire or null:

$$^{12}_{6}C \; ; ^{16}_{8}O \; ; ^{32}_{16}S \qquad\qquad I = 0$$

$$^{14}_{7}N \; ; ^{2}_{1}H \qquad\qquad I = 1$$

$$^{10}_{3}B \qquad\qquad I = 3$$

$$^{11}_{3}B \; ; ^{35}Cl \; ; ^{37}Cl \qquad\qquad I = 3/2$$

$$^{1}H \; ; ^{19}F : ^{15}N \; ; ^{13}C \; ; ^{31}P \qquad\qquad I = \frac{1}{2}$$

Let us consider the nucleus $^{A}_{Z}X$, where A is the mass number and Z is its atomic number. Table 1 links the spin number I to the characteristics of the nucleus X.

Table 1: Spin Number I and Characteristics of the Nucleus.

A	Z	I
Odd	Odd or even	$\frac{1}{2}$, 3/2, 5/2
Even	Even	0
Even	Odd	1,2,3

Hence, the proton has a spin number $I = \frac{1}{2}$. The nuclei $^{12}_{6}C$; $^{16}_{8}O$; $^{32}_{16}S$ ($I = 0$) do not give rise to any direct effect. As a great consequence, the N.M.R. spectra of organic molecules, which obligatorily possess plenty of carbon atoms, are relatively simple. One particularly studies and uses the N.M.R. spectra of the proton, of ^{19}F, ^{13}C, ^{31}P and ^{15}N.

^{1}H, ^{19}F and ^{31}P are easily observed by N.M.R. because these isotopes constitute 100% of the corresponding element in its natural abundance, whereas other isotopes exist in very weak proportions (^{13}C 1%: ^{15}N 0.37%).

Several other isotopes are possible candidates for N.M.R. studies but today, they are not used because of their natural abundance which is too weak to create a useful interest.

Magnetic Moments—Magnetons

For the study of the phenomenon of nuclear resonance itself, the notion of the magnetic moment must be dealt with. It leads to that of magnetons and is an important parameter that permits it to go further into the explanation of the phenomenon of nuclear resonance.

Magnetic Moment

One knows that we can consider a magnet as being constituted by a pair of magnetic masses +m and –m separated by a distance R. It is a magnetic dipole. It possesses the magnetic moment μ defined by the expression:

$$\mu = m\mathbf{R}$$

The μ is a characteristic of the magnet.

---

[36] Wolfgang Pauli, American physicist from Austrian origin (1900–1958). Nobel prize in physics (1945).

(*Remark*: Let us recall that the notion of magnetic mass is fictitious. It is, however, convenient for the study of the actions suffered or produced by magnets.)

The Magnetic Moment of the Electron

In order to initiate this study, let us consider the case of one electron which is moving in a circular orbit with the angular rate ω (Figure 2). It is convenient to begin like this because the behavior of an electron in this respect is better known than that of a nucleus.

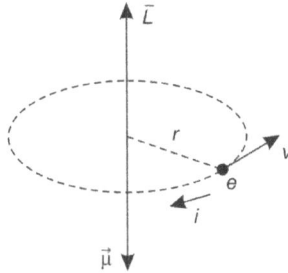

Figure 2: Electron Moving on a Circular Orbit. (Reprinted with Permission from Gwenola and Jean-Louis Burgot, Paris, Lavoisier, Tec et Doc, 2017, 413).

Let us consider an electric current i moving through a circulate orbit of radius r. The magnetic moment of the circuit is given (in a module) by the expression:

$$\mu = iS$$

Here, S is the internal surface of the circuit. Therefore, this displacement is equivalent to an electric current in an opposed direction to that of the displacement (see electromagnetism). The intensity is given by the general definition of one intensity:

$$i = q/t$$

Where q is the charge e of the electron and ω the angular rate of the particle, given by the expression:

$$\omega = v/r$$

The *kinetic moment* or kinetic momentum is by definition (in the module)

$$L = m_e v\, r$$

The m_e is the mass of the electron.

It results, from the handling of these expressions, that:

$$i = e\, L/(2\pi m_e r^2)$$

The relation between the angular and magnetic moments is the following:

$$\mu = (-e/2m_e)\, \mathbf{L} \tag{4}$$

This relation is established by writing:

$$I = q/t \text{ with } q = 1e \quad \text{and} \quad \omega t = 2\pi r$$

The sign minus comes from the fact that one electron is negatively charged. Vectors μ and $\mathbf{L}$ are antiparallel. μ is the magnetic moment (orbital) of the electron and $\mathbf{L}$ is its kinetics (orbital) moment. The kinetic moments are quantified at the microscopic scale.

The quantity β

$$\beta = e\, \hbar/2m_e$$

It is named the *electronic Bohr[37] magneton*. Its value is: 9.2740 10^{-24} JT^{-1} (T is the tesla). Its other symbol is μ_B:

$$\mu_B = \beta$$

Nuclear Magnetic Moment

It is the same for nuclei with mass m_N, of charge number 1, having a spin, which rotates around charged particles. They also possess a magnetic moment so-called nuclear magnetic moment μ. With the help of analogous reasoning as the one followed just previously, one finds that their nuclear magnetic moment μ is given by the relation $(e/2m_N)L$. One introduces one unit of magnetic moment corresponding to the proton; the *nuclear magneton β_N* in unities $\hbar$ by the expression:

$$\beta_N = e\,\hbar/2m_p$$

Where m_p is the mass of a proton. β_N is the *nuclear Bohr magneton*.

$$\beta_N = 5.0505 \ 10^{-27} \ J.T^{-1}.$$

Its other symbol is μ_N.

(*Remark*: The tesla[38] is the unit SI of the density of magnetic flux. In terms of basic units of the SI system, the tesla is expressed in kg s^{-2}A^{-1}. Likewise, 1 tesla = 1 Wbm^{-2}. A smaller unit is the Gauss[39]; 1 gauss (G) = 10^{-4} T. The weber[40] (Wb) is, as to itself, the unit of magnetic flux. In terms of basic units SI, it expresses in m^{2}kgs^{-2}A^{-1}.)

Kinetic and Magnetic Moments: Magnetogyric Ratio

According to what is previously said, it is evident that vectors kinetics moment **L** and magnetic moment μ are proportional and coaxial (Figure 3).

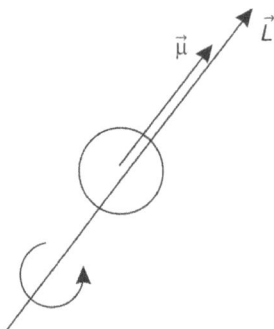

Figure 3: Angular and Kinetic Moments of Active Nuclei. (Reprinted with Permission from Gwenola and Jean-Louis Burgot, Paris, Lavoisier, Tec et Doc, 2017, 416).

Usually, nuclear magnetic moments are expressed by using the factor γ and is called *the magnetogyric factor* or *ratio*, which is defined by the expression:

$$\mu = \gamma\, L \quad \text{or}$$

$$\gamma = g_N \mu_N \, /\hbar$$

The magnetogyric ratio commands the frequency of the resonance. Table 2 mentions the I values.

[37] Niels Bohr, Danish physicist (1885–1962). Nobel prize in physics (1922).
[38] Nikola Tesla, American physicist, from Yugoslav origin (1856–1943).
[39] Karl F. Gauss, German astronomer, mathematician and physicist (1777–1855).
[40] Wilhelm E. Weber, German physicist (1804–1891).

Table 2: Values of Magnetogyric Ratios γ of Some Nuclei (10^7 rad. $T^{-1}s^{-1}$) and Resonance Frequency (MHz) ν of the Nuclei for a 2.11 T Magnetic Field.

Isotope	I	Natural Abundance	γ	Resonance Frequency
^{1}H	½	99.98	26.752	90 (60 pour 1.41T, 100 pour 2.35T, 220 pour 5.20T, 500 pour 12.00T)
Neutron	½		−18.325	
^{13}C	½	1.11	6.726	22.63
^{14}N	1	99.6	1.933	6.5
^{15}N	½	0.36	−2.711	9.12
^{19}F	½	100	25.167	84.67
^{31}P	½	100	10.829	36.43

The resonance frequency ν is calculated after the relation:

$$\nu = \gamma\, B_0/2\pi$$

B_0 is the magnetic field. The *magnetogyric factor* is a quantity that only depends on the nature of the nucleus. μ_N is about 10^3 smaller than μ_B. The nuclear magnetic moments are about 10^3 smaller than the electronic moments.

Behavior of the Active Nuclei in a Magnetic Field

In the Absence of Magnetic Field

The energy of a magnetic dipole is independent of its orientation. The magnetic dipole acquires some potential energy solely when it is placed in a magnetic field.

Let us suppose that the dipole is placed in a uniform magnetic field of induction $\mathbf{B}_0$ (Figure 4). The action of the field on the magnetic mass + m is that of a force + $\mathbf{F}$ parallel to the induction $\mathbf{B}_0$ and equal to − $m\mathbf{B}_0$. The action on the mass − m is a force parallel to the induction $\mathbf{B}_0$, but in the opposite sense. It is equal to $m\mathbf{B}_0$ in absolute value. Both forces are parallel, equal and of contrary sense. No neat force acts on the dipole, but a couple of forces do with respect to the point O. The couple tends to bring in alignment the dipole and the field (Figure 4).

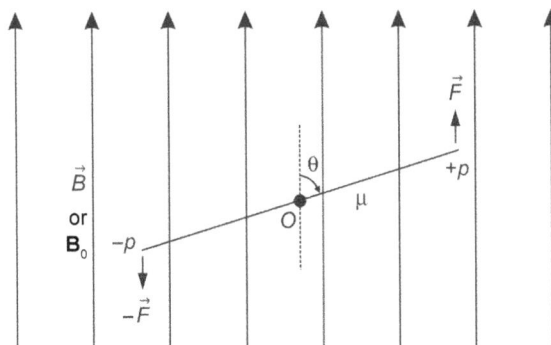

Figure 4: A Magnetic Dipole in a Magnetic Field. (Reprinted with Permission from Gwenola and Jean-Louis Burgot, Paris, Lavoisier, Tec et Doc, 2017, 417).

In a Magnetic Field

The work must be done by the surroundings in order to change the orientation of the dipole. This work is kept under the form of the potential energy E. Any value of θ could be chosen as corresponding to the value 0 of the potential energy E. Usually, it is the value $\theta = \pi/2$, which is adopted. As a result:

$$E = \int_{\pi/2}^{\theta} \mu B_0 \sin\theta d\theta$$

$$E = -\mu B_0 \cos\theta$$

Hence, the potential energy stored by the dipole is expressed by the scalar product:

$$E = -\mathbf{\mu} \, \mathbf{B}_0 \qquad\qquad (5)$$

Behavior of Active Nuclei Localized in a Magnetic Field

When active nuclei are immersed in a magnetic field, two phenomena occur:

- The rotation axis of nuclei takes position so that they form an angle θ with the direction of the magnetic field.
- The nuclei enter in precession around the latter.

The Rotation Axis of Nuclei

It has already been described. When one applies the field $\mathbf{B}_0$ in a direction z'z, a magnet of magnetic moment μ accepts energy E given by the expression (6):

$$E = -\mathbf{\mu}_z \cdot \mathbf{B}_0 \qquad\qquad (6)$$

Here μ_z is the component of the nuclear magnetic moment μ in the direction z'z (Figure 5).

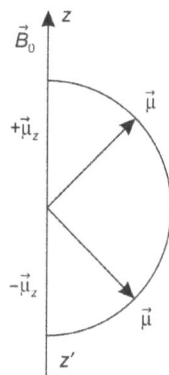

Figure 5: Component μ_z. (Reprinted with Permission from Gwenola and Jean-Louis Burgot, Paris, Lavoisier, Tec et Doc, 2017, 418).

Since:

$$\mathbf{\mu} = \gamma \mathbf{L}$$

$$\mathbf{\mu}_z = \gamma \mathbf{L}_z$$

And since:

$$L_z = m_I \bar{h} \quad \text{(viz., relations (1) and (2))}$$

$$\mu_z = \gamma m_I \bar{h}$$

The m_I defines the component of this moment on the axis z'z. As a result:

$$E = -\gamma m_I \bar{h} B \qquad\qquad (7)$$

The m_I can take the values $l, l-1, \ldots -l+1, -l$. The energy can take $2l+1$ levels corresponding to the $2l+1$ values of m_I. Each value of the quantum number m_I corresponds to a different orientation of the nuclear magnetic moment. For $I = \frac{1}{2}$ (for example, for the proton), there exist only two possible levels which correspond to the values $m_I = \frac{1}{2}$ and $m_I = -\frac{1}{2}$ (Figure 6).

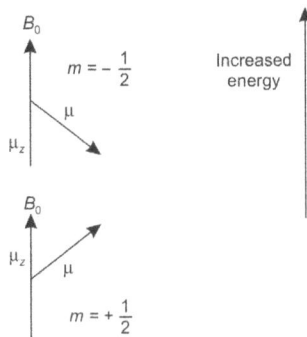

Figure 6: Components μ_z for a nucleus whose $I = \frac{1}{2}$. (Reprinted with Permission from Gwenola and Jean-Louis Burgot, Paris, Lavoisier, Tec et Doc, 2017, 419).

The Second Phenomenon

It operates when the nuclei rotation axis enters in precession around the direction of the magnetic field with the angular rate ω (Figure 7). The phenomenon is called Larmor's precession.

In both preceding phenomena once become effective, the system gets ready to give rise to resonance under the action of electromagnetic radiation.

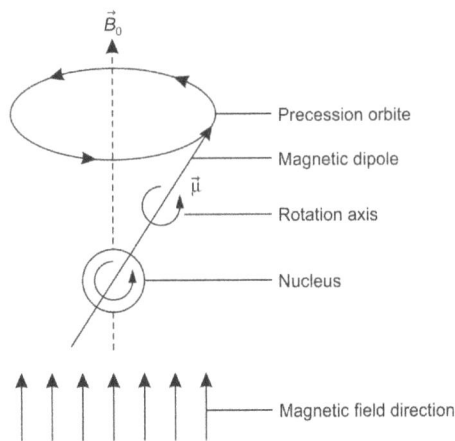

Figure 7: Precession of μ Around B_0. (Reprinted with Permission from Gwenola and Jean-Louis Burgot, Paris, Lavoisier, Tec et Doc, 2017, 420).

From the point of view of classical mechanics, with this phenomenon of precession, we are faced here with an analogous system which is that of the gyroscope system in the field of terrestrial gravitation. It is the result of the couple exerted by $\mathbf{B}_0$ on $\mathbf{\mu}$. The gyroscopic effect is found each time that a force of any nature is applied to a revolution solid animated by a quick rotation around its axis.

Energetic Transitions in N.M.R.

Before studying the phenomenon of resonance in N.M.R., it is judicious to specify the transitions which can play a part during an N.M.R. experiment.

The transitions, as we shall see them, are located in the domain of radio frequencies, that is to say in the range $10^7 < v < 10^{10}$ Hz.

According to relation (7), the energy levels are separated by the quantity ΔE:

$$\Delta E = \gamma\, \bar{h} B_0\, \Delta m_I$$

Here, Δm_I is the change in the quantum number m_I accompanying the transition and B_0 is the intensity of the field. According to the selection rules, the only permitted transitions are such that:

$$\Delta m_I = \pm 1$$

As a consequence, the corresponding change in energy is:

$$\Delta E = \gamma\, \bar{h} B_0 \qquad (8)$$

As in other spectroscopic methods, the absorbed or emitted energies obey Bohr's condition:

$$\Delta E = hv$$

ΔE is the difference of energy between the final and initial levels, v is the frequency of the radiation and h is the Planck constant. According to the two preceding relations, one finds:

$$v_0 = \gamma\, B_0/2\pi \qquad (9)$$

By using magnetic fields from 1 to 5T (10 to 50 kilogauss) and giving us the data in Table 2, we deduce the frequencies of the transition radiations which are of the order of ten MHz, that is to say in the domain of radio frequencies (M = Mega = 10^6). For example:

- For a frequency of 100 MHz, the energy change between the two levels is about 0.0418 J. mol^{-1}. It is very weak.

- Dealing now with the possible orientations of the nuclear magnetic moment in the case of the proton and the neutron, for example (I = ½), we see that increasing θ or changing m_I from ½ to – ½ corresponds to an increase in energy and that, in these conditions, $\cos\theta = \pm 1/\sqrt{3}$. That is to say, $\theta = \pm 54°44'$.

The relation (9) shows that in order to detect the resonance, the scanning can be carried out either by varying the frequency while the principal field remains constant or by varying the principal field while the frequency remains constant. The last alternative is the easiest method to apply. Figure 8 summarizes these considerations.

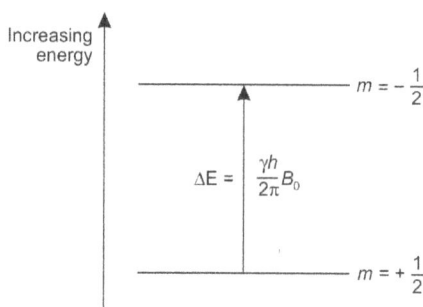

Figure 8: Magnetic Moments and Energy Levels of a Nucleus Possessing a Spin Quantum Number of Spin I; I =1/2. (Reprinted with Permission from Gwenola and Jean-Louis Burgot, Paris, Lavoisier, Tec et Doc, 2017, 421).

The angular velocity of the precession at the resonance ω_0 is found with the help of the following reasoning. We know that in a circular uniform motion, the frequency v of the corresponding periodic motion is related to the angular velocity ω through the relation:

$$v = \omega/2\pi$$

At the resonance, the relation (9) is verified. Therefore, we can deduce:

$$\omega_0 = \gamma \, B_0$$

This relation is called Larmor's relation. Any change in the field changes the precession rate, but θ remains constant.

It is important to notice that the resonance frequency is in linear relation with the magnetic field, as it is indicated by relation (9). This point constitutes the great difference between the other spectroscopic methods and N.M.R. For the former ones, the energy differences depend only on the studied compound.

The Nuclear Magnetic Resonance

In an N.M.R. experiment, the nuclei of the element to study are submitted to a magnetic field and jointly to electromagnetic radiation which produces a change in their energy levels. There is resonance when the frequency of the electromagnetic radiation is such that it induces a change in the energy level. Its frequency is given by expression (9).

One can pre-view the phenomenon of resonance by reconsidering the vectorial model of the process (Figure 7). The change of the level of energy may be considered as resulting of the application of a magnetic field $\mathbf{B}_1$ perpendicular to $\mathbf{B}_0$, $\mathbf{B'}_1$ (viz., under the justification of $\mathbf{B}_1'$). $\mathbf{B'}_1$ exercises a couple on μ which tends to modify θ. If $\mathbf{B}_1'$ turns around $\mathbf{B}_0$ at the same frequency as that of precession while remaining perpendicular to the plane $(\mathbf{B}_0, \mu)$, its effect on μ always tends to increase θ until its value corresponds to different energy permitted (Figures 9 and 10). In any way,

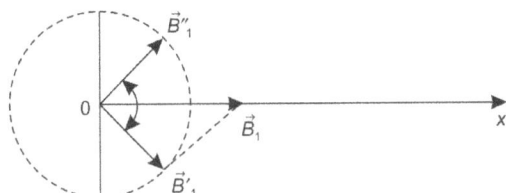

Figure 9: Decomposition of a Linear Sinusoidal Field Into Two Fields Turning in Opposite Senses. (Reprinted with Permission from Gwenola and Jean-Louis Burgot, Paris, Lavoisier, Tec et Doc, 2017, 422).

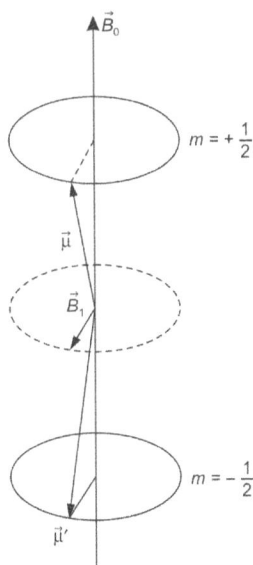

Figure 10: Vectorial Model of the Absorption of the Electromagnetic Radiation by the Precessing Nucleus. (Reprinted with Permission from Gwenola and Jean-Louis Burgot, Paris, Lavoisier, Tec et Doc, 2017, 423).

μ transforms itself into **μ'**. One obtains the resonance when the frequency of the turning field is equal to Larmor's frequency. The absorbed energy at the resonance is provided by **B$_1$**.

In practice, on uses a linear sinusoidal field **B$_1$** given by the expression:

$$\mathbf{B}_1 = \mathbf{B}_1\text{'}\cos(\omega t) + \mathbf{B}_1\text{''}\cos(\omega t)$$

That we can decompose into two turning fields in inverse senses one from the other and of the same period. Only the one possessing the suitable sense **B$_1$'** (and not **B$_1$''**) plays a noticeable part. The radiation coming from the emitting coil in the radiofrequency domain must, hence, be plane, i.e., polarized (Figure 9) (see Chapter 27).

Exchanges of Energy: Saturation

We have seen that in a field **B$_0$**, a nucleus of spin ½ possesses two levels of energy E(−1/2) and E(1/2) located as it is indicated in Figure 11.

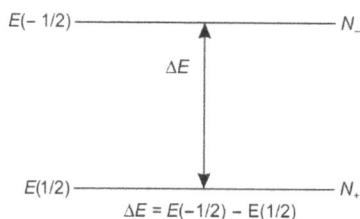

$$\Delta E = E(-1/2) - E(1/2)$$

Figure 11: A transition Between Two Energy Levels (Nuclei of Spin ½). (Reprinted with Permission from Gwenola and Jean-Louis Burgot, Paris, Lavoisier, Tec et Doc, 2017, 423).

According to the relation (9):

$$\Delta E = |\gamma \, \bar{h} B_0|$$

Let us consider a set of N nuclei of spin ½ and N$_+$, N$_-$ the populations of the above nuclei. Let r be their ratio:

$$r = N_-/N_+ \tag{10}$$

According to Bolzmann's relation:

$$r = e^{-\Delta E/kT}$$

$$r \approx 1 - \Delta E/kT \tag{11}$$

After a series of development:

$$\Delta E = h\nu$$

At the value of 60 MHz:

$$\Delta E = 6.626. \ 10^{-34}.60.10^6 = 3.97 \ 10^{-26} \text{ J}$$

At 300K:

$$kT = 1.381.10^{-23}. \ 300 = 4.14 \ 10^{-21} \text{ J}$$

$$r = 1 - 1. \ 10^{-5} \text{ J}$$

With the hypothesis that N$_-$ = 10^6, N$_+$ = 1.000010 10^6. There only exists an excess of 10 particles in the lowest level. The excess of nuclei in the state the most occupied is very weak. From relations (10) and (11), we find:

$$N_-/N_+ = 1 - |\gamma \, \bar{h} B_0|/kT \tag{12}$$

This relation shows that the excess of nuclei in the weakest level of energy is linearly related to the intensity of the magnetic field. It incites to work with magnetic fields the stronger as possible.

From a general standpoint, the transitions in spectroscopy may correspond to several phenomena:

- Absorption of energy (Way A);
- An induced or simulated emission (Way B);
- A negligible spontaneous emission in the domain of radio-frequencies, as is shown by the theory of electromagnetic radiations and hence negligible in N.M.R. (Way C) (Figure 12).

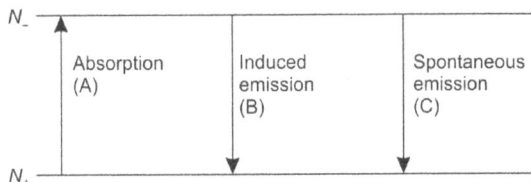

Figure 12: Spectroscopic Transitions. (Reprinted with Permission from Gwenola and Jean-Louis Burgot, Paris, Lavoisier, Tec et Doc, 2017, 424).

The probabilities of processes A and B are identical. Let p be this probability. The number of nuclei that are the cause of absorption by time unit is pN_+. The one which gives rise to an induced emission during the same interval of time is pN_-. The effective absorption is given by the expression:

$$(N_+ - N_-)\, p$$

The excess of nuclei in the lowest energy level is responsible for the absorption. During the time it is more filled than the upper one, the absorption occurs and one observes a signal. But the absorption tends to make equal the populations of both levels. In this case, we would obtain a saturation state. The intensity of the signal would become worthless. During the N.M.R. experiment, we would obtain a limit of time, from which once reached, it would become impossible to observe the absorption of energy. However, it is not the case. This is due to the *phenomenon of relaxation.*

Relaxation Phenomena

The different kinds of transition, which permit the coming back from the state of high energy to that of low energy without any emission of radiation, are called relaxation phenomena. One distinguishes the relaxations *spin-spin* and *spin-lattice*.

The relaxation spin-spin is due to the mutual exchange of spins between two nuclei that are in precession close to one other. When two neighboring nuclei of the same kind have the same precession velocities but have different spin-states, the magnetic fields due to each of them may enter into interaction and induce an exchange of their spin state. This kind of relaxation does not contribute to the keeping of the required excess of nuclei but reduces the lifetime of the excited state. The relaxation spin-spin is characterized by a relaxation time T_2. T_2 is the time constant that quantifies the loss of magnetization (see next section as well as Chapter 25) in the plane XY when there is no inhomogeneity in the static field $\mathbf{B}_0$. One also introduces the constant T_2^*, the so-called effective constant of loss of magnetization. It takes into account the lack of homogeneity of $\mathbf{B}_0$ and permits quantifying the spectral resolution. The relation spin-spin is sometimes called *spin-transverse relaxation.*

In the relaxation spin-lattice, it is the medium where the nucleus (which is in precession) is immersed and induces the fall back to the state of lowest energy. The medium is composed of molecules in solution, in the gaseous, liquid and solid states. All these molecules possess magnetic properties due to the fact that they are animated by translation, vibration and rotation movements. As a result, there exist in the medium small magnetic fields, the transition of which could be induced. During them, the whole energy of the system remains unchanged since that of transition is transferred to the medium as additional energy for translation, rotation and vibration. This kind of relaxation participates in the keeping of the excess nuclei possessing the weakest energy level and hence the phenomenon of N.M.R. It is characterized by a relaxation time T_1, which is a measurement of the average lifetime of nuclei in the state of the uppermost energy. T_1 is the time constant of falling back of the magnetization vector to its equilibrium (see following chapter). In other words, it corresponds to the total length along the axis z'z, which is traveled.

The width of a spectral ray is inversely proportional to the average time during which the system remains in the excited state. Hence, the fine rays are due to long lifetimes in the excited state whereas the large rays are due to weak lifetimes.

The behavior of nuclei in N.M.R. is described by Bloch's equations. In principle, that of each nucleus might be described by the formalism of quantum mechanics. However, for one set of nuclei that do not interact, one can adopt a classical description considering that *nuclear magnetization* is equal to the vectorial sum of individual magnetic moments. Bloch's equations permit the calculation of nuclear magnetization as a function of time. They involve the relaxation times T_1 and T_2.

(*Remark*: We shall come back on the nuclear magnetization in the chapter devoted to the N.M.R. with Fourier transform(s). (See Chapter 26).)

Apparatus

High resolution apparatuses are of two kinds:

- Those with continuum waves with which one follows the absorption signals of the radiation as a function of the frequency of the electromagnetic radiation that is changed slowly.
- Those with Fourier transforms are now the most used.

Figure 13 is a scheme of a continuous wave N.M.R. apparatus. With this kind of equipment, the magnetic field $\mathbf{B}_0$ of the electromagnet can vary, such as from 0 to 5 T. The radio transmitter at high frequencies is essentially composed of one winding traversed by a sinusoidal current.

Evidently, the heart of the apparatus, like with other N.M.R. ones, is a magnet. It can be permanent as it must be for an electromagnet. Their field must be homogenous and reproducible. Let us recall that for reasons of limits of detection and resolution (see the section on 'Energetic Transitions in N.M.R'), it is advantageous to work with a field as strong as possible.

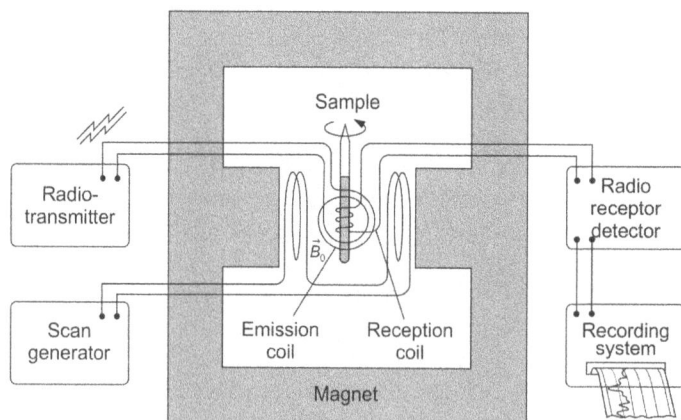

Figure 13: Scheme of a Continuous-Wave N.M.R. Apparatus. (Reprinted with Permission from Gwenola and Jean-Louis Burgot, Paris, Lavoisier, Tec et Doc, 2017, 427).

Some permanent magnets may generate magnetic fields practically of 2.1 T. That is to say, in proton N.M.R., the frequency of resonance of the order is 90 MHz. In more recent apparatus functioning with the help of supra-conductors magnets, fields extremely powerful such as 21 T can be used.

We notice that the coil transmitting radio frequencies located on the x-axis is disposed at the perpendicular of the radio-receiving one (y-axis) and the perpendicular of the field of the electromagnet (Figure 14).

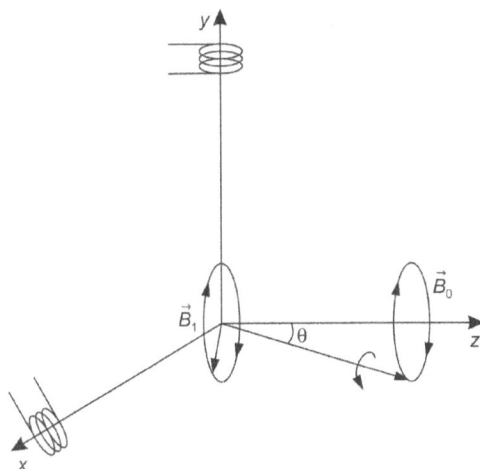

Figure 14: Relative Dispositions of the Radio-Transmitting and Radio-Receiving Coils. (Reprinted with Permission from Gwenola and Jean-Louis Burgot, Paris, Lavoisier, Tec et Doc, 2017, 427).

When transitions occur in the sample, the oscillations of the magnetic field, which are subsisting after the absorption, induce an alternative tension (that can be amplified and detected) in the coil of the radio receiver. Given the very weak energy difference between both levels of transition, the source of the radio frequencies must not be too powerful. However, the detector must be very sensitive.

Let us also mention the fact that for compensating the fluctuations of magnetic fields, devices that block ratios of field/frequency ("field frequency lock system") are used. They are used in order to maintain constant the ratio of the intensity of the magnetic field and the resonance frequency for a kind of nucleus in order for relation (9) to be satisfied.

Forms of the Peaks, Widths of the Rays and Intensity of an Absorption Signal

The resonance appears as induction of an electromotive force in a coil. According to the fact that this induction is in phase or out of phase with the field $\mathbf{B}_1$, the signal of resonance appears in the form of a curve of dispersion (Figure 15a) or absorption (Figure 15b).

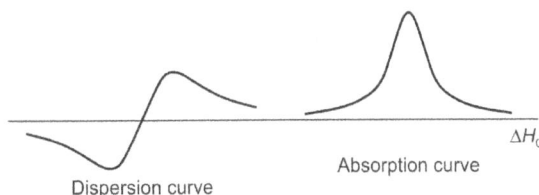

Figures 15a and b: Dispersion Curve a and Absorption Curve b. (Reprinted with Permission from Gwenola and Jean-Louis Burgot, Paris, Lavoisier, Tec et Doc, 2017, 428).

The N.M.R. signals in absorption are the most often Lorentz's curves which are mathematical functions of the type $y = 1/(1 + x^2)$ (Figure 16). The curve is symmetrical with respect to the absorption frequency v_0. Larmor's relation (9) allows us to foresee a resonance ray extremely fine. The absorption takes rise in the interval of frequency of $2\Delta v$ around v_0, that is to say in the interval $v_0 \pm \Delta v$. The width of the ray is in relation to the relaxation times. It is all the more important that these relaxation times are weak. There exist more practical reasons why the ray may be enlarged, such as the magnetic field $\mathbf{B}_0$ is not homogenous (Figure 16).

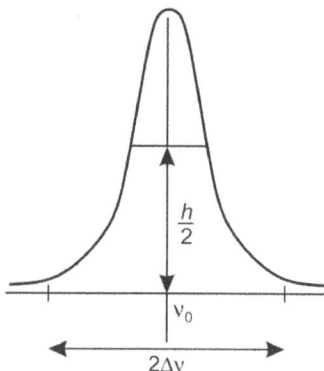

Figure 16: N.M.R. Signal of Absorption. (Reprinted with Permission from Gwenola and Jean-Louis Burgot, Paris, Lavoisier, Tec et Doc, 2017, 429).

The surface under the absorption curve is proportional to the number of resonating nuclei at the same frequency and can be determined by integrating the curve with respect to time at a slow scanning rate. The spectrometers provide a measurement of the surface S under the form of a curve so-called *integration curve* (Figure 17). The distance h between two floors on one side of the absorption signal and the other is proportional to S. This property is very important for two reasons:

- Firstly, it provides fundamental information for the interpretation of spectra.
- The second reason is that it permits the quantitative analysis by N.M.R.

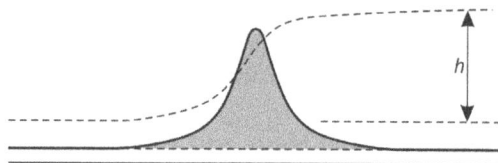

Figure 17: Integration Curve. (Reprinted with Permission from Gwenola and Jean-Louis Burgot, Paris, Lavoisier, Tec et Doc, 2017, 429).

Kinetic Processes—Changes in Configurations and Exchanges of Atoms

The aspect of an N.M.R. spectrum may evolve when the active nuclei can go from one molecule to another. This is the phenomenon of exchange. When the rate of exchange is weak at the scale of time of N.M.R., the spectrum shows the signals of the two molecular forms which can exchange the active nucleus. When the exchange is rapid at the scale of N.M.R., the spectrum no longer exhibits the two molecular forms but only one located at the average value of the *chemical shifts* (see next chapter) of the initial forms.

The phenomenon is analogous when both conformations of the same molecule become "interconverted" and give rise to equilibrium under the effect of the temperature. When the system tends toward equilibrium, there exists a time for which the signals of both initial forms are no longer separated. The time τ necessary to attain the coalescence is given by the expression:

$$\tau = \sqrt{2}/\pi\Delta$$

Here, Δ is the difference (Hz) between the two peaks, when there is no "interconversion". (An example of exchange is given in the next chapter).

Chapter 25

N.M.R. Spectra Fundamentals
Characteristics and
Analytical Applications

There are several kinds of N.M.R. spectra. Distinction criteria can be, for example, the type of the used apparatus, the nature of the active nucleus, the physical state of the analyzed sample and so forth.

The tendency is to distinguish two kinds of spectra, i.e., wide-line or large band spectra and high resolution spectra. High resolution spectra are obtained with apparatus which are able to differentiate the signals, the differences of resonance frequencies of which are very weak.

The large band spectra are obtained when the source of possible frequencies possesses a bandwidth sufficiently large, which precludes the vision of the fine structure of the studied compound. This kind of N.M.R. is used to quantitatively determine isotopes together with the study of the surroundings of the species which absorbs. Spectra are obtained with magnetic fields rather weak. We shall not study this sort of N.M.R. further.

In this chapter, we shall define the principal grandeurs which characterize high resolution spectrum and we shall infer the information concerning the studied system from them. Then, we shall give some examples of applications essentially provided by the study of molecular structures. Then, we shall consider examples of quantitative analysis by N.M.R.

In this chapter, we are only interested in the high-resolution N.M.R. of proton ^{1}H and to a less extent that of ^{13}C and, in an anecdotic way, that of ^{19}F. These N.M.R. are the most practiced.

Definition of N.M.R. Spectrum: Its General Form

N.M.R. spectrum is the diagram intensity of the absorbed radiation I(ν) as a function of the frequency ν of the radio frequency. Figure 1 shows the ^{1}H N.M.R. spectrum (continuum wave) of ethylbenzene. It is a very simple spectrum.

We notice some points. They constitute the essential features of the N.M.R. spectrum. They are:

* The existence of two scales of abscissa. One is located at the top of the figure. It permits noticing the frequency of resonance of the different protons of the molecule. A second is mentioned at the bottom of the spectrum. It is the scale of ppm (parts per million) (symbol δ) defined further under 'Chemical Shift'.

* The existence of three groups of protons numbered 1, 2 and 3. Manifestly, the protons of each group are in resonance in different frequency zones. This fact gives rise to the notion of electronic screens and to that of *chemical shift* (see the section on).

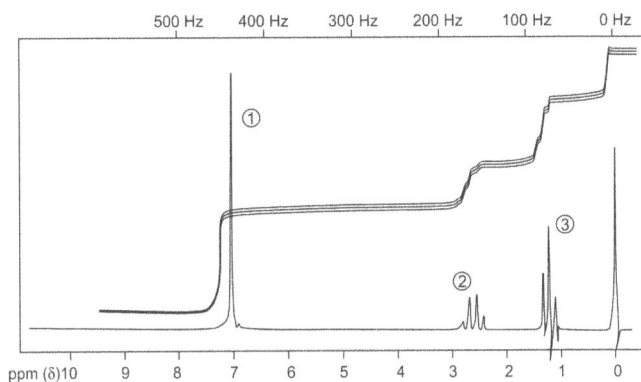

Figure 1: N.M.R. ^{1}H at 60 MHz of Ethylbenzene in CDCl$_3$ With the TMS as an Internal Standard. (Reprinted with Permission from Gwenola and Jean-Louis Burgot, Paris, Lavoisier, Tec et Doc, 2017, 434).

- Some groups of protons appear under the form of multiplets. In this example, they are groups 2 and 3. This point is based on the notion of *spin-spin coupling* (see the following section on 'Spin-Spin Coupling').

The Effect of Electronic Shielding Effect

A naïve vision of N.M.R. may induce the thinking that all the atoms of a given element are in resonance at the same frequency. In other words, this vision might induce the thinking that, in the studied molecule, the static magnetic field is not perturbed by the surroundings of the active nucleus. As it is shown by the above example, this is false.

A substance situated in a magnetic field acquires an induced magnetization. In order to explain this point, it is necessary to consider two magnetic effects due to the electronic surroundings of the active nucleus:

- A global effect is related to the magnetic susceptibility χ of the substance constituting the sample, which generally (diamagnetic substances) reduces the magnetic field inside the substance. It is poorly interesting in N.M.R.
- A very interesting local effect that gives rise to the notions of *shielding effect* and of *chemical shift*.

Let us consider the hydrogen atom immerged in the magnetic field $\mathbf{B}_0$ (Figure 2). The electron acquires a movement in its orbit. This phenomenon can be assimilated to an electric current circulating in a winding.

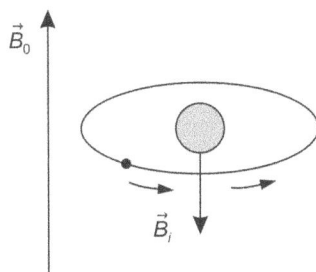

Figure 2: Origin of the Induced Current. (Reprinted with Permission from Gwenola and Jean-Louis Burgot, Paris, Lavoisier, Tec et Doc, 2017, 435).

There is the creation of an induced magnetic field $\mathbf{B}_i$ antiparallel to the applied field $\mathbf{B}_0$. The latter is on average opposed to the former.

(*Remark*: The notion of average is justified by the fact that for one majority of liquids, the rotation time of one molecule is very short: of the order of 10^{-10} s.)

As a result, the active nuclei are submitted to an effective field **B** weaker than $\mathbf{B}_0$. The relation between both is:

$$\mathbf{B} = \mathbf{B}_0(1 - \sigma) \tag{1}$$

σ is named screening constant.

Chemical Shift

As the result of the effect of electronic screening, the values of σ a given nucleus depend on its electronic surroundings. Given the relation (see preceding chapter):

$$v = |\gamma| \, B/2\pi$$

The resonance frequency of a nucleus j is given by the expression:

$$v_j = (|\gamma|/2\pi) B_0(1 - \sigma_j) \tag{2}$$

Where B_0 is the applied field. It changes with the position of the nucleus in the molecule. Let us take the case for example of chloroacetone (chlorodimethylketone) as a molecule $ClCH_2C(=O)CH_3$. In N.M.R. ^1H, one obtains two signals (Figure 3):

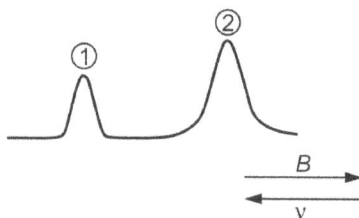

Figure 3: Signals of the Protons of Chloroacetone. (Reprinted with Permission from Gwenola and Jean-Louis Burgot, Paris, Lavoisier, Tec et Doc, 2017, 436).

These signals are due to the protons of the rest methylene and those of the rest methyl. Manifestly, the latter possesses different surroundings from that of the former. Signal 1 is due to the methylene and signal 2 to the methyl. This assertion is notably justified by the integration curve.

Relation (2) shows that resonance frequencies remain in linear relation with B_0. The differences in the chemical shifts measured in Hz change with B_0 as it is demonstrated by the following relation coming directly from (2):

$$v_i - v_j = (\gamma/2\pi) \, B_0(\sigma_j - \sigma_i)$$

The fact that the difference in the resonance frequency depends on the field makes it a changing one with the different apparatus. This has for consequence the impossibility to build an absolute scale of resonance frequencies. One refers to a relative scale, dimensionless, measured with respect to a reference compound, which is chosen arbitrarily and gets dissolved in the sample. It is called an *internal reference*. The position of the signal, symbolized by δ, is defined by the expression:

$$\delta = 10^6(v_x - v_{ref})/v_{ref} \tag{3}$$

The difference $(v_x - v_{ref})$ is very small with respect to v_{ref}. This is the reason why coefficient 10^6 is assigned to the numerator. It permits handling simple numbers. δ is the *chemical shift* of the nucleus x. It is evident that it is independent of B_0. δ is expressed in ppm (parts per million).

Usually, one compares the resonance frequencies to those of nuclei of an internal reference substance in preference to those of nuclei of an external reference in order to avoid the corrections of magnetic susceptibility.

The wished properties for a reference substance are the existence of only one kind of proton in its structure, a weak reactivity and a non-polar character of its molecule (given the solubility properties of the majority of organic compounds). Today, the reference substance the most used for protons is tetramethylsilane (TMS) $Si(CH_3)_4$. As a result, according to (3):

$$\delta = 10^6 (v_x - v_{TMS})/v_{TMS} \tag{4}$$

The solvents the most used in proton N.M.R. are carbon tetrachloride, deuterochloroform $CDCl_3$, carbon disulfide CS_2, deuterated sulfoxide $(CD_3)_2SO$, hexa-deutero-dimethyl-ketone $(CD_3)_2CO$ and deuterated trifluoroacetic CF_3COOD. In water, one can use as an internal reference the sodium sulfonate of formula $(CH_3)_3 -Si-(CH_2)_3-SO_3Na$ (2,2 dimethyl-2-silapentane-5-sulfonate or DSS) whose chemical shift of the methyl signals is located at 0.00 ppm/TMS.

One can say further that there exists another scale permitting to locate the position of a signal. It is called the scale τ. The conversion of one unit to the other is straightforward:

$$\tau = 10 - \delta$$

Today, this scale is little used.

Spin-Spin Coupling

In some conditions, the resonance rays of a nucleus appear in the form of a multiplet (see the groups of signals 2 and 3; Figure 1). For a molecule having two coupled nuclei of no-nul spins I_1 and I_2, the resonance signal of nucleus 1 appears under the form of $2I_2 + 1$ rays and that of nucleus 2 under the form of $2I_1 + 1$ rays (for example, if $I_1 = I_2 = \frac{1}{2}$, each nucleus appears under the form of two rays). The distances between two rays of each multiplet are equal. These characteristics suggest that the separations result from molecules exhibiting different values of m_i (magnetic moments) owing to interactions between the spins.

The spin-spin couplings are characterized by a constant J called *coupling constant*. They depend on the chemical surroundings. Let us consider the hexachloro 1.1.1.2,3,3-propane and the ethylbenzene.

The Hexachloro 1.1.1.2,3,3-Propane

The formula of hexachloro 1.1.1.2,3,3-propane is given in Figure 4.

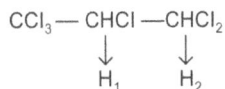

Figure 4: Formula of the Hexachloro 1.1.1.2,3,3-Propane. (Reprinted with Permission from Gwenola and Jean-Louis Burgot, Paris, Lavoisier, Tec et Doc, 2017, 438).

Let us name the proton brought by carbon 2 by H_1 and that brought by carbon 3 by H_2. Both are coupled. Both possess the spin $I = \frac{1}{2}$. Each one appears in the form of a doublet (Figure 5). One of the doublets appears at $\delta_2 = 6.67$ ppm and the other at 4.95 ppm. The distances between the two rays of each doublet are measured in Hz the value of the coupling constant. It is the same, that is to say, J = 1.4 Hz.

(*Remark*: Notice that both coupling protons are separated by three bonds. We shall come again on the fact that there exists a weakening of the coupling constant with the number of atoms between the two coupled nuclei in the coming section.)

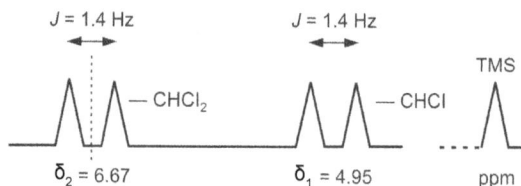

Figure 5: Schematic ^{1}H Spectrum of the Hexachloropropane. (Reprinted with Permission from Gwenola and Jean-Louis Burgot, Paris, Lavoisier, Tec et Doc, 2017, 438).

Ethylbenzene (see Figure 1)

One notes that the protons of the methylene and those of the methyl groups are coupled. It is the same case as the preceding one. However, there are between them several differences. Let us designate by H_1 and H_2 the protons of the methylene and those of the methyl:

- The protons H_1 give rise to a signal which is a quadruplet whereas the two protons H_2 appear as a triplet.
- Although that is not evident at first sight at the examination of the spectra, the two protons H_1 are equivalent. They give the same signals which are exactly superposing. It is the same for the three protons of the methyl CH_3 which are superposing (the three ones) at a different shift from that of the two preceding. We shall come back to the notion of chemical and magnetic equivalences in the coming sections.
- The measurements of the area under the peaks (see the preceding chapter) show that the quadruplet (Group 2) corresponds to two protons and the triplet (Group 3) to three protons.

Supplementary Description of a N.M.R. Spectrum

Usually, N.M.R. spectra are presented with two scales of abscissa, the chemical shift δ (ppm) and the resonance frequency ν (see Figure 1). The term frequency is very often raised. It is related to two kinds of frequencies. The first one is linked to Larmor's frequency which concerns the considered nucleus. The second one is the frequency at which resonates a kind of nuclei in a narrow domain of frequencies.

The first kind of frequency is given by the relation (9) of the preceding chapter:

$$\nu_0 = \gamma B_0/2\pi$$

This frequency evidently depends on the nucleus via the magnetogyric ratio γ. It also depends on the magnetic field B_0. Therefore, concerning the proton, the frequencies 60 MHz, 100 MHz, 220 MHz and 300 MHz correspond to the magnetic fields $1.41T - 2.35T - 5.17T - 7.05T$. This is the reason why one speaks of ^{1}H spectra scanned at 60 MHz, that is to say for a magnetic field of $1.41T$.

The second type of frequency corresponds to the retained interval of frequencies of scanning which depends on the nucleus. Thus, with the organic compounds the most often encountered, protons absorb in a band of frequencies of 1,200 Hz for a field of $1.41T$, a band centered on the frequency of 60 MHz. One of the two used kinds of abscissa brings about the value of the frequency at which the nucleus located in the band resonates. In ^{1}H N.M.R., one essentially works in the scanning frequencies going from 0 to 500 Hz and from 0 to 1,000 Hz.

Concerning now the chemical shifts δ, it must be noticed that they are identical for the same compound, whichever the working frequency is chosen. For example 60 MHz or 100 MHZ or, if we prefer, whichever the chosen magnetic field is $1.41T$ or $2.35T$. This is not astonishing if we consider the definition (3) of the chemical shift in which we see that the magnetic field no longer appears in the numerator and the denominator, and it can be understood that the frequencies v_x and v_{ref} are in linear relation with the field B_0. This is not the case for the resonance frequencies in the scanning scale.

For historical reasons, the signals are registered in such a way that the applied field B_0 increases when the spectrum is read from the left to the right (see Figure 1). It is said that while we are studying a spectrum we are going more towards the right, we are going towards more strong fields, or (what is equivalent) towards low frequencies. The domain of low frequencies has also been named the region of strong *shielding*. Inversely, when we consider the left part of a spectrum, one is situated in the region of the weak fields or weak shielding or that of high frequencies. The protons resonating in this region are said *deshielded*. For example, the TMS which is the most used reference resonates in the strong fields. The origin of this vocabulary is given in the following paragraph devoted to the chemical shift.

Generally, most protons resonate in the domain 1 to 15 ppm, that is to say in an interval of about 1,200 Hz. For other nuclei, the domains of resonance are markedly different.

The coupling constants are measured in Hz. Concerning the ^1H N.M.R., they are rather weak. They do not depend on the strength of the magnetic field. In this behavior, we have a very interesting means to distinguish one peak due to a chemical shift from a signal due to a coupling spin-spin.

A Comeback into the Domain of Chemical Shifts

The major interest of N.M.R. in analysis lies in the fact that all the nuclei of a given isotope do not resonate at the same frequency because of the occurrence of their electronic surroundings, even in the same molecule. Figure 6 mentions the values of the chemical shifts for different kinds of protons.

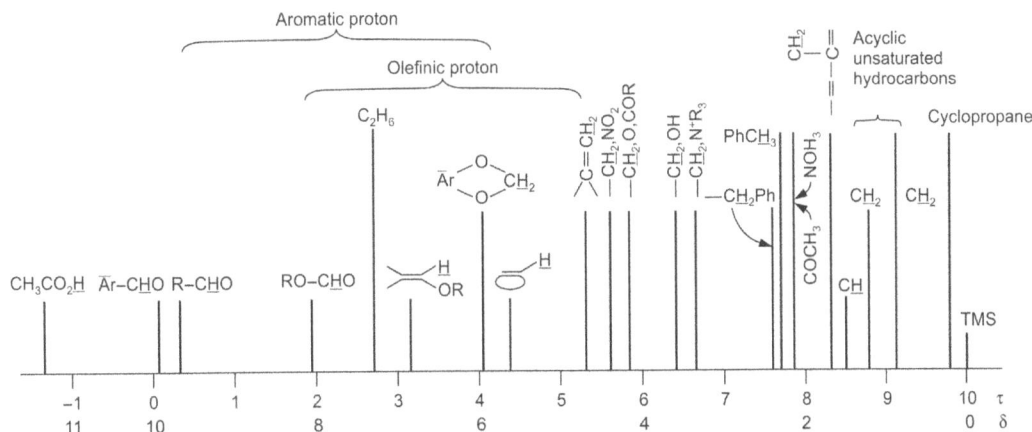

Figure 6: Chemical Shifts of Protons of Different Organic Functions (Recall: The Chemical Shifts Depend on the Nature of the Solvent). (Reprinted with Permission from Gwenola and Jean-Louis Burgot, Paris, Lavoisier, Tec et Doc, 2017, 440).

The value of the chemical shift is a major criterium of identification of the kind of proton which is resonating and, as a result of this possibility, a criterium of elucidation of the structure of the molecule to which they belong. Furthermore, in some conditions, the presence is easy to identify and easy to distinguish as a group of protons can permit knowing the concentration of the molecule in a mixture (see the section on quantitative analyses by N.M.R. via the chemical shift).

(*Remark*: In passing, let us point out that chemical shifts can be calculated in some cases. The hypothesis on which is grounded their calculation is that chemical shifts due to the groups which are components of the molecule are additive.)

From the theoretical point of view, as has been said in the preceding sections, the chemical shifts are due to secondary magnetic fields produced by the displacements of the electron in the molecules induced by the primary magnetic field.

Effect of the Shielding

We have already seen (see the section on 'The Effect of Electronic Shielding Effect') that under this influence, the electrons which are around the proton, tend to give rise to a phenomenon of precession all around it, in a plane perpendicularly to that of the principal magnetic field. As a consequence, a secondary magnetic field is forming which is opposed to the primary one. The nucleus is, hence, under the influence of a total field that is weaker than the principal. The nucleus is said masked or "shielded" with respect to the sole principal field. As a result, the applied field must be stronger in order to attain resonance.

The effect of shielding suffered by a nucleus is directly linked to the electronic density, which is all around it. Therefore, it is conceivable that the shielding effect depends on the electronegativity of the atoms and groups of atoms that are adjacent to the nucleus. This hypothesis is perfectly baked up with the chemical shifts of the methyl halogenides, the values of which are $\delta = 2.16 - 2.68 - 3.05$ and 4.26 ppm, respectively, for the iodide, bromide, chloride and fluoride as it is shown in Figure 7.

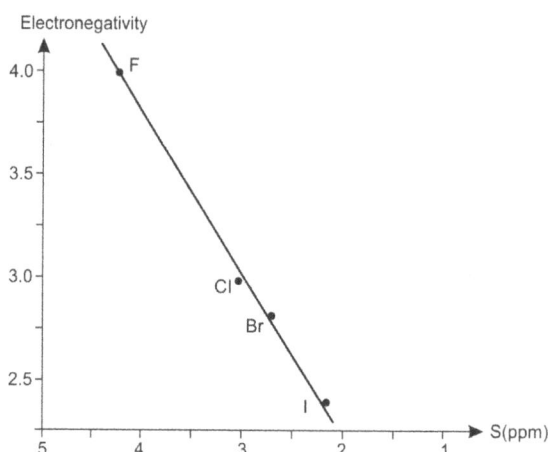

Figure 7: Relation Between the Chemical Shift of the Methyl of Methyl Halogenides as a Function of the Electronegativity (Pauling Scale) of the Halogen Atom. (Reprinted with Permission from Gwenola and Jean-Louis Burgot, Paris, Lavoisier, Tec et Doc, 2017, 441).

Evidently, it is in the methyl iodide that the methyl group remains surrounded by the stronger electronic density since the iodide atom is the least withdrawing halogenide atom. It is in this molecule that the protons are the more shielded.

The Anisotropic Effect

The electronegativity of the adjacent group is not, however, a sufficient factor in order to explain all the phenomena. Magnetic effects due to some particular chemical groups sum up, indeed, with electronegativity. Let us mention the anisotropy effects which appear in the form of anomalous chemical shifts even when eventual electronegativity effects must be taken into account.

Let us consider an aromatic nucleus, such as benzene. The magnetic effect it brings is a function of the orientation of its plan with respect to the principal field. When they are perpendicular, there occurs a current of the cycle that is opposed to the principal field (Figure 8). However, this induced field carries out a magnetic effect of the same direction and same sense as the principal field on the protons located on the aromatic nucleus (that is to say in its plane). Here is the explanation of the values of the chemical shifts of the protons located in the region of 6 to 10 ppm. These protons are strongly deshielded. It is the same for ethylenic protons (Figure 9). Their chemical shifts are located at fields markedly weaker than those of alkanes (from 4.8 to 8 ppm) although both groups of molecules are composed of the same elements, besides being very weakly electronegative. (Alkanes give signals at high fields: 0.2 to 1.5 ppm). There exist numerous other examples of deshielding but also of shielding due to particular groups. In the 1-alkynes, in contrast, there exists a shielding effect due to the current of the cycle all along the triple bond. It induces an opposite field to the principal one (Figure 9).

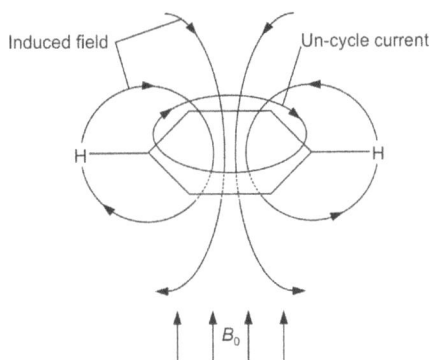

Figure 8: Deshielding of Protons Located on the Aromatic nucleus. (Reprinted with Permission from Gwenola and Jean-Louis Burgot, Paris, Lavoisier, Tec et Doc, 2017, 442).

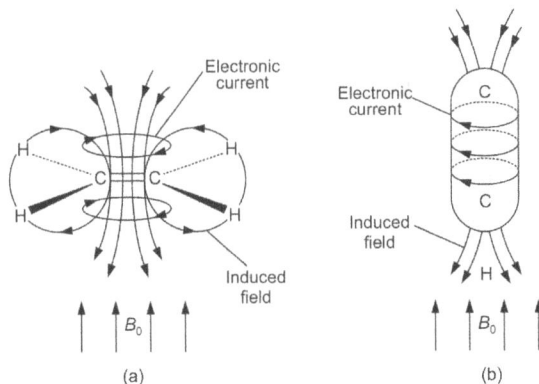

Figures 9a and b: Deshielding of Ethylenic Protons (a) and Shielding of Protons of Alkynes (b). (Reprinted with Permission from Gwenola and Jean-Louis Burgot, Paris, Lavoisier, Tec et Doc, 2017, 438).

Changes in Chemical Shifts Under the Influence of Some Reactants

When the N.M.R. spectra are too complicated to be explained, one can use reactants that permit displacement of some signals towards weak fields. Therefore, it is possible to enhance its comprehension. They are named "N.M.R. shift reagents". In the majority of cases, they change the chemical shifts without changing the coupling constants. A majority of them are complexes of lanthanides. One example is the following complex of europium; $Eu(DPM)_3$ (Figure 10). Most often, they are neutral complexes of the 2,4-pentanedione (and derivatives) and paramagnetic ions. The metallic ion possesses one or several sites of coordination that are not blocked by the pentanedione and with the help of which the reactant can associate itself to the analyte. The latter must be a neutral electron-donating species, such as amines, alcohols, ketones, aldehydes, thiols, thiocarbonylated

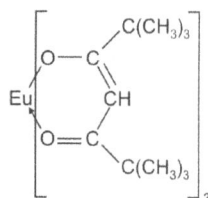

Figure 10: Formula of Eu(DPM)$_3$. (Reprinted with Permission from Gwenola and Jean-Louis Burgot, Paris, Lavoisier, Tec et Doc, 2017, 443).

derivatives and so forth. With these ligands, a labile complex forms. Because it is anisotropic, the lanthanide ion Ln^{III} displaces the signal of resonance of protons according to their distance to the ion. Moreover, the effect of the reactant is only noticeable in the approach to the coordination site.

Beyond the magnetic properties, the choice of a lanthanide complex is justified by the double fact that it does not intervene in the relaxation dipole-dipole and that it possesses important anisotropies which induce important displacements.

A Comeback on the Spin-Spin Couplings

The occurrence of coupling constants in a spectrum provides very important elements for the elucidation of molecular structures.

The origin of the spin-spin couplings is attributed to the fact that the spin of a group of nuclei influences the resonance of another group of nuclei. Hence, it is due to an interaction of nuclei more or less removed from each other by the intermediary of bonding electrons.

Example of Coupling in the Ethyl Group

Let us study the effect of the methylene protons on those of the methyl group, for example in the ethylbenzene (see Figure 1). We noticed that the methyl group is represented by a triplet. The spins of both protons of methylene can be grouped according to Figure 11.

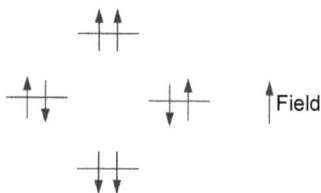

Figure 11: Pairing of the Spins of the Nuclei of the Group Methylene in the Ethylbenzene. (Reprinted with Permission from Gwenola and Jean-Louis Burgot, Paris, Lavoisier, Tec et Doc, 2017, 444).

In the low scheme, the nuclear spins are parallel and opposite to the magnetic field whereas in the scheme of the top, the spins remain parallel but they are in the opposite sense to previously. In the lower, the principal field is slightly weaker than the one applied. Therefore, the field for which there is resonance is slightly enhanced. In the second case, there is resonance at a field slightly weaker. The two median configurations have no influence on the resonance of the protons of the methyl. The area under the central component is equal to two times that which is under the two remaining combinations. From these considerations, it results that the signals of the protons of the methyl (in this molecule) appear under the form of a triplet.

[*Remark*: Let us remark that in the reasoning we have just followed when we calculated the number of combinations corresponding to the different possible pairings. That is to say, in the area under the components we have supposed that the ratio of the numbers of protons in each state of spins is equal to unity. It was legitimate because the excess of the number of protons in the spin state of weakest energy is extremely weak (see the previous chapter).]

Let us now study the influence of the three protons of the methyl on those of the methylene. The possible combinations of the spins of the three protons of the methyl group are represented in Figure 12.

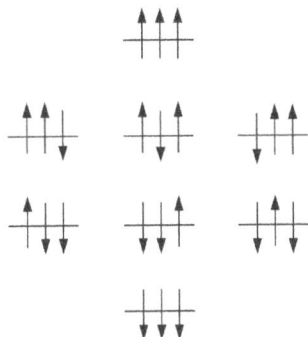

Figure 12: Pairing of the Spins of the Protons of the Methyl Group in the Ethylbenzene. (Reprinted with Permission from Gwenola and Jean-Louis Burgot, Paris, Lavoisier, Tec et Doc, 2017, 445).

There are eight possible combinations in the proportions 1,3,3,1. The reasoning is the same as previously, except for the fact that there are two groups of median combinations, which this time contributes to modifying the applied magnetic field besides being in the opposite senses. The methylenic protons appear in the form of quadruplets.

Coupling Constants

The interval between two consecutive rays is named *coupling constant*. It is symbolized by J (Figure 13) and is measured in Hz.

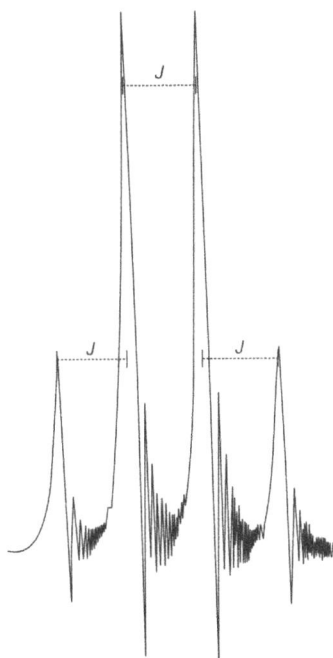

Figure 13: Coupling Constants in a Quadruplet. (Reprinted with Permission from Gwenola and Jean-Louis Burgot, Paris, Lavoisier, Tec et Doc, 2017, 438).

The preceding quadruplet is due to the presence of a neighboring methyl. If one makes the hypothesis that it is the signal of one methylene (see the example of ethylbenzene above), the methyl appears then under the form of a triplet; the rays of which are separated by the *same value J* of the coupling constant.

Values of J are a measurement of the force of the coupling. They are extremely variable. They increase with the atomic number of the coupled elements. In the case of couplings J_{HH}, the value is of the order of 0 to 20 Hz. For J_{FF}, it is of the order of 20 to 200 Hz.

Spectra of First and Second Orders

Let us name Δv the difference in resonance frequencies between the nuclei mutually coupled. Spectra of the first order are those for which:

$$J/v \leq 1/10$$

They are of second order when:

$$J/\Delta v \approx 1$$

The interpretation of the coupling constants is relatively easy in the case of spectra of the first order. It is not the same for the spectra of second order. This case is not investigated here.

Chemical Equivalence—Magnetic Equivalence

Because there may exist some elements of symmetry in some molecules, several nuclei may have their electronic surroundings identical. They have identical chemical shifts, that is to say, the same Larmor's frequency. Nuclei having identical Larmor's frequencies are named *isochrone nuclei, symmetrically equivalent* or *chemically equivalent*. (The accidental isochrony which does not come from the symmetry of the molecule does not enter this nomenclature). In the symmetric molecules, the equivalent nuclei are those which can be "interchanged" by transformations that have certain symmetries.

On the other hand, two or several nuclei having the same *screening constants* are said to be equivalent. Isochrone nuclei, if they are chemically equivalent, are not obligatorily magnetically equivalent. For example:

- The 1-2-3-trichloro-5-fluorobenzene (Figure 14) possesses two protons magnetically and chemically equivalent by symmetry; also because they are coupled identically to the fluor atom ^{19}F (spin ½).

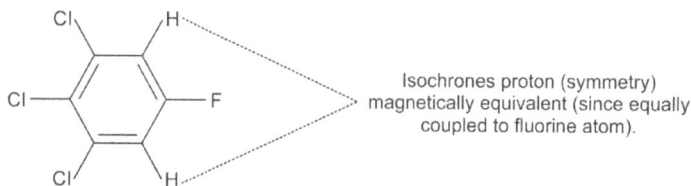

Figure 14: 1-2-3-trichloro-5-fluorobenzene. (Reprinted with Permission from Gwenola and Jean-Louis Burgot, Paris, Lavoisier, Tec et Doc, 2017, 438).

- The 1,2-dichloro-3,6-difluorobenzene possesses two isochrone protons, but they are not magnetically equivalent. The coupling constants J_{ab} and J_{ac} are not identical (Figure 15).
- The cyclobutene (Figure 16), protons a, b, c and d are isochrone, but they are not magnetically equivalent since $J_{ac} \neq J_{bc}$.

Generally, in cases of chemical or (and) magnetic equivalence, spectra are simplified.

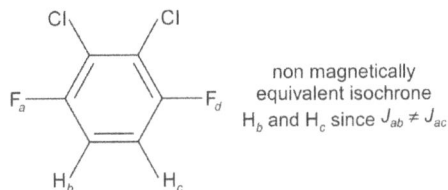

Figure 15: 1,2-Dichloro-3,6-Difluorobenzene. (Reprinted with Permission from Gwenola and Jean-Louis Burgot, Paris, Lavoisier, Tec et Doc, 2017, 447).

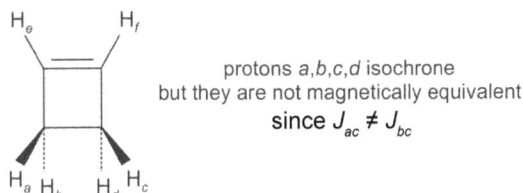

Figure 16: Cyclobutene. (Reprinted with Permission from Gwenola and Jean-Louis Burgot, Paris, Lavoisier, Tec et Doc, 2017, 447).

Homotopic, Enantiotropic and Diastereotopic substances (see Appendix 6)

Characteristics of the First Order Spectra

• Equivalent nuclei do not interact between them. Hence, they do not give several peaks. For example, the three protons of the methyl group of the ethylbenzene only give one signal, even if this is under the form of a triplet. The fact that there is a triplet has nothing to do with the presence of three protons. It is due to the coupling with the adjacent methylene.

• The coupling constants decrease in absolute values with the number of bonds separating the nuclei. Generally, the coupling is null for the proton after four bonds. However, if among these bonds there are double bonds or when some stereochemical structures do exist, a non-null value of the coupling constant may persist after 4 or 5 bonds.

• The multiplicity of the rays of a signal is determined by the number n of protons magnetically equivalent brought by the neighboring atoms and is given by the value n + 1.

• When the protons of an Atom B "interact" with those (nonequivalent) brought by atoms A and C, the multiplicity of the signal given by B is given by the factor.

$$(n_A + 1)(n_C + 1)$$

Here, n_A and n_C are the numbers of equivalent protons, respectively, brought by A and C.

• Approximatively, the relative area of the components of the multiplets are symmetrical with respect to the center of the ensemble of the signal and are proportional to the terms of the expansion of the signal $(x + 1)^n$, that is to say to the coefficients of Pascal's binomial relation.

• The values of coupling constants are independent of the applied magnetic field. Here, a property lies that permits distinguishing the ray of a multiplet from a chemical shift (see the previous section).

• An interesting case that is frequently encountered is that in which a central group of protons is influenced by two (or several) other groups of protons. Let us consider three protons that we respectively designate by A, M and X. We say that it constitutes an A, M and X system. From the standpoint of the nomenclature, we notice that the used alphabet letters are rather remote from each other. This means that the three protons (A, M, X) are sufficiently different so that the system they constitute is of the first order. The three protons influence each other. After the application of the relation $(n_A + 1)(n_C + 1)$, we see that each one appears in the form of a quadruplet. They can be built by degrees by successively considering the three types of possible interactions; AM, AX and MX (Figure 17).

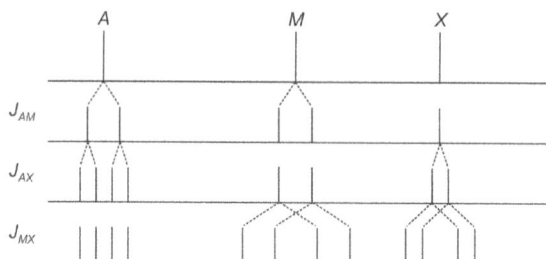

Figure 17: Multiplicity in the System A, M and X. (Reprinted with Permission from Gwenola and Jean-Louis Burgot, Paris, Lavoisier, Tec et Doc, 2017, 449).

This example shows how a spectrum of the first order can be interpreted by grounding ourselves on the chemical shifts, the coupling constants and the integration.

Simplification of the N.M.R. Spectra: The Technique of the Double Resonance

Often, it happens that there is an overlapping of the signals of chemical shifts and rays resulting from spin-spin couplings coming from the resonance of other nuclei than those giving the preceding shifts. As a result, spectra may be difficult to interpret, even when they may be considered as being of the first order. Spectra may be simplified by adopting one of the following protocols:

- One can increase B_0. The resonance frequencies increase but not the coupling constants.
- One can proceed by changing the solvent. Some protons may appear more shielded, some others less with respect to their chemical shift in their initial solvent.
- One can proceed to a decoupling by irradiation of a proton at its resonance frequency by a sinusoidal field of high intensity. If the latter is sufficient, there is a saturation of this proton and its coupling with similar nuclei disappear from the spectra. This is the technique of the double resonance (*homonuclear spin decoupling*). The technique of decoupling between different nuclei (*heteronuclear decoupling*) may also be carried out with a relatively recent apparatus. The technique is realized with the help of auxiliary radiation. It is overall used in N.M.R.^{13}C. (see section on N.M.R.^{13}C).

Values of Some Coupling Constants

We give in Table 1 the values of some coupling constants.

Table 1: Some Values of Coupling Constants J_{H-H} (Hz).

CH_4	Carbon sp^3	+6 to −20
C_2H_6	Carbon sp^3	+3 to +11
=CH_2	Carbon sp^2	+2
H_2C=CH_2 CIS	Carbon sp^2 cis	+5 to +12
H_2C=CH_2 TRANS	Carbon sp^2 trans	+10 to +20
Cyclohexane		(See the section on 'Determination of a quantity of matter (or of a concentration)'
Benzene Ortho* Meta para		6–9 1–3 0–1

(*In the ethylbenzene, the couplings between the protons *o*, *m*, and *p* are not observable because they have the same chemical shift and because the couplings are visible only when the chemical shifts are different.)

These values possess an unquestionable diagnostic power. It is interesting to notice that, in some cases, it is possible to calculate the values of some coupling constants as a rule and reasonably.

Coupling Constants and Exchange of Atoms

In some conditions, the couplings are not visible. It is the case when there is an exchange of atoms between two molecules. A classic example is when there is the exchange of a proton between water and ethanol:

• When ethanol is very pure, the awaited couplings appear. That is to say, the proton brought by the hydroxyl group appears in the form of a triplet because of its coupling with the protons of the adjacent methylene, whereas the protons of the methylene are coupled at once with the protons of the methyl group and with that of the hydroxyl. Hence, its signal is that of a complex multiplet.

• In alcohol, that is to say, in the ethanol-water mixture and in the presence of some acid or basic impurities which react as catalyzers, the spectrum exhibits one singlet only for the group hydroxyl and for water. This is due to the exchange of protons between them. The rate of the exchange depends on the temperature. One could determine that, at the coalescence of the peaks of water and ethanol, the proton of the radical hydroxyl changes in the molecule about 11 times per second (See Chapter 24).

^{13}C N.M.R.

Let us recall that the carbon atom ^{13}C has for its spin number the value ½, whereas that of the carbon ^{12}C is null. Its magnetogyric ratio (see Table 2 Chapter 24) is $6.726 \ 10^{-7}$, whereas that of the proton is $26.752 \ 10^{-7}$ rad.T^{-1} s^{-1}. As a result, ^{13}C resonates at frequencies about 4 times as weak as ^1H. For a magnetic field of 2.11 T, the resonance frequencies of the two kinds of nuclei are respectively 90 MHz and 22.63 MHz. Moreover, the natural abundance of ^{13}C is only 1.1%, whereas it is 99.8% for ^{12}C.

Its weak abundance and the small value of its magnetogyric factor make the "sensibility" obtained by handling this nucleus about 6,000 times weaker than that conferred by the proton. It is so weak that the conventional standard apparatus cannot be used for this use. This difficulty can be surmounted by proceeding with multiple repeated scanning, the signals of which are stocked in electronic memory and averaged. Today, the methodology of choice is to use apparatus with Fourier transforms (see the following chapter) in which, in addition, very stark magnetic fields are used.

^{13}C N.M.R. presents some advantages with respect to ^1H N.M.R.:

• The chemical shifts in ^{13}C N.M.R. are more "sensitive" to the structures than those obtained in ^1H N.M.R. They are concerned indeed by the framework of the organic molecules, that is to say by the carbon chain rather than by protons that can be qualified as being "peripheric atoms".

• On the other hand, signals of most organic compounds are located in the interval of 0 to 14 ppm in ^1H N.M.R., and they are found in the domain of 150 to 250 ppm in ^{13}C N.M.R. (with respect to the T.M.S.). The consequence is that there is less overlapping of peaks in ^{13}C N.M.R.

• In ^{13}C N.M.R., one cannot observe rays resulting from the homonuclear coupling of the type ^{13}C—^{12}C because ^{12}C is inactive magnetically.

• Finally and overall, there are several methods of the decoupling of atomes ^{13}C and ^1H, which is the most frequent in N.M.R. ^{13}C (and that we do not see in ^1H N.M.R.). The evoked decoupling is all the more interesting as the ^{13}C atoms generally adopt the form of a singlet after decoupling.

All these points contribute to facilitating the study of spectra.

There exist several processes which permit to decouple of the atoms ^{1}H and ^{13}C in the bond ^{13}C—^{1}H. They are the decoupling by broad bands, rupture of resonance and by applying frequent impulsions.

• The decoupling by broad-band consists in irradiating the sample by a band of radio frequencies that is larger than the resonance domain of the protons. In these conditions, the NOE effect (Nuclear Overhauser Enhancement) appears. It is expressed by a significant increase in the area of the ^{13}C signals. This is a general effect once there is a decoupling. It comes from a direct magnetic interaction between a decoupled proton and a neighboring ^{13}C carbon. There is an increase in the population in the lowest energy state of ^{13}C with respect to the distribution of Boltzmann.

The process by breaking the resonance named "off-resonance decoupling" is based on the irradiation of the sample by radiation of frequency less broad than the previous one, brought to play for values located between 1,000 and 2,000 Hz stronger than the spectral region of protons. In these conditions, one obtains spectra that are only partially decoupled.

Therefore, only structural information which can be brought only by some types of couplings cannot disappear and can bring still information. Before applying this process, the carbon atoms bringing three protons (primary carbons) appear in the form of quadruplets, the secondary ones as triplets, the tertiary as doublets and the quaternary as singlets. After application of the process, they can appear as singlets. Hence, the methyl iodide, which normally appears under the form of a quadruplet, can in these conditions give one singlet only.

The third type of decoupling is based on the intervention of a system of impulsions of frequencies judiciously chosen. They permit the improvement of the ratio signal/noise with respect to the preceding processes.

We give the chemical shifts of some groups in ^{13}C N.M.R./TMS in Table 2 and some coupling constants ^{1}H—^{13}C in Table 3.

Table 2: Chemical Shifts δ (ppm) of some Groups in ^{13}C N.M.R./TMS.

Alkanes	10 to 55	Carbonyl (aldehyde)	205–230
Alkenes	120 to 150 (C sp^2)	Carbonyl (ketone)	210–230
Alkyne	60 to 100 (C sp^1)	Carbonyl (esters)	180–200
Aromatic	90 to 150	Carbonyl (amides)	180–200

Table 3: Some Coupling Constants ^{1}H—^{13}C.

H-C sp^3	120–130 Hz
H-C sp^3 with electro-attractor substituents	125–210 Hz
H-C sp^2	150–170 Hz
H-C sp^1	250 Hz

As in 1H, N.M.R. some values of coupling constants can be approached, as a rule, by calculations.

^{19}F N.M.R.

^{19}F possesses the spin number I = ½. Its natural abundance is 100% and its gyromagnetic ratio is $\gamma = 25.167 \ 10^{-7}$ rad T^{-1} s^{-1} (see Table 2 Chapter 24). Hence, ^{19}F resonates at frequencies near those of ^{1}H. At 2.11 T, ^{19}F resonates at 84.67 MHz, whereas ^{1}H resonates at 90 MHz. The chemical shifts are located in a region close to 300 ppm. The reference is often the fluoro-trichloro-methane. A majority of fluorinated species resonate in fields that are stronger than that of this reference. The coupling constants $J_{H—F}$ are of the order of 50 Hz.

Quantitative Analyses by N.M.R.

N.M.R. permits quantitative analyses. These analyses are founded on the linear relation which exists between the area of the peaks and the number of nuclei responsible for the peak (see Chapter 24). If the peak is a multiplet, the whole rays must be taken into account in the proportionality relation. The proportionality constant depends on numerous experimental parameters, some of which are instrumental ones, but does not depend on parameters inherent to the sample, except its concentration. The consequence of these facts is that N.M.R. is an absolute method of analysis because one is not obliged to handle a sample of the pure analyte in order to carry out calibrations. Every substance of known purity and giving an exploitable N.M.R. signal can be used to carry out these calibrations.

Therefore, if we can identify a signal that only emanates from the derivative to determine, which only emanates from only one kind of protons in a spectrum, one can directly determine the concentration of the compound, provided we know the linear relation between the area of the peak and the number of protons. Here, we mention two kinds of quantitative determinations. They are:

- The determination of the quantity of substance (or its concentration) present in the N.M.R. tube. It can, naturally, be that of an external solution;
- The determination of the composition in molar fractions of a mixture permitting also to determine the concentrations.

Determination of a Quantity of Matter (or of a Concentration)

The analysis is carried out by putting the solution containing the analyte in a suitable solvent into the N.M.R. tube and also by putting an exactly known quantity of the standard compound. The latter permits to set up the linear relation. Let A_a and A_s the area of the signals chosen after the spectra have been registered. Let n_a and n_s be the numbers of the equivalent nuclei responsible for the peaks of the sample and the standard, and M_a and M_s be their molar masses. The number of mg of the substance contained in the sample is given by the expression:

$$mg_a = (A_a/n_a)(n_s/A_s)(M_a/M_s)mg_s$$

(*Remark*: The above relation is found in the following manner. The number of moles of each substance is:

$$m_a/M_a \quad \text{and} \quad m_s/M_s$$

The numbers of protons per signal are:

$$m_a n_a/M_a \quad \text{and} \quad m_s n_s/M_s$$

The proportionality relation is:

$$A_a = k\, m_a n_a/M_a \quad \text{and} \quad A_s = k\, m_s n_s/M_s$$

By eliminating the proportionality constant between these two relations, one obtains what we seek.)

Measurements require concentrations of the analyte and the standard of the order of 0.5 mol L^{-1}. N.M.R. is known as being poorly "sensitive".

Determination of the Molar Fractions of the Constituents of a Mixture

Let us consider the following mixture by way of example. The mixture contains tetralin, naphthalene and n-hexane (respectively x, y and z moles). Their structures are represented in Figures 18, 19 and 20.

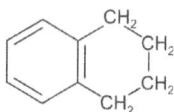
Figure 18: Tetralin.

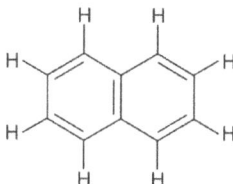
Figure 19: Naphthalene.

$$CH_3(CH_2)_4CH_3$$
Figure 20: n-Hexane.

The N.M.R. spectrum of the mixture is given in Figure 21.

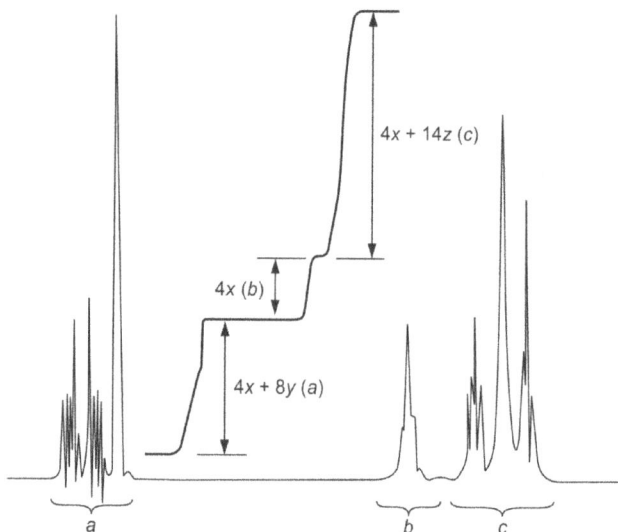

Figure 21: ^{1}H N.M.R. Spectrum of a Mixture of Tetralin, Naphthalene and n-Hexane. (Reprinted with Permission from Gwenola and Jean-Louis Burgot, Paris, Lavoisier, Tec et Doc, 2017, 455).

According to the very numerous known data of the chemical shifts, the peak marked b (on the figure) can only belong to the benzylic protons α and α' of tetralin, that is to say, four protons. According to the basic theory of proportionality, the heights of the levels, (measured for example in mm on the spectrum) correspond to the peaks:

- a at $4x + 8y$ protons. The peak marked a points to the four aromatic protons of the tetralin and the eight ones of naphthalene;
- b at $4x$ benzylic protons of tetralin;
- c at the $14z + 4x$ protons of the n-hexane plus the four remaining methylenic protons β and β' of the tetralin.

One can deduce the following molar fractions:

$$x = 0.39 \quad y = 0.25 \quad z = 0.36$$

Knowing the total mass of the mixture in the N.M.R. tube (for example, $60 \ 10^{-3}$ g) and the molar masses (tetralin 132 g L^{-1}, naphthalene 128 g L^{-1} and n-hexane 86 g L^{-1}), one can deduce successively that:

- The total number in the tube is ($5.2328 \ 10^{-4}$ moles) whose $n_x = 2.056 \ 10^{-4}$, $n_y = 1.3188 \ 10^{-4}$ and $n_z = 1.8570 \ 10^{-4}$ moles;
- The concentration in mol L^{-1} of the three substances in the tube. Taking for granted that the volume of the solution in the tube is 0.5 ml, we obtain $C_x = 0.41$, $C_y = 0.26$ and $C_z = 0.37$ mol L^{-1}.

It should be understood that, for these determinations, the solvent must not present signals located near those of the analyte and the standard. The solution must be not too viscous in such a way that the peaks remain sufficiently sharp. The solute and the standard must be sufficiently inert not to provoke a chemical reaction with every substance present in the solution.

Applications of N.M.R.

The applications of N.M.R. are innumerable. Let us mention some examples.

Identification of the Compounds and Eventual Impurities

It is not surprising that with N.M.R. one has a very good means of identification of a compound when one takes into account its power to distinguish the carbon chains and the presence of functional groups in different molecular surroundings. For example, the literature entails very numerous descriptions of ^{1}H and ^{13}C of medicines. They are capable, for the same reasons, to distinguish the impurities accompanying another substance, despite its relative lack of sensitiveness. It seems that, approximately, the impurity must have a concentration at least equal to 0.5% of that of the analyte.

Qualitative and Structural Analysis

Numerous examples of the application of N.M.R. in organic chemistry have already been given here. It is our opinion that organic chemistry would not be what it is today without the advent of N.M.R. methods, in particular the advents of ^{1}H and ^{13}C N.M.R.

Beyond the structural analysis in the usual meaning (this is nothing from the standpoint of applications). Let us mention the very interesting fact that diastereotopic nuclei are not magnetically equivalent and isochrone, even in achiral media. In principle, they give chemical shifts and coupling constants that are different from one to the other, even if these differences can be very weak. Here, there is in principle, a means to verify the optical purity of a compound. Now, this is an according to regulations exigence in the pharmaceutical field because of the unexpected arrival in the past of very severe accidents in the field coming from the prescription of racemic mixtures and racemates.

N.M.R. is also valuable in the domain of structural analysis in inorganic chemistry. For example, the ^{13}C N.M.R. permits demonstrating equivalence or not of carbon-ligands of some complex.

Conformational Analysis

(*Remark*: Let us recall that a conformation is a spatial arrangement of the atoms of one molecule in a mobile equilibrium with other conformations. A particular conformation, although it exists, is not isolable in the usual conditions contrary to a configuration, that is isolable. Examples of conformations are the forms "chair and boat" of the cyclohexane and examples of configurations are the forms Z and E of one olefine.)

For kinetic reasons of mutual interconversions between conformations as a function of temperature (see Chapter 24), N.M.R. permits keeping evidence of some of them. At ambient temperature, the conformations of different compounds that can be named "rotation isomers" cannot, in general, be detected by N.M.R. Only the averaged characteristics of the considered protons are accessible because of the quick rotations around the simple bonds. When the temperature decreases, the rotation rate also decreases and it is possible to observe the absorptions of rotation isomers, at least in some conditions.

Therefore, one can evidence the conformations "chair and boat" of the cyclohexane. According to the coupling constants, one can distinguish the coupling H–H/axial-axial/axial-equatorial/equatorial – equatorial. The values of the coupling constants are given in the Figure 22.

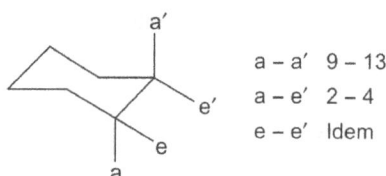

$$
\begin{array}{ll}
a - a' & 9 - 13 \\
a - e' & 2 - 4 \\
e - e' & \text{Idem}
\end{array}
$$

Figure 22: Coupling Constants in the Cyclohexane (Hz). (Reprinted with Permission from Gwenola and Jean-Louis Burgot, Paris, Lavoisier, Tec et Doc, 2017, 458).

Quantitative Analysis

This point has been developed in the previous paragraph. In the domain of the quantitative determination of medicines, the examples are very numerous. Here, we confine ourselves only to describing the determinations of the components of mixtures. Let us mention the simultaneous determination in the pharmacy field:

- Of the trimethoprim and the sulfamethoxazole used in tablets. The internal standard used is the 1,4 – dinitrobenzene;
- Of the quinidine and the hydroquinidine with the internal standard the 2,3,5-tribromothiophene;
- Of the theophylline and the ethylenediamine in the aminophylline with as internal standard, the t- butanol and so forth.

One example particularly interesting from the analytic standpoint is that provided by the determination of barbiturates possessing a chain of 4 or 5 carbon atoms in Position 5.

They possess several isomers due to some unexpected and abnormal substitutions. No chromatographic system is capable to solve this problem. In contrast, the ^{13}C N.M.R. is capable to do that.

N.M.R. is also capable to analyze excipients. For example, it permits to determine the content in rests hydroxypropyl of some modified starches.

Analysis of Functional Groups

The chemical shifts of nuclei in similar molecular surroundings are close to each other. ^{1}H N.M.R., for example, is capable at first sight making it possible to distinguish some from others among aromatic, vinylic, methyl, methylene and methine in a sample. This possibility is rare with other methods. One example is provided by the determination of the ratio of methyl-methacrylate in the chirurgical cements coming from its polymerization. It is obtained after measurement of the area due to the vinylic protons, which no longer exist after polymerization and of the area coming from the group methoxy which exists in all the species, polymerized or not. Let us also mention the method of determination of molecules possessing "mobile hydrogens" of the kind RXH. They react with the hexafluoro-dimethyl-ketone ^{19}F (hexafluoroacetone ^{19}F) quantitatively according to the reaction:

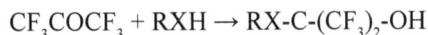

$$CF_3COCF_3 + RXH \rightarrow RX\text{-}C\text{-}(CF_3)_2\text{-}OH$$

The formed derivative is quantified by ^{19}F N.M.R. Another possibility in^{1}H N.M.R., close to the previous one, consists in obtaining the number of "mobile hydrogens" of one molecule by simply putting it in contact with heavy water. There is disappearing or weakening of some peaks and the disappearance of some couplings by the replacement of the hydrogen atoms with those of deuterium.

N.M.R. in Physical Chemistry

According to the foregoing, it is evident that N.M.R. is very useful in physical chemistry. Let us recall only two possibilities:

- The determination of the electronegativity of a group of atoms is considered as being a substituent by the measurement of the chemical shifts (effect of deshielding);
- The measurement of the kinetics of transformation or exchange of atoms and through these experiments and that of the activation thermodynamics parameters.

The principal disadvantage of N.M.R. is the high prizes of the apparatus and also its detection limit which is rather high (method poorly "sensitive").

CHAPTER 26

Fourier Transforms and Fourier Transforms N.M.R.

We know that there are now two general types of N.M.R. spectrometers being currently in use. They are the *continuum-wave* and the *pulsed or Fourier[41] transform* spectrometers. It seems that FT transform has now considerably supplanted the C.V, apparatus except perhaps for some routine analysis. Moreover, for reasons of convenience, apparatus based on the principle of Fourier transforms are used now in other types of spectroscopies such as infra-red and electron spin spectroscopies.

It is interesting, before considering this method, to give a brief mathematical overview of Fourier transforms. Actually, considering Fourier's series and Fourier transforms constitute two different subjects, although they are soundly related to each other from the mathematical viewpoint. For our purpose, we are overall interested in Fourier's transforms. Then, we shall study the N.M.R. with Fourier transforms.

Fourier's Transform

A Fourier's transform is from the strict mathematical standpoint a particular integral. The Fourier's transform F(t) of the function f(ω) is defined by the expression:

$$F(t) = 1/2\pi \int_{-\infty}^{\infty} f(\omega)e^{i\omega t}d\omega \tag{1}$$

Provided that the function $f(\omega)$ is defined $-\infty < t < +\infty$ and that the integral exists.

(*Remark*: The writing of the integral can differ according to the authors. For example, some authors use the factor $1/\sqrt{2\pi}$ instead of $1/2\pi$.)

Recall that t and ω are two variables. t is the time and ω is the circular rate of the periodic movement. i is the symbol of the imaginaries. In what is following, F( ) and f () are functions of t and ω (see just below).

A point that is quite remarkable is that the relation between F(t) and f(w) can be qualified as being *reversible*. One demonstrates, indeed, by taking into account the mathematical properties of Fourier's series (not given here) that the following equation is also verified:

$$f(\omega) = 1/2\pi \int_{-\infty}^{\infty} F(t)e^{-i\omega t} dt \tag{2}$$

We can imagine easily that the function Fourier Transform F(t) is finally a limit case of the development of Fourier's series which would be the integrand in relation (2) or inversely. A reason

---

[41] Joseph Fourier, French mathematician (1768–1830).

(among others) is the fact that an integral is a sum infinitely large of infinitely small terms. The two following points are important:

- Let us consider a waveform that repeats itself continuously as does the sine wave. Such a waveform is called periodic. It is the case for example of the square wave function and that of the sine wave one. It turns out that the only sine waves needed to "generate" any such repeated wave form are those which are harmonic or multiple of some fundamental frequency f_0. However, not all periodic functions are Fourier's functions. They cannot be described by a function as simple as a sine wave;

- Concerning the case of truncated waveforms or even of a pulse, they can be synthesized by the superposition of sine waves in the condition that the frequencies of their components will be a continuous spread of frequencies.

The mathematical concept of Fourier's transform is founded on the two relations (1) and (2). We shall see that the reversibility permits a backward and forward motion between a kind of variable or the other. However, there is a difficulty. The calculation of one of these integrals from the other is seldom easy. It is only the case when the two functions f(t) and F(ω) are simple. If not, their calculation involves using computers. However, the difficulty is quasi-disappeared since the advent of very powerful computers and algorithms. Calculations are now quasi-instantaneously carried out.

N.M.R. and Fourier Transforms

Recent developments of N.M.R. are due on one hand to the use of Fourier transforms and, on the other one, to the carrying out of very powerful magnetic fields. They constitute salient technological progress that has taken its origin only at the beginning of the years 1970. FT N.M.R. is N.M.R. with impulsions; both names designate the same technology and must be now understood as synonymous.

As we shall see, the FT N.M.R. is N.M.R. in which the sample remains submitted to a principal magnetic field, as in conventional N.M.R. But, among the differences between them, it is also submitted to pulses of frequencies judiciously chosen instead of continuously changing radio frequencies. From a general viewpoint, techniques involving the use of impulsions, although of older conception than those involving Fourier's transforms (1950), have been developed in the domain of the N.M.R. since the use of the latter. The use of pulses in N.M.R. has been widely facilitated by Fourier transforms. Therefore, as we have said earlier, the FT N.M.R. is synonymous with pulse N.M.R.

The Successive Steps of the Carrying Out of FT N.M.R. Experiment

In FT N.M.R. experiment:

- The nuclei are for the first time immersed in a static magnetic field. Afterward, they are periodically submitted to the action of a series of very brief impulsions in the domain of radio frequencies;

- After absorption of the latter ones, the signals emitted by the nuclei, which have just been excited by the impulsions and are relaxing, are registered as a function of time and are stocked;

- Finally, they are transformed in order to be described in the domains of the frequencies.

Principal Steps of the Registering of FT N.M.R. and the Future of Nuclei

Applying the Static Magnetic Field

This step is not different from the corresponding one in conventional N.M.R. in which was applied the static field $\mathbf{B}_0$. The interactions of nuclei/magnetic field establish the precession around the latter of the global magnetic moment **M** of the set of nuclei, named "magnetization". (Handling the global magnetic moment permits considering the system from the standpoint of classical mechanics and not from quantum mechanics).

Applying the Impulsions

In this methodology, the nuclei are periodically excited by several salves of grouped radio frequencies (impulsions) located in the domain in which they resonate, according to the sequence shown in Figure 1.

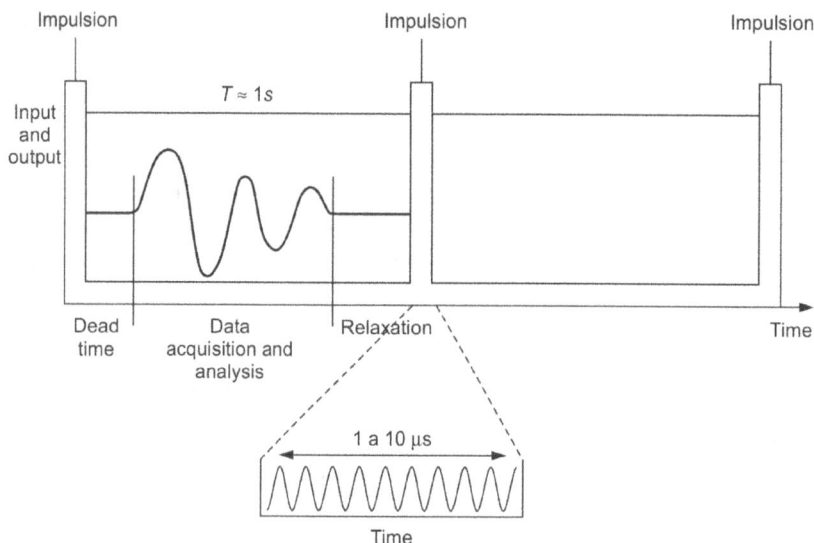

Figure 1: Sequence of the Impulsions and Scale of Times in FT N.M.R. (Reprinted with Permission from Gwenola and Jean-Louis Burgot, Paris, Lavoisier, Tec et Doc, 2017, 464).

The impulsions result from the application of an oscillating magnetic field B_1 perpendicularly to the static one B_o (see relation (9) in Chapter 24). This is obtained by passing a radio frequency current in a coil located, such as along the *x*-axis. Let us place ourselves in the hypothesis for which only one impulsion is applied (Figure 1).

When B_1 oscillates exactly with the Larmor's frequency of the nuclei, there is resonance. There is a transition of a given number of nuclei toward the superior energetic level as in conventional N.M.R.

It is important to notice, for this purpose, that all the nuclei do not resonate exactly at the same Larmor's frequency. (This can be due, for example, to a slight lack of homogeneity of the static field B_o). Here is located the interest to strike the sample with impulsions constituted by numerous radio frequencies. There are more nuclei excited than in CV N.M.R. As a result, a maximum of nuclei will enter simultaneously into resonance. To make sure it will be true, it is necessary that the band of frequencies conveyed by the impulsion covers the domain in which a majority of the nuclei resonate.

It is essential to notice that the imposition of one impulsion finally induces the knowledge of the same information as those obtained with CV N.M.R.

We have just seen what is happening after one impulsion. Actually, in this methodology, we accumulate the data coming from the application of several successive impulsions. The time interval T between two impulsions is of the order of one or several seconds (Figure 1).

Relaxation and Signal FID

After a brief moment of application of the impulsion, the relaxation becomes effective (Figure 1). In the interval of time T (of the order of some seconds) separating the application of two impulsions, signals are emitted by the nuclei (previously excited) when they are relaxing. These signals are named "free induction decay signal – FID". The relaxation phenomena are the same as in CV N.M.R. (see Chapter 24). They are "spin-lattice (relaxation time T1)" and "spin-spin relaxations (relaxation time T2)".

The signal FID emitted during the relaxation is detected by a receptive coil located perpendicularly to the static magnetic field on the axis x or y (the same coil may, besides, be also used as the emitting coil mentioned previously). The origin of the phenomenon of FID is complex and its study is not mentioned here. Let us only make clear that when the FID phenomenon occurs the oscillating potential at the terminals of the receptive coil decreases since the magnetization vector is taking again its initial position before resonance or before applying the following impulsions (see the section on 'Principal Steps of the Registering of a DT N.M.R. and the Future of Nuclei'). After the absorption of the radio frequencies brought by the impulsions, the emitted signals by the nuclei, which have been just before excited and relaxed, are registered as a function of time passed from the beginning of the relaxation and not as a function of the frequency.

About this, we can say that for the first time, there is an answer which occurs in the domain of the time in contrast to classic N.M.R. But, except for the very simple cases, it is necessary next to transform the obtained signal according to Fourier's transformation in order to be utilizable.

Obtaining the Data

Digitalization

The signal is digitalized. An analogical signal is transformed into a digital signal by an automatic selection at regular intervals of time of the information brought continuously by the analogical one.

During the process of obtaining the data before digitalization, the frequency of the spectrophotometer v_S is automatically subtracted from that of Larmor v_L of the nucleus which resonates. The Larmor frequencies are, indeed, too high to be handled with apparatus which are easily accessible. Hence, they are systematically reduced from the value of the oscillator frequency by electronic subtraction. This *modus operandi* induces no problem in chemistry, which is given the domains of frequencies corresponding to the chemical shifts which are by far weaker than those evoked before.

(*Remark*: Let us recall that digitalization is one of the manners permitting coding of some electrical data, such as potential differences, currents, charges and powers with devices, which can catch only two possible states.)

In the digital domain, the data are stocked accordingly in a binary manner. The information is coded with the help of binary numbers which represent the numerical and alphanumerical data of the process.

Stock and Treatment of Data

After treatment in the digital domain, the FID signal is stocked in the electronic memory of a computer in order to be treated after. Most often, signals in the time domain that result from the application of several successive impulsions are added in order to increase the ratio of signal/noise. Hence, in this methodology, one accumulates the data coming from several FID. The sequence is applying the impulsion, relaxation and data acquiring, and this is repeated several times and later the results are averaged.

Going Into the Domain of Frequencies

It is at this step when the data are summed, a step which is still in the time domain, the summed data are changed into data in the frequencies domain by Fourier's transformation. This substantially simplifies the spectrum. So far, indeed, there are several kinds of nuclei to enter into resonance and the FID (obtained in the domain time) become quickly inextricable because of the phenomena of beatings. (The beating phenomenon results from the superimposition of two vibratory phenomena of the same period and direction. If the periods, instead of being equal are close, there appears the phenomenon of beatings.) This is due to the fact that the impulsion applies a large domain of radio frequencies simultaneously to all the active nuclei, although all of these do not absorb exactly at the same frequency. In contrast, after having passed into the domain of frequencies, the obtained spectrum becomes workable. One example is provided by the spectrum FT ^{13}C N.M.R. of cyclohexene (Figure 2).

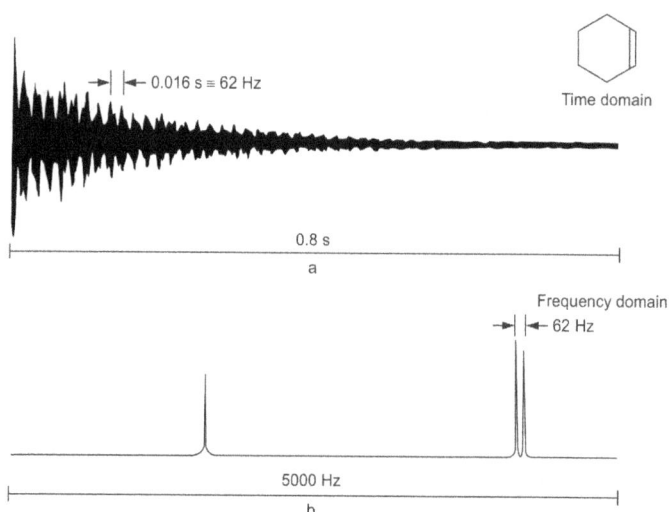

Figure 2: FT ^{13}C of Cyclohexene. (Reprinted with Permission from Gwenola and Jean-Louis Burgot, Paris, Lavoisier, Tec et Doc, 2017, 468).

Fourier's transformation is achieved as follows. One starts by noting that the signal is in the time domain. The total FID curve is the integral over all the contribution frequencies:

$$f(t) = \int F(v)e^{-2i\pi vt}dv \qquad sum\ from - \infty\ to + \infty$$

The result F(v) is:

$$F(v) = 2Re\int f(t)\ e^{-2i\pi vt}dt \quad sum\ from\ 0\ to + \infty$$

Here, Re is the real part of the expression. F(v) is the spectrum in the frequency domain. It is the first integration that is carried out here. It is carried out numerically with the help of a computer, according to Cooley and Tukey's algorithm.[42] It is an undeniable fact that the achievement of this algorithm has strongly contributed to the development of the FT N.M.R.

The profile of the signals of the FID is particularly complex. In contrast, after Fourier's transformation into frequencies, the spectrum only exhibits three peaks corresponding to three kinds of carbon atoms.

---

[42] An algorithm for the machine calculation of complex Fourier series. Math. Comput., 1965, 19(90): 297–301.

Concerning the complexity "in the time domain" of the spectra and their simplification by Fourier's transformation, let us mention that:

- When there is only one kind of protons (protons magnetically equivalent), there exists no possibility to be faced with beatings phenomena since their resonance frequency is strictly identical. The FID is a simple exponential and its transform presents a sole peak only;
- When the frequency of the spectrometer is slightly higher than Larmor's frequency of the proton, there is a battement phenomenon. Then, the FID consists of sinusoidal oscillations, the envelope of which is a decreasing exponential. After transformation into frequencies, the spectrum only exhibits a peak but it is slightly shifted with respect to the preceding one.

Last Filtering

Finally, digital filtering is again carried out in order to again increase the ratio signal/noise.

A Brief Comparison Between CV N.M.R. and FT. N.M.R.

In a conventional N.M.R. experiment, the sample is immersed in a magnetic field and is subjected to radio frequency radiation, the frequency of which is continuously varying. The spectrum is a diagram intensity of the absorbed radiation $I(\omega)$ versus the frequency ω. Each used frequency is experienced by the sample. Hence, there is an accumulation of noise in addition to the signals of interest.

In Fourier's transform N.M.R., a strong impulsion of radio frequency radiation (of about 10 μs duration) is applied to the sample and repeated several times. The frequencies of the radiations constituting the pulse are not necessarily equal to any of the resonance frequencies in the spectrum. The spectrum is a plot of the decay of the N.M.R. signal following the radiofrequency impulsion versus time. The obtained curve $G(t)$ is known as "the spectral response pattern". The function $G(t)$ is also known as the "Fourier transform of the frequency intensity $I(\omega)$". The results of several pulses are accumulated in a digital computer. The Fourier's transform is then calculated. At the end of this process, the obtained spectrum is the diagram $I(\omega)/\omega$. It is frequency-domain spectroscopy.

[*Remark*: Fourier's transforms are also used in other types of spectroscopies. Let us, for example, mention the "time-domain spectroscopies" in which the obtained spectra are diagrams of changes of radiant power (emitted or absorbed)/time. [Such diagrams offered technological difficulties of obtention before the advent of Fourier's transforms. Due to them, they have been surmounted (see 'Infrared Spectroscopy')].

Advantages and Applications of FT. N.M.R.

Transformations of Experimental Data

- Using Fourier's transforms permits converting experimental data, which are a function of an independent variable, into information that is dependent on another kind of independent variable. For example, the variable time can be replaced by the variable frequency and inversely.
- Using Fourier's transforms is well indicated to carry out this process. One says that the use of Fourier's transforms converts the data presented in the domain of time into the domain of frequencies and inversely. The grandeur's time and frequencies are said to be conjugated. They are related to each other mathematically. The transformation is, in the great majority of

cases, difficult from a mathematical standpoint. It is realized with the help of very efficient algorithms. They are the object of numerous routine ones.

- It is an experimental fact that it is easier, once the transformation of the experimental data according to Fourier is realized, to extract the sought information in the new working domain rather than in the ancient. Here is the advantage of the transformation. This point is all the more interesting as the presented data in one domain bring the same information as those which are brought in the other.

The phenomenon of beatings in physics which appears when there is a superposition of periodic phenomena well demonstrates the interest to change the independent variables in order to simplify the interpretation of the data.

We know that two sources A and B animated by a vibratory sinusoidal as a function of time, the same amplitude and periods T and T' and neighboring frequencies f and f' give phenomena of beatings. They appear under the form of a movement the amplitude of which periodically takes a null and a maximal value. The Amplitude Y of the resultant movement as a function of time is shown in Figure 3. The maximum amplitude reproduces periodically as seen in Figure 3.

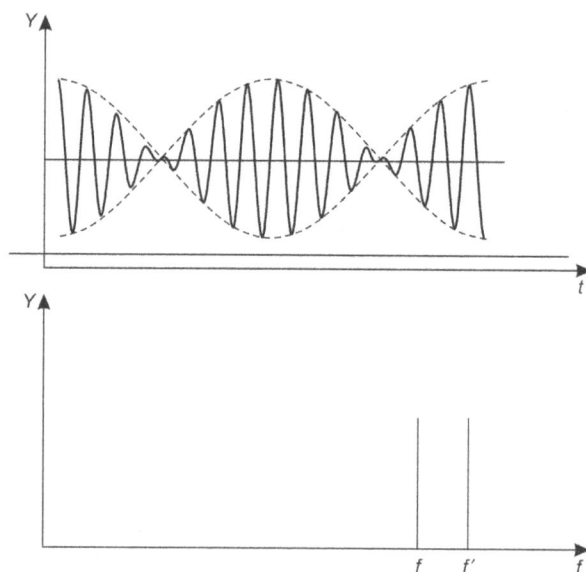

Figure 3: Phenomena of Beatings in the Domain of Time. (Reprinted with Permission from Gwenola and Jean-Louis Burgot, Paris, Lavoisier, Tec et Doc, 2017, 405).

The frequency of the beatings is equal to the difference of those of the vibratory movements. We also represent, by way of example, the diagram of Y as a function of frequencies f and f'. It is evident that the diagram of frequencies can only involve vertical right lines. The simplification of the interpretation of the diagram after the crossing from the time domain to the frequency one is striking.

- Among other methods, those which are spectroscopic and hence involve the measurement of a radiative power or intensity $P(v)$ as a function of the frequency of the radiation or other possibility, the radiative power $P(t)$ as a function of time, are subject to use Fourier's transforms.

Other Advantages (of Technological Order) of the Use of Fourier's Transforms

Handling Fourier's transforms is one of the processes emanating from the "software" permitting to increase in the ratio signal/noise. Let us recall that the hardware of a computer is constituted by the physical elements constituting it. Its software is constituted by the programs and instructions commanding it, plus the devices permitting to stock them. Varied electronic components, such as high and low frequencies filters, choppers and some types of detectors are pieces of the "hardware".

An example of the enhancement of the signal/noise ratio by using Fourier's transforms is schematized in Figure 4. Initially, one is in presence of a spectrum f(t)/t contaminated by the fundamental noise (frame a). A first Fourier's transformation transforms it into the spectrum g(f)/f (frame b). One knows, from the foregoing, that the latter spectrum may be easier to interpret than the former. In process c, finally one carries out a second Fourier's transformation. One comes back in the domain of the time (frame c). It is conceivable that the last transformation can in some cases "purify" further the spectrum Figure 4.

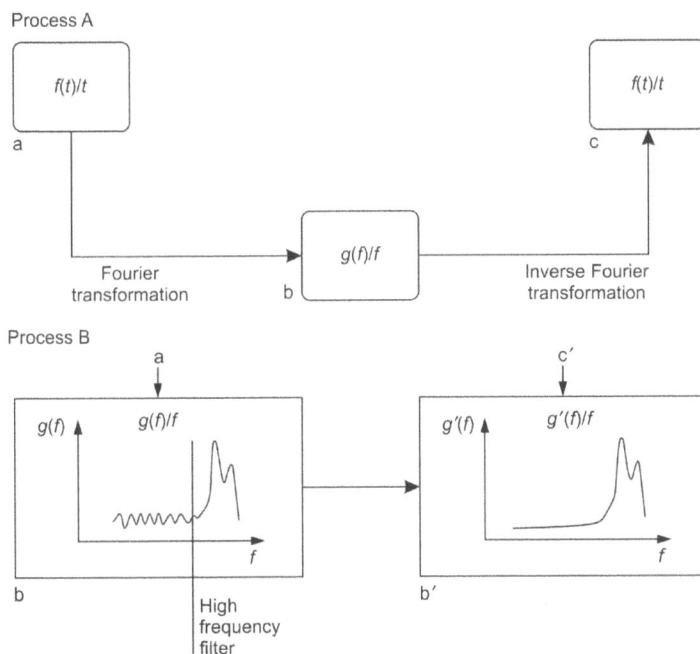

Figure 4: Using Fourier's Transforms in Order to Enhance the Ratio Signal/Fundamental Noise. (Reprinted with Permission from Gwenola and Jean-Louis Burgot, Paris, Lavoisier, Tec et Doc, 2017, 407).

The following alternative is still further effective (Figure 4; process b). It consists of submitting the spectrum g(f)/f (already in the frequency domain) to the action of a filter high or low frequency, according to the domain we want to keep. Once the filtering is realized, the obtained spectrum g'(f), still in the frequency domain, is purified with respect to g(t). Finally, the spectrum may submit a second Fourier's transformation to come back into the time spectrum. The latter and the preceding became not only poor in noise, but also in the time domain corresponding to the filtered frequency.

Another advantage of an FT apparatus lies in the fact that it contains few optical components, notably the slits which lower the power of the radiations. As a result, the power of the radiation, which strikes the detector, is by far stronger than that obtained with an apparatus functioning with slits. Hence, with Fourier's transforms, the ratio signal/noise is improved.

A supplementary advantage is the resolution power extremely high of FT apparatus. It permits the analysis of very complicated spectra, notably when there exists a superimposition of individual spectra. The advantage is due to the number of resolution elements which is higher than with the "classical" apparatus. The resolution elements correspond to the intervals of wavelengths or frequencies necessary for two successive measurements to be independent of each other. The quality of the spectrum is all the better as the number of resolution elements is higher.

Finally, a very important interest presented by these apparatus lies also in the fact that the signals coming from all the elements of resolution are measured simultaneously, contrary to the traditional apparatus. As a result, the time necessary to obtain a spectrum is considerably shortened. This advantage is called the "Multiplex Advantage" or "Advantage of Felgett" from the name of the author who has put it in evidence.

CHAPTER 27

Electron Spin Resonance (E.S.R.) and Electron Paramagnetic Resonance (E.P.R.)

The E.S.R. (or E.P.R.) is a method involving the magnetic character of electromagnetic radiations. It is a very sensitive method that permits the studying of molecules, products intermediates radicals, etc., that contain an odd number of electrons (that we shall name now "paramagnetic center" or "PC"). It consists of noticing the magnetic fields with which these species enter in resonance with monochromatic radiations. It is grounded on the fact that electrons possess a magnetic moment, that is to say, a spin angular momentum.

As a first comment which can be made, one can advance that the already studied basic N.M.R. theory applies to electron spin resonance, at a list in its great basic principles, despite some differences. The method provides spectra very rich in information based notably on the magnetic interactions between the constitutive elements of the studied structures.

A Brief Overview of ESR

In E.S.R., as in N.M.R., the sample (PC) is immersed in a strong magnetic field and is submitted to the action of electromagnetic radiation, the frequency of which is in principle changing regularly. The apparatus emits a signal when the resonance between the system and the radiation is attained. The half-integral *spin angular momentum value* of an electron gives rise to the two possible spin orientations, which are distinguished by the two quantum numbers $m_s = \frac{1}{2}$ and $m_s = -\frac{1}{2}$. One can say that the strong magnetic field removes the degeneracy of the two orientations of the electron magnetic moment.

There exists a great analogy between E.S.R. and N.M.R. In particular, in both methods, two actions operate; that of electromagnetic radiation and the magnetic field in which both materialize different energy levels of the PC. (However, for purely technical reasons, the frequency of the radiation is maintained constant, and it is the value of the magnetic field which is changed).

Some Paramagnetic Centers

They are free radicals, some transition metal ions and all other molecules in triplet states containing unpaired electrons. If electrons are paired, their magnetic moments cancel each other and they are not observable.

Generally, electrons of internal shells atomic or molecular of species are complete. They do not play a significant part. This is a consequence of the *Pauli exclusion principle*, which requires that whenever two electrons occupy the same orbital and their spins must be paired. In contrast, in atoms and molecules possessing an odd number of electrons, there necessarily exists a non-paired electron. The species is a free radical. As examples, let us cite the following radicals: $OH\bullet$, $CN\bullet$, $NO\bullet$, $NO_2\bullet$, and also other numerous organic and inorganic molecules.

Principle of E.S.R.

Recalls

We have seen in Chapter 24 that an electron, the movement of which is characterized by a kinetic moment $\mathbf{L_l}$, does possess a magnetic moment $\boldsymbol{\mu_l}$ given by the relation:

$$\boldsymbol{\mu_l} = -(e/2m_e)\mathbf{L_l} \tag{1}$$

In this relation, e is the charge of the electron and m_e its mass. l is a quantum number (which can only take even positive values). l is one of the three quantum numbers characterizing the orbital of the electron. Let us also recall the expression of the factor β:

$$\beta = e\,\bar{h}\,2m_e$$

$$\beta = \mu_B \text{ (in modules)}$$

β is a universal constant. It is named "Bohr's electronic magneton". Its value is $9.274\ 10^{-24}$ J T^{-1}. Let us also recall that according to the demonstration of the relation giving the expression of Bohr's electronic magneton, the moments $\boldsymbol{\mu_l}$ and $\mathbf{L_l}$ which have been aforementioned, are *orbital* moments and that they are quantified.

We also recall that the kinetic orbital moment of a mobile on a circular orbit is quantified. If one chooses the axis z'z for which m_l is the corresponding quantum number:

$$L_z = \bar{h}\,m_l$$

Spin Moment of the Electron

In supplement to its kinetic orbital moment, the electron does possess a spin L_s since it is animated by a proper rotation motion. The theory which provides the parameters describing the phenomenon is strictly analogous to that followed in the orbital case. In this way, we define the following relations and parameters:

$$L_s = \bar{h}\,s$$

Here, s is the number characteristic of the spin of the electron and L_s is the kinetic moment of spin of the electron. At the moment $\mathbf{L_s}$ correspond to the magnetic moment of spin $\boldsymbol{\mu_s}$ through the relation:

$$\boldsymbol{\mu_s} = -g_e(e/2m_e)\mathbf{L_s} \tag{2}$$

The value $s = \frac{1}{2}$ implies that in the relation $L_{zs} = \bar{h}\,m_s$, analogous to the "orbital relation" $L_{zl} = \bar{h}\,m_l$, the quantum number m_s can take only the values $\pm \frac{1}{2}$.

The total magnetic moment of an electron is:

$$\mu = -(e/2m_e)(\mathbf{L_z} + g_e\mathbf{L_s}) \tag{3}$$

The g_e is a constant for the free electron. Its value is 2.0023. It can vary by a few percent for other compounds such as free radicals, transition-metal ions and other bodies. In these cases, the symbol is g. The g is called the *splitting factor*. The subscript e of g_e means that g_e is the value for the electron.

The expression (3) can be simplified by putting $\beta = e\bar{h}/2m_e$ in common factor in expression in it:

$$\mu = -\beta(\mathbf{L} + g_e\mathbf{s}) \tag{4}$$

L and **s** are called reduced kinetic moments. They are dimensionless. The μ remains the total magnetic moment of one electron. β is Bohr's electronic magnet which has already been introduced.

Paramagnetic Centers in a Magnetic Field

In order to explain the phenomena, we begin with simple experiments. Later, we shall complete these explanations by saying a few words about the "hyperfine structure" that this method provides.

Let us consider a paramagnetic center that possesses given energy only due to its spin. Its magnetic moment can be written:

$$\mu = -g\beta\mathbf{S}$$

(Notice that the symbol g replaces g_e, since it is not the free electron that is under study). **S** is its total spin moment. It obeys the usual rules of quantum mechanics. Its projection S_z on any axis z'z can take any value m_S of the $2S + 1$ possible, that is to say:

$$m_S = -S, -S + 1, ...S \tag{5}$$

The interaction energy E_m of the center and the magnetic field is given by the dot product:

$$E_m = -\mu\mathbf{B}$$

$$E_m = g\beta\mathbf{S}\mathbf{B}$$

If the axis z'z is chosen in the direction of **B**, the interaction energy is given by the expression:

$$E_m = g\beta B m_S \quad \text{with} \quad m_S = -S, -S + 1, ...S$$

Here, B is the module of **B**. Hence, there exist $(2S + 1)$ levels of energy equidistant from each other and separated from the value $\Delta E_M = g\beta B$. This ensemble of $2S + 1$ levels of energy is located at the level of the paramagnetic center in absence of the field.

Figure 1 shows an example for which $S = 1/2$.

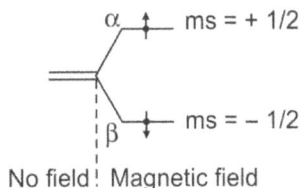

$$\alpha \quad \uparrow \quad ms = + 1/2$$
$$\beta \quad \downarrow \quad ms = - 1/2$$

No field   Magnetic field

Figure 1: Interaction of the Magnetic Moment with the Field **B** ($S = 1/2$).

This is the case of a free electron without a magnetic field (no magnetic field is applied). In this case, the interaction energy E_m is given by the relation. There is the appearance of 2 $(2S +1)$ levels of energy separated from each other by the energy $\Delta E = g_e\mu_B B$. The set of these levels (for this example, 2 levels) is centered on the level of energy of the paramagnetic center. From a general standpoint, these levels of energy intervene in the ESR transitions.

Figure 2 provides another example. It shows electron spin levels in a magnetic field in the case for which $S = 3/2$. In this case, the interaction energy E_{mS} is given by the relation:

$$E_{(mS)} = g\beta B m_S \quad \text{with} \quad m_S = -3/2, -1/2; +1/2, +3/2$$

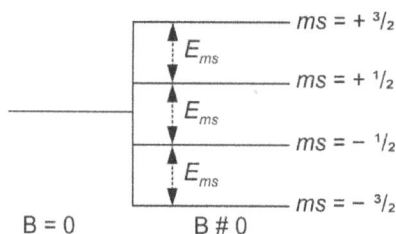

Figure 2: Different Spin Levels of PC in a Magnetic Field: S = 3/2.

We note that there are 4 levels of energy after the application of a magnetic field.

Transitions in E.S.R.

They occur when the paramagnetic center, immersed in field **B**, is submitted to electromagnetic radiation of a particular frequency v. Transitions occur for some particular values of B and v.

According to the transition rules (dependent on quantum mechanics), they can only occur when:

$$\Delta_{mS} = \pm 1$$

Given the fact that the levels of energy of an electron spin in a magnetic field **B** is:

$$E_m = g\mu_B B$$

And that $\Delta_{mS} = \pm 1$, the condition of resonance is therefore:

$$h\nu = g\mu_B B$$

Transitions occur when **B** takes the particular value **B**$_0$ such as:

$$h\nu = g\beta B_0$$

The Parameter g

The parameter *g* is a positive number. It characterizes the paramagnetic center. Its value can give some information about its electronic structure. It is the parameter that is measured. Its main applications are the facts that it can aid in the identification of the paramagnetic center. It also can bring some information on the character radical or not of an organic reaction. As in N.M.R., the spin magnetic moment interacts with the local magnetic fields. This fact corresponds to a variation of the parameter g which now is different from g_e. The resonance condition becomes:

$$h\nu = g\mu B_{local}$$
$$h\nu = g_e \mu B(1 - \sigma)B$$

Thus:

$$g = g_e(1 - \sigma)$$

The *g* is the g-factor of the radical. The difference between g and g_e depends on the ability of the applied field to induce local currents in the radical and as a result to bring some information about its electronic structure. The part played by the difference $g - g_e$ plays the part of σ in N.M.R. in some way.

For a free electron, the value of g is $g_e = 2.0023$. The g value of an unpaired electron in a radical is different. The cause is the fact that there exist local magnetic fields due to the framework of the radical. In this condition, it is the parameter g rather than g_e which is used. Values of g have often some values close to g_e. Thus, many organic radicals exhibit the value $g = 2.0027$, whereas inorganic radicals have values typically in the range of 1.9 to 2.1 g values of reach the value 6 for some d complex.

There is another factor that permits to study further of the radicals. It is the notion of the *density of spin*. It will be briefly tackled after having considered the hyperfine structure.

Hyperfine Structure

One knows that ESR measurements are based on the absorption of radiation of frequency v by PC contained in a magnetic field H. Spectra are recorded by measuring the absorption as a function of H while the magnetic field strength is scanned. But, in E.S.R., a hyperfine structure exists. It is due to the magnetic interactions between a paramagnetic center (for example one electron) and the magnetic dipole moments of the nuclei present in the radical (neighboring protons and other nuclei – N^{14}, P^{31}-) which have magnetic moments that interact with the unpaired electron. Hyperfine structure means a splitting up of individual resonance lines into some components.

Let us consider a single H nucleus present in the radical. Its spin is a source of the magnetic field. Depending on its nuclear spin, the field it gives rise to add or subtract from the applied field. The total local field is given by the relation

$$B_{loc} = B + am_1 \quad \text{with} \quad m_1 = + \text{ or} - 1/2$$

Here, a is the *hyperfine coupling constant*. Half the radicals in the sample resonate according to the relation.

$$B_{loc} = B + 1/2a \qquad m_1 = + 1/2$$

And
$$B_{loc} = B - 1/2a \qquad m_1 = -1/2$$

For the other half

$$B_{loc} = B - 1/2a \quad m_1 = -1/2$$

$$B_{loc} = B + 1/2a \quad m_1 = +1/2$$

Hence, instead of a single line, the spectrum shows two lines of half the original intensity separated by a and centered on the field determined by g (Figure 3). It results in four energy levels in place of the original two.

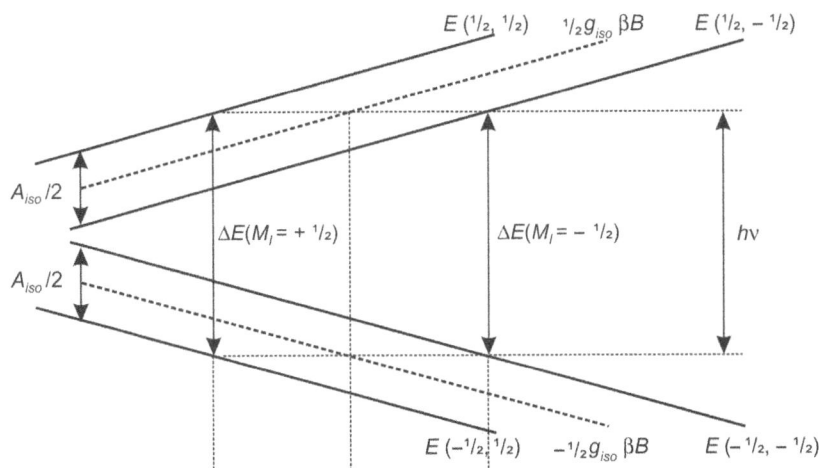

Figure 3: The Hyperfine Interaction Between 1 Electron and a Spin ½ Nucleus. (Reprinted with Permission from Patrick Bertrand (2010), EDP Sciences).

Finally, the spectrum consists of two lines of equal intensities instead of one. The intensity distribution is obtained by simple addition.

When there are several magnetic nuclei in the radical, each one contributes to the hyperfine structure. If the radical contains N equivalent protons, there are N + 1 hyperfine lines with a binomial intensity. The E.S.R. signal for the radical $CH_3\bullet$ consists of four peaks in a 1/3/3/1 ratio because an electron spin can couple to the spins of three equivalent protons.

Instrumentation

In a continuum wave E.S.R. experiment, magnetic fields of about 0.3 T correspond to resonance with an electromagnetic field of frequency of 10 GHz (giga: 10^9). Most spectra are carried out at a constant frequency of the electromagnetic sinusoidal wave.

In these conditions, the resonance occurs in the domain of microwaves (wavelength of about 3 cm). It is considered simpler to operate by maintaining a constant frequency and varying the field. The microwave source is a klystron or a semi-conductor device. There is an electromagnet with a fixed direction whose field may vary in the region of 0.3 T. The sample is inserted into a glass or quartz container. Cells have been devised which permit the formation of radicals into them, either with U.V., gamma, x-rays or by electrochemical reactions in solutions.

It is possible now to do impulse E.S.R. By a judicious choice of the parameters which command this methodology, some information supplementary to those obtained with a continuum wave apparatus can be obtained. They concern, for example, the getting relaxation times.

In E.S.R., the obtained peaks can be absorption peaks or more usually, peaks obtained in which is used in N.M.R. In this case, the obtained peak is the first derivative of the absorption one, as it is indicated in Figure 4.

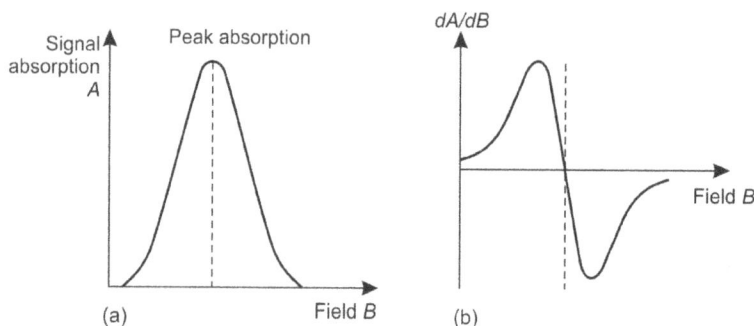

Figure 4: Absorption Signal (Scheme a) and First Derivative of the Absorption Peak (Scheme b). After a Phase Sensitive Detection.

It is interesting to notice that the obtained rays are not infinitely narrow as could be inferred from a given theory.

In E.S.R., the obtained peaks can be absorption peaks or more usually, peaks obtained in *phase-sensitive detection* which is used in N.M.R. In this case, the obtained peak is the first derivative of the absorption one, as it is indicated in Figure 4. It is interesting to notice that the obtained rays are not infinitely narrow as could be inferred from a given theory.

Let us say briefly that in quadrature phase-sensitive detection, it is possible to sense frequency differences notably between that of the signal and that of the carrier.

Spin Density

The hyperfine structure of an ESR spectrum is a kind of fingerprint permitting the identification of a radical present in a sample. Moreover, since the magnitude of the splitting depends on the distribution of the unpaired electron near the magnetic nuclei present, the spectrum can be used to map the molecular orbital occupied by the radical. For example, the hyperfine splitting in the benzene anion $C_6H_6^-$ is 0.375 mT and one proton is close to a C atom with one-sixth the unpaired electron *spin density* ρ (because the electrons are spread uniformly around the ring). Neither hydrogen atom is privileged. Each possesses the same part of spin density, that is to say, 1/6 of it.

The hyperfine structure of the ESR spectrum of naphthalene can be interpreted as arising from two groups of four equivalent protons, one at the α position in the ring and the other in the β position (P. Atkins and J. de Paula, 2014). They give for spin density $\rho = 0.22$ for the α-position and $\rho = 0.08$ for the β positions.

The spectrum of the benzene radical anion $C_6H_6^-$ is shown in Figure 5. In addition to the fact that it is a first derivative curve, we find that the curve exhibits a septuplet (6 + 1 lines). This means it is perfectly symmetric. This proves that the six protons are equivalent.

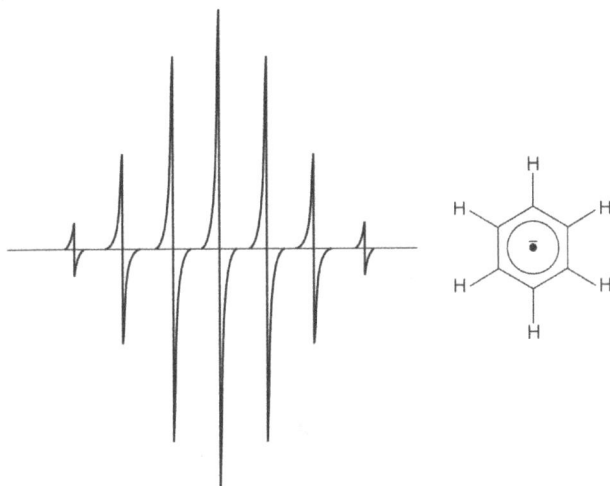

Figure 5: E.S.R. Spectrum of the Benzene Radical Anion.

Figure 6 shows the spectrum of a semiquinone intermediate. We note that the semiquinone free-radical anion exhibits a five-line pattern, which is the consequence of the magnetic spin interaction between the odd electron and the four protons on the ring. The five lines of the hyperfine structure are in the ratios: 1-4-6-4-1.

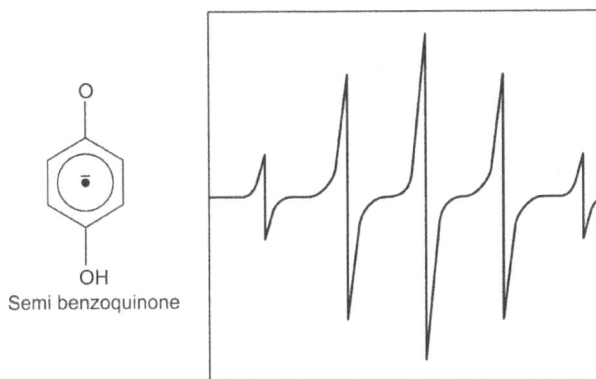

Semi benzoquinone

Figure 6: The E.S.R. Spectrum of the Quinone-Hydroquinol Redox System.

It proves that a semiquinone free-radical anion is an intermediate of the transformation. We also give in Figure 7 the spin density of naphthalene.

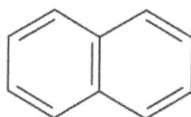

Figure 7: Structure of the Naphthalene.

Some rules concerning the hyperfine structure can be drawn. In the case of *isotropy* into the PC itself, interactions between the non-paired electrons of a molecule with a nucleus of spin I create one ensemble of 2I + 1 equidistant rays on the E.S.R. spectrum. They are centered on the position the ray would have in the absence of any interaction. When the hyperfine interaction is *anisotropic*, the position and the disposition of the rays have changed. In particular, they are dependent on the orientation of the field with respect to the molecule.

Anisotropy of the nuclear hyperfine interactions can be observed when a radical is immobilized in a solid for example. An example is provided by the E.S.R. spectrum of the di-tert-butyl nitroxide radical which changes with temperature (Figure 8).

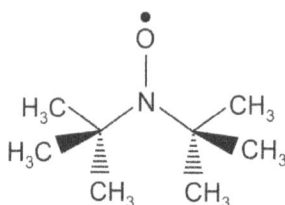

Figure 8: Structure of the Di-Tert-Butyl Nitroxide.

At 292 K, the molecule tumbles freely and isotropic hyperfine coupling gives rise to the three *sharp* awaited signals. At 77 K, the motion of the radical is restricted. The anisotropic and isotropic hyperfine couplings are visible on the spectrum. Thus, appear three *broad* peaks. The phenomenon is comparable with that of the broadening of the proton N.M.R. spectrum due to chemical exchange. The electron exchange between two radicals can broaden an E.S.R. spectrum, both radicals might be *spin probes*. This effect is used in some biochemical studies, notably with the following reactant; i.e., 2,2,6,6-tetramethylpiperidine-1-oxy-4-amino-4-carboylic acid (Figure 9).

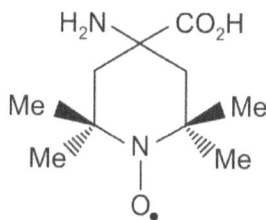

Figure 9: Structure of the 2,2,6,6-Tetramethylpiperidine-1-Oxy-4-Amino-4-Carboxylic Acid.

Applications: Spin Label

As in N.M.R. a standard substance of reference is convenient in E.S.R. A proposed one is the 1,1-diphenyl-2- picrylhydrazyl free radical. Its splitting factor is g = 2.0036. It cannot be used as an internal standard for a reason of distinction from the samples. However, it can be used consecutively (Figure 10).

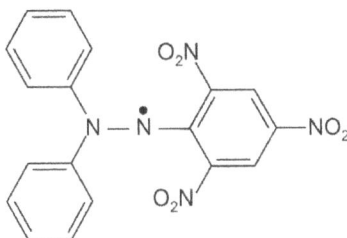

Figure 10: Structure of the 1,1-Diphenyl-2-Picrylhydrazyl Free Radical.

Free radicals can be readily studied by E.S.R. even at very low concentrations.

ESR spectroscopy allows the detection of radical ions at about 10^{-8} mol L^{-1}. It has found extensive application in electrochemistry, especially in studies of aromatic compounds in non-aqueous solutions. E.S.R. can also be used in the estimation of trace amounts of paramagnetic ions. It is the case of Mn^{2+} ions.

CHAPTER 28

Infrared Spectroscopy I
Fundamentals of Infrared Spectra

Infrared spectroscopy is one of the spectral methods most used in the organic analysis, in particular in functional analysis. In this chapter, essentially devoted to some aspects of the theory of infrared spectra, we shall study successively the spectra of rotation pure and those of vibration-rotation of diatomic molecules. The case of polyatomic molecules is also equally evoked.

As a short explanation, it is possible to say that the photons involved in the obtention of these spectra do not have sufficient energy to change the electronic states of the molecules.

Domains of the Infrared Spectroscopy

One can divide the domain into three regions; the near IR, the normal IR and the far IR. The part the most used is the normal-infrared or mid-IR.

Figure 1 gives the limits of the different domains.

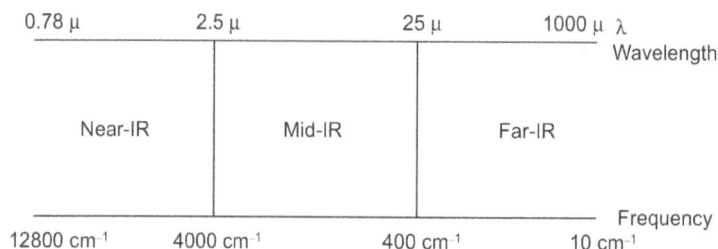

Figure 1: Domains of the Infrared IR Spectroscopy.

From now on, it is important to know that in the domain of infrared, one only finds spectra of rotation pure in the region of microwaves and spectra of vibration-rotation in the region of normal and near-infrared. The photons involved in the obtention of these spectra do not have sufficient energy to change the electronic states of the molecules. Figure 2 roughly indicates the levels of the electronic (E" and E'), vibrational (v", v') and rotational (J", J') energies of a diatomic molecule. One can see that the intervals of the different levels of electronic energy are higher than those between the vibrational energies. It is the same for the intervals existing between the vibrational and rotational levels of energy. Besides, it is important to notice that there exist several levels of vibrational energy per level of electronic energy and that it is the same case for vibrational and rotational energies. In general, the orders of magnitude of the differences between electronic levels are of the order of 420,000 J/mol, that between vibrational levels is of the order of 21,000 J/mol and

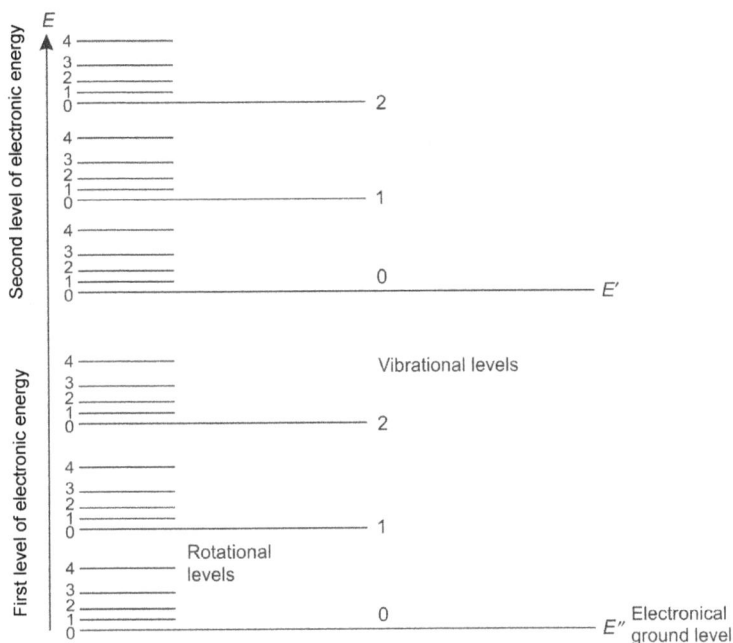

Figure 2: Levels of Electronic, Vibrational and Rotational Energies of a Diatomic Molecule.

that between rotational levels is 42 J/mol. Since the average thermal kinetic energy of a molecule at room temperature is about 4,200 J/mol, the use of Boltzmann law shows that molecules in general are in the lowest vibrational level of the ground electronic state but that they possess several quanta of rotational energy.

(These results have been obtained with the help of the study of thermal capacities of diatomic molecules.)

Origin of the Absorption Spectra IR

According to quantum mechanics, every molecule exists in some number of discontinuous states. All these molecules are in perpetual agitation. They are animated by three kinds of motions, i.e., translation, rotation and vibration. It is impossible to directly observe these movements but their levels of energy are quantified and the molecules can be indirectly studied (at least in part) through their spectra of absorption IR. This means that:

- Only the radiation IR which has a frequency exactly equal to that which is necessary to raise the energy level of a bond is absorbed;
- The amplitude of a particular vibration is suddenly enlarged.

Born-Oppenheimer Principle

Let us recall that it is possible, with a good approximation, to separate the vibration and rotation energies (see 'Introduction to the Spectral Methods') from the nuclear and electronic energies.

It is possible, with some approximation, to separate the electronic energies from the nuclear and vibration-rotation energies. This approximation is sound given the huge energy difference that exists between the masses of one electron and one nucleus. From another standpoint, the motion of the nuclei is very slow and the internuclear distances play the part of parameters.

Motions of Molecules

Before investigating the origin of the absorption spectra IR, it is necessary to specify what are the different motions that animate molecules. All the molecules are in perpetual agitation. The motions of some molecules together with their chemical bonds look like a system of springs and balls (Figure 3).

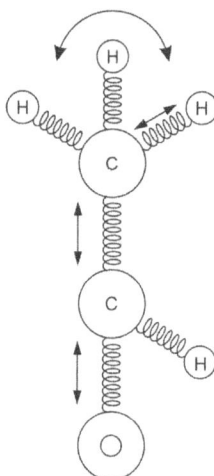

Figure 3: Representation of "Springs and Balls" of Acetaldehyde.

A non-linear molecule whatever its shape, constituted by *n* atoms, can be animated by three motions:

• A motion of translation that can occur along the orthogonal axes;
• A motion of rotation that can be sustained by the three orthogonal axes;
• The motions of vibration of the atoms.

The number of the intra-molecular vibrations corresponding to these motions is 3n -6 when the molecule is formed by 3 atoms and when it is non-linear.

The number 3n comes from the fact that each atom does possess three degrees of liberty (1 degree of translation,1 degree of rotation and 1 degree of vibration) from which one must deduce 6 motions of the whole system; that is to say, 3 global motions of translation and 3 global motions of rotation. These normal vibrations of atoms are very important. In order to make the calculations easier, one always supposes that one atom is immobile with respect to the others. A linear molecule shows 3n – 5 normal vibrations.

There are two kinds of vibrations:

• The valence vibrations;
• Valence or stretching vibrations.

The atoms vibrate but remain on the axis of the bond adjoining the atoms. There are two valence vibrations, a symmetric and an asymmetric (Figure 4).

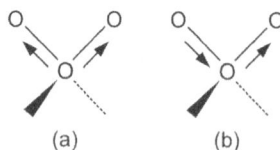

(a)   (b)

Figure 4: Symmetric (a) and Asymmetric (b) Valence Vibrations.

This kind of vibration involves a continuous change in the interatomic distance along the axis of the bond between two atoms.

• Distortion or bending vibrations. They are of two kinds, in the plane and out of the plane, where they can be symmetric and asymmetric. The schematic representation of the different types of bending vibrations is given in Figure 5.

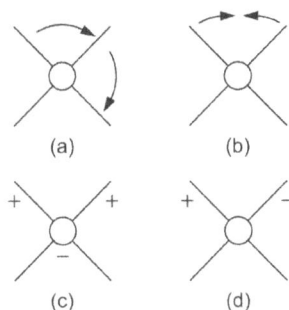

Figure 5: In-Plane (a and b) and Out of the Plane (c and d) Distortion Vibrations.

Bending vibrations are characterized by a change in the angle between two bonds and are of four types (Figure 5).

Characteristics of the Absorption IR Spectra

The frequencies of the vibrations of valence and deformation are quantified. When the molecule receives IR radiations, it absorbs energy. The motions of the system "spring-balls" amplify and the amplitude of the vibrations increases. Nevertheless, the important point is that the energy levels of the molecule are quantified. Only the IR radiations having a frequency equal to that necessary to increase the level of energy of a bond are absorbed. Then, the amplitude of a particular vibration is abruptly enlarged. There is another phenomenon when the frequency of the radiations regularly increases, some are absorbed since they are used in the valence and deformation vibrations. Therefore, some parts of the spectrum may partly disappear. The emergent corresponding IR radiation is weakened.

The come back to the fundamental state relates this energy under the form of heat. Contrary to their spectra UV which comprise relatively few absorption peaks, a vast majority of organic compounds exhibit numerous bands. Most often, they cannot be attributed but they give however some precious information concerning the absorbing molecules.

The number of theoretical peaks cannot be observed for various reasons. This is the case, for example, if they do not belong to the studied domain or if they are badly resolved or if they are of too weak intensity. In contrast, there can exist supplementary bands due to the harmonics of weaker intensity or due to combinations corresponding to the sum or the difference of several wavenumbers.

There are restrictions on the allowed changes in molecules. They are known as selection rules. There are two sorts of selection rules:

• The particular selection rules;
• The gross selection rules.

The particular selection rules concern the conditions that govern changes in the individual quantum numbers. They are discussed when the various types of spectra are considered. Gross selection rules specify the general features a molecule must have if it is to have a spectrum of a given sort. They are used, for example, to decide whether a molecule will give a vibration or rotation spectrum at all. Notice that there are no gross selection rules governing spectra in the visible or ultraviolet since all the molecules have an electronic spectrum. Both types of selection rules are based on symmetry.

For example:

• In the case of diatomic molecules, the gross selection rules say that when a heteronuclear molecule such as HCl vibrates, there is a net displacement of charges. There is a net interaction with the incident radiation which entails an oscillatory electric component. The vibration is allowed and is active in infrared. This is not the case for a homonuclear diatomic molecule such as Cl_2 for a reason of symmetry which demands that every displacement in one half of the molecule shall be balanced by an equal but opposite change in the other half. There is no interaction with the incident radiation. The vibration is "forbidden" and is inactive in the infrared;

• For polyatomic molecules, similar arguments apply, but there are several vibrations to be considered. For a frequency to be active in the infrared, the motion of nuclei which can be associated with a vibrational frequency must produce a change in the dipole moment μ of the molecule. This does not require the molecule to have a permanent dipole moment since the vibration may change the dipole moment from zero to any finite value. In the occurrence, the gross selection rule for a frequency to appear in the infrared can be symbolized by the relation:

$$d\mu/dq \neq 0$$

The q represents either one coordinate or several coordinates defining the motion of the nuclei.

   The conditions are similar for a molecule to show a rotation spectrum in the far infrared or microwave region; the dipole moment of the molecule must change with the direction of the incident radiation as the molecule rotates. The molecule must possess a permanent dipole moment. The gross selection rule for the existence of a pure rotation spectrum is, therefore:

$$\mu \neq 0$$

In summary, the gross selection rules governing activity in infrared are:

Spectrum	*Gross Selection Rule*
Vibrational Infrared	Vibration must involve a change in the dipole moment of the molecule
Rotational Infrared	Molecule must possess a permanent dipole moment
Microwave	

Obtaining an IR Absorption Spectrum: Example and Tables

An IR absorption spectrum is obtained with the help of a spectrophotometer (see Chapter 30). Most often, the spectrum, once determined, is recorded on a standardized paper. One can determine the spectrum of a compound either in a solid or in the liquid or gaseous states. Here, we present the spectrum of acetaldehyde recorded under the form of a pellicle in Figure 6.

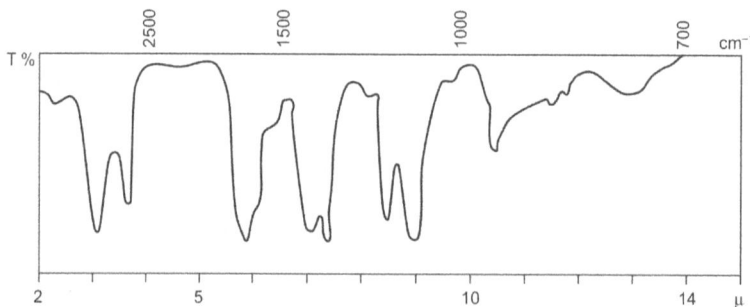

Figure 6: IR Absorption Spectrum of Acetaldehyde (Pellicular Film) (Thin Layer).

The first commentary which can be done is that an IR absorption spectrum is quite complicated even when the molecule is rather simple. The spectrum cannot be interpreted entirely. Let us, however, annotate that the imputations drawn in Figure 3 can be made.

Spectra Infrared of Diatomic Molecules

Absorption in the Far Infrared

Gases formed by two different atoms absorb in this region of the spectrum. They give a simple spectrum that consists of a series of rays, the wavenumbers of which are equidistant. For example, it is the case of hydrochloric gas (Figure 7).

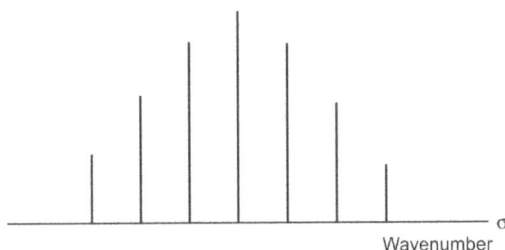

Figure 7: Spectrum in Far Infrared of Hydrochloric Gas.

It will be seen that it is a spectrum of pure rotation.

Absorption in the Normal Infrared

With HCL and Spectrograph of Weak Resolution

This way one obtains a gaussian peak ill-resolved (Figure 8).

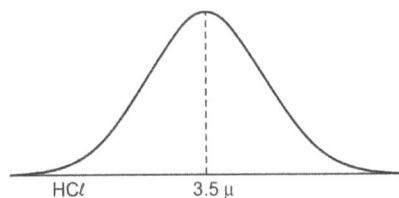

Figure 8: Structure in Normal Infrared of Hydrochloric Gas (Apparatus of Weak Resolution).

With an Apparatus More Powerful, One Obtains the Following Spectra (Figures 9a, 9b, 9c and 9d)

The absorption band consists of two series of equally spacing rays and is symmetrical with respect to a radiation σ which does not correspond to any ray.

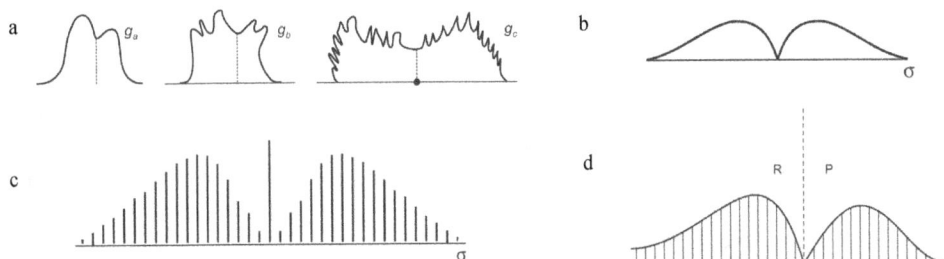

Figures 9a, b, c and d: Normal Infrared Spectra of Hydrochloric Acid Obtained with an Apparatus more Powerful.

Rotation and Vibration Spectra: Theoretical Studies

Pure Rotation Spectra

Let us first recall that in order to show a pure rotation spectrum, the molecule must possess a permanent dipole moment according to the gross selection rule. It is the case of the molecule AB (Figure 10).

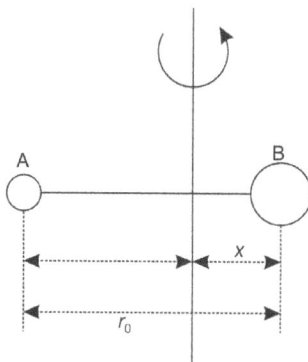

Figure 10: Parameters of a Diatomic Molecule AB.

Let m_A and m_B the masses of atoms A and B located at a fixed distance r_o apart. This system is known under the name "rigid rotor model". If this molecule rotates about an axis perpendicular to the internuclear axis through the center of gravity O, the moment of inertia I_o is given by:

$$I_o = m_A(r_o - x)^2 + m_B x^2$$

Here, x is the distance from atom B to the center of gravity. If one takes moments about the centers of gravity, one gets:

$$m_A(r_o - x) = m_B x$$

Therefore:

$$x = m_A r_o/(m_A + m_B)$$

Hence:

$$I_o = m_A[r_o - m_A r_o/(m_A + m_B)]^2 + m_B[m_A r_o/(m_A + m_B)]^2$$

$$I_o = [m_A m_B/(m_A + m_B)]r_o^2$$

The reduced mass M of the molecule is defined as:

$$1/M = 1/m_A + 1/m_B$$

Hence:

$$M = m_A m_B/(m_A + m_B)$$

And from the preceding equation:

$$I_o = Mr_o^2$$

This means that one can regard the molecule AB spinning about its center of gravity as equivalent to a single particle of mass M and describing a circle r_o. The motion of a particle in a circle is a well-known problem in wave mechanics. An exact solution of the corresponding Schrödinger equation has been obtained. It is found that only certain discrete values of the energy of the system are possible. These values are given by the relations:

$$E_{rot} = h^2 J(J + 1)/8\pi^2 Mr_o^2 = h^2 J(J + 1)/8\pi^2 I_o = BhJ(J + 1)$$

With

$$B = h/8\pi^2 I_o$$

B is called the rotational constant. h is Planck's constant. J is the quantum number of rotation. It can have integral values including zero. If $J = 0$, $E_{rot} = 0$. There is no rotational energy. This energy can be changed by addition of energy. The excess of the received energy must correspond to an increase of 1 unity of the quantum number J, that is to say to $\Delta J = \pm 1$. The energy difference between two successive levels is given by:

$$\Delta E_{rot} = Bh[J'(J' + 1) - J''(J'' + 1)]$$

Here, J' refers to the upper rotational state and J" to the lower. Since the selection rule demands that $J' - J'' = 1$

$$\Delta E_{rot} = B2hJ'$$

Although the spacing between the energy levels increases linearly with J, the spectrum consists of a number of equally spaced lines (Figure 11).

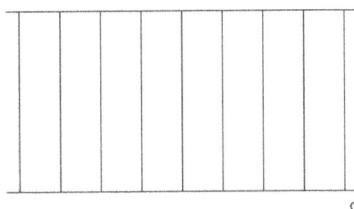

Figure 11: Energy Differences Between Successive Levels.

The result is in agreement with the experience.

Rotation-Vibration Spectra

Change of the Potential Energy of Two Atoms as a Function of their Distance

A diatomic molecule consists of two positively charged nuclei and a number of electrons. The stable molecule corresponds to a minimum in the potential energy of the system. The minimum is the result of the part played by the repulsion of the two positively nuclei and that played by the attraction between the nuclei and the electrons. The existence of a stable molecule is characterized by a minimum value of this potential intra-molecular energy. Since the molecule is stable, the potential energy passes by a minimum M for the equilibrium value. When this distance changes, the two atoms are recalled towards the equilibrium position. The potential curve is the Figure 12:

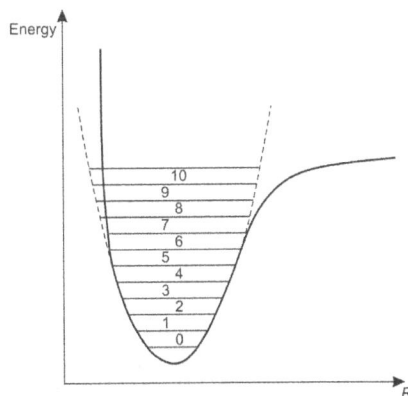

Figure 12: Potential Energy Curve of a Diatomic Molecule.

The zero potential energy is the potential energy when the two atoms are separated by an infinite distance. In order to study these phenomena, one assimilates the chemical system into a vibrating spring. Let us consider a spring, without mass vertically hung, bringing a mass M (Figure 13):

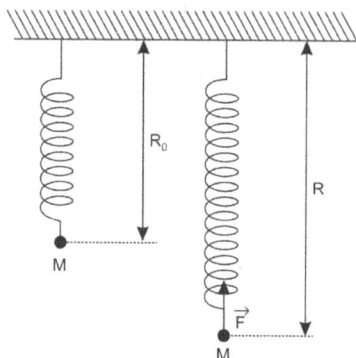

Figure 13: A Vibration of Diatomic Molecule (Schematic).

Not in use, the set possesses a length of R_0. If, by traction, one causes the lengthening of the spring, its length becomes R, and it appears an elastic force of recovery **F** which tends to bring back the system to its equilibrium position, according to:

$$\mathbf{F} = -k\,(\mathbf{R} - \mathbf{R}_o)$$

Where k is the force constant. Finally, a motion of oscillations results from this system. For the weak amplitudes, the system behaves as a harmonic oscillator and the frequency ν of the periodic motion is equal to:

$$\nu = (1/2\pi)(k/M)^{1/2}$$

M is the reduced mass of the extremity of the spring in the mechanical model. The potential energy of the system E_p is given by the relation;

$$Ep = 1/2k\,(R - R_o)^2$$

The total energy, the sum of the potential energy and the kinetic energy, is constant for a harmonic oscillator and is equal to the value of the potential energy at the position which corresponds to the maximal spreading. Hence, the total energy depends on the amplitude of the vibration, according to a function of the second order of the form;

$$y = kx^2$$

Hence, the graphical representation of the system is a parabola. This is true only when $R \approx R_o$. If R becomes markedly different from Ro, the curve is no longer a parabola. Wave mechanics say that the vibration energy Ev is quantified. It is given by the relation:

$$E_v = (v + \tfrac{1}{2})\,h\nu$$

ν is the wavenumber of the vibration and v the quantum number of vibration. It is any integral number. The fact that the vibrational energy is quantized is represented in Figure 11 by the horizontal lines which are labeled with the vibrational quantum number v = 0,1,2, 3…

Energy of Rotation Vibration

The energy of rotation can vary by quanta very smaller than the energy of vibration, the frequency of vibration of which is ten times higher than that of rotation. As a result, every cause which may enhance the vibration of the molecule also enhances their rotation. One cannot speak vibration without evoking the rotation. The energy of rotation vibration is, hence, the sum:

$$E_{vib,rot} = E_{vib} + E_{rot}$$
$$E_{vib,rot} = J(J+1)h2/8\pi^2 I_o + (v + \tfrac{1}{2})hv$$

Non-Linear Molecules

The spectra of non-linear molecules are much more complex than those of linear ones. It is rare that can give a simple interpretation of such a molecule. It is necessary to address the group theory in order to elucidate the nature of the vibrations. We shall not discuss these mathematical derivations. Let us only say that the satisfactory frequency assignments have been made most stable.

Molecules that possess some symmetry and which have less than about twelve molecules. A full frequency assignment remains a formidable task. It remains to be said, however, that certain groups of atoms always absorb at roughly the same frequency, no matter in what molecules they are located. Such observations enable us to detect the presence of certain groups of atoms in the most complicated molecules (see Chapter 29).

CHAPTER 29

Infrared Spectroscopy (II)
Apparatus, Qualitative and Quantitative Analysis MID, Infrared Reflection and Near-Infrared Spectrometries

After having recalled specific particular points concerning the apparatuses used in infrared spectroscopy, we essentially develop the contribution of this instrumental method to the qualitative analysis and especially to the functional analysis which constitutes its great interest. Its application in quantitative analysis is, indeed, weak. We finish this chapter with a few words about diffuse-reflectance spectrometry.

A Brief Overview

Infra-red spectroscopy is essentially spectroscopy by absorption. In infra-red spectroscopy, a collimated beam of radiation from a body at red heat is passed through a sample of the substance under investigation and is then dispersed by a prism or grating. The prism must be transparent to the radiation and usually, the infra-red sodium chloride or potassium bromide prisms and windows are used.

The dispersed beam then passes to a detector that looks at each small portion of the spectrum separately. The smaller the portion of the spectrum that falls on the detector at any one moment, the greater the resolution of the instrument. The detector finds that certain parts of the continuous spectrum of the sources have been weakened compared to others. The frequencies of these attenuated parts correspond to the different absorption phenomena of the molecule.

Techniques and Apparatus

Spectrometers IR are similar but not identical as those used in visible-UV. They differ by the nature of some constitutive and essential pieces.

Sources of Radiations

The sources of radiation IR can be a Nernst filament or a globar for mid-infrared radiation ($200 \ cm^{-1}$–$400 \ cm^{-1}$). The Nernst filament consists of a ceramic filament containing oxides of zirconium, yttrium, thallium and cerium. They are heated from 1,200 K to 2,000 K. The globar consists of a rod of silicon carbide heated electrically to about 1,500 K. Many apparatuses use some materials which emit black-body radiations when heated (35 cm^{-1}–200 cm^{-1}). A tungsten filament gives an intense light in the range (320 nm–2,500 nm) when heated at about 3,000 K.

The Dispersing Element

One uses:

- A diffraction grating is operating from the near IR until UV. It is a ceramic or a glass plate in which fine grooves have been cut and covered with a reflective aluminum coating. The grating causes interferences between waves reflected from its surface. There is a formation of constructive interference (see Chapter 36).
- A monochromator is a narrow slit allowing a narrow range of wavelengths to reach the detector.

The Receptor

This is the essential piece of the spectrometer. It was the case of the oldest apparatus because of the smallness of the energy to measure of the order of 10^{-9} to 10^{-13} watts. The receptor was a thermocouple constituted by very fine wires of 100 to 500 μm^2 of section. The weld is blackened and located in a vacuum at the point of focalization of the radiations (see Appendix 5 for the description of a thermocouple).

Fourier Transform Techniques

At the present time, the reception is made with the help of the Fourier transform technique (see Chapter 26). There is, indeed, a problem with proceeding according to conventional spectroscopy. It is elegantly solved due to this technique. The problem is the following one. Conventional spectroscopy can be qualified as being *frequency-domain spectroscopy*, given the fact that radiant power data are recorded as a function of the frequency or the inversely related wavelength of the absorbed radiation. In contrast, *time-domain spectroscopy* is concerned with changes in radiant power with time. Hence, we can say that IR spectroscopy is not conventional spectroscopy. Time-domain signals cannot be acquired experimentally directly given the range of frequencies associated with this spectroscopy (10^{12} to 10^{15} Hz) because there are no transducers that can respond to these enormous frequencies. However, frequency-domain spectroscopy can be transformed into time-domain spectroscopy by a Fourier transform technique with the help of a device called Michelson's interferometer[43] (see Appendix 7). It works by splitting the beam from the sample into two parts and by introducing a supplement parameter p called *retardation* in relations, describing this new system. When the two components recombine, a phase difference between them may appear. If the radiation has now the average wavenumber $\bar{v}$, the intensity is:

$$I(p) = I(\bar{v}) \cos 2\pi\bar{v}p$$

Notice that the detected signal depends on p. It remains to find $I(\bar{v})$ to go into the frequency domain. This function represents the spectrum we require. We use the Fourier transform technique. We find:

$$I(\bar{v}) = \int_0^\infty I(p)\cos 2\pi\bar{v}\, p\, dp$$

Thus, at each value of *p*, we take the signal I(p), multiply by cos2π$\bar{v}$p and add the products together (in order to integrate). The first expression which gives I(p) as a function of I($\bar{v}$) is legitimated by Fourier's theorem which permits writing a periodic function as a function of the phase difference of its harmonics.

The apparatus is described in Appendix 7.

---

[43] Michelson. Albert, American physicist from Polish origin, 1852–1931, Nobel Prize in Physics, 1907.

Detector: Thermal Transducers

The usual photo transducer is generally not applicable in the infrared domain because photons in this region lack the energy to cause photoemission of electrons. It is used in this case thermal transducer in which the radiation strikes a small blackbody and is absorbed by it. The resultant temperature rise is measured. It must be noticed that the power of a typical infrared beam is very weak (10^{-7} to 10^{-9} W). As a result, the heat capacity of the absorbing element must be as small as possible and the element is designed according to this fact. The temperature change of the element can amount to a few thousand of 1 kelvin. A parasitic thermal effect must be exclusively measured, it is the thermal noise of the surroundings. One way to avoid it is to chop the beam from the source. In these conditions, the analytic signal exhibits the frequency of the chopper after treatment. It can then be easily electronically isolated. Among thermal transducers, thermocouples, bolometers and pyroelectric transducers are used.

Thermocouples are the most widely used infrared detectors. They are manufactured in a number of ways. The principle of their functioning is given in Appendix 5. They are generally equipped with a tiny bit of blackened gold foil which is the actual absorber of radiation.

The bolometer is another kind of thermal change detector (see Appendix 5). Finally, we can also mention the use of pyroelectrics crystals. The pyroelectrics crystals develop a potential difference across their opposed faces when they are heated. Among them, let us mention triglycine sulfate, barium titanate and lithium niobate; they have been used successfully as infrared detectors.

Let us also mention the mercury-cadmium-telluride MCT detector for the mid-infrared region. It is a voltaïc device for which the potential difference changes upon exposure to infrared radiation; the deuterated triglycine which is a pyroelectric device in which the capacitance is sensitive to temperature.

Qualitative and Quantitative Analysis

Qualitative Analysis

Absorption IR spectra may be used to identify pure substances or to detect and identify impurities. Infrared spectroscopy is more used in organic chemistry than in inorganic chemistry. Inorganic substances often exhibit large bands whereas organic compounds show narrow bands. Water, the frequent solvent of inorganic species does not facilitate the analysis of the latter ones because of its very strong absorption occurring at about 1,5 μ. The spectrum of a mixture is, roughly speaking, the same as that of the sum of spectra of the individual compounds. Exceptions occur when there are formations of associations and dissociation or polymerization compounds. Such phenomena may provide displacements of bands or the occurrence of supplementary bands due to hydrogen bonds. We shall give some examples when we consider the functional analysis.

Functional Analysis of Polyatomic Molecules

Elementary analysis and the determination of the molecular mass permit knowing the molecular formula of an organic compound. In most cases, in order to identify it, it remains to establish its developed formula. However, in several cases, there exist several developed formulas corresponding to the same molecular mass. They differ by the nature and by the position of their functional groups. The functional analysis has the goal to evidence these ones and permitting the knowledge of the structure of the investigated compound.

Functional analysis is carried out by chemical and physical methods. Today, quasi-only physical methods are used in functional analysis and among them (which can be varied); let us mention spectral methods and especially infrared spectroscopy. (Functional analysis is an old term and tends to disappear).

Basis of the Identification

Figure 1 recalls the different interesting domains of analysis by IR.

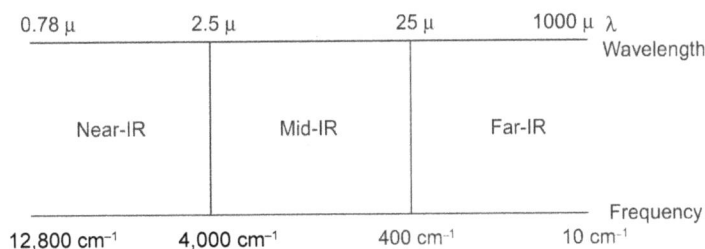

Figure 1: Different Domains of Analysis by IR.

One can say that in organic chemistry, only are interesting the spectra recorded between 2 and 16 μ, the domain of absorption which is very indicative of the molecular structures. As a result, the infrared spectra are very sensitive to the modifications of molecular structures.

Theoretically, in order to confirm a proposed structure compound, it should be interesting to compare the experimental IR spectrum of the unknown product and the spectrum calculated according to the proposed structure or with some calculated frequencies of the latter structure. Unfortunately, it is too complicated to calculate IR spectra, hence we must address ourselves with an empirical interpretation of the spectrum of the unknown product.

In this way, we compare the IR spectrum of the unknown compound and that of another (existing) product, whose (some) frequencies of vibration have been calculated and interpreted by starting from experimental data. The entire studied molecule participates in the normal vibrations but only some atoms show motions of important amplitudes; the others remain practically at rest. This comparison may be more easily understood if we consider a thought experiment that would consist in isolating two directly linked atoms of the molecule and comparing them and a diatomic molecule containing the same atoms. By introducing this vibrating system into a complex molecule, it will be coupled with other systems.

This possibility is grounded on the hypothesis that in a molecule most often complex, an association of two atoms constituting a functional group (such as cyano group –CN) or a part of the functional group (CO of COOH, etc.) can vibrate with an important intensity whereas the other atoms stay rather immobile. It results from this point that the frequency of vibration of these associations of two atoms remains statistically the same. This hypothesis is largely verified excepted for resonance phenomena. Hence, the functional analysis by IR spectroscopy is based on the comparison of spectra of a great number of products more or less homologous.

Functional Analysis Via the Study of Vibrations De Valence

We have seen that the spectra of bands are usually those that are usually obtained for which the maximum corresponds to a vibration transition. Thus, it is the value of the wavenumber of vibration that is based on the identification. One comeback on the *approximate* relation, already seen:

$$\bar{\nu} = (1/2\pi C) \sqrt{k/\mu}$$

It permits justifying the identification of some bands to some functional groups. The two only variables are the constant of recovery k, and it is also called the constant of the force of bond and the reduced mass μ.

Concerning the influence of μ, that is to say, the influence of the two masses of both atoms, one can see that the more the reduced mass importance, the weakest the wave number is. If one of the two atoms still possesses a smaller mass, the reduced mass is weak:

$$\mu = M_1 M_2/(M_1 + M_2)$$

In the case of bonds in which intervenes a hydrogen atom, the reduced mass is near the unit:

$$M_2 = 1 \quad \mu = M_1/(M_1 + 1) \quad \mu \approx 1$$

It results that the reduced mass is neighbor the following vibrations of valence;

$$OH: \mu = 16/17 \quad NH: \mu = 14/15 \quad CH: \mu = 12/13$$

The bands of vibration of valence corresponding to the bonds in which intervene H are located in the domain of the great wavenumbers between 4,000 cm^{-1} and 2,800 cm^{-1} (2,500 cm^{-1}), that is to say in the near I.R.

In the case where the bonds are not between hydrogen atoms, the reduced masses considerably increase. For a bond between two carbons:

$$\mu = 12 \times 12/(12 + 12) = 6$$

As a result, with respect to a bond CH ($\approx 3,000$ cm^{-1}), a vibration de valence band is located at $\sqrt{6} = 2.45$ times weaker, close to 1,200 cm^{-1}. More generally, the greater the mass of one (at least) of the two atoms is, the weaker the wavenumber is (see the following table).

Table 1: Vibrations of Valence of the Bonds Between the Carbon and Principal Heteroatoms.

Bond	Wave Numbers cm^{-1}
C-H	3,300–2,800
C-F	1,400–1,000
C-O-	1,300–1,050
C-N	1,360–1,050
C-Cl	800–600
C-S-	800–600
C-P	750–650
C-Br	650–550
C-I	600–500

With valence vibrations, the atoms are more and less moving along the bond axis. These are the cases of the vibration C-H of the chloroform and the two vibrations of water in which both hydrogen atoms symmetrically or asymmetrically approach the oxygen atom. As a rule, valence vibrations are poorly sensitive to the surroundings. Hence, they are rather characteristic of a group of atoms. Therefore, one can define valence vibration frequencies for several couples of atoms. For the zone:

- 1.6 to 3 μ, one encounters the valence vibrations of groups containing one hydrogen atom. They are N-H; C-H; O-H.
- 4 to 6 μ: This is the region of the multiple bonds of organic compounds; C=C; C=O; C N; C C; —C(H)=O: —C(=O)O-H.
- 7 to 16 μ: This is the region of simple bonds between more heavy atoms, such as the groups containing one halogen atom.

Influence of the number of multiple bonds

It is linked to the number of bonds established between the two atoms implicate and also to the bond length (both are linked). The force constant k is perceptibly proportional to the number of bonds (1 newton $= 10^5$ dynes). If we consider the case of the bonds between carbon atoms, we find:

Table 2: Values of Force Constant k for Different Number of Bonds.

	k (dynes/cm)	$\bar{v}$ cm^{-1}
C C	15 10^5	2,250–2,100
C C	10 10^5	1,700–1,600
C C	5 10^5	1,260–1,100

As it is the square root of the constant which plays a part, the wavenumber of an ethylenic bond is close to $\sqrt{2} = 1.414$ higher than that of the simple ether. That of an acetylenic $\sqrt{3} = 1.731$ times higher. The influence of the force constant is preponderant when the reduced mass is slightly different from a couple of atoms to another. It is the reason why we find in the same domain:

Alkyne-derivatives and nitriles $\approx 2,200$cm^{-1}
Olefins and carbonyl derivating compounds ≈ 160 to$1,800$ cm^{-1}

The length of the bond is important. The force constant k is all the greater as the distance between the atoms is weak. Hence, the biggest is one of the atoms, the less important the force constant is. Thus, $\bar{v}_{OH} = 3,600$ cm^{-1} and $\bar{v}_{SH} = 2,250$ cm^{-1} for a reduced mass quasi-identical.

Other Factors Playing a Part in the Location of the Band of Valence Vibration

- Considering intramolecular associations, the spectroscopy IR permits detecting intramolecular associations such as those which are formed through hydrogen bonds. Let us consider the alcohol R-OH.
- When it is in a very dilute solution, it is constituted by molecules very far from each other. They vibrate without being constrained. Then, we can observe an IR absorption band at 2.76 μ which is the band of valence vibration of the molecule named monomer.
- In solutions less diluted, the molecules are sufficiently near each other to be associated two by two by forming dimers.

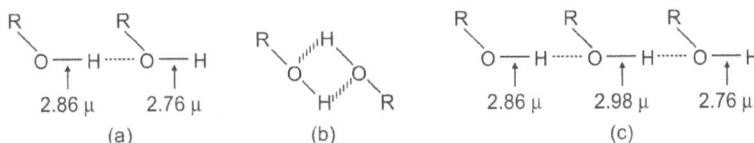

Figures 2a, b, c: Formation of Intermolecular Associations through Hydrogen Bonds.

In dimer a, one observes an ordinary valence vibration at 2.76 μ and a slower valence vibration at 2.86 μ due to the fact that the hydrogen atom is made heavy by the other molecule. In *dimer b,* one has only one band at 2.85 μ, both atoms being made heavy identically. Finally, if the solutions are still more concentrated. It appears a *chain c.* We notice three bands at 2.86 μ, 2.76 μ and 2.98 μ. The first two are already described. The third corresponds to all the bonds O-H extremely hindered in their vibrations.

There occur other factors which play a part in the location of the band of the valence vibration. They are the conjugation; if a bond by conjugation does possess some double bond character, its force constant is:

- re-inforced relatively to a simple bond ($\bar{v}_{OH}$ increases);
- weakened with respect to a double bond. For example, one knows that benzene has been written under the false form 3a whereas it must be written under the form 3b which is correct:

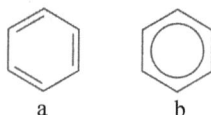

Figures 3a, b: Effect of the Conjugation on the Location of the Band of Vibration.

The valence vibration of the bonds carbon-carbon ($\overline{v}_r = 1,500$ cm^{-1}) possesses a wavenumber lower than that of a classical double bond (1,600–1,700 cm^{-1}) and markedly greater than for a simple bond ($\approx 1,200$ cm^{-1}). This note is true for every conjugation. Inversely, the study of the infrared spectrum may permit to confirm that in a given structure, a group cannot enter resonance.

Functional Analysis According to its Study of Deformation Vibrations

The necessary energy in order to obtain a deformation vibration is always weaker than that which corresponds to a valence vibration. The domain of the spectrum which is attributed to it is located in "the weak wavenumbers". They are very numerous and the most sensitive to their atomic neighbors. It is difficult to assign them to a specified deformation vibration. However, some of them are used for a structural analytical goal. It is the case of deformation vibrations out of the plane of the bonds C-H of the aromatic derivatives. They permit distinguishing the presence of substituents in ortho, para and meta positions:

v_{CH} cm^{-1}	monosubstituted	disubstituted
	770–730	ortho 770–730
	710–690	meta 810–750
		para 860–800

Functional Analysis After Study of the Combination Bands

In the case of aromatic derivatives, the combination of two deformation vibrations out of a plane is the cause of the appearance of several bands located between 2,000 cm^{-1} and 1,600 cm^{-1}. The band near 1,600 cm^{-1} is characteristic of the aromatic nuclei. The aldehydes exhibit several bands located between 2,820 and 2,700 cm^{-1} in supplement of $\overline{v}_{OH}$ which are combination bands. Hence, the examination of the infrared spectrum of an organic compound permits asserting or not the presence of some functional groups in its structure such as a carbonyl.

Identification of a Compound

It is the application most important of infrared spectrometry. Its principle consists in verifying the identity of each of the bands of the reference spectrum and of the spectrum of the compound under study, spectra registered in the same conditions of sampling and with the same apparatus. The smallest structural difference appears particularly at the level of the vibration deformation bands which are the most sensitive to their atomic neighboring (domain 1,000 to 1,300 cm^{-1}). This domain is often compared to the fingerprint of a human. The identity of the infrared spectra of two compounds provides a very strong assurance that they are identical and reciprocally if the two spectra differ.

However, one must know that this test is not infallible for some solid compounds. Some polymorph species and some particles of different tails may present different spectra. The fabrication of pastilles, necessary for the registering of the IR spectra, may command the use of important pressures which can induce a polymorphism. Sometimes, there is an exchange of ions between the ions Br- from the KBr necessary usually to prepare the pastilles; in turn necessary to register the infrared spectra with an anion of the product under study. It is the case of the thiamine hydrochloride which exchanges its ion chloride Cl$^-$ with the ion Br$^-$.

On the other hand, some racemic compounds may have different IR spectra in the solid state. They are called racemates. In these mixtures, each enantiomer possesses a more marked affinity (in its solid state) for the opposite enantiomer rather than for any other compound. This is no longer the case in the liquid state, where the optical isomers have the same infrared spectrum.

For the identification of IR in compounds, there exist catalogs of IR spectra. Moreover, there exist informatic systems where are registered the position and the intensity of the peaks of some importance in the memory of a computer. The profile of the studied derivative is compared to other profiles already registered.

Quantitative Analysis

Quantitative measurements by infrared spectroscopy are by far less frequent than those by UV-visible. They are of less good quality, in particular when they are determined with the help of dispersive apparatuses. This is due to the fact that the bands of IR are much narrower than the UV-visible ones; also that the sources are less intense and the detectors are of weak power. To come back to the narrowness of the bands IR, it results from this fact that the band-width of the gratings in slits is superior to the bands. As a result, there are deviations from the Beer-Lambert law. This is why calibration curves are, more than ever, necessary. One does not use a double ray as is the case in UV-visible because of the difficulty one has to obtain measurement cells whose transmissions are identical. Their lengths are by far too weak to be exactly duplicated. Furthermore, their windows can be differently attacked by contaminating agents. Generally, one solely operates in solution. One uses two techniques:

- The first one consists in working relative to the ray which does not pass through a cell. One measure successively:
    * The power transmitted by this one Pr;
    * The power transmitted by the pure solvent Po;
    * The power transmitted by the solution with the sample P.

The transmittances of the solvent T_o and of the sample T_s are:

$$T_o = P_o/P_r$$

$$T_s = P/P_r$$

The transmittance of the sample only is:

$$T = T_s/T_o$$

$$T = P/P_o$$

One operates with respect to the basic line.

- The technique of the basic line.

One considers that the transmittance of the solvent linearly changes between the shoulders of the peaks of absorption P, whence the absorbance A of the solute is given by the relation:

$$A = \log T_o/T_s = \log P_o/P$$

Given the relationships and, notably, a lot of similarities existing between the infrared and UV-visible spectroscopies, it is not surprising that Beer-Lambert's law also applies in infrared spectroscopy. In this case, the methodology is the same as in the former.

Unfortunately, it is difficult to apply it. One of the major encountered difficulties in IR spectroscopy is the presence of diffused energy. It makes false the application of Beer Lambert's law. Another problem is that the energy at our disposal in the domain of the IR wavelengths is very weak and the spectrophotometers must have large slits and the bands become, therefore, difficult to study.

Some Examples of Applications

Among the countless applications in functional analysis widely exploited by organic chemists are:

- Concerning the identification, let us recall that there are several hundred tests of identification in the world pharmacopeias.
- In quantitative analysis: They are based on the bands of vibration (stretching) C=O; NH and OH. The most used band is the carbonyl band. Let us cite for example:
  - ∗ The analysis of penicillin from the band located at 1,760 cm^{-1} of the carbonyl of the cycle lactam, the products being in solution in the chloroform;
  - ∗ The simultaneous determination of the components of mixtures; for example the mixture of aspirin, phenacetine and caffeine.

Other Methodologies of IR Spectroscopy

Let us mention:

• The mid-infrared reflection spectroscopy:
When a beam of radiation passes from a denser to a less dense medium, reflection occurs. It is shown that during the reflection process, the beam acts as if it penetrates a small distance into the less dense medium before reflection occurs. The depth of the penetration depends upon the wavelength of the incident radiation, the index of refraction of the two materials and the angle of the incident beam. The methodology has found a number of applications, particularly for dealing with solid samples. Mid-infrared reflection spectra are not quite identical to the corresponding absorption spectra, but they are similar in their general appearance and, finally, bring the same information. In this methodology, the radiation undergoes multiple internal reflections before passing from the crystal to the detector.

• The near-infrared spectroscopy:
The near infra-red region extends from 770 nm to 2,500 nm (13,000 to 4,000 cm^{-1}). Absorption bands of this region are overtones or combinations of fundamental stretching vibrational bands that occur in the region 3,000 cm^{-1}–1,700 cm^{-1}. It is overall used for the routine quantitative determination of species, such as water, proteins, low-molecular-weight-hydrocarbons and fats. Instrumentation is similar to that used for UV-visible absorption.

In near-infrared, the determinations are located between 1,700 and 3,000 cm^{-1}. The used bands are combinations and the harmonics of the bands C-H, N-H, and O-H. It is interesting to notice that the quantitative applications of the near-infrared are more numerous than its applications in qualitative analysis.

In IR by reflection, analysis of powders and of suspension can be made in usual IR as in near IR. Some words have already been devoted to this methodology under the heading "Fiber optics" in Chapter 21. In near IR, the methodology brings the name "near reflectance spectroscopy". The solid sample is irradiated with bands of radiation of wavelengths located in the interval ranging from 1 to 10,000 to 4,000 cm^{-1}. Near reflectance, the methodology is its speed and the simplicity of preparation of the sample it permits.

The chromatography /IR coupling has been the subject of some studies. The difficulty is the following one; the constituents of the mixture are emergent too quickly versus the answer of the spectrophotometer.

Finally, some applications of IR emission spectroscopy have been published. They implicate the use of Fourier transforms via the help of an interferometer.

Applications in Physical Chemistry

Infra-red spectra come from varied modes of vibration and rotation. As it can be seen, pure rotation spectra of molecules are located at very high wavelengths, beyond the value of 25 μm which is often considered as being the limit of the far infrared. For the values weaker than 25 μm, the radiation IR becomes sufficiently energetic to induce changes in the two states, i.e., the vibrational and rotational ones.

CHAPTER 30

Raman Spectroscopy

In 1928, Raman[44] has placed a scattering of the light phenomenon in a prominent position. This phenomenon has very narrow affinities with IR spectroscopy. The two combined effects permit the study of molecular structures.

Principle: Raman and Rayleigh Effect

When an intense luminous monochromatic ray goes through a transparent medium (liquid, solid and gas) a fraction of the light is scattered by the molecules which are present. The molecular scattering of a monochromatic ray without any change of the wavelength has been known for a long time and is called Rayleigh's effect[45] (Figure 1).

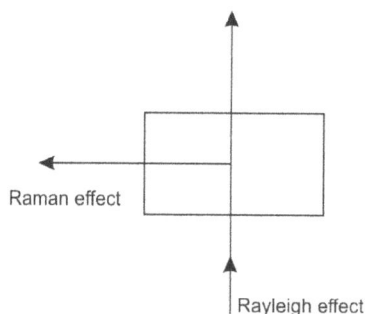

Figure 1: Rayleigh and Raman Effect.

Raman discovered that a part of the light could be scattered with changes in its wavelength (or of its frequency).

Apparatus

Let us consider any medium (gas, liquid or solid) under the same source of light as in IR spectroscopy (Figure 2).

The light is scattered laterally. The beam goes through a lens and goes towards the spectroscope. One observes perpendicularly to the incident beam.

An interesting point is that Raman spectroscopy is based upon visible and near-infrared radiation. Both have the great advantage that they can be transmitted for a very long distance of the order of 100 m. Thus the sampling can be made with a *fiber-optic* probe.

---

[44] Raman Chandrasekhara Venkata, Indian physicist, 1888–1970, Nobel prize in physics, 1931.
[45] Rayleigh, John William, English physicist, 1842–1919, Nobel prize in physics 1904.

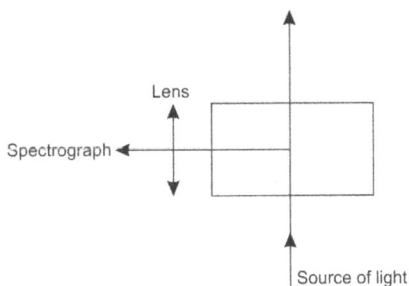

Figure 2: Apparatus to Observe Rayleigh and Raman Effects.

Experimental Facts

Now, let us consider a beam of monochromatic light, the wavenumber of which being σ_0. Its frequency is ν_0. These values may be whichever is it or any other one provided they are located outside the zone of absorption of the molecule. (In practice, it must be very intense and as monochromatic as possible if the Raman lines are to be sharp. Raman spectroscopy is always carried out in the visible region. Raman was prior to the introduction of lasers and now the latter ones are universally employed since they produce great power and monochromaticity).

In the direction perpendicular to that of the incident beam, one observes in the obtained spectrum:

- A ray of frequency ν_0 identical with that of the preceding;
- A series of other rays located symmetrically on both sides of the preceding ray. Their frequencies are:

$$\nu_0 \pm \nu_1: \quad \nu_0 \pm \nu_2: \quad \nu_0 \pm \nu_3: \quad \ldots \ldots \nu_0 \pm \nu_n$$

These rays are named Raman's rays. If one changes the wavenumber or the frequency of the incident line (which for example becomes σ' and ν_0'), one observes other rays surrounding symmetrically the central ray. As in the previous ones, the intervals between the rays are characteristic of the chemical nature of the substance and its physical state.

- The intensity of Raman's rays is much weaker than that of Rayleigh. Their observation necessitates an intense illumination. The frequencies of Raman's rays are expressed in cm^{-1}. Thus, their spectrum spreads from 60 to 3,600 cm^{-1}.
- Based on the same category of phenomena as those obtained by absorption of IR rays, the elementary vibrations of the atoms in the molecules are also interpreted by the variations ΔE of the energy which is linked to the frequencies by the usual relation:

$$\Delta E = h\nu$$

Raman's rays have not the values $\nu_0 \pm \nu_1: \quad \nu_0 \pm \nu_2: \quad \nu_0 \pm \nu_3: \quad \ldots \quad \nu_0 \pm \nu_n$ for proper frequencies. The frequencies which are characteristic of the substance are $\nu_1, \nu_2 \ldots \nu_n$. These frequencies remain, indeed, identical when the molecule is submitted to another initial frequency ν_0'. Let E_o be the initial value of the energy of the molecule. One quantum $h\nu_0$ of the incident radiation excites the molecule by bringing it to the level of energy (E) such as:

$$(E) = E_o + h\nu_0$$

This level is fictitious since the molecule cannot absorb the radiation of frequency ν_0. (E) is not a quantified level of energy of the molecule. From this level (E), the molecule comes back to a permitted level E_1 (defined by the values of the quantum numbers of rotation and vibration) by emitting the radiation of frequency ν such as:

$$h\nu = (E) - E_1$$

By taking into account the preceding relation:

$$h\nu = E_0 + h\nu_0 - E_1$$

or

$$E_1 - E_0 = \Delta E$$

$$\Delta E = h(\nu_0 - \nu)$$

ΔE is the difference of energy between the final and initial states of the molecule.

Three cases may now be distinguished:

- $\Delta E = 0$, that is to say $E_1 = E_0$

  $\nu_0 = \nu$. This is simple diffusion.

- $\Delta E > 0$ $E_1 > E_0$

The molecule goes to an upper level of energy. The scattered ray shows a frequency lower than the lower initial one $\nu < \nu_0$. The rays produced in the same manner are Raman's negative rays of strong intensity. These rays are called Stokes rays.

- $\Delta E < 0$

The molecule goes back to a lower energy level. The scattered ray has a frequency higher than the initial. The rays produced in the same manner are Raman's positive of weak intensity. These rays are called anti-stokes rays.

Let us notice that the fact $\Delta E < 0$ is possible only if the initial state of the molecule is an excited state. We represent a Raman spectrum in Figure 3.

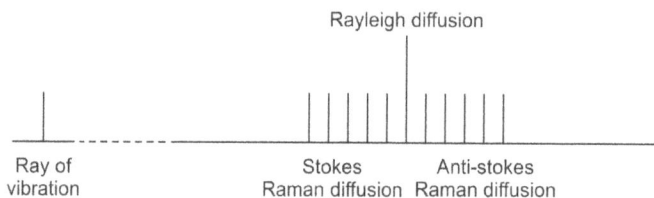

Figure 3: A Raman Spectrum (Schematic).

Selection Rules

In infrared, one condition is necessary so that an absorption of radiation should be possible. There must be a change in the value of the dipolar moment of the molecules of the sample accompanying the absorption.

In Raman's effect, the rule is different. The gross selection rule says that in order to have an absorption, there must be a change in the *polarizability* of the sample molecules. The polarizability is defined as follows. Let us consider any molecule at rest. It exhibits a permanent dipolar moment μ:

$$\mu = q\,l$$

q is the electric charge separated from the other charge $-q$ of the electric dipole constituted by the dipolar molecule, l being the distance between both charges. When this molecule is submitted to an electric field, there is a displacement of the positive and negative electric charges in the sample molecule. There happens a distortion of the molecule. A new dipolar moment μ^* of the sample molecule appears. As long as the applied electric field $\mathbf{E}$ is not too strong, one can write:

$$\mu^* = \alpha \mathbf{E}$$

α is the *polarizability*. The unity of polarizability is unusable. One prefers handling the polarizability volume α' defined by the expression:

$$\alpha' = \alpha/4\pi\varepsilon_0$$

Here, ε_0 is the *vacuum permittivity*. α' has the dimensions of volume (cm^3). Hence, the gross selection rule edicts that when the polarizability of a homonuclear diatomic molecule does change as the length of the bond is altered, such a molecule will give a vibrational Raman line. Hence, the polarizability of the molecule must be altered by the vibrations, which means the shape of the molecule is changed but without generating a dipole moment. Since the production of a dipole moment is the requirement for vibrational absorption of infrared, it follows that frequencies observed in the infrared are probably not to be active in the Raman spectra and vice versa.

Rotational Raman Spectra: Gross Selection Rules and Energy Spacing Successive Levels

Concerning now a rotational Raman spectrum, the polarizability of the molecule perpendicular to the axis of rotation must be different in different directions. A molecule need not necessarily possess a permanent dipole moment in order to show a rotational Raman spectrum. In any molecule, the polarizability in any direction must be equal to the polarizability in the opposite direction. As a result, if a diatomic molecule rotates with a frequency v about an axis at right angles to the internuclear axis, the polarizability must vary with a frequency 2v. The particular selection rule governing the appearance of rotational Raman lines of a diatomic molecule is:

$$\Delta J = \pm 2$$

The energy spacing between two successive levels is, as before:

$$\Delta E_{rot} = B_0 h \, [J'(J' + 1) - J''(J'' + 1)]$$

With J' – J" = 2

$$\Delta E_{rot} = B_0 h \, (4J' - 2)$$

The spectrum will consist of a series of equally spaced lines on each side of the exciting line, but the frequency separation, in this case, will be:

$$\Delta v = 4B_0$$

A Brief Comparison of Infrared and Raman Spectroscopy

As it has been already said, infra-red spectroscopy is essentially spectroscopy by absorption. The Raman effect is frequently complementary to that which has been described just above. In Raman spectroscopy, a monochromatic beam of light of any convenient wavelength is sent on the sample, and observations are made on the scattered light at right angles to the incident beam. If the scattered light is dispersed by means of a prism or by a grating, a spectrum consisting of discrete lines may be observed. These are the Raman lines.

The principal difference between both spectroscopies is that IR spectroscopy is spectroscopy by absorption whereas Raman spectroscopy is spectroscopy based on the study of the scattered light dispersed by a convenient means.

Applications

They are essentially concerned about the functional analysis. The Raman spectroscopy completes the information provided by the spectroscopy IR.

• Identification of Substances

A Raman spectrum can be useful in fingerprint identification as the infrared spectra. It permits, in some cases, to identify a compound in mixtures by a procedure far quicker than by one of the chemical analyses. One can easily recognize the constituents of a mixture of hydrocarbons. This methodology is particularly interesting when we have little material at our disposal. It is as it is mentioned that the spectra of the three xylenes are sufficiently different to be distinguished in a mixture.

• Functional Analysis

As the spectroscopy IR, the spectroscopy Raman constitutes a means to recognize functional groups. These groups are C-Cl, C=C, -C=C-, etc. They produce characteristic rays in Raman spectra, whose frequencies present few changes whatever the investigated molecule is. One example is given by the alkynes the Raman frequency of valence which is at 4.9 μ and the Raman frequency of deformation is at 33 μ.

Concerning now the influence of the nature of the neighboring groups, let us investigate some molecules of the type $X_1(C=O) X_2$.

The frequency of vibration of group C=O depends on the nature of X_1 and X_2. It can change from 6.9 μ to 5.5μ. When $X_1 = X_2 = H$, the vibration is located about 5.5 μ. The groups X_1 and X_2 modify the feature of the double bond C=O because of their property of electron-withdrawing or electron donor by the mesomeric and (or) inductor effects they exercise. The C=O bond can take a character more and less marked of the simple or double bond according to their nature. A donor group leads to a lowering of frequency. In the carboxylate ion, the value of λ is 6.3 μ. A group that withdraws the electrons leads to an enhancement of the frequency. For example, for an acyl chloride, R–(C=O)-Cl, $\lambda = 5.6$ μ.

Raman spectroscopy has also received some biological applications. It has not received many applications in quantitative analysis.

CHAPTER 31

Polarimetry

One knows that many transparent substances characterized by a lack of symmetry in their crystalline or molecular structure have the property of rotating the plane of polarized radiation. This is the phenomenon of *rotatory polarization*. Such types of materials are said to be *optically active*. Polarized light and optical rotation are given a considerable part in the fields, among others, of analytical chemistry and organic chemistry, particularly stereochemistry.

Polarimetry is rather concerned with the analytical aspects that procure the occurrence of the optical activity. Optically rotatory dispersion and circular dichroism (which are studied in the next chapter) are more concerned than polarimetry by the different aspects of stereochemistry.

Definition

The term polarimetry can be defined as the study of the rotation of the plan that is polarized or linearly-polarized (both terms are synonymous) light by transparent substances. The knowledge of the sign ± of the rotation and the value of the rotation angle is interesting in qualitative and structural analysis and also, in quantitative analysis.

Structure of a Luminous Wave

Refraction

The index of refraction of an optical medium is the ratio of the velocity of radiation of a particular frequency in a vacuum to that in the medium. The variation of refractive index of a substance with wavelength is called its refractive dispersion or simply its dispersion.

Monochromatic Luminous Wave

We recall (see the 'Introduction of Spectral Methods') that a monochromatic luminous wave can be represented by two perpendicular vectors giving rise to transversal vibrations, that is to say to vibrations located in a plan which is perpendicular to the propagation direction:

- One vector is the vector electric field **E**
- The other is the vector magnetic field **H**

The two vectors define the *plane of the wave*. It is represented in Figure 1.

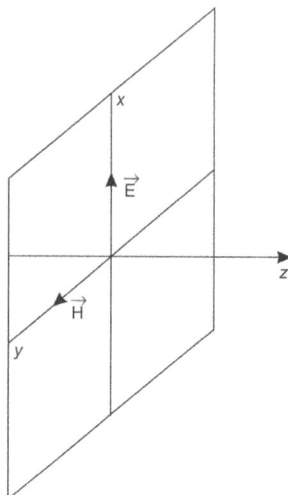

Figure 1: Schematic Representation of Vectors **E** and **H** of a Luminous Monochromatic Wave.

We name:

- Vibration plane, the plane normal to the plane of wave and containing the luminous vibration;
- Polarization plane, the plane normal to the plane of wave and containing the vector magnetic field.

Natural or ordinary light consists of a bundle of electromagnetic waves in which the elongation of vibration is the same as an infinity of planes passing through the direction of light (Figure 2).

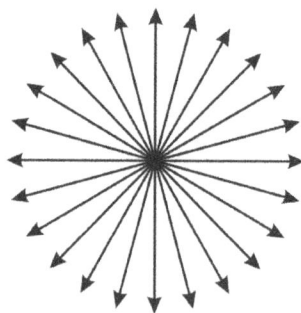

Figure 2: Natural or Ordinary Light.

A beam of monochromatic natural light viewed from the end of its axis of propagation looks like an infinity set of electric vectors that fluctuate in length. Each arrow (Figure 2) is changing (with regularity) of the area which is a function of the amplitude of the vector **E**.

Each arrow is circular looking like a spot. This is because it vibrates in all directions perpendicular to that of propagation. Figure 3 shows the evolution of the different vectors in every plane in which is located the propagation axis.

Each vector in a plane, for example, the plane of the wave can be reduced to the two vectors mutually perpendicular which are the components of the initial vector. In this example, they are the components AB and CD. We shall see that the polarizing material has the property of eliminating one of these components of the vibrations (say CD) and passing the other AB. The emergent beam consists of vibrations in one plane only and is said plane-polarized.

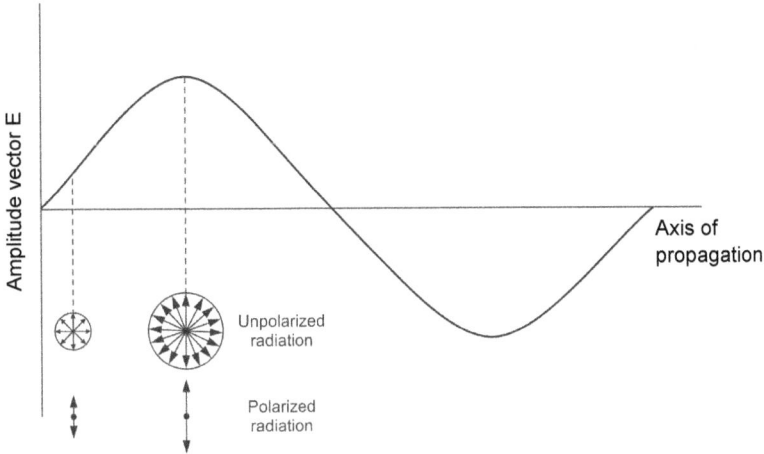

Figure 3: View of Polarized Radiation (Cross Section).

This result can be studied from a standpoint more mathematical. Every luminous vibration is a sinusoidal vibration, which is a function of time t. It can be represented by an equation of the form:

$$x = a \cos(\omega t - \varphi)$$

Here, a is the amplitude, ω is the pulsation related to the period T by the relation $T = 2\pi/\omega$ and φ is the difference of phases. Let us seek, now, how two rectangular sinusoidal vibrations can combine and what is their resultant. The solution can be found by carrying out the geometrical sum of two vectors **OM$_1$** and **OM$_2$** directed along two perpendicular axis Ox and Oy (Figures 4a and 4b), that is to say how to carry out the sum (Figures 4):

$$x = a \cos \omega t$$

$$y = b \cos(\omega t - \varphi)$$

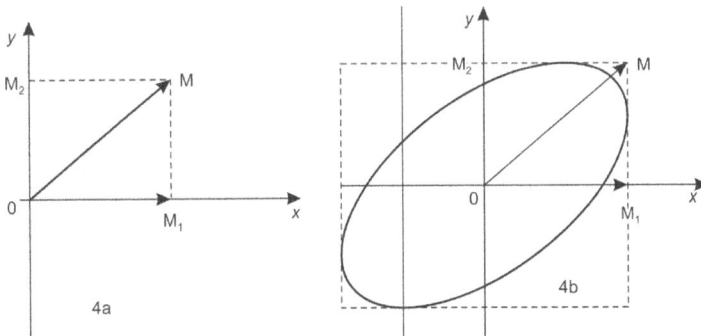

Figure 4: 4a) Composition of Both Vectors OM$_1$ and OM$_2$. 4b) Ellipse of the Extremity M of the Vector OM.

The handling of these relations and some usual other trigonometric ones permits obtaining the following equation which are those of one ellipse:

$$\sin^2\varphi = y^2/b^2 - 2xy \cos\varphi/ab + x^2/a^2 \quad \text{where } \varphi \text{ is a constant.}$$

The ellipse is inscribable into a rectangle of dimensions 2a and 2b. The Point M, the extremity of the vector sum **OM**:

$$\mathbf{OM} = \mathbf{OM_1} + \mathbf{OM_2}$$

This is an *elliptic vibration*. We shall see (see next chapter) that an elliptical light can be considered as the vectorial sum of two unequal inverse circular light components.

Fundamental Experience: Polarizer and Analyzer

Let us consider the following apparatus constituted by (Figure 5):

- A cylindrical horizontal tube T brings a mirror M_1 and M_2 at each extremity; each of them having its bottom area blackened by a glass lame in order to avoid any reflection on this face.

- The mirror M_1, owing to collar A, can rotate around the horizontal axis XY of tube T. Moreover, it is movable around the axis x_1y_1 perpendicular to XY. Likewise, the mirror M_2 can take two motions of rotation; one around XY due to collar B and the other around x_2y_2 perpendicular to XY.

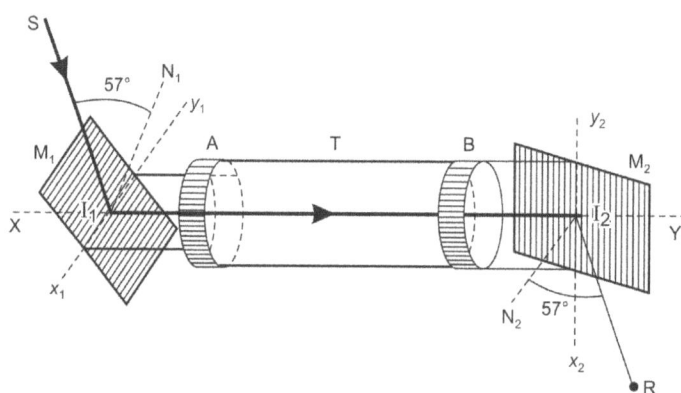

Figure 5: Fundamental Experience on Light Polarization.

Let us make the following experience (Figure 5):

- Let us settle the orientations of the two mirrors M_1 and M_2 by rotating them around x_1y_1 for M_1 and x_2y_2 for M_2 in such a manner that the normal I_1N_1 or I_2N_2 to each mirror makes an angle of 57 degrees with the axis XY of the tube T. The apparatus being settled as it is said above, let us a beam of parallel rays under the incidence of 57 degrees. They are reflecting in a parallel direction of the axis of the tube and encounter M_2. Finally, the beam reflected by M_2 on a screen gives a luminous splash.

- If we turn collar B, the splash describes a circle centered on XY and its illumination is variable. This proves that the properties of the incident ray I_1I_2 are not the same in all directions.

- If we start from the position M_2 parallel to M_1, let us rotate collar B. The splash disappeared when the collar turned 90 degrees. At this moment, the incidence planes $S_1I_1N_1$ and $I_1I_2N_2$ are perpendicular.

- Let us continue to rotate B. The luminous splash appears again. Its illumination progressively increases again and is maximum when the collar has rotated 90 degrees with the respect to the extinction. Then, the planes of incidence $S_1T_1N_1$ and $I_1T_2N_2$ are parallel.

- If one continues the rotation in the same sense, one obtains the same sequence as above.

- As a result, the beam reflected by M_1 along the axis I_1I_2 does not possess the same properties in all the directions around the direction of propagation. One says that it is a *beam of polarized lines*.

The device which has produced this beam, that is to say, mirror M_1 is called the *polarizer*. The mirror M_2 which permitted discovering this property of the beam is called the *analyzer*.

One can ask the question; quid of these results if the incidence angle on M_1 was not exactly 57 degrees?

By rotating, judiciously, collar B, we could notice that the luminous splash reflected on M_2 has still a variable brightness and that also the minimum of illumination is no longer null. One says the light is partially polarized. Under the incidence of 57 degrees, the light is wholly polarized. This incidence is called *brewsterian*.

Plane of Polarization

The study of the reflected beam with the mirror M_2 shows that the properties of this beam are symmetric with respect to the symmetry plan SI_1N_1. One says that the ray I_1I_2 admits the plan SI_1N_1 as the *plane of polarization*. From another viewpoint, one also notes that the mirror M_2 gives a reflected beam of intensity null when the incidence plan $I_1I_2N_2$ is normal to the plane SI_1I_2. The plan of polarization is perpendicular to the plan of incidence corresponding to the extinction of the reflected beam.

Nature of the Polarized Light

These last developments permit to specify of the nature of the polarized light. One can say that the luminous vibrations of a monochromatic polarized ray are rectilinear, and sinusoidal and transversal vibrations are perpendicular to the plane of polarization. These vibrations can be characterized by a vector **V**. It is perpendicular to the plan of polarization. Its length is proportional to the amplitude of the vibrations (Figure 6).

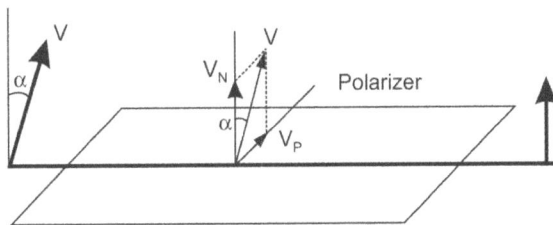

Figure 6: A Polarized Ray in a Horizontal Plan with Vertical Luminous Vibrations.

A polarizer that receives a polarized ray permits only the vibrations corresponding to the component V_N of the luminous vector **V** to pass according to the normal plan of polarization of the apparatus (Figure 6). The polarizer only transmits the vibration V_N and stops the horizontal vibration.

Polarizers and Analyzers

A *polarizer* is a device that receives natural light and gives polarized light. An *analyzer* is a device that receives a polarized line and permits to determine of its plane of polarization.

However, nothing can distinguish one polarizer from an analyzer from the standpoint of their constitution. Every polarizer can be used as an analyzer. In this chapter, we only mention a few words about the polarizers.

When a luminous ray strikes the surface of an isotropic medium, it is conveyed into it at the same speed, whichever the followed direction. Contrary to the preceding case, in a medium anisotropic, one generally observes the phenomenon of *birefringence*. The beam of incident unpolarized light gives rise to two refracted rays; an "ordinary" ray and an "extraordinary" ray. Both rays go away from each other, even in monochromatic light. They are polarized at right angles. The phenomenon of birefringence is taken into account in order to obtain polarized light. However, preliminary, it is necessary to eliminate one of the two refracted rays. For this purpose, one can use a Nicol. It is a prism of Spath of Island (CO_3Ca) (calcium carbonate), cut judiciously in a particular form (Figure 7).

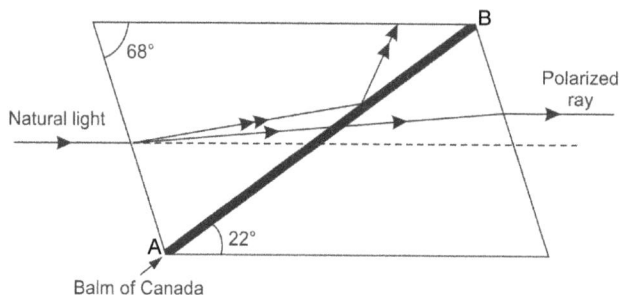

Figure 7: The Prism of Nicol.

The prism has been cut with a saw along the line AB and the two parts are glued with the balm of Canada. One of the refracted rays is wholly reflected on AB. When one works with a Nicol, one notices that there correspond two refracted rays to one incident ray. (One says that there is double refraction). Their study shows that they are both polarized and that their plans of polarization are rectangular. One is eliminated by total reflection.

Polarimetry

In polarimetry, the medium under study is traversed by a plane-polarized light coming from the polarizer. If the medium is optically active, the polarization plane turns with respect to the initial direction. This experimental fact is evidenced by the analyzer (Figure 6). There is the value of the rotation angle and its sign which are interesting.

Typically, monochromatic radiation from a sodium lamp is polarized by a calcite prism which plays the part of a polarizer. The radiation then passes through the sample which is contained in a glass tube of known length and then through the analyzer and finally to the device permitting the visual observation. A rotating analyzing prism permits the detection of the optical rotation caused by the sample solution and a graduated scale permits measuring the angular rotation. In order to make an accurate visual measurement of the degree of rotation and analytical polarimeters incorporate a half-shade device by which the matching of two half-fields gives the balance point.

Variables Affecting the Optical Activity

The rotation of the vibration plane may change from several hundred degrees to several hundreds of degrees above. The variables affecting the optical activity are the wavelength of the radiation, the length of the cell measurement, the density of the substance and the concentration of the solute.

The Wavelength

We postpone the treatment of this item to the next chapter devoted to optical rotatory dispersion and circular dichroism.

Specific Rotatory Power

It is a parameter largely used in order to describe the properties of a liquid (pure or solvent – viz., V–IV) permitting calculation of the rotation of the plane of polarization. It is defined by the relation:

$$[\alpha]^T_\lambda = \alpha/lc$$

Here, α is the observed rotation (degrees). l the length of the measurement cell (dm) and c is the number of grams of solute in 100 cm^3 of solution. The wavelength λ and the temperature T of the measurement are usually specified. The most numerous specific rotations are measured at 20°C with the sodium ray. This, for example, explains the symbolism $[\alpha]^{20}_D$.

For a pure liquid, the concentration c is replaced by its density. By convention, a levogyre rotation (l) is anticlockwise.

Molecular Rotation [M] or [φ]

It is given by the relation:

$$[\phi] = [M] = M[\alpha]/100$$

Here, M is the molecular mass of the solute.

.

Influence of the Solvent

It is frequently necessary to measure the rotatory power in the solution. Unfortunately, the specific rotation of a compound changes with the nature of the solvent. It is not possible to define a standard solvent, given the very important variability of the solute solubilities. Moreover, the specific rotation is not fully independent of the concentration, although, in diluted solutions, the change is weak. Because of this fact, it is convenient to notice the name of the solvent and the concentration of the solute in it.

Influence of the Temperature

The change in the specific rotation with temperature is approximately linear but the linearity coefficient (the slope of the line) may change widely from one substance to another.

The True Angular Rotation

From the practical standpoint, it is important to realize that compounds giving the following rotations of + 50°, + 230°, 410° or –130° will appear as having the same rotation. In order to know the true angle, the measurement must be twice carried out with two concentrations, one with a (weak) concentration that is equal to that of the other.

Some Applications

Qualitative Analysis

The optical rotation of a pure compound in specific conditions is a physical constant that is useful to know for the identification of the substances under study. It has the same virtue as the fusion point, the ebullition point, and the refraction index for this determination. By way of recall, amino acids, sugars, steroids and alkaloids exhibit optical activity.

Structural Analysis

It is not used a lot in this domain since it can be carried out easier and more soundly by optical rotatory dispersion and by circular dichroism (see next chapter).

Quantitative Analysis

It is based on the approximative relation giving the specific rotation;

$$c = \alpha/l[\alpha]^T_D$$

Hence, it is sufficient to know the constant $[\alpha]^T_D$ by a previous calibration. A practical and interesting example is the polarimetric determination of sucrose (saccharose) in the presence of other sugars. More precisely, it is the estimation of sucrose in the presence of inverted sugar in molasses and syrups. The invert sugar contains equivalent amounts of dextrose and fructose. It is obtained by acidic hydrolysis of sucrose. During hydrolysis, the inverted sugar concentration is followed by the measurement of the rotatory power. The specific rotations are at about 20°C. They are mentioned under the reaction:

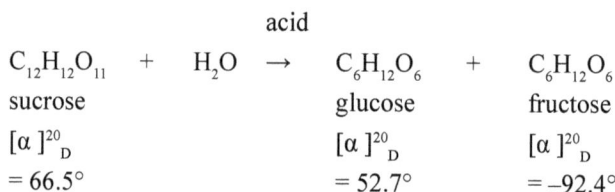

$$
\begin{array}{ccccccc}
& & & \text{acid} & & & \\
C_{12}H_{12}O_{11} & + & H_2O & \rightarrow & C_6H_{12}O_6 & + & C_6H_{12}O_6 \\
\text{sucrose} & & & & \text{glucose} & & \text{fructose} \\
[\alpha]^{20}_D & & & & [\alpha]^{20}_D & & [\alpha]^{20}_D \\
= 66.5° & & & & = 52.7° & & = -92.4°
\end{array}
$$

One demonstrates that the concentration of remaining sucrose during the hydrolysis reaction is directly proportional to the difference in rotations before and during the hydrolysis.

CHAPTER 32

Rotatory Dispersion
Circular Dichroism

Rotatory dispersion and circular dichroism are phenomena that are observed when solutions of some organic compounds (rarely inorganic substances) are traversed by a beam of polarized light. After having defined rotatory polarization and circular dichroism, we explain these phenomena and give some of their applications, notably in organic chemistry. The circular dichroism is registered in the major Pharmacopeia. It is a spectroscopic method essentially permitting stereochemical analyses in the domains of chemistry, biochemistry and even pharmacology. Above all, analyses concern chiral molecules.

Introduction

For several years, the study of the interactions between optical substances and polarized light was limited to the measurement of the rotatory power at 589 mµ with the wavelength of the ray D of sodium. However, phenomena of rotatory dispersion and of circular dichroism have been known for about one century due to the works of Biot,[46] Fresnel[47] and Cotton.[48] The first applications to the studies of organic compounds were done about the years 1930. Rotatory dispersion took flight in 1955 and circular dichroism in 1960 when the pharmaceutical firm Roussel-UCLAF put its apparatus in a running position.

Rotatory dispersion and circular dichroism are today's techniques of experimental investigations in the domains of research concerning substances possessing an optical activity. They are complementary to the spectroscopies UV-visible, infrared, N.M.R. and so forth in other domains. However, it must be already noticed that optical activity in solution is only placed in prominence when the substance under study:

- Has neither a center nor a plane of symmetry;
- Receives a light of helicoidal symmetry i.e., a circularly polarized one. It is this kind of light that must be employed in the studies of the interactions of electromagnetic radiations with the matter in these conditions.

---

[46] Biot, Jean-Baptiste, French physicist, 1774–1862.
[47] Fresnel, Augustin, French physicist, 1788–1827.
[48] Cotton, Aimé, French physicist, 1869–1951.

Theoretical Aspects

Optical Rotation

We know that at each luminous beam are associated two fields, one electric and the other magnetic which vibrate perpendicularly one with respect to the other and propagate along the direction of the light with the speed of the latter. For natural light, the vector electrical field **E** can occupy every position in the vibrational plane (see Figure 1; Chapter Introduction). On the other hand, if the light is polarized, **E** can only vibrate in a well-determined direction, the plane formed by the direction of propagation of the light and the direction of the vibration itself. This vibration is straight. This representation is that of a rectilinear polarized light. It is a simplified one of reality, but it is sufficient for the practice of polarimetry.

The theory of Fresnel goes further. The rectilinear polarized light must be considered as the resultant of left circularly polarized light and of a right circularly light (Figures 1a, b and c).

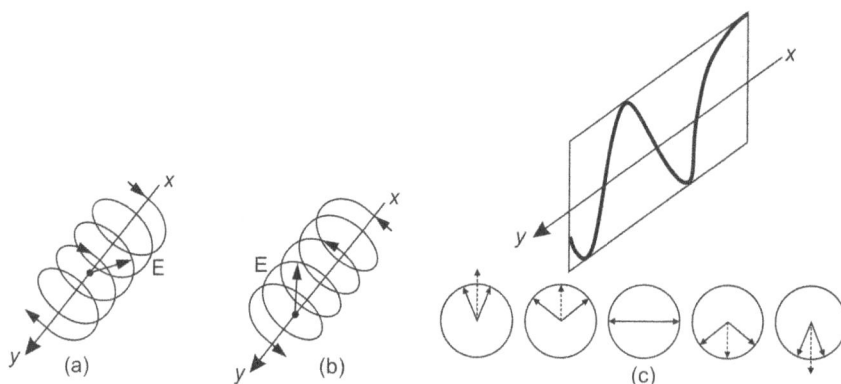

Figures 1 a, b and c: Circularly Polarized Light.

The resultant electric field **E** is the resultant of the two vectors E_l and E_r which are its circular components. This theory explains why **E** always propagates in the same plane and why its length is a sinusoidal function of time (Figure 2).

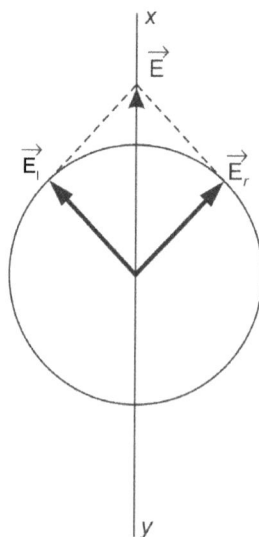

Figure 2: Electric Field as a Sum of Two Oppositely Rotating E_l and E_r.

A ray of linearly polarized (chromatic) light may be regarded as equivalent to two circularly polarized beams of equal amplitude but with opposite senses. In the passage through an inactive material, the velocities of transmission of both components are affected to the same extent, the position of the line in which they meet, and hence the plane of polarization, remains unchanged (Figure 2).

In an optically active medium, however, it may be supposed that the two circularly polarized rays have *different velocities* in the direction of the beam of light. The result is that the components now meet on a line that is inclined to the original and hence the plane of polarization is rotated through a definite angle (Figure 3).

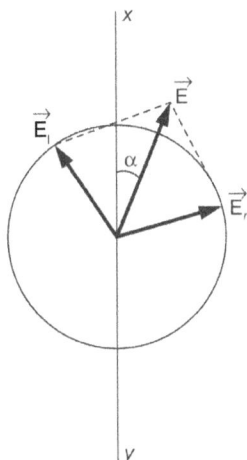

Figure 3: Decomposition of a Linearly Polarized Line into Two Circularly Polarized Beams. The Velocities of Transmission of which Being Different Because of Interactions with the Material of the Medium.

It appears that optical rotation may be regarded as due to a difference in the refraction of right- and left-circularly polarized light since the refractive index of a given medium is related to the velocity of light in it. This interpretation of optical activity was proposed by A.J. Fresnel (1825). He showed that the angle of rotation α per cm of the medium was given by:

$$\alpha = \pi/\lambda \, (n_l - n_r)$$

Here, λ is the wavelength of the light and n_l and n_r are the refractive indices for left and right, circularly polarized light, respectively. If n_l exceeds n_r, the angle of rotation is in one direction, but if n_r is greater than n_l, it will be in the opposite sense. In view of the relatively small wavelength of visible light, the difference between the two refractive indices need only be very small; e.g., 10^{-5} 10^{-6} to give appreciable angles of rotation.

The specific rotatory power increases in absolute value when λ decreases. It can be positive or negative. Figure 4 shows the curves obtained with the two isomers of the 17-hydroxy androstane. It is important to notice that these curves represent the phenomenon only when the domains of wavelengths they represent do not possess a band of absorption. Such curves are in general weakly interesting except for the fact that they can permit to differentiate of two enantiomers.

Generally, these curves do not show irregularities (maximum, minimum, inflection point and change of sense of the rotatory power). However, some exhibit anomalies. One characteristic of these curves is to show a very high maximum of its rotatory power followed by an abrupt break of its value down to a minimum. This behavior is called the *Cotton effect* (see the section on 'Circular dichroism: The Cotton effect').

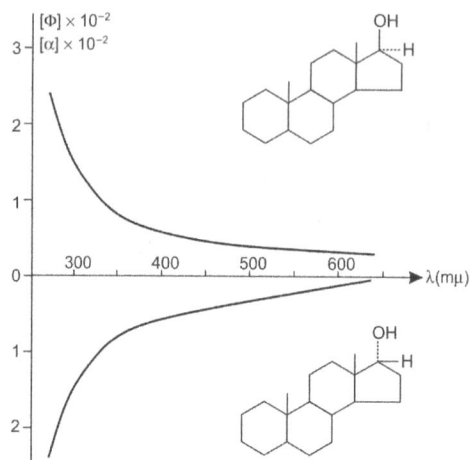

Figure 4: Dispersion Rotatory Curves of 17-Hydroxy Androstane (Transparent Domain).

Rotatory Dispersion

Since the refractive index of light depends on the wavelength, it is legitimate to expect that the rotary power of a substance will change in about the same manner as the wavelength. The variation of the angle of rotation with the wavelength is called *rotatory dispersion*. An important contribution to the theory of rotatory dispersion is due to *Drude*. We confine ourselves to say that in this book an equation has been found by him, the expression of which is;

$$[\alpha] = k / (\lambda^2 - \lambda_o^2)$$

λ_o and k are characteristic of the given substance. Close and in the absorption band, the Drude equation does not apply. Some more sophisticated equations can be satisfactory. Often, there are several terms analogous to that on the right hand of Drude's equation.

Circular Dichroism: The Cotton Effect

A. Cotton (1896) discovered interesting facts which are named the "Cotton Effect". The first point is the fact that the optical rotation within the band was found to be anomalous. It increased to a maximum near the absorption band and then decreased to zero within the band and finally increased in the other sense to a second maximum (Figure 5):

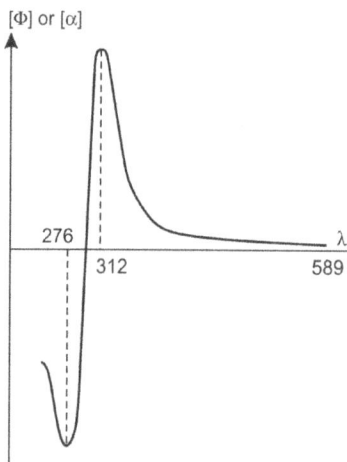

Figure 5: The Cotton Effect.

Secondly, it was observed that the right and left-circularly polarized light are absorbed differently by an optically active substance. As a result, a linearly polarized ray, the wavelength of which is located within an absorption band, is converted into an elliptically polarized one. Instead of finding any rotation angle (Figure 3), it is another one that is found, on passage through the material. If there exists a difference in absorption, there is of course a change in the amplitude of the two electrical whence the elliptic polarization (Figure 6).

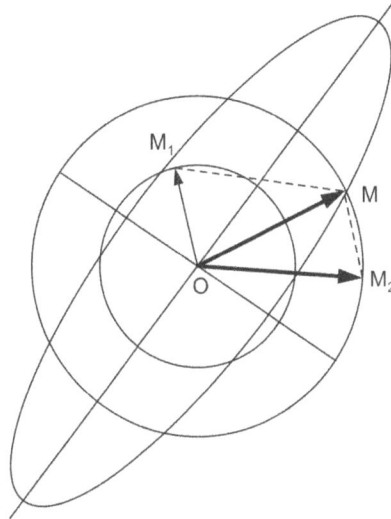

Figure 6: Vector Diagram Obtained After Traversing a Sample for which the Right-Polarized Component is Absorbed More to the Left.

This phenomenon is known as *circular dichroism*. Its magnitude is expressed by the ellipticity ϕ such as;

$$\phi = \pi/\lambda \; (\varepsilon_l - \varepsilon_r)$$

Here, ε_l and ε_r are the absorption indices for left and right-circularly polarized lights.

The ellipticity increases numerically as the absorption band is approached, reaches a maximum and then decreases. The wavelength at the maximum corresponds approximately to that of the center of the band and to that at which the rotation changes sign. The maximum dichroic absorption measures the amplitude of the Cotton effect. In the ideal case, it is exactly located at the maximum of absorption. But the superimposed effects of supplementary chromophores may cause shifts.

There are two Cotton effects:

- The positive Cotton effect for which the maximum is located at wavelengths higher than the minimum.
- The negative Cotton effect shows an inverse behavior.

Figure7 represents a positive Cotton effect and a negative Cotton effect. Usually, one notices the two following parameters in order to characterize these curves. They are:

- The amplitude a. It is the vertical distance between the maximum and the minimum
- The width b. It is the length of the horizontal line between the maximum and the minimum.

One says that the medium presents the phenomenon of circular dichroism and the corresponding luminous wave is said elliptically polarized. Circular dichroism is an *anisotropic absorption*. Circular dichroism takes place essentially in the UV or visible part of the spectrum. The difference $\Delta\varepsilon$ in molar absorption coefficients ε_l and ε_r of an optically active medium for left and right-circularly polarized light is termed *differential absorption dichroic*. It is a measurement of the intensity of the circular dichroism of the medium at the absorption wavelength.

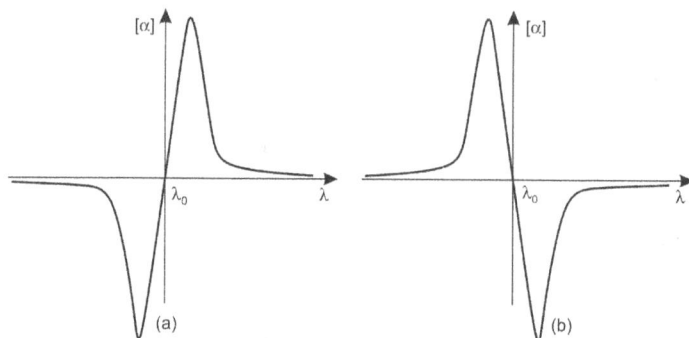

Figure 7: Positive (a) and Negative (b) Cotton Effect.

The difference $\Delta\varepsilon$ between the molecular absorption coefficients of the circularly polarized components left and right is given by:

$$\Delta\varepsilon = \varepsilon_r - \varepsilon_l$$

And one calls molecular ellipticity $[\theta]$ the grandeur defined by the relation:

$$[\theta] = \phi/100cd = k\,\Delta\varepsilon$$

Here, c is the concentration (mol L^{-1}) and d the thickness of the cuvette in dm is called differential dichroic absorption. It is a measurement of the intensity of the circular dichroism. The angle of ellipticity ϕ is directly proportional to $\Delta\varepsilon$.

$\Delta\varepsilon$ is only sufficiently important to permit measurements of circular dichroism close to the wavelength of maximum absorption λ_o. The curve of circular dichroism is obtained by registering $\Delta\varepsilon$ as a function of λ. Most often it is bell-shaped. The maximum dichroic absorption measures the amplitude of the Cotton effect. In the ideal case, it is exactly located at the maximum of absorption. But the superimposed effects of supplementary chromophores may cause shifts.

Values of Parameters

In order to observe the phenomenon of circular dichroism, the medium must be not only optically active but also it must be absorbing.

The value of $\alpha = \pi/\lambda\,(n_l - n_r)$ can be measured directly with a polarimeter but θ can only be indirectly determined through the circular dichroism from the difference in the molar absorptions of the circularly polarized lights. This difference can be positive and negative. It is bell-shaped. When the beam passes through a dichroic medium, light generally changes into another form of elliptical light with a different ellipticity. In the special case where the ellipticity of the incident elliptical vibration is so chosen that, following unequal absorption by the dichroic medium, the two circular components become equal, we obtain plane-polarized light.

We can notice the presence of the concentration of the solute under study in most of the given quantitative relationships. This is the reason why quantitative analysis can be carried out by these methods.

Origin of the Phenomena

The origin of the phenomena is related to the presence in the molecule of chromophores. This is not surprising since chromophore groups have sufficient labile bonding electrons for having characteristic absorptions in the spectral region from the visible to the near UV. Active chromophores can be classified into two types:

- Those which are active by nature;
- Those that are becoming as a result of the presence of asymmetric centers in their vicinity.

The chromophores, which are active naturally, do not possess a plane of symmetry nor a center of symmetry in the group of atoms participating in the optical transition. These chromophores are rare. They are very active but rarely encountered. One example is provided by the hexahelicene (Figure 8):

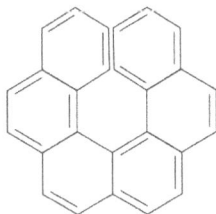

Figure 8: Structure of Hexahelicene.

• Other chromophores possess these symmetry elements, but they may become optically active by induction. All the members of these groups do possess at least one mirror plane, even when they are considered without substituents. These molecules may undergo chiral perturbations arising in the chromophore. These perturbations are exerted by substituents located in the vicinity of the chromophore. These chromophores are very often encountered but their activity is often poor. It is the case of chromophores that are inherently achiral by symmetry such as carbonyl and carboxyl groups, ordinary alkenes and sulfoxides. For example, cyclohexanone is not active. It has a grouping possessing a plane of symmetry. In contrast, 2-methylcyclohexanone is active from the center of asymmetry in position 2 and that induces a certain degree of asymmetry in the chromophore (see 'the octant rule').

Apparatus

A brief description of a polarimeter is given in the preceding chapter. Concerning the spectropolarimeter, one can say that the commercialized apparatuses are recording ones and permit measurements in the domain 200 mμ–600 mμ. They continuously give the rotatory power as a function of the wavelength. A prism called a polarizer permits the production of an extraordinary polarized beam. A second one, called the analyzer, permits determining the rotation brought about by the medium. The beam of light from a conventional monochromator passes sequentially through a polarizer, the sample and an analyzer to a multiplier tube. The beam is modulated with respect to its state of polarization at some frequency which causes the polarizer to rock back and forth through an angle of ± 1°. The amplifier responds only to this frequency and causes the apparatus to adjust the analyzer continuously to the extinction. The recording pen is also automatically positioned.

Concerning now a dichrograph, the monochromatic beam linearly polarized and coming from the polarizer, goes through a lamina of monoammonium phosphate cut perpendicularly to its optical axis. This crystal under the action of an electric field parallel to the axis becomes birefringent (this result from an effect called the Pockel's effect). Due to a convenient orientating of the lamina and by applying an alternative electric field on its two faces, one obtains a beam at its way out which during a period of the alternative voltage goes from a right-circularly polarized light to a left-circularly polarized light and inversely. The luminous flux, obtained after having passed through the solution produces at the terminals of the photomultiplier a voltage formed by a continuous tension which characterizes the average absorption of the solution and an alternative component indicating the difference of absorption between the two circular waves. The information is stocked in an electronic device and the apparatus registers the changes of a quantity which is proportional to the "dichroical" absorption as a function of time.

(Pockel's device is a circular resolver. Its functioning principle is the following. A high potential is applied across a plate of potassium dihydrogen phosphate cut perpendicular to its optical axis. The retardation of one phase it induces can be programmed by the choice of potential).

Applications

Circular dichroism is one of the most fundamental and useful physical methods of study of stereochemistry. But it tends to be also used in biochemistry and also in quantitative analysis. Its interest appears clearly when one considers its applications not only in the field of organic chemistry but also in those of biological sciences and inorganic chemistry where it has been employed to elucidate the structure of some metal complexes. Circular dichroism is used in conjunction with optical rotatory dispersion in establishing the configurations of different chemical substances.

The apparatuses used in these methods possess several practical advantages:

- The circular dichroic apparatuses are easy to handle.
- They permit to work in a diluted solution.
- The results are obtained quickly. This facilitates stability studies.
- There is no limit to the size of molecules (the analysis of macromolecules is possible).
- There is compatibility with an elution gradient DC-HPLC.
- They provide some chiral and structural information.
- The choice of the solvent and the buffers of dilution does not give rise to great problems. Let us cite borate or ammonium salts.

They also present some disadvantages:

- They seem to bring less information on the structure of proteins than do diffractometry of X-rays and N.M.R.
- Interferences with dioxygen must be avoided by using devices fed with nitrogen:
  i. The linearity range is limited;
  ii. The sensibility is moderate.

Structural Chemistry

Its interest, in this domain, appears clearly when one considers its applications not only in the field of organic chemistry but also in those of biological sciences and inorganic chemistry where it has been employed to elucidate the structure of some metal complexes. Circular dichroism is used in conjunction with optical rotatory dispersion in establishing the configurations of different chemical substances. The practical interest of the curves of rotatory dispersion and circular dichroism is notably demonstrated by the examples given under.

They show that there exist relations between the positions of substitutions of molecules, their orientation in series of analogous compounds which may help to elucidate their structures by comparison of their curves of rotatory dispersion and of circular dichroism.

Concerning the curves of rotatory dispersion, in a region where the wavelength is far from the absorption band, the molecular rotatory power is related to the wavelength of the incident radiation by Drude's equation:

$$[\phi] = \text{constant}/(\lambda^2 - \lambda_0^2)$$

The obtained rotatory dispersion curve $[\phi]$ or $[\alpha]$ as a function of λ is then a *normal curve*. $[\phi]$ increases (in absolute value) when λ decreases and $[\phi]$ might be positive or negative. Such curves are generally poorly interesting but can help to distinguish two enantiomers. In Figure 4, we have reproduced the dispersion rotatory curves of the two isomers of the 17-hydroxyandrostane. Figure 9 shows the UV spectrum and the dichroic absorption of 3β-hydroxyandrostane-17-one.

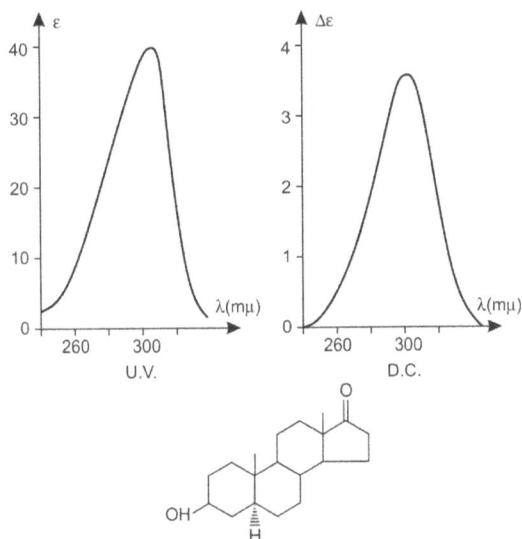

Figure 9: UV Spectrum and Dichroic Absorption of 3β-Hydroxyandrostane-17-One.

In contrast, when the measurements can be drawn through the absorption band of the active substance, the maximum of which is located at the wavelength λ_0, the rotatory power and the dichroic absorption exhibiting their most characteristic and explicit curves. One says that the substance exhibits a Cotton effect. The dispersion rotatory curve then shows curves called *abnormal curves*. But near λ_0, circular dichroism permits discovering hidden absorption maxima more easily than with the curves of rotatory dispersion.

If we compare now the circular dichroism and the rotatory dispersion, the curves of the former method are less disturbed than those of the second by the presence of supplementary chromophores in the molecule under study and are easier to explain.

It is not possible, here, to review all the applications of the rotatory dispersion and of the circular dichroism devoted to stereochemical analysis. We only present the interest of these two techniques in the case of derivatives possessing a group carbonyl. The chromophore C=O has been particularly studied in this domain. There are several reasons:

- Their absorption band is located near 300 mµ in the domain of wavelengths covered by the apparatuses of circular dichroism.
- Their absorption is relatively weak and do not disturb the measurements.
- There exist numerous carbonylated derived.

From the structural viewpoint, the Cotton effect, the circular dichroism and the rotatory dispersion permit easily distinguishing the geometrical and optical isomers.

It is probably in the steroid series that the interest of these methods is the most remarkable. In a typical steroid molecule, the 17-substituted androstane, there exists a chain of 7 asymmetric carbons (Figure 10). Whatever the position where a group ketone may be located, it cannot be located more than one carbon atom from this chain. The synthesis of these ketones has been carried out. We only

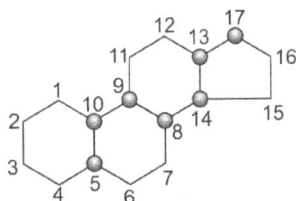

Figure 10: Structure of 17-Substituted Androstanes with its 7 Asymmetric Carbons.

investigate two types of modifications; those concerning the hydrogen atom in 5 and the substituent in 17.

From another standpoint, by way of example, we are also interested by the rule of octant.

Steroid Series

Figure 4 concerns with the 17-hydroxyandrostane, which is a normal curve of rotatory dispersion. It shows a means to distinguish two enantiomers. It also shows that these two enantiomers have a stereochemistry of the kind trans-decalin.

One knows that cyclohexane can exist without any tension. There are the two forms chair I and boat II. Although it is impossible to isolate these two forms, it is well-established that it is the form chair which is the most stable from the thermodynamic standpoint in the absence of any constraint.

Concerning now other steroids, it is interesting to recall that in this series the cyclohexane rests can take the forms chair I and boat II as those which are the most stable. This has for result the fact that at the joining between cycles A and B, there occurs a possible isomerism cis-trans. In this case, it is named an isomerism of the kind trans-decalin and cis-decalin (Figure 11).

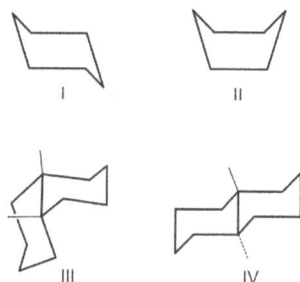

Figure 11: Forms Chair-Boat I and II and Chair Form in the Cis-Decalin III and the Trans-Decalin IV.

The nomenclature β corresponds to a joining trans-decalin, whereas α corresponds to the joining cis-decalin. The joining of two cycles cyclohexane does not induce constraints. This permits to both cycles to adopt the chair form in the cis-decalin III and in the trans-decalin IV (Figure 11).

But, in the steroid series, we encounter two kinds of structural isometry:

- The optical isomery;
- The geometrical isomery.

The studied steroids are classified into two groups according to their decalin form and their type of Cotton effect:

- Joining Trans-Decalin:

   a: Cholestane 1 – one, Cotton effect positive;

   b: Cholestane 2 – one, Cotton effect positive;

   c: Cholestane 3 – one, Cotton effect positive;

   h: 5α-Androstane 17 β-ol 3-one, Cotton effect positive.

   d: Cholestane 4 – one, Cotton effect negative;

   e: Cholestane 6 – one, Cotton effect negative.

- Joining cis-decalin:

   f: Coprostane 4 – one, Cotton effect positive;

   g: 5β-Androstane 17 β-ol 3-one, Cotton effect negative;

Structures and curves of rotatory dispersion of steroids are given in Figures 12, 13, 14 and 15.

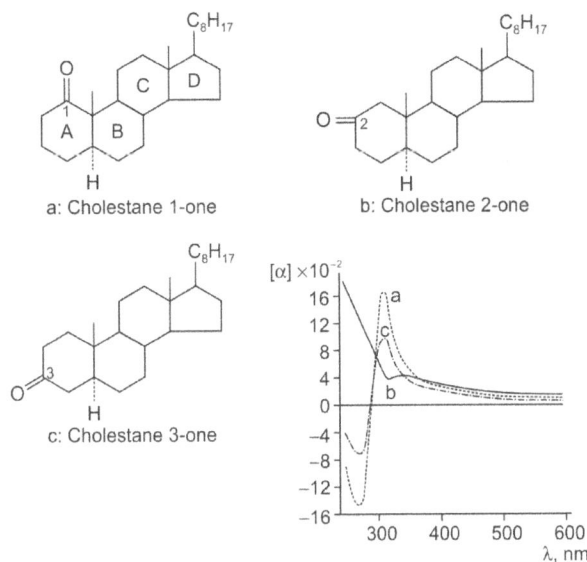

Figure 12: Structures of the Steroids a, b, c and their Curves of Rotatory Dispersion.

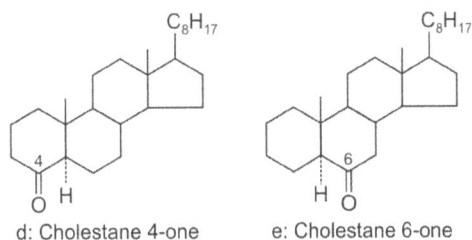

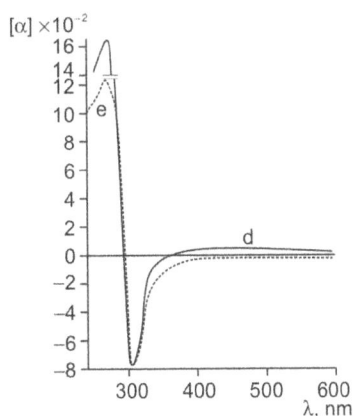

Figure 13: Structures of the Steroids d and e and their Curves of Rotatory Dispersion.

Rule of the Octant: Study of Monocyclic Ketones

Numerous monocyclic ketones are optically active. Many have been studied by rotatory dispersion and by circular dichroism. Here, we give some results concerning their stereochemistry and we show how the octant rule justifies them. We only consider the case of cyclohexanones. This rule can be applied to other kinds of series of compounds. We shall see with this rule that these methods are able not only to distinguish different isomers (which are configurations) but also to distinguish among different conformations.

C_8H_{17} C_8H_{17}

f: Coprostane 4-one   d: Cholestane 4-one

$[\alpha] \times 10^{-2}$

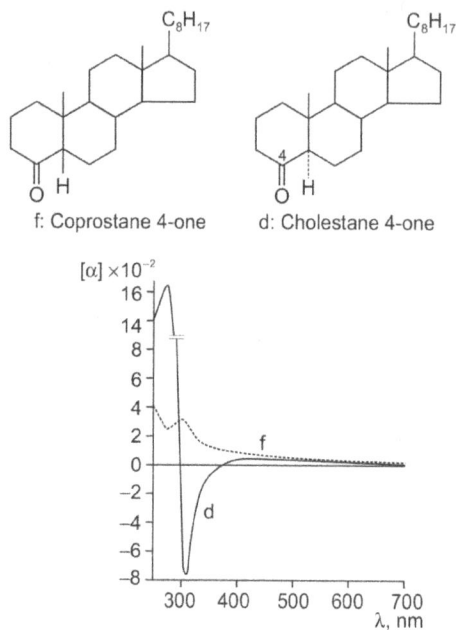

Figure 14: Structures of the Steroids d and f and their Curves of Rotatory Dispersion.

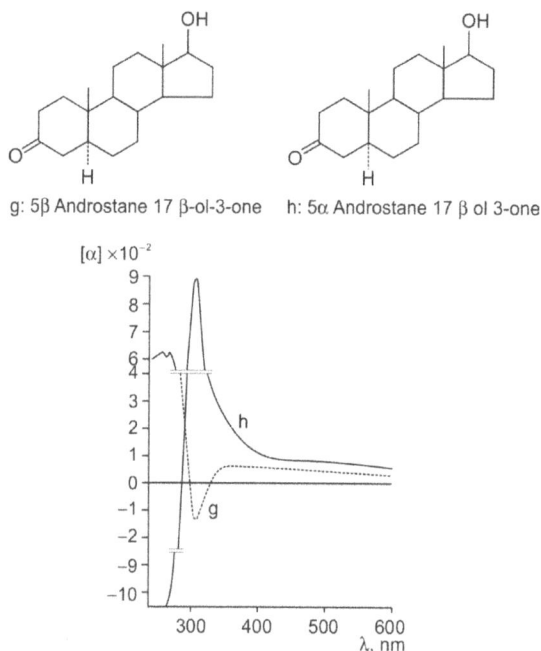

OH   OH

g: 5β Androstane 17 β-ol-3-one   h: 5α Androstane 17 β ol 3-one

$[\alpha] \times 10^{-2}$

Figure 15: Structures of the Steroids g and h and their Curves of Rotatory Dispersion.

The Geometry of the Cyclohexanones

Let us consider the cyclohexanone (Figure 16) and chose the carbonyl group as the reference.

- Plane A is vertical. It contains the carbon 1 of the carbonyl and the carbon 4;
- Plane B is horizontal. It contains carbon 1 and the two carbons C2 and D2 adjacent to the former (Figure 16). The substituents of these two carbons located in the equatorial position are practically in the B plane;

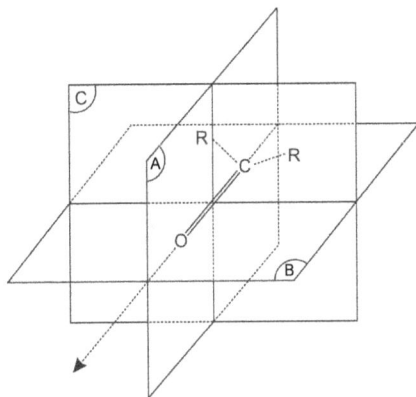

Figure 16: Structure of Cyclohexanone.

- Plane C is normal to planes A and B. It intersects the carbonyl group. In the plane projection on C of the plane B, the trace of the latter is constituted practically by the line C1G2D2 and contains the equatorial substituents of G2 and D2. Plan A is constituted by the axis C1C4 and contains the substituents in C4 (Figure 17).

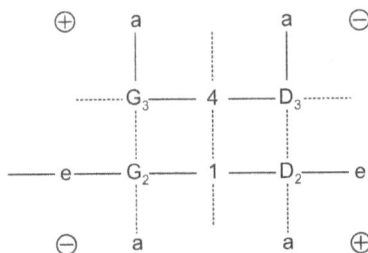

Figure 17: Placing a Cyclohexanone with Respect to the Octant Rule. Octant Rule and the Three Planes of Symmetry of Cyclohexanone.

It is interesting to locate the eight parts of the space which constitute the octant. When we frontal look the system along the axis C1 → C4 (Figure 18).

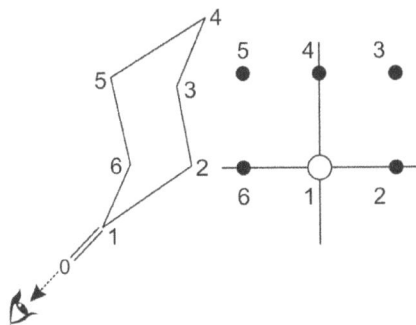

Figure 18: The Eight Parts of the Octant.

Statement of the Octant Rule

Let us only consider the space located at the right of C_1. The rule of octant can be expressed as follows:

- The substituents contained in planes A and E eq (that is to say the substituents of C4 and the equatorial substituents of D2 and G2 do not contribute to the Cotton effect);

- The substituents located above the left octant (substituents of G3) and under the right octant (axial substituent of D2) tend to create a positive Cotton effect;
- The other substituents located in the under left octant (axial substituent of G2) and in the above right octant tend to create a negative Cotton effect.

In the case of compounds more complex than a cyclohexenone, the octants are located on the left of the plane C. Then, the signs of the effects are reversed. By example, a substituent located in the right under octant gives a negative effect (Figure 18).

Applying of the octant rule to others cyclohexanones such as the case of the (+)3-methyl-6-chlorocyclohexanone.

Some physical measurements have shown that the (+)3-methyl-6-chlorocyclohexanone is under one of the two following conformations. The conformational analysis foresees that the methyl radical is located in an equatorial position (a position). The octant diagram shows that conformation *a* is connected to a positive cotton effect whereas a negative effect is connected to conformation *b*. A positive effect is observed. It is consistent with the conformation *a* (Figure 19):

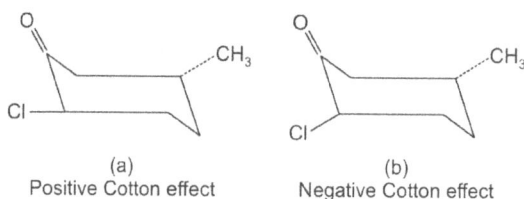

(a)
Positive Cotton effect

(b)
Negative Cotton effect

Figure 19: Dominant Conformation of the (+)3-Methyl- 6-Chlorocyclohexanone. Conformation (a) is Connected to a Positive Cotton Effect and (b) to a Negative Cotton Effect.

Biochemistry

Circular dichroism has also applications in biochemistry. They notably concern the determination of the secondary structure of proteins.

- It can be used when molecules under study are optically active; they can be biological molecules such as proteins, nucleic acids, polysaccharides, some medicines with their primary structure, the identification of the isomers and the analysis of all the chiral molecules.
- Macromolecules, the secondary structure of which plays a part on their dichroism: the structures in helix α the conformations β of proteins and in double helix of nucleic acids (Figure 20).

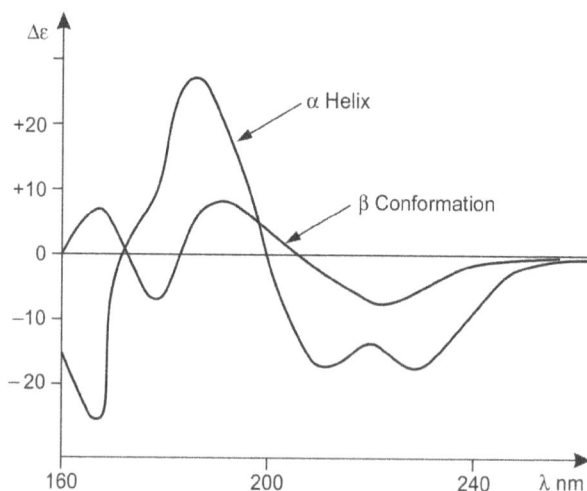

Figure 20: Secondary Structures of Proteins: Helix α and β Conformation.

- The study of the interactions between and active principles inside the cells.
- The effect of the environment on the structure of a molecule.
- The study of the interactions protein/protein and nucleic acids/proteins.
- The study of denaturations.

CHAPTER 33

Introduction to the Atomic Absorption and Emission Spectroscopies

In this chapter, we investigate some points that spectrometries of atomic absorption and emission have in common. We are interested in:

- The invoked transitions in both atomic emission and absorption.
- The different steps of the analysis according to the principle of the chosen method.
- The criterion of the choice of one of these two methods.
- Their quantitative and practical aspects.

In this chapter, when we speak of atomic absorption, we only think atomic absorption, whereas when we speak atomic emission, we think flame emission but also arc, spark spectrometries and inductively coupled plasma–atomic emission spectrometry (ICP-AES inductive coupled plasma-atomic emission spectrometry spectra). Some further points concerning all these methods are given in the next chapters.

The atomic mass spectrometry, whose relationships with present methods are evident, will be studied in one chapter still further.

Generalities

The spectroscopies or spectrometries of atomic absorption or emission are qualitative and quantitative methods essentially applicable to inorganic analysis. They are based on the absorption or emission of luminous radiations by a population of atoms in the state of vapors. The improvement of these techniques in particular with the advent of the emission spectroscopy that is named ICP-AES. In addition to lasers, it has permitted the:

- Extending the field of applications to the analysis of numerous elements under the state of traces;
- Favoring the *automatization* of these methods.
- Improving the studies of speciation of the elements, that is to say the determination of the valence state under which the element is in the studied sample.

(*Remark*: It is interesting to know that the elimination of some toxic elements from some surrounding matrices cannot be conceived if they are not under a given valence. For example, the trivalent chrome is easily eliminated from an aqueous medium (one says aqueous matrix) contrarily to the chrome hexavalent.)

The atomic spectral methods obligatorily entail two steps:

- The *vaporization* of the ions or the atoms forms atomic vapors.
- The exposure of the atoms (formed in this manner) to a source of *energy* (heat, light) to bring the vaporized atoms in the state of electronic excited atoms in the case of the spectrometry by emission. The study of the obtained spectra permits characterizing the element by the discovery of its absorbed (or emitted) luminous rays and to determine its concentration according to the intensities of the latter ones.

Origin of the Transitions

Experiments of Kirchhoff and Bunsen

The experiments of G. Kirchhoff and R.W. Bunsen (originally carried out on glowing gases, more than one century ago; 1859–1860) explain the principles of the absorption and emission atomic spectroscopies (Figure 1). They show that atoms can absorb the radiations that they can emit.

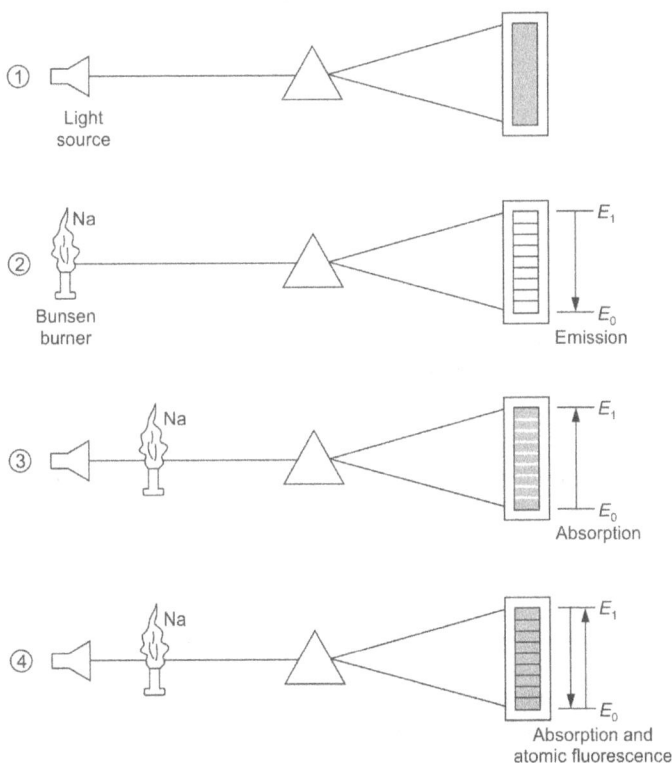

Figure 1: Experiments of Kirchoff and Bunsen.

In order to interpret this experiment, let us start an experiment of atomic absorption in which the element sodium is implicated. In brief, for doing it, we;

- Produce a beam of radiations whose formation is due to the element to study, (which in the occurrence is the element sodium);
- Produce an atomic vapor of this element;
- Make the luminous beam pass through the atomic vapor which absorbs a part of the beam;
- Determine the percentage of absorption by the measurement of the weakening of the luminous intensity which is a function of the concentration in the atomic vapor, at least in some well-

defined conditions. (The weakening is, on principle, only measured in the case of atomic absorption and not in the case of emission spectrometry).

Experimentally, the production of atomic vapor is obtained with the help of a flame in which is pulverized a solution of the element to determine. (One uses the equipment sprayer-burner in photometry by emission).

The similarity (Figure 1) of steps of both methods may indicate that their mechanism is the same. It is not true. Specifically, the phenomena which are the same are those occurring in the flame. Indeed, in photometry by emission, the excited atoms emit radiations and return to a state of lowest energy than the excited one from where they are coming (the intensity of the emitted radiations is measured).

Moreover, in photometry by absorption, some atoms at the fundamental level of energy, (that is to say non-excited but which are in the vapor state) absorb the radiation given by Bohr's law which is the difference of energy of the two levels of electronic-energy between the fundamental and the first excited one.

- The dispersion of the light emitted by a polychromatic source (glow lamp or electric arc, every system producing an electric current visible in an isolating medium) by a prism furnishes a continuous spectrum 1. (In this spectrum, there are the rays D of sodium).

- The substitution of the luminous source by a Bunsen burner with projection into the flame of a few quantities of sodium chloride gives the emission spectrum of the element sodium formed by rays (flames emission) with at 589 nm a very bright line 2, named ray D by Franhäufer (strengthened drawings in the figure).

- By placing the flame of the Bunsen burner (where the sodium chloride was projected initially) in the optical path of the initial polychromatic radiations, one obtains on the screen one spectrum comparable with that obtained during the Experiment **1** with, however, some dark rays in the places of the emission of the element sodium **3**. This phenomenon is called *reverse of the rays of emission of the element sodium*, which results from the presence in the flame of a large proportion of atoms remaining in the fundamental state which absorb the light emitted by the same excited atoms.

- Finally, with the same assembling, it is possible to obtain, if there are sufficiently excited atoms in the flame, the spectrum of atomic fluorescence of the element (the emitted wavelength corresponds exactly to the absorbed energy 4).

Kirchhoff has thus shown that a compound submitted to some conditions of excitation can emit the radiations that it can absorb.

Interpretation: Fine Structure of the Atomic Emission and Absorption

We know that, in atoms where there are several electrons, the origin of the rays is much more difficult to attribute than in the case of hydrogenoides or hydrogen-like atoms. In particular, the multiplicity is higher in the first case, given the different energy levels of the sub-shells and spin effects present in the first case.

(*Remark*: By definition, hydrogenoides are ions which only contain one electron on their valence shell as does the hydrogen atom. Only, for these species, the Schrödinger equation can be solved analytically.)

In order to interpret the spectra, the coupling of the *orbital kinetic moments of electrons* and those of *spins of electrons* must be considered. The motion of an electron can be represented by a vector called orbital kinetic moment. Its module and its orientation in space are quantified and are under the dependence of the atomic quantum numbers n, l and m, n is the quantum number principal, l is that of the angular moment and m is the magnetic quantum number. Its length is equal to

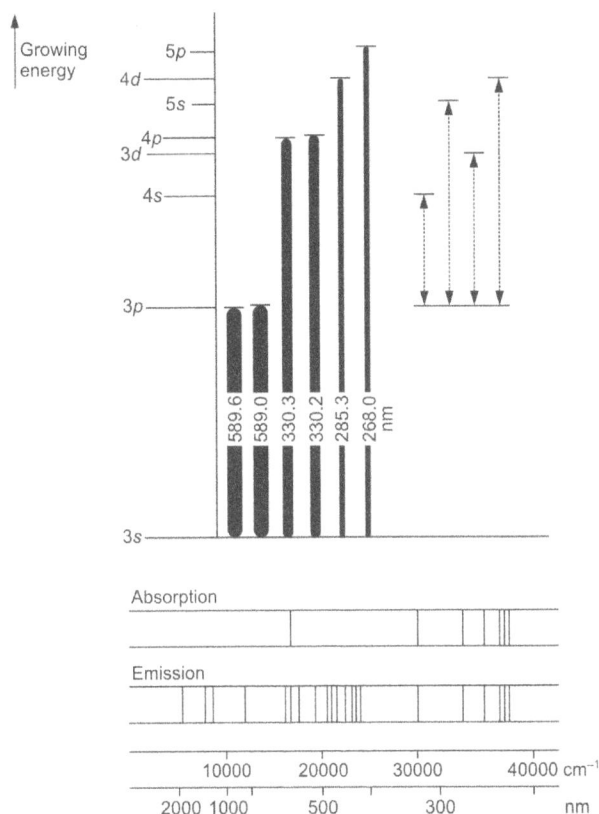

Figure 2: Electronic Transitions of the Sodium Atom and Spectra of Absorption and Emission (According to the German Pharmacopoeia). (Reprinted with Permission from Gwenola and Jean-Louis Burgot, Paris, Lavoisier, Tec et Doc, 2017, 476).

$[l(l + 1)]^{1/2}$ in units $\bar{h}(= h/2\pi)$. Likewise, a satisfactory representation of spin (for our purpose) is the fact that the electron is animated by a motion of rotation about itself.

One connects the kinetic moment of spin with this motion. It is quantified. Its length is equal to $[s(s + 1)]^{1/2}$ in units $\bar{h}$. s is the quantum number of spin. It can take only the values $+1/2$ and $-1/2$. Two vectors representing the same physical phenomenon (displacement of an electrical charge) can be added together. One says that there is a coupling spin-orbit. The resulting vector is called *total angular moment of the electron*. It can also only take quantified values. The quantum number governing it is J. Hence, the total angular moment of the electron is defined by the quantum numbers l and *j*.

$$J = l + s$$

$$J = l + \tfrac{1}{2} \quad \text{and} \quad J = l - \tfrac{1}{2}$$

In the first case, both kinetic moments have the same sign. In the second, they are opposite. For example:

- $l = 0$ (electron s): $J = s$:
- $l = 2$: $J = 5/2$ and $J = 3/2$ (see later)

- Case of One Electron.

Let us take the case of sodium. Its electronic configuration in the fundamental state is: $[Ne](1s^2 2s^2 2p^6 3s^1]$. Now, let us be interested by the transition $3p^1 \leftarrow 3s^1$. It is important from the analytical standpoint. Firstly, we notice that this transition exists because the sub-shells 3s and 3p have different levels of energy. This transition is objectified by two rays called D_1 and D_2. The theoretical interpretation, legitimately, begins by the neglecting of the electrons filling the shells the most profound. This is a good approximation. Thus, we only consider the electron s (l = 0) and spin s = ½ and J = ½. It is the electron 3s, the celibate electron. In the excited state 3p, l = 1, there are two possibilities for J = 3/2 and J = 1/2. The term corresponding to the highest value of the total kinetic moment (J = 3/2) is that of the highest energy. Therefore, there are two possible transitions (Figure 3). They occur at 589.8 nm and 589.2 nm. Generally, the difference of energy between the different levels J is weak for the light elements, but it becomes important for the heavy ones, more precisely for those which possess an important nuclear charge.

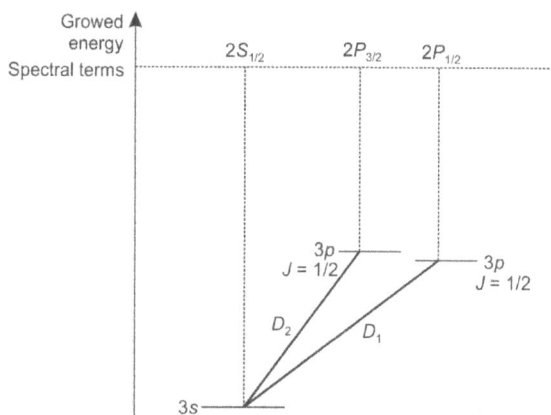

Figure 3: Rays D_1 and D_2 of Sodium and Spectral Terms. (Reprinted with Permission from Gwenola and Jean-Louis Burgot, Paris, Lavoisier, Tec et Doc, 2017, 477).

There exists a symbolism markedly more efficient in order to designate the transitions than that involving J and the electronic levels. This new symbolism introduces new parameters called *spectral terms*. It is based on the coupling so-called Russel-Saunders coupling according which one begins by coupling the orbital kinetic moments of all the peripheric electrons. Therefore, one obtains a global orbital kinetic moment L (a vector). One also couples the kinetic moments of spin of the peripheric electrons. One obtains the kinetic moment of spin global S (a vector). Then, one proceeds to the coupling of L and S. For example, for two peripheric electrons with the orbital kinetic moments l_1 and l_2 and with the kinetic moments of spin s_1 and s_2:

$$L = l_1 + l_2 \quad \text{(in modules)}$$

$$S = s_1 + s_2 \quad \text{(in modules)}$$

When L = 0, the corresponding term spectral is symbolized by S.

Do not confuse the symbol S with the total spin S. Here, S comes from sharp.

When L = 1, L = 2, the spectral terms are P, D (by analogy with the layer s, p, when l = 0,1,2 ...). The "multiplicity" of the transition is given by the formula 2S + 1 where S is the total spin. When S = 0 (full shell- noble gases), there is only one term spectral possible since the multiplicity is one. For S = 1, the multiplicity is 3. From a general standpoint, the total angular moment J is given by the relation (in the module):

$$J = L + S, L + S - 1 \ ...., |L - S|$$

Each spectral term comprises:

- The letter S, P and D indicate the total orbital moment;
- One exponent on the left gives its multiplicity;
- One index on the right gives the total value of J.

For sodium,

- In the fundamental state ($3s^1$), there is only one electron to consider, $l = 0$, $L = 0$ (from which the symbol S). $s = 1/2$, value of the spin total $S = 1/2$ and multiplicity $2(1/2) + 1 = 2$. Finally, $J = 0 + 1/2 = 1/2$. The spectral term is $^2S_{1/2}$.
- In the excited state ($3p^1$), $l = 1$, $l = 1$ (P), spin total value $S = 1/2$, multiplicity 2, $J = 1 + 1/2$, $J = 3/2$ and $J = +1/2$. The two transitions (rays D_2 and D_1) are then:

$$^2P_{3/2} \leftarrow {}^2S_{1/2} \quad \text{and} \quad {}^2P_{1/2} \leftarrow {}^2S_{1/2} \quad (\text{viz., Figure 3})$$

The simple case in which the outer electron is raised by one energy level and then returns into its initial level is known as corresponding to the *resonance absorption*. The atomic absorption is based on this phenomenon.

- Case of Several Electrons

Another interesting example in atomic absorption is the magnesium and more particularly the transition $3p \leftarrow 3s$. Let us recall that in the fundamental state, the electronic structure of magnesium is $[Ne]3s^2$.

In the fundamental state, $l_1 = 0$, $l_2 = 0$, $L = 0$ (S), $s_1 = 1/2$, $s_2 = -1/2$, value of the spin total $S = 0$, multiplicity 1; $J = 0$. The spectral term is $^1S_0$.

In the excited state ($3s^1, 3p^1$):

on one hand $s_1 = 1/2$, $s_2 = 1/2$, value of the spin total $S = 1$, multiplicity $= 3$, $l_1 = 0$, $l_2 = 1$, $L = 1(P)$: $J = 2, 1, 0$ from which the spectral terms $^3P_0, ^3P_1, ^3P_2$

on the other $s_1 = 1/2$, $s_2 = -1/2$, spin total $S = 0$, multiplicity $= 1$, $L = 1(P)$, $J = 1$.

The transitions $3p \leftarrow 3s$ are, hence, in terms of spectral terms:

$$^1P_1 \leftarrow {}^1S_0 \quad \text{and} \quad {}^3P_0 \leftarrow {}^1S_0$$
$$^3P_1 \leftarrow {}^1S_0$$
$$^3P_2 \leftarrow {}^1S_0$$

From the three transitions singlet/triplet which should be envisaged, only the transition $^3P_1 \leftarrow {}^1S_0$ is permitted given the selection rule which imposes $\Delta J = \pm 1$ when $j = 0$. Definitively, there are two rays $3p \leftarrow 3s$; the rays $^1P_1 \leftarrow {}^1S_0$ and $^3P_1 \leftarrow {}^1S_0$. The transitions $^3P_0 \leftarrow {}^1S_0$ and $^3P_2 \leftarrow {}^1S_0$ are forbidden by the selection rule. Notice also that among the two remaining authorized transitions, $^1P_1 \leftarrow {}^1S_0$ and $^3P_1 \leftarrow {}^1S_0$, the singlet/triplet one is by far less probable than the other which does not necessitate a change in the spin state. Figure 4 summarizes these considerations.

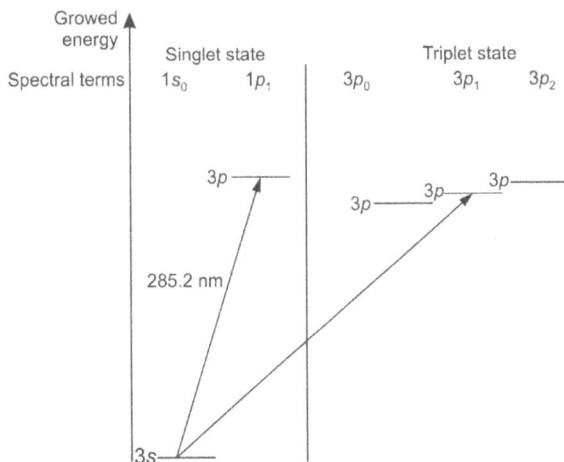

Figure 4: Transition $3p \leftarrow 3s$ for the Magnesium. Notice that $^3P_1 < {}^1P_1$ From the Standpoint of Energy. (Reprinted with Permission from Gwenola and Jean-Louis Burgot, Paris, Lavoisier, Tec et Doc, 2017, 479).

The Different Steps of an Analysis by Atomic Spectroscopy

The methods implicate the entire destruction of the molecules containing the element to be determined in order to obtain the free atoms which must be in the state of vapors. Let us consider for example sodium chloride. It must undergo the following steps before obtaining both elements sodium and chlorine under the form of atomic vapors;

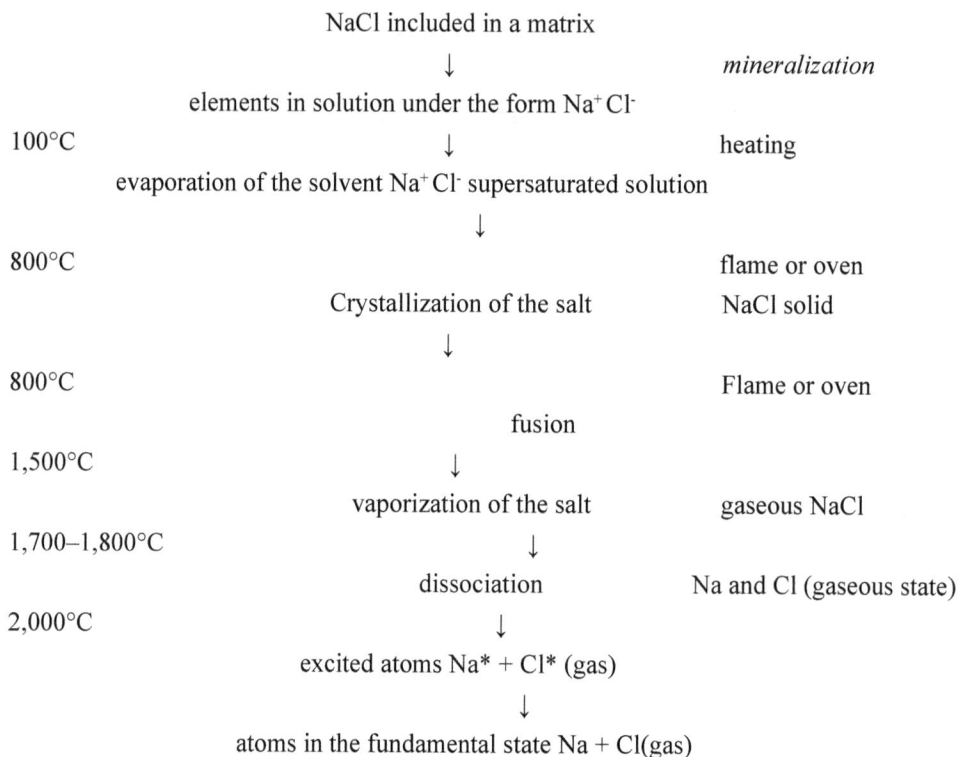

NaCl included in a matrix

$\downarrow$ *mineralization*

elements in solution under the form $Na^+ Cl^-$

100°C $\downarrow$ heating

evaporation of the solvent $Na^+ Cl^-$ supersaturated solution

$\downarrow$

800°C   flame or oven

Crystallization of the salt   NaCl solid

$\downarrow$

800°C   Flame or oven

fusion

1,500°C $\downarrow$

vaporization of the salt   gaseous NaCl

1,700–1,800°C $\downarrow$

dissociation   Na and Cl (gaseous state)

2,000°C $\downarrow$

excited atoms Na* + Cl* (gas)

$\downarrow$

atoms in the fundamental state Na + Cl(gas)

It is important to notice the double part played by the flame; it plays a part in the process of vaporization of the element and in the process of its excitation once it is vaporized.

Criteria of Choice Between the Atomic Absorption and Emission

The choice of the method (absorption or emission) is a function of the ratio number of excited atoms/number of atoms remaining in the fundamental state (once vaporized). This ratio is given by the law of Maxwell-Boltzmann:

$$N_e/N_o = g \exp(-\Delta E/kT)$$

Here, N_e is the number of atoms in the excited state, N_o the number of atoms in the fundamental state, T the absolute temperature (K), g ratio of the statistical weights of the excited and fundamental states, ΔE is the difference of energy between the excited and the fundamental levels, k is the Boltzmann's constant (= $1.380658 \ 10^{-23}$ J K^{-1}). g is given by the expression:

$$g = p_e/p_o$$

p_e and p_o are also statistical factors. Their values are determined by the number of states having equal energies for each quantum level. These parameters are systematically encountered in statistical thermodynamics.

This ratio changes with the working temperature and the nature of the elements. Thus, for example:

Na (λ = 589 nm)   T = 2000 K N_e/N_o = 1.305 10^{-5}
Cu (λ = 325 nm)   T = 2000 K N_e/N_o = 4.77 10^{-10}

The obtained values indicate that the atoms remain in their major part in their fundamental state. This fact would justify the quasi-systematic use (in this case) of atomic absorption. The measurement of the absorbance of the radiations is delicate since locating the absorption rays is difficult because of the presence of numerous interferences existing at these high temperatures. The monochromators have not yet sufficient resolution to eliminate them. This is the reason why one prefers working in emission. However, in order to obtain a good sensitivity, a ratio $N_e/N_o < 10^{-7}$ may be retained. This result is easily obtained in flame photometry with sodium, potassium and lithium at temperatures of 2,000 K. As a result, the choice between emission and absorption can be made according to the following criterion:

$N_e/No > 10^{-7}$ $\rightarrow$ Flame emission,
$N_e/No < 10^{-7}$ $\rightarrow$ Atomic absorption.

Quantitative Aspects

Flame Atomic Absorption (Atomic Absorption Spectrometry)

The quantitative measurements are based on a law that formally looks like that of Beer-Lambert. It relates the incident I_0 and transmitted I intensities. It is an empirical fact that the transmitted intensity obeys an exponential relation of the type:

$$I = I_0 \exp(-K/N_0)$$

This law of the type of Beer-Lambert is only valuable in rather narrow domains of concentrations. From the existence of this relation, three techniques are used:

- The external calibration which consists of relating the absorbance A = log(I_0/I) to a concentration through a calibration curve A = F(C) ; (F ( ) : function of) with a standard. This function can be a straight line in a narrow domain (of concentrations). This technique supposes a good knowledge of the medium to analyze. Obtaining sufficiently specific results implicates that the standards should be diluted in media of compositions as close as possible to that of the matrix in order to limit the interferences;

- The method of the standard addition: According to the principle of this method, known and increasing quantities of the element to dose are added to the sample. The straight-line (absorbance/added quantities) intersects the x-axis at a distance from the origin equal (absolute values) at the sought concentration. For example, let us take three tubes containing the same volume v of the solution which must be titrated. In volume v, there are q_o moles of the compound. We add increased and known volumes v_1, v_2 and v_3 of a standard solution of the compound to be measured. In other words, we add quantities q_1, q_2, and q_3 moles of the compound to be dosed. We complete to the same total volume V with the same solvent as that used in the addition of v_1, v_2, and v_3. The concentrations in Tubes 1, 2 and 3 are:

$$C_1 = q_1/V + q_0/V$$

$$C_2 = q_2/V + q_0/V$$

$$C_3 = q_3/V + q_0/V$$

One measures the absorbances of A_1, A_2, and A_3 (Figures 5 and 6). Since the equation of the straight line is:

$$A = q/V + q_o/V$$

When A = 0

$$-q/V = q_o/V$$

- The method of the internal standard: One carries out one calibration curve of the element to determine. One adds a concentration C' of another element judiciously chosen into the solutions which will serve to trace the calibration curve. One also adds the concentration C' to the sample.

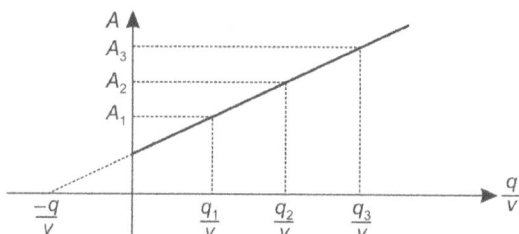

Figure 5: Absorbances as a Function of the Known Concentrations q_1/V; q_2/V; q_3/V of the Element to Determine it. (Reprinted with Permission from Gwenola and Jean-Louis Burgot, Paris, Lavoisier, Tec et Doc, 2017, 483).

Then, one draws the curve (a straight line as given in Figure 5) of calibration A/A' = F(C_x/C') where A_x is the absorbance of the product and A' that of the internal standard and C_x and C' the concentrations of the sample and the standard. Carrying the value of the ratio A_x/A' obtained with the studied solution over the calibration curve directly leads to the ratio:

$$C_x/C'$$

As a result, C_x is known, since C' is known. In this case, the reached precisions can be very good if the internal standard is well-chosen: neighboring wavelengths of absorption and likewise for their redox potentials for example (Figure 6):

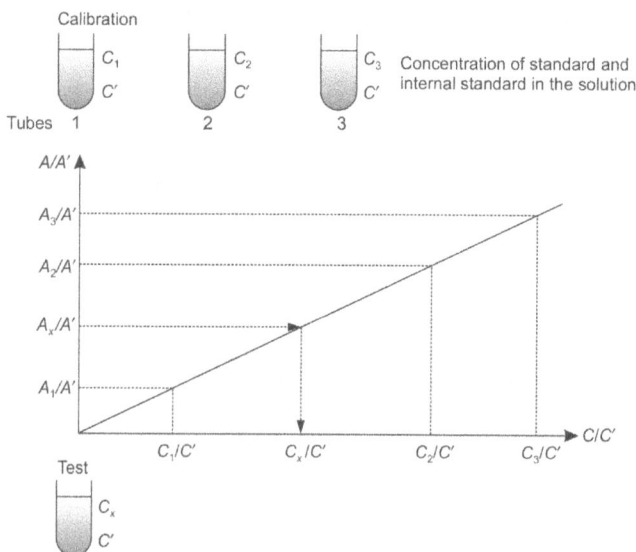

Figure 6: Method of the Internal Standard. (Reprinted with Permission from Gwenola and Jean-Louis Burgot, Paris, Lavoisier, Tec et Doc, 2017, 484).

The method of the internal standard is based on the same basic principle as the technique of the pilot-ion in polarography and that of the internal standard in chromatography.

Imperatively, the luminous source must emit the two wavelengths specific of both elements since it is impossible to envisage the change of lamp during the measurement. Unfortunately, the technique is applicable with difficulty since the luminous source remains specific of each element. This is why one rather uses the method of standard additions.

Quantitative Aspects in Atomic Emission

As previously, one supposes that the element to determine is in solution. It is an experimental fact that the intensity of light emitted is proportional to the concentration of the element (at least in a narrow domain of concentrations):

$$I = KC$$

As in absorption, it is possible to use the methods of the external calibration, of standard additions and that of internal standard.

Let us mention a variant of the calibration curve when one has one order of magnitude of the concentration of the sample. One prepares two standard solutions of concentrations C_1 or C_2 framing C_0. Let I_1, I_2, and I_0 the corresponding intensities. Given the relation of proportionality existing between the intensity and the concentration, it is possible to write:

$$I_2 - I_1 = K (C_2 - C_1)$$

$$I_0 - I_1 = K (C_0 - C_1)$$

And $(C_0 - C_1) = (C_2 - C_1)(I_0 - I_1)/(I_2 - I_1)$

C_0 is calculated *via* the relation:

$$C_0 = C_1 + (C_2 - C_1)(I_0 - I_1)/(I_2 - I_1)$$

Practical Aspects

The preparation of the samples is changing as a function of the nature of the matrix and the elements to be determinate. For example, one can carry out:

- A decomposition of the organic matter in acidic medium (HNO_3, HCL and H_2SO_4) and action of oxidative reactants (H_2O_2). The decomposition is never complete. This point can be tricky with some techniques such as ICP-AES with ultrasonic nebulization and with ICP-MS (terms defined in the following chapter). It is also reserved to the case for which the carrying out of the mineralization is impossible. It is important beforehand to validate the ratio volume of reactant/ volume of the sample. Nitric acid does not pose any problem. In contrast, hydrochloric acid provokes chemical interferences in atomic absorption with furnace Electrothermal Atomic Absorption (ETAAS) and sulfuric acid, liquid, viscous, leads to risks of errors in the techniques in which the sample is introduced by a system of nebulization. The preparation of the sample by wet process must be avoided when there is nebulization by ultrasonic process.

- One mineralization, that is to say, a calcination of the sample of organic nature at 450°C followed by a taking back of the ashes under a weak volume of an acid solution. The addition of some agents such as the nitrate or the oxide of magnesium makes easy the calcination of arsenic, selenium or iron by decreasing their volatility under the form of a salt. The dry process offers the advantage of wholly mineralizing the sample but some elements volatilize at temperatures above than 450°C.

- One extraction in an organic phase under the form of a chelate in order to eliminate the interfering substances and also in order to concentrate these elements often initially present in the state of traces.

Some experimental conditions must be respected in order to carry out a good analysis. Therefore, the samples in solutions (waters, beverages, biological liquids) are acidified to avoid the adsorption of the elements-trace on the sides of the vessel.

The use of the acids is one part of all the protocols of preparation of the samples. Therefore, it is necessary to achieve the standard solutions in the same conditions. This implicates to deal with the quantity of acid consumed for the mineralization. When this information is difficult to obtain, it is better to preferentially chose a technique which is weakly influenced by the acidity of the medium as the spectrometry of atomic emission in an argon plasma ICP-AES with an ultrasonic nebulizer.

Performances of the Analytical Techniques

The performance of the atomic spectroscopy depends on the element and the medium to analyze. It is possible to give a rough estimate of the value of limit detection.

FAAS: Flame absorption atomic spectrometry 1 to 100 $\mu g.L^{-1}$

ICP-OES: Inductively coupled plasma optical emission spectroscopy 0.1 to 100 $\mu g.L^{-1}$

GFAAS: Graphite furnace absorption atomic spectrometry 0.001 to 5 $\mu g.L^{-1}$

ICP-MS: Inductively coupled plasma coupled to the mass spectrometry 0.00001 to 1 $\mu g.L^{-1}$

These techniques are developed in the following chapters.

CHAPTER 34

Atomic Absorption Spectrometry

Atomic absorption is a method used for the quantitative analysis of a great number of chemical elements. It is the oldest spectral atomic method of analysis. It permits the analysis at less expense than the techniques ICP-AES or ICP-MS. This method is registered in the major pharmacopeia. In this chapter, we shall successively study:

- Its principle
- Its apparatus
- Some problems to which it is faced
- Some of its applications
- The advantages it presents

Principle

A monochromatic radiation, emitted by one lamp and corresponding to the ray of the element to determine it (see the preceding chapter) is sent to a population of atoms of the same element in the state of vapor. The measurement of the weakening of the luminous intensity due to its absorption is, in some conditions, a function of the concentration of the element to determine.

Apparatus: Some Components

There exist some analogies between the apparatus of atomic absorption and that of spectrophotometric absorption (Figure 1).

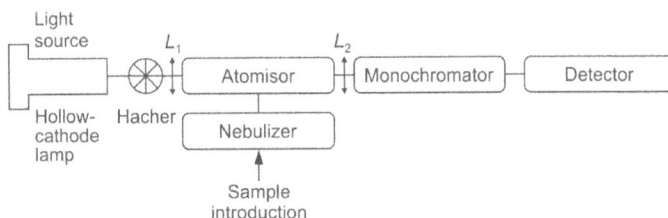

Figure 1: Scheme of an Apparatus of Atomic Absorption. (Reprinted with Permission from Gwenola and Jean-Louis Burgot, Paris, Lavoisier, Tec et Doc, 2017, 488).

Systems of Nebulization

The systems of nebulization transform the sample in a fine and homogeneous mist. Only the finest drops reach the source of energy and thus permit the transformation of the element to be determine into atoms. This means that a great proportion of the sample is lost. The phenomenon constitutes a limit for the analysis of elements trace in weak contents in the sample.

There exist two types of nebulizers.

Pneumatic Nebulizers

The compressed air is coming through a narrow orifice and provokes a depression which sweeps away an aspiration of the solution and its fragmentation (Figure 2).

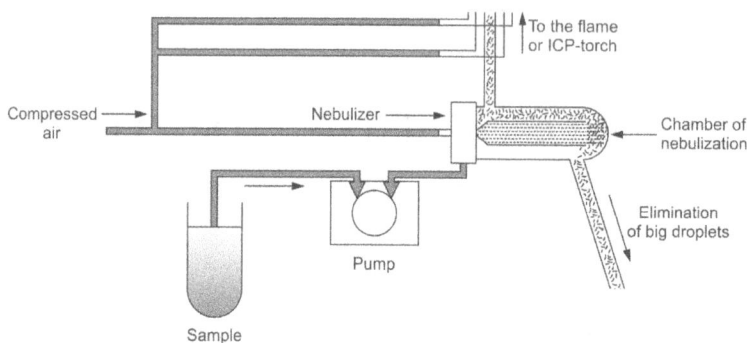

Figure 2: Pneumatic Nebulizer.

However, the device remains crude. The big droplets condense in the chamber of nebulization and only the finest ones (< 10 µm) are swept toward the flame. The general result is that the yield in the element reaching the energy source does not extend beyond 5%.

Ultrasonic Nebulizers

The solution is deposited on a vibrating plate. The ultrasonic nebulizer is more efficient than the pneumatic one since it permits to obtain an aerosol finer than the preceding. But its aerosol is richer in water. This necessitates a particular device in order to avoid a cooling of the source of energy. For a frequency of vibration of the lame of 1 MHz the average diameter of the droplets is lower than 4 µm. This system sweeps away ten times more product than a pneumatic nebulizer but also ten times more constituents of the matrix and impurities.

Atomization of the Element

Flame

It is used in atomic absorption spectrometry in flames (FAAS) for the analysis of samples in solutions. The thermal energy must be sufficiently high in order to atomize the elements; however, without exciting and ionizing them.

The flame is maintained by:

- A source of carburant (gas in bottle under pressure), the outflow of which is regulated by a pressure-reducer.
- A source of comburant (dioxygen in a bottle or compressed air with the help of a compressor the output of which must be stabilized).
- A burner where the gases are mixed and inflamed.

The flame is generally isolated from the surroundings by a kind of chimney which avoids fluctuations due to phenomena of convection of the heat.

The mixtures comburant/carburant the most used are:

- 2,500°C: Air /acetylene flame for the analysis of the following elements Mg, Ca, Fe, Cu and Pb;
- 3,100°C: Nitrous-oxide/acetylene flame for Al, Si and Ta. The nitrous oxide increases the temperature of the flame up to 3,000°C.

The cold flames lead to less parasitic phenomena of ionization but favor the chemical interactions. The reducing flames obtained with the mixtures N_2O/acetylene decrease the formation of oxides or complex, refractory to the atomization. For example, the use of a flame N_2O/acetylene suppresses the interaction between the calcium and phosphate ions.

In order to increase the probability of collision of a photon and of an atom, the produced flames must be long (5 cm to 15 cm) and thin (1 cm). The apparatus is regulated so that the luminous rays pass through its hottest part. All the flames absorb under 230 nm. This can be an inconvenient for the analysis of arsenic (193.7 nm) and of selenium (196 nm). Flame techniques present problems when dealing with either small or solid samples.

(*Remark*: A flame possesses an emission of base with a maximum located between 600 and 700 nm. This phenomenon must be taken into account in the carrying out of the quantitative measurements, even in the absence of a metal in the sample.)

Furnace in Graphite

It is used in electrothermal atomic absorption (ETAAS). It permitted to lower the limit of detection down to the part *per* billion (ppb, part per billion: $\mu g\,L^{-1}$). It is presented in the form of a graphite tube into which the sample is injected with the help of a syringe. The tube is cut longitudinally by a very luminous ray (Figure 3). The atomic vapor is obtained by the joule effect at a temperature between 800°C and 3,000°C. The heating of the tube can be "programmed" according to the operations which must be done. Thus, in this chamber:

- The sample can be dried at 100°C;
- Organic matter can be decomposed at 400°C;
- Mineral compounds can be volatilized at 2,500°C.

An inert gas such as argon flows into this closed chamber to carry the vaporized sample into the atomizer.

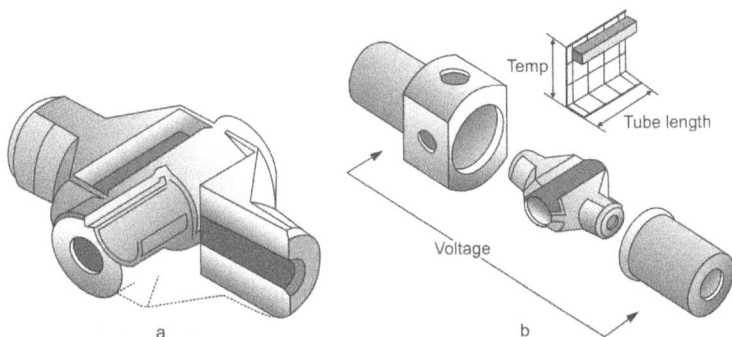

Figure 3: A Furnace in Graphite with a Platform of L'vov (a), Which is Heated Transversally (b). (Reprinted with Permission from of the Society Perkin-Elmer Instrument, 12 Avenue de la Baltique, 91140 Villebon sur Yvette).

The usual system is heated longitudinally but the observed differences in temperatures between the central parts and the extremities have led to ameliorating the system by the introduction of the sample on a platform (so-called L'vov platform) heated transversally.

It is the sole atomic spectral technique in which the sample is integrally introduced into the source of energy. Hence, using a furnace increases the "sensitivity" by permitting the handling of test samples of only some microliters whereas in a flame, it is necessary to introduce several milliliters since the majority of the volume is lost.

However, the furnace presents the drawback to give interferences in relation with:

- The proper emission of graphite starting at about 2,500°C;
- The presence of some elements which react with the carbon of the furnace giving carbides which are responsible for a parasitic emission.

Chemical Vaporization System

This methodology applies to the elements present in a high oxidation state, which are reduced and also vaporized by the flame with difficulty (arsenic, bismuth, tin, selenium and mercury). It is frequently the case for the elements such as trace (some ppb) of environment samples. The process consists in reducing preliminary them and in transforming them into volatile hydrides at ambient temperature.

In order to do that, the sample is acidified and then placed into a reactional flask. Then, a stream of an inert gas eliminates the dioxygen which would destroy the hydride and the elements are reduced by a solution of sodium tetrahydroborate $NaBH_4$ which, according to some authors, would produce atomic hydrogen:

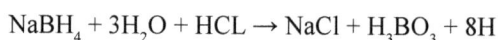

$$NaBH_4 + 3H_2O + HCL \rightarrow NaCl + H_3BO_3 + 8H$$

Atomic hydrogen, a very reactive species, forms hydrides. These ones are swept away into a quartz tube heated at about 1,000°C at which they are decomposed in the atomic state. The tube is lighted by the luminous source (Figure 4). A particular process is used for mercury which, directly (without heating), passes in the atomic state after addition of tetrahydroborate.

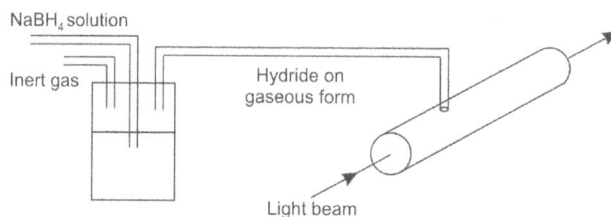

Figure 4: System of Chemical Vaporization. (Reprinted with Permission from Gwenola and Jean-Louis Burgot, Paris, Lavoisier, Tec et Doc, 2017, 491).

(*Remark*: Some recent apparatuses are equipped with two sources of atomization by flame or furnace with an optical system permitting to pass from one to the other without any dismantling and any adjustement.)

Optical Part

Luminous Source

The continuous luminous sources (lamp xenon or deuterium) used in molecular spectrophotometry are not suitable in atomic absorption. Their luminous intensity is dispersed on the whole spectrum, hence their name. This intensity becomes too weak at the characteristic wavelength of the element for a quantitative application. Moreover, in order that all photons coming from the source should be absorbed, it is necessary to have a spectral width $\Delta\lambda$ (a) of the ray of emission inferior to that of the ray of absorption $\Delta\lambda$ (b) (Figure 5). The difficulty is the very strong spectral narrowness of the absorption ray and the order of 10^{-3} nm.

It only removes a minute fraction of the energy passed by a conventional monochromator. The area corresponding to the energy passed by the conventional monochromator is only slightly diminished by the absorption of the narrow line due to the absorption.

Given this fact, an apparatus of atomic absorption does not use continuous sources. A clever device, called *hollow-cathode lamp*, emits the ray with which the element can precisely be determined. A hollow-cathode lamp emits the rays which are exactly the same as those the element must absorb to be determined. These lamps contain the same element as that which is to determine and it is their emission which are the rays that the sample absorb. Therefore, the emitted and absorbed area are identical. The hollow-cathode lamp lam are enclosures (Figure 6 and the

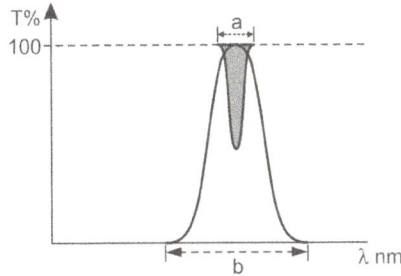

Figure 5: Atomic Absorption of an Element from a Continuous Source.

Photo 7) filled with a noble gas (neon and argon) at weak pressure containing a hollow cylinder built with the element to determine.

The cylinder constitutes the cathode. The window of the lamp is in appropriate material in order to transmit the emitted radiation. It is either quartz or silica. A voltage of the order of 100 to 400 V provokes the ionization of the noble gas, the ions of which go to the cathode and strike it. These ones tear away the atoms from the cathode, which although they stay confined in the enclosure, are excited by the ions Ne^+ by exchange of energy. By returning in their fundamental state, they emit their specific ray (Figure 6 and Photo 7).

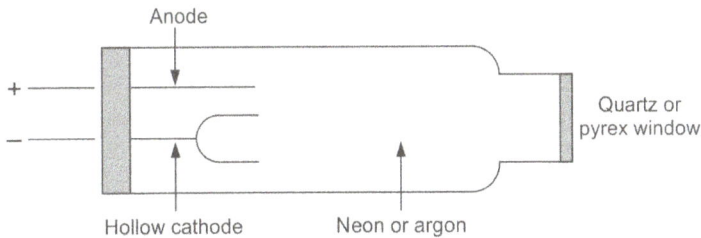

Figure 6: A Hollow Cathode Lamp. (Reprinted with Permission from Gwenola and Jean-Louis Burgot, Paris, Lavoisier, Tec et Doc, 2017, 492).

Photo 7: Hollow Cathod Lamp.

Its working principle necessitates to change the lamp according to the element that we want determinate but this operation is considerably simplified with today's apparatus equipped with several handrails with several lamps. The hollow lamps have a length of life of the order of 1,000 hours. These are also used lamps at discharges without electrode, which supply a luminous radiation 10 to 100 times more intense than those given by hollow lamps but with a widening of the rays which decreases the sensitivity of the method. The electrodes are replaced by a quartz tube containing a salt of the element to determinate and a noble gas at weak pressure, encompassed by a coil transmitting an intensive electromagnetic field. This one provokes the ionization of the gas noble the ions of which excite their specific ray.

Monochromators

The monochromator, placed between the atomizer and the detector, helps to eliminate all the luminous rays the wavelengths of which do not correspond to that of the element. Its constitution is identical to that of monochromators used in molecular spectrometry (see Chapter 20). For example, the usual spectral domain is located between 190 nm and 900 nm (193.7 nm for arsenic and 852.1 nm for the cesium).

Background Effects: Modulator or Hacher

Background effects are the parasitic lights, called interferences or "chemical noises" which come from:

- The excited atoms, in spite of the precautions, which "re-emit" the same ray by falling back into the fundamental state. Such a case would give the "false" impression there is too much beam transmitted. That is to say, one would make an error by default:
- Of the flame under the form of continuum background. It superimposes on the transmitted light.

The goal is to eliminate these parasitic lights. In order to do that, it is necessary to introduce a device playing the part of a constant reference signal. Such a device is called a modulator. There are several kinds of modulators. This part can be played by a double beam optical system. In the double beam device, the radiation is divided by two, one beam is passing through the sample and the other through the blank. Then, both strike the detector. The modulator also contains a chopper, that is to say a rotating wheel and holes. The electronic system in relation to the chopper, sorts out the signals from the two sources and compares them. Both beams are sent through the optical system at the same time but chops differentiate them.

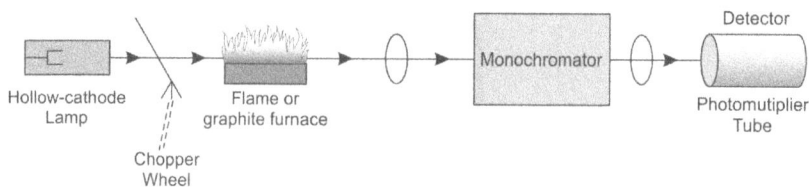

Figure 7: Atomic Absorption Spectrometer with a Chopper for Corrections.

Another possibility is the following (Figure 7). The chopper transforms the incident light, initially continuous in alternative light whereas the electrical modulator sends a pulsed light into the vapor. The amplifier makes the difference between the continuous signal related to the parasitic emissions and the alternative signal related to the presence of the sample in the flame. The sample disappears in the flame whereas the alternative light does not.

- The rays of the element to determine emitted by the lamp which is of no concern in the analysis.
- The rays of the filling gas.
- The rays of eventual impurities.
- The proper emission of the flame.

Configurations of the Apparatus

There exists apparatus with a configuration single-beam according to the scheme described previously and apparatus double-beam (Figure 8) that takes into account the variations of the light source.

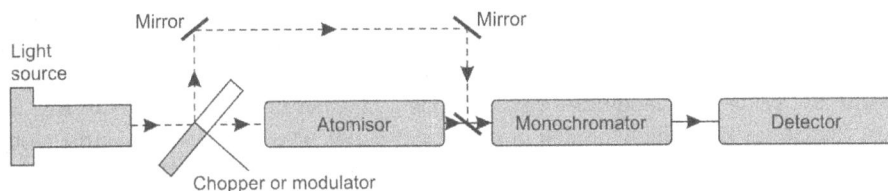

Figure 8: Configuration Double-Beam. (Reprinted with Permission from Gwenola and Jean-Louis Burgot, Paris, Lavoisier, Tec et Doc, 2017, 493).

Both beams are directed towards the entrance of the monochromator.

Problems with Correction of the Background Noise

The observed background noise has several origins. That is to say:

• *Spectral or non-specific interferences* which come from the atomizer. They appear as modifications of the luminous intensity transmitted to the detector. It is the case of:

  • A *superimposition of rays* in which these interferences can be corrected by the Zeeman effect.
  • *Molecular absorptions* due to the existence in the source of atomization of molecules coming from the matrix or products of transformation. For example, the calcium forms radicals CaOH˙, which, even at weak content, disturb the determination of the barium (the presence of 1% of the calcium in a solution absorbs 50% of the light if one considers the phenomenon at 553.6 nm where the calcium absorbs).
  • The diffusion of the incident light on solid or liquid particles coming from the matrix or its decomposition. These interferences can be offset by using deuterium lamps or by the Zeeman effect.

These are background effects that are called also *non-specific effects*.

• *Chemical interferences* or *effects of the matrix* result from chemical parasitic reactions (redox, ionization or dissociation) or the presence of particles in suspension in the source of atomization. They appear as a change of slope of the straight-line absorbance/quantities added in the method of addition standard. The new systems of atomization (with the platform in the furnaces and the addition of a modifier of medium limit this kind of interference).

• *Physical interferences* related to the viscosity of the samples which may be different from that of the standards. This is a problem essentially encountered when the sample is introduced under a liquid form into a flame.

Correction of the Spectral or Non-Specific Interferences

Background effects are the parasitic lights, called interferences or background noise, which come from

  • The excited atoms, despite the precautions, which "re-emit" the same ray by falling back into the fundamental state. Such a case would give the "false" impression there is too much transmitted. That is to say, one would make an error by default.
  • It also comes of the flame under the form of continuum background. It superimposes the transmitted light. The goal is to eliminate these parasitic lights.

Deuterium Lamp

The coupling of the hollow-cathode lamp with a deuterium lamp permits us to eliminate the problem of spectral interferences. The principle consists in the alternative lighting of the flame with the hollow lamp (discontinuous source) and with the deuterium lamp (continuum source). In order to do that, a modulator is inserted between the luminous source and the atomizer. It is constituted by a continuous source (hydrogen or deuterium lamp in supplement to the line-source (the hollow lamp). There are several kinds of modulators. One is that the modulator should be both a mechanical and an electrical modulator. The modulator also contains a chopper or hacher, that is to say, a wheel and holes. Radiations from the hydrogen or deuterium lamps pass through the sample along with the radiations coming from the hollow lamp. The electronic system sorts out the signals from the two sources and compares them. Both beams are sent through the optical system at the same time but chop them at different rates. A significant fraction of metal atoms in the flame or vapor will be raised to excited levels. These atoms will emit resonance radiations in all directions, at the same wavelength as the monochromator can transmit it. If the radiation from the hollow cathode lamp were continuous, there would be no way to distinguish between the spurious radiation and that transmitted from the lamp. Both beams are equally attenuated with respect to background effects, but, the line-source radiation is the only appreciably attenuated in the sample. This result comes from the extreme narrowness of the line of absorption of atomic vapor with respect to that passed by a conventional monochromator.

Another possibility is the following one. The chopper transforms the incident light, initially continuous in alternative light whereas the electrical modulator sends a pulsed light into the vapor. The amplifier makes the difference between the continuous signal related to the parasitic emissions and the alternative signal related to the presence of the sample in the flame. The sample disappears in the flame whereas the alternative light does not hash, is inserted between the luminous source and the atomizer (Figure 9). It is constituted by a continuous source (hydrogen or deuterium lamp) in supplement to the line-source, the hollow lamp. There are several kinds of modulators. One is that the modulator should be both a mechanical and electrical modulator. Modulator, also contains a chopper, that is to say a wheel and holes. Radiations from the hydrogen or deuterium lamps pass through the sample along with the radiations coming from the hallow lamp. The electronic system sorts out the signals from the two sources and compares them. Both beams are sent through the optical system at the same time but chops them at different rates. Both beams are equally attenuated with respect to background effects, but the line-source radiation only is attenuated in the sample.

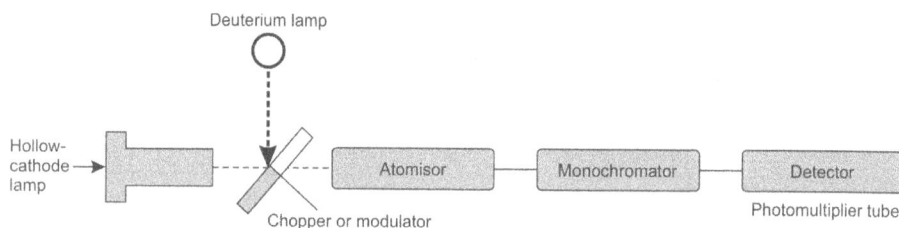

Figure 9: Correction of Spectral Interferences by Lamp Deuterium.

Correction by Application of Zeeman's Effect

With a furnace, even by working with a powerful deuterium lamp, it is not possible to wholly correct the interferences. Then, it is necessary to appeal to another process the theoretical basis of which is Zeeman's effect.

The application of an intense magnetic field (1 T) on a luminous source provokes a division of the spectral rays into several polarized components. The two exterior rays μ (or σ) are circularly polarized and the central ray π is plane-polarized. This phenomenon has for origin in the perturbation of the energy states of the electrons and has the name of normal Zeeman Effect (1). Only the atoms are affected by the Zeeman effect.

During the absorption of energy in the absence of a magnetic field, there is only one electronic transition. It is not the case in presence of a magnetic field. Then, the level p is degenerated. It is divided into three sub-levels according to the value of the magnetic quantum number m (Figure 10).

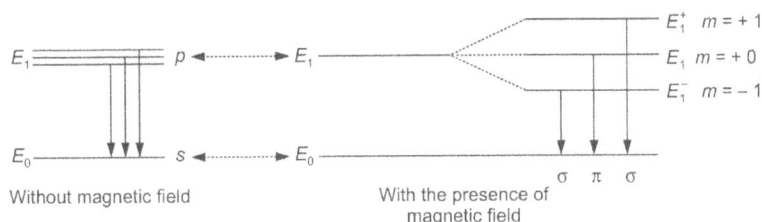

Figure 10: Correction by the Normal Zeeman Effect. (Reprinted with Permission from Gwenola and Jean-Louis Burgot, Paris, Lavoisier, Tec et Doc, 2017, 496).

The phenomenon of division of the rays is accompanied by a polarization of the three rays. In the absence of a magnetic field, there is no polarization. Therefore, specific and non-specific intensities are detected. In the presence of the field, the rays are polarized. The central ray is, therefore, eliminated. There remain the peripheric rays that cannot be absorbed by the element. As a result, one obtains the measurement of the non-specific absorbance.

Correction of Chemical Interferences

At the Level of the Flame

At the moment of the vaporization, some compounds form refractory compounds, the kinetic of vaporization and the decomposition of which become very slow. It is the example of calcium which provides pyrophosphate of calcium $Ca_2P_2O_7$ with phosphates:

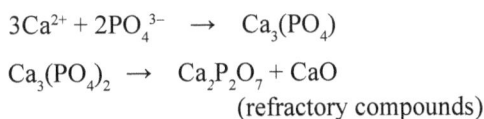

$$3Ca^{2+} + 2PO_4^{3-} \rightarrow Ca_3(PO_4)$$

$$Ca_3(PO_4)_2 \rightarrow Ca_2P_2O_7 + CaO$$
$$\text{(refractory compounds)}$$

The addition of spectroscopic buffers such as the complexing agents (the ethylene-diamino-tetraacetic acid and its salts; e.d.t.a.) or the addition of the precipitation agents, such as lanthanum is useful to limit the phenomenon. Hence, the phosphate of lanthanum, weakly soluble, precipitates during the vaporization of the solvent. On the other hand, with a flame temperature too high, the elements are ionized. This provokes a loss of sensitivity to the method. The addition of cesium (which is a very electropositive element) which ionizes itself preferentially, eliminates this kind of problem by regression of the equilibria of ionization of the last ones.

At the Level of the Furnace

The problems of condensation on the walls of the furnace are widely limited with the use of platforms and the transverse heating mode. One also adds modifying agents such that mixtures of palladium (2% Pd in HNO_3 5% m/m) with ammonium nitrate. The first favors the volatility of the compounds and the second favors the calcination.

Performances of the Techniques

These techniques are endowed with very nice analytical possibilities. Numerous elements are justifiable of them. Let us cite Ca, Cd, Cr, Co, Cu, Fe, Mg, Hg, Ni, Pt, Pb, and Zn. It is possible to fabricate hollow cathode lamps with a mixture (or alloy) of several metals lining the cathode cup so that a number of elements can be determined without the necessity of changing lamps. Examples are Ca, Mg and Al; Fe, Cu and Mn; Cu, Zn, Pb and Sn; Cr, Co, Cu, Fe, Mn and Ni.

These methods are endowed with very low detection limits for:

- The atomic absorption with flames (FAAS): 1 to 100 $\mu g.L^{-1}$;
- The atomic absorption with furnace (GFAAS): 0.001 to 5 $\mu g.L^{-1}$.

In terms of limits of detection, the atomic absorption with furnace offers performances comparable to those of electrochemical methods (polarography and anodic redissolution).

Applications

There exist tables in the literature which give wavelengths of the rays used in the analysis for the elements the most usual. Let us mention some applications in the following domains:

- Pharmaceutical (the spectrometry of atomic absorption is inscribed in the major pharmacopeia as an example in French (now European) pharmacopeia since 1975 with "determination of various elements in different medicines"; for example, some micro-nutrients, such as Zn, Co, Hg, Al, Mg and Ca.
- Determination of various metals as impurities, Al, Cu, Fe, Cd, Cr, Hg, Ni, Pb, Zn, Ag and Pt.
- In the analysis of *foods*, determination of micro-nutrients and toxic residues.
- In *clinical* biochemistry, analysis of Fe in case of anemia.
- Determination of Pb in the research of lead-poisoning.
- In hydrology, for a quick analysis of drinking water.
- Analysis of metallurgic products, such as alloys. Notice in passing that speciation is more and more wanted in order to have a better knowledge of the impact of the detected elements, especially in the products of synthesis where some catalysts or some of their immediate derivatives may remain. This can be, for example, some remnants of palladium, silicon, etc.

Interest in the Atomic Absorption

The great interest in atomic absorption is its *specificity* since there exist very few examples of rays of frequencies identical emitted by two different elements. This property avoids proceeding to a preliminary separation of the element to determine. Only sodium and magnesium exhibit the same resonance ray at 285 nm. In this case, it is indicated to choose another ray, even if it is less luminous than the preceding.

CHAPTER 35

Atomic Emission Spectrometry

An experiment of atomic emission spectrometry is carried out by introducing the sample into a flame or a plasma. Emission can also be realized from arc and spark sources.

Then, one speaks of arc and spark spectroscopies. The major part of this chapter is devoted to the emission in flames together with those in plasmas of rare gases. We shall also briefly study atomic fluorescence.

General Principle

A sample containing the element of interest (analyte) is introduced into the atomic emission source. The latter, typically a flame or a hot ionized gas (a plasma), is used to completely decompose the sample into atoms. Then, energy coming from the source is absorbed and electrons of the analyte are promoted to higher-energy orbitals. They are in *excited electronic states*. Atomic emission spectrometry has the vocation of the measurement of the electromagnetic radiations (photons) produced when these electrons drop to lower-energy orbitals.

An important point deserves further attention. Atomic orbitals have specific energy levels. Therefore, the resulting photon has a specific wavelength. Because the energies of these orbitals are different from one element to another, each element has a single set of atomic emission wavelengths. As a result, one can differentiate between several elements for both analytical viewpoints, that is to say, the qualitative and quantitative ones.

Emission Spectra

Among all kinds of spectra, we can distinguish:

Flame Spectra

They are obtained by throwing a trace of a metallic salt into one flame. The simple flame of a Bunsen burner is sufficient for the alkaline or alkaline earth salts. It is a well-known experiment for a sodium salt that gives rise to the double D-ray of sodium. For other metals, flames are hotter than the previous one (flame dioxygen-dihydrogen or dioxygen-acetylene). Flame spectra only contain a few numbers of rays (see Chapter 33).

Arc Spectra

If some current passes through a resistant wire, its temperature rises. As the current continues to pass, the joule effect permits the dispersion of the energy into the wire under the form of heat. But, the hotter the wire is, the higher the calorific wastes by radiation, convection. Finally, the wire will reach one temperature, called the equilibrium temperature, at which the calorific wastes exactly counterbalance the produced heat by the current. The system remains stationary as the conductor will not be destructed by the rising temperature.

From the practical standpoint, one searches to isolate thermically the conductor as well as possible, one enquires for a good material conductor which resists well to high temperatures. The conductor the most refractory is the carbon in the form of graphite. Its temperature of volatilization is 3,500°C. Hence, one chooses two electrodes in graphite at different potentials and one makes explode one electrical ark between them. They are isolated from the ambient medium by blocks of refractory material (magnesia). The temperature in the ark can reach 3,500°C (Figure 1):

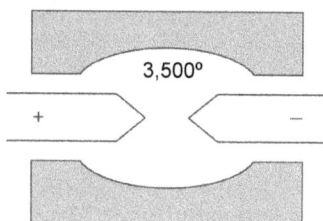

Figure 1: A Device to Obtain Emission Ark Spectra.

We must remark that the ark rushes out between the two solid electrodes. Between them, the current is conducted by the vapors of carbon. Between these solid electrodes and the vapors of carbon, there exists a true equilibrium solid vapor. With a voltage of 110 to 220 V, an intensity of 3 or 4 amperes can be reached. The obtained spectra are some lines spectra. The number of rays is greater than in the case of flame spectra.

The obtained spectra are spectra of rays. They possess more rays than flame spectra (see Chapter 33).

Spark Spectra

These spectra are produced by the passage of a spark between two metallic electrodes or between two drops of a salt solution, formed at the extremities of two capillary tubes built-in quartz. The spark comes from the discharge of a condenser charged at a high potential. The spark is very brief but its intensity can reach several hundred A. By working in a vacuum, the rays coming from oxygen and nitrogen disappear. Moreover, the absorption in the far ultra-violet is avoided.

The obtained spectra are some lines spectra but the rays are different from the arc rays. This source is well adapted for the analysis of low-melting materials since the heating effect is weak (see Chapter 33).

Spectra of Discharge of Gases

A discharge in one gas permits obtaining the spectrum of the gas. The gas is contained in a Geissler tub under a weak pressure (about 1 mm Hg). Two electrodes ensure the passage of the current coming from the discharge of condensers. The obtained spectra are composed of rays and bands. A continuous spectrum is often superimposed on the preceding ones.

Induction Spectra at High Frequencies: Plasmas-Inductively Coupled Plasma Source and ICP Torch

They also are *gases* or *vapors spectra*. The discharge is obtained due to metallic coil wrapping the container of the gas into which passes the discharge of a group of condensers. One finds these devices in plasmas. An inductively coupled plasma source is also called a torch.

A plasma of rare gas is constituted in variable proportions of atoms, free radicals, ions and free electrons. It permits to reach temperatures of the order of 10,000 K, permitting all atoms to be emissive. It is the case for example of the refractory elements: boron, phosphorus and tungsten.

Atoms are excited using the large energy levels associated with the inductively coupled plasma (ICP). This method of excitation is far more effective and permits the analysis of elements beyond the scope of simple flame emission techniques.

A plasma of a rare gas is a gaseous medium heated at a high temperature, thermodynamic equilibrium, electrically neutral but conductor of the electricity. Recall that, here, we only study the capacitive or inductive plasmas working in the domain of high frequencies (ICP). They are the most used.

Figure 2 represents the principal devices participating in the functioning of a torch of the kind ICP-OES. (OES full form is 'Optical emission spectroscopy'. In principle, the part played by the plasma is to raise the element of the sample to the state of an atom in an excited electronic state).

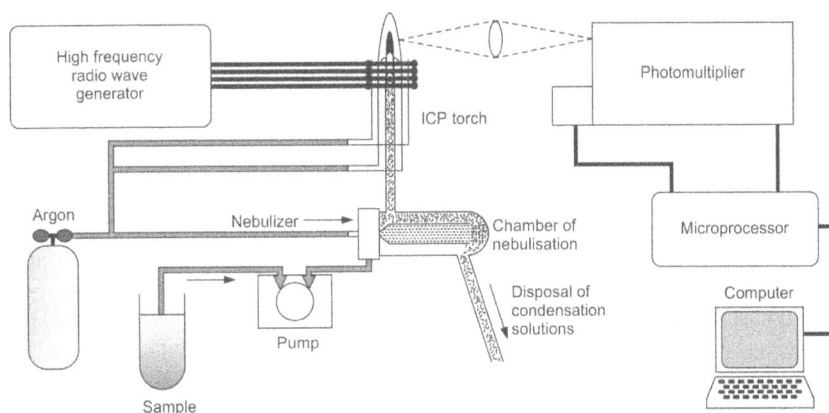

Figure 2: Scheme of a Plasma-Torch of the Type OES. (Rreprinted with Permission from of the Society Perkin-Elmer Instrument, 12 Avenue de la Baltique, 91140 Villebon sur Yvette).

Its functioning is as follows.

A stream of argon emerges at the extremity of a quartz tube surrounded by a coil traversed by an electric current coming from a generator of radiofrequencies (5 to 100 MHz). Ionization of the flowing argon is initiated by a spark from a Tesla coil. The resulting ions and their associated electrons interact with the variable created magnetic field. They are confined into an annular path (Figure 3).

The medium (plasma) becomes hotter because of the heat dissipated by the Joule effect. The plasma is maintained isolated from the side of the ensemble by a second flux of argon, non-ionized, which circulates in the circular wending concentric to the quartz-tube.

Finally, the third internal tube conveys the sample, in the form of an aerosol towards the plasma. The obtained temperature is very high as it is indicated.

This methodology has the advantage to be quicker than the atomic absorption spectrometry. Hence, it is more interesting in routine.

The high temperature eliminates many of the interference effects. The method can permit the multi-elements analysis. Usually, the plasmas used in atomic emission spectroscopy (AES) are formed from noble gases such as helium or argon. An external source of power is required to maintain the functioning state of plasmas. From the theoretical standpoint, it can be said that the heating according to this method is due to the radiofrequency field on argon gas flowing through a quartz tube. The high power frequencies cause a changing magnetic field in the gas and this in turn results in a heating effect. For an inductively coupled plasma, a radiofrequency generator is used with an induction coil to concentrate electrical energy into a small area.

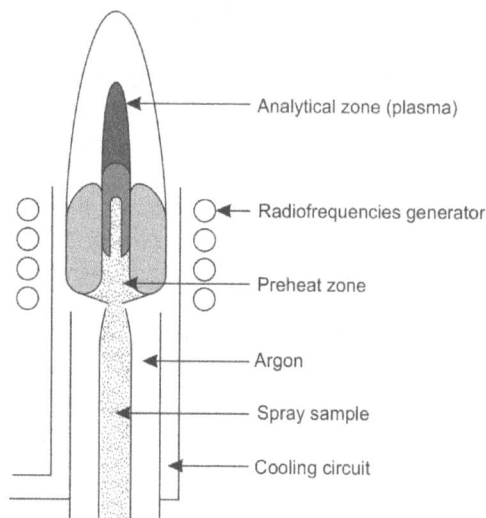

Figure 3: A Typical Inductively Coupled Plasma (Head of the Torch). (Reprinted with Permission from of the Society Perkin-Elmer Instrument, 12 Avenue de la Baltique, 91140 Villebon sur Yvette).

To ignite the plasma, one provides "seed" electrons to a flowing stream of gas in the electrical field. These seed electrons collide with atoms of the gas and initiate an ionization effect in cascade in presence of a high-voltage spark from a coil. There are several means of preparation for plasmas. The process which is the most usual is the plasma induced by high frequency (P.I.H.F) given their high sensitivity limits and the weak interferences they give. In this case, the coil which is traversed by an electric current of high frequency (20 to 40 Mhz) creates an intensive magnetic field the pulsation of which is in linear dependence on its frequency. A current of gas that can engender plasmas traverses the inductor coil and, after ignition, creates an "inductively coupled plasma". The argon is generally used since it is easily ionizable and is endowed with good electrical conductivity.

The mechanism of the production of rays is not fully understood. There exist several kinds of plasmas:

• Plasmas d.c. (direct current) or plasmas-arc,
• Plasmas microwave,
• Capacitive plasmas or inductive plasmas working in the domain of high frequencies (ICP). They are the most used. They are the only ones studied here.

It can be said that the use of plasmas which permit working at high temperature on numerous elements explain why, recently, measurements in emission have got a renewal of interest and practice.

Helium plasmas seem to have special properties. For example, excited-state electronic populations are not described accurately by the Boltzmann equation.

Spectra of Discharge in the Gases

In a gas, the discharge permits obtaining the spectrum of the gas. The latter is obtained at a weak pressure ($\approx$ mmHg) in a Geissler tube (Figure 4).

The spectra can be constituted by rays and bands (see Chapter 36). A Geissler tube is a glass cylinder under partial vacuum, the extremities of which are sealed to two electrodes, a cathode and an anode. When a sufficient voltage is applied between the two electrodes, some molecules of gas present in the tube are ionized and an electrical current appears. Often a color is emitted. The Geissler tubes are the first tubes with gaseous discharge.

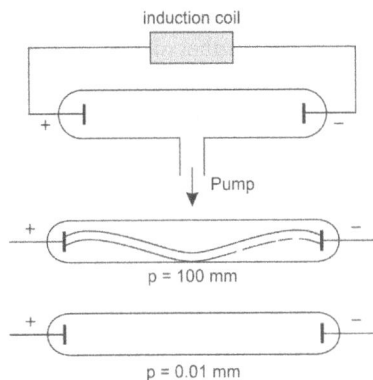

Figure 4: A model of Discharge Tube: (a) Pressure = 100 mm Hg and (b) = 0.01 mm Hg.

Sources of Energy

Flame

The flame is fed by mixtures of carburant-comburant such as:

- *Air/acetylene* provides a temperature of 2,300°C–2,500°C. This flame permits the determination:
  - Of calcium and magnesium. The measurement is often disturbed by the presence of the oxygenated ions PO_4^{3-} which form refractory compounds.
  - Of copper, iron and manganese which form oxides in the flame which are not refractory.
  - Of tin, barium, chrome and aluminum which can be excited at this temperature but form refractory oxides.
- *N_2O/acetylene* which permits to reach temperatures of the order of 3,200°C. This flame, a very reducing one, is used for the elements giving refractory oxides in the presence of dioxygen.

The temperatures of the flames are limited by the presence in their superior part of numerous chemical species (H_2O, CO_2, CO, H_2 and N_2) whose dissociation consumes a part of the thermal energy. The sensitivity (limit of detection) depends on the temperature but also on the presence of anions in the sample, such as Cl^-, SO_4^{2-}, NO_3^- and PO_4^{3-}, responsible for the formation of refractory compounds. Water is an obstacle to an increase of sensitivity. It enhances the presence of radicals and oxides in the medium and has a non-negligible emission spectrum of its own.

Plasmas

(See the section on 'Induction Spectra at High Frequencies').

Instrumentation for Optical Emission Spectroscopy (OES)

As in atomic absorption, the sample is nebulized before its introduction into the flame or into the device providing energy.

Major Atomic Emission Sources (Flames-Plasmas)

Firstly, the sample introduction device is used to convert the sample into an analytically useful form. As an example, a nebulizer which converts an aqueous sample to a fine mist is commonly used. The emission source is hot and permits a fraction of the sample to be atomized and to be electronically excited. The nature of the emission source is an important factor in AES.

For example, some flames are good sources for alkali and alkaline earth elements but they do not provide the energy needed to excite many other metals. As other examples, plasmas can produce intense emissions but non-metals are excited especially well excited with helium plasmas. (Further other point concerning plasmas are studied in a following paragraph).

The Optical System

It permits the analysis of the radiations emitted by the elements in the sources of energy. It contains:
- A system intended to choose the wavelengths;
- A detector;
- A measurement system of the amplified current.

Selection of the Wavelengths

One uses:
- Optical Filters
  - With colored filters. The passant bands are broad (60 to 100 nm)
  - With interference filters: the resolution is better (1 to 10 nm)
- monochromators (prism or gratings) whose passant band varies from 0.1 to 1 nm.

Detection of the Radiations

One uses:

Photon transducers. These include semiconductor photo-transducer. The most generally useful for light-measuring purposes is the silicon diode transducer which consists of a reversed-biased pn junction formed on a silicon chip. The reverse bias creates a depletion layer that reduces the conductance of the junction to nearly zero (see Chapter 19).

More precisely, in this type of diodes, the semi-conductor element *p* of the diode is related to the negative pole of the generator, the element semi-conductor *n* (n is negative) being connected to the positive pole of the generator. The negative pole withdraws the elementary charges of very mobile positive electricity (holes) toward itself of the element p. The holes move away from the junction *p-n*. Simultaneously, one observes the attraction by the positive pole of the generator of electrons, elementary negative charges. Electrons, also, move away from the junction. Hence, both electrons and holes are moved away from the junction.

Because of these facts, the exchanges of electricity charges are not facilitated. Only, a very weak electrical current circulates. For this reason, the zone around the junction is named the depletion zone. This carrying out of the diode is its reverse-bias functioning. Figure 5 summarizes it.

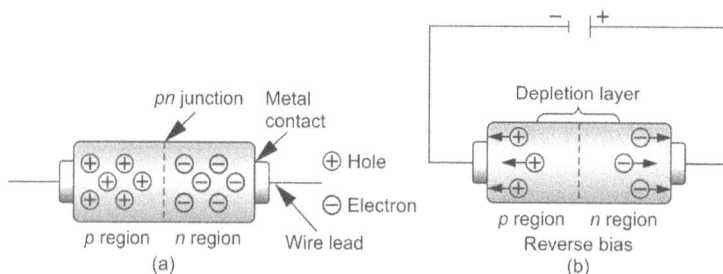

Figure 5: Principle of the Reversed Biased *p-n* Junction. From "Principles of Instrumental Analysis", Skoog et al., 2018, Chapter 7, p 179, Figures 7–32, 2018 with Permission of Cengage Editor.

Photomultipliers: The intensity of rays is weak, overall, when they are selected by monochromators. As a consequence, it is necessary to amplify the current. This is done with the help of amplifiers.

With apparatus of the type ICP-AES, technical improvements have permitted the simultaneous observation of several wavelengths, thanks to the multichannel detection.

These systems permit the simultaneous analysis of the continuous emission and the luminous rays. These points present the following advantages:

- Optimization of the standardization by using several rays of the same element. This possibility permits the identification of the existence of a matrix effect.
- Evaluation and lowering of the background.

At wavelengths lower than 190nm, the rays cannot be detected in presence of dioxygen and humidity. Some apparatuses have an optical system definitively and exclusively under argon.

Finally, the light from the source is focused into the entrance slit of a spectrometer. The latter disperses the light as a function of wavelength, allowing the choice of a particular working wavelength. At the exit of the spectrometer, a photodetector converts the light to an electrical signal.

Characteristics of Atomic Emission Spectroscopy

Line Emission

One of the characteristics of the atomic emission spectrometry is to produce narrow rays. Consequently, interferences caused by overlapping emission lines from other elements are generally avoided.

Populations in the Different States of Energy and the Temperature

Analytical plasmas populate higher energy levels as efficiently as possible. At higher temperatures, there is more energy to excite the atoms.

These methods are useful for the determination of the alkaline and alkaline-earth elements, sodium and potassium particularly. The sensibility may reach 0.01 ppm with good apparatus. For the device in flame, the precision depends much on the uniformity of the pulverization of the flame. It is often better than 5%.

Interferences

Although they are weak, the interferences in AES are numerous (spectral and chemical ones).

Choice of SAA or ES

The use of all these techniques is dependent on disturbing phenomena of several origins such as spectral, physical and chemical origins, already described elsewhere. Let us particularly mention the formation of complexes with anions, such as sulfate, phosphate and silicate which lower the number of atoms which can be excited. Moreover, these methods are particularly sensitive to the viscosity of the samples. In order to correct these effects, it is necessary to add a spectral buffer as a derivative of the metal lanthanum in concentrations 20 to 50 times stronger than those of the disturbing elements or, in the second case, to dilute the sample. The addition of lanthanum chloride $LaCl_3$ to the phosphate of the element to be determined displaces the latter, which remains available for the luminous absorption whereas the phosphate of lanthanum precipitates.

It is because both methods are founded on the use of flames that sometimes the atomic absorption spectrometry is called "flame photometry by absorption". The spectrometry by emission studied in this chapter might rather be called "flame photometry by emission". The similitude of these two definitions would mean that the mechanism of both methods is the same. It is not true. There are the phenomena occurring in the flame in both cases, which are the same.

- In photometry by emission, there are the atoms in electronic excited states which emit radiations by dropping down to their fundamental level of energy.

- In photometry by absorption, contrary to the preceding case, there are the atoms the electrons of which are in their fundamental levels, which absorb external radiations.

The principal advantage of the spectrometry by atomic absorption comes from the fact that the absorption is a function of the number of non-excited atoms. This ratio is in favor of the spectrometry by absorption. In disfavor of emission spectrometry is the fact that, in its conditions, the ionization of the atoms in the excited electronic state displaces the equilibrium excited/their ionization in the favor of the latter (Saha's equilibrium).

It can be said that the use of plasmas which permits to work at high temperature on numerous elements explain why, recently, measurements in emission have got a renewal of interest and of practice.

An interesting domain of comparison of the two methods is that of *clinical biology*. For example, let us mention the determination in the plasma and in urine of sodium (589 nm) and potassium (766–769 nm; the ray 404–405 nm is not used) and of lithium (670.8 nm).

Concerning the lithium, the sensitivity of the technique is quite comparable to that of the atomic absorption, that is to say of the order of 10^{-2}, 10^{-3} µg ml^{-1}. Now, the flame photometry competes with specific electrodes at the level of the laboratories of clinical biology.

As internal standard, one uses the lithium for the determination of sodium and potassium and the cesium for the determination of the lithium. Standard solutions must always be kept in polyethylene flasks since glass loosens sodium ions in solutions.

In a flame air-acetylene, the calcium shows several characteristic rays: 422.2 nm, 554 nm, 570 nm and 622 nm. One exploits the latter since it is well-different from that of sodium.

Applications

Numerous applications can be envisaged overall by inductively coupled plasma emission since the latter practically concerns all the elements. One can distinguish two categories of applications:

• Those concerning the element itself, Al, Pt, Zn, Cu, Fe and Be in biological liquids, and those concerning the metallic hydrides easily formed, such as those of As, Sb, Se, Te and Pb.

• The use of all these techniques is dependent on disturbing phenomena of several origins such as spectral, physical and chemical origins, already described elsewhere. Let us particularly mention the formation of complexes with anions such as sulfate, phosphate and silicate which lower the number of atoms which can be excited. Moreover, these methods are particularly sensitive to the viscosity of the samples. In order to correct these effects, it is necessary to add a spectral buffer as a derivative of the metal lanthanum in concentrations 20 to 50 times stronger than those of the disturbing elements or, in the second case, to dilute the sample. The addition of lanthanum chloride $LaCl_3$ to the phosphate of the element to be determined displaces the latter which remains available for the luminous absorption whereas the phosphate of lanthanum precipitates.

Coupling ICP-MS

The matter, here, is the study of the coupling plasma-torch and the mass-spectrum technologies. It is talked about more in Chapter 42.

CHAPTER 36

Analysis by X-Rays

X-rays are short wavelength electromagnetic radiations. They were discovered in 1895 by Röentgen.[49] They are, by far, more energetic than the luminous rays. This property explains the particularities of their applying.

From an analytical standpoint, it is convenient to distinguish:

- The emission spectrography X;
- The absorptiometry of X rays;
- The fluorometry of X rays.

X-ray fluorescence and X-ray absorptiometry methods are widely used for the qualitative and quantitative determination of all the elements in the periodic table having atomic numbers greater than that of sodium. Even, with special equipment, elements with atomic numbers 5 to 10 in the range of atomic number can also be determined.

General Characters of X-Rays

X-rays are notably formed when a beam of electrons impinges on a target material (see the section on 'Apparatus'). X-rays are short wavelength *electromagnetic radiations*. The wavelength range of X-rays is situated from about 10^{-5} Å to 100 Å. Usually, X-ray spectroscopy is located in the range 0.1 Å–25 Å.

- All the classical experiments of optical spectroscopy such as reflection, refraction, interferences can be carried out with X-rays, since they are electromagnetic radiations.
- X-rays are constituted by photons possessing a great energy w. The photons, whose wavelength is about 5 Å, possess an energy one thousand times greater than an optical radiation of wavelength about 5,000 Å does possess.
- A beam of X- rays loses a part of its intensity (or power) when it is passed through a thin layer of matter. It is a consequence of its *absorption* and of its scattering.

X-rays also exhibit the following properties:

- They provoke the fluorescence of numerous substances.
- They are easily absorbed by some substances. This property is particularly studied in the next paragraph.
- They act on photographic plates.
- They possess the property of destruction of some biological tissues.
- They may modify the properties of some matters.

---

[49] Röentgen, Wilhelm, Conrad, German physicist, 1845–1923, Nobel Prize in physics 1901.

Apparatus

One method of preparation of X-rays consists in sending a beam of thermo-ionic electrons on a target material. The electrons are slowed down by multiple interactions with the electrons of the target. This induces one very energetic excitation of the atoms. The energy lost is converted into a continuum of X-radiations. This is a process of bombardment of a metal target with a beam of electrons.

One uses a Coolidge 's tube or an X-ray tube which is a modification of Crookes[50] tube with which Röentgen discovered X-rays (Figure 1). The Coolidge's tube[51] contains one anticathode M under a block of tungsten which may be refrigerated. In such a tube, the electrons impinge on the anticathode constituted by the metal target. They are obtained by a thermo-electronic effect in a perfect vacuum from a tungsten filament F brought to incandescence by an electric current produced, such as by a small transformer. The filament is placed at the bottom of a cavity in order to direct the electrons on the anticathode.

A Coolidge's tube also possesses a window through which escape X-rays. In general, it is constituted by a fine shell of beryllium. Finally, it is connected to a transformer with a redressor. It gives a high voltage in order that the electrons should possess a sufficient speed. The fixed voltage is often located between 50,000 and 200,000 volts, the value most often chosen being 120,000 volts. (There are other processes of preparation of X-rays; see section on 'Other Modes of Generation and X-Rays').

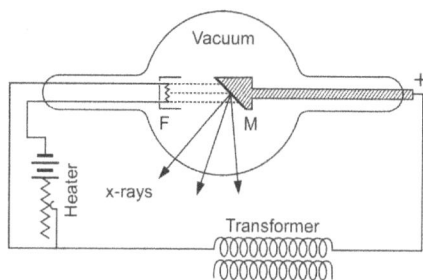

Figure 1: Coolidge's Tube.

X-Ray Spectroscopy by Emission

Form of a X-Ray Spectrum by Emission

Spectra X are usually diagram intensities of the peaks emitted by the anticathode as a function of the wavelengths at which they appear. They are constituted:

- By a continuous part (or continuum grounding)
- By a ray spectrum possessing discrete lines superimposed to the preceding part.

These rays are characteristic of the elements of the target. The emission of X-rays of a target material resembles to that shown in Figure 2.

---

[50] Crookes William, 1832–1878, English physicist.
[51] Coolidge William, 1873–1975, American physicist.

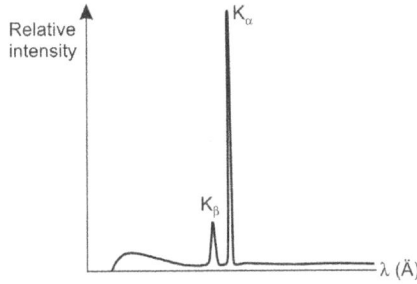

Figure 2: Type of X: Rays Spectrum Emitted by a Target.

Continuum Grounding

It has for origin the transfer of a kinetic energy into a radiative energy by an inverse photoelectric effect when a cathodic electron is suddenly braked in the electrical field of the atoms of the anticathode.

Recall that the photoelectric effect is the emission of electrons by the enlightened matter, in particular by some metals such the alkaline ones (potassium, sodium, rubidium and cesium) and negatively charged zinc. It is evidenced for example by use of a photoelectric cell. The latter can be a glass bulb. A part of its inner surface is recovered by the metal. This sheet is linked to the negative terminal of a battery which constitutes the cathode. The positive pole of the battery is linked to a ring of tungsten which is the anode. There is a microamperometer in series with the battery. In absence of light, no current is circulating but, if there is lighting, there is a electric current. The luminous energy pulls out the electrons from the metallic surface.

They are withdrawn by the anode and they circulate from the cathode to the anode. In order to explain this phenomenon and events in direct connection with it (Einstein, 1905), it has been necessary to admit that light is formed from particles called photons. For a light of frequency v, each photon possesses the energy $w = hv$. The fact that the intensity of the current was proportional to the lighting and the speed of electrons independent of the luminous intensity induces the reasoning that the energy w permits the photon to both

- Pull up the electron from the metal.
- Give to the electron its kinetic energy.

That is to say:

$$hv = w + 1/2\ mv^2$$

The energy of the photon is:

$$w + 1/2\ mv^2 = Ve$$

Here, V is the accelerating voltage (of the electron beam), e the charge of electron and h is the Planck's constant.

In order there is expulsion of one electron, the photon must possess the energy such as:

$$hv \leq Ve$$

Since $v = C/\lambda$

Here, C is the rate of the light and λ the wavelength of the photon:

$$\lambda \geq hC/Ve$$

The energy hv of the produced photons X in that manner is at the maximum equal to that of the incident electrons, that is to say:

$$hv \leq Ve$$

Here, c is the speed of the light. λ is given by the relation:

$$\lambda = 12{,}400/V$$

by taking into account the values of the constants and in which λ is expressed in angstroms and V in volts. Given the fact that the cathodic electrons unequally lose their energy during their braking, X radiation is formed by a continuous series of rays, beyond a threshold value of λ.

Characteristics of the Elements of the Anti-Cathode

The discrete lines which superimpose the continuum are characteristic of the elements of the anticathode if the voltage of acceleration of the electrons is greater than the critical potential of excitation. The emission of the characteristic rays follows the same principle as that of the optical spectra. In these series labeled K, L, M, etc., the characteristic frequencies of the rays are regularly moving according to their increasing wavelengths. However, here, the excitation is produced by the shock of the cathodic electrons with the internal electrons of the elements of the anticathode. The ultimate state of interactions between the cathodic electrons and the atoms of the target (anticathode) is the wrench of an electron K the latter. It is only possible when the energy eV of the incident electrons is, at least, equal to the bond energy U_K of this electron in the atom. The *notion of critical potential* V_K results from this point. It is defined by the relation:

$$eV_k = U_k$$

The ionized atom recovers its stability by transition of the electrons toward levels of stronger energy K, L and M from levels of lesser energy, i.e., L, M and N and emission of series of rays K, L, M. In a same series, each ray is labeled by an index α, β and γ in order to specify the starting level of the transfer (Figure 3):

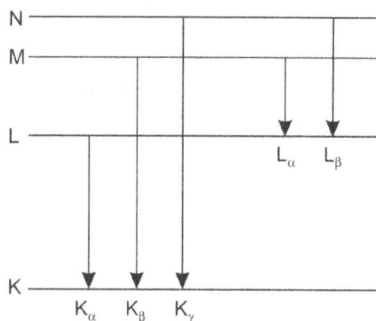

Figure 3: Mechanism of the Emission of X-Rays.

Spectra are characteristic of the radiant element. They are formed by monochromatic radiations as are the spectra of incandescent gases. They are continuous and simpler than the luminous ones. The intensity of the X-rays is proportional to the number of photons emitted *per* second by the anticathode. Thus, it is proportional to the number of electrons the latter receives. The electrons emitted by the cathode arrive on the anticathode (which is in the metal or which contains the sample under study) with the kinetic energy given by the relation:

$$1/2mv^2 = eV$$

Here, e is the charge of the electron and V the tension applied to the Coolidge's tube. The wavelength of the X-rays decreases when the tension V applied to the tube is increased.

If the whole kinetic energy of the electron is located only in the photon, one can write:

$$hc/\lambda_{min} = eV$$

λ_{min} is the minimum wavelength of the radiation that can put the tube under the tension V.

Moseley's Law—Rays of the Series K

The frequency of the ray K_α, $\nu_{K\alpha}$ emitted by an element of the anticathode of atomic number Z is defined by the expression of the energy of the electron in its principal level of order n:

$$U_n = -A\,Z^2/n^2$$

According to Moseley's law:

$$(\nu_{K\alpha})^{1/2} = aZ$$

A and *a* are constants. The square root is a linear function of the atomic number of the element. Moseley's law can be written for ray L and K:

$$h_{\nu K\alpha} = U_L - U_K$$
$$= -AZ^2/2^2 + AZ^2/1^2 \quad \text{(principal level of order 1)}$$

The rays of the series K are the most important since they are the most intensive once and since their rays K are easily detectable. Moseley's law is a kind of spectral series such as those encountered in optical spectra (type Balmer, etc.). It exhibits two terms and finally obeys the Ritz combination principle. Nevertheless, this law is only approached. Each ray only appears when the tension is superior to a critical value.

Analysis by Spectroscopy X of Emission

The identifications consist in measuring the wavelengths of the characteristic radiations emitted by the substance to be analyzed, located at the focus of the T Tube. Its atomic number can be immediately deduced by application of Moseley's law.

The principle of the quantitative determinations is the same as that of the optical spectroscopy of emission. The intensities of the rays of the product under study is reported to the values of a calibration curve established in the same conditions by working with samples of known compositions.

X-ray emission permits to analyze high-alloy steels together with heat-resistant alloys of the chromium- nickel-cobalt type. Analysis of low percentages is limited by the absorption of the emitted radiation of the element by the materials of the specimen. Samples in which the main constituent is an element of high atomic weight will absorb a higher percentage of the radiation than would be absorbed by a light element. Thus, nickel in an aluminum alloy can be determined with a greater sensitivity than is possible for nickel in steel or in a silver or a lead alloy where the absorption of the Ni K_α radiation is high. In favorable circumstances accuracy of the order of 0.5 percent of the element present can be achieved. The limit of detectability may be as low as a few parts per million.

Emission methods have also been applied successfully in trace analysis following a preconcentration step. Such a preconcentration step has been an electrolysis onto a cathode of pyrolytic graphite and subsequent X-ray examination of the cathode surface.

Absorptiometry of X-Rays

A beam of X-rays loses a part of its intensity or power when it is passed through a thin layer of matter. It is a consequence of its *absorption* and of its scattering. The losses of intensity by absorption obey the general Beer-Lambert's law. Let us consider a monochromatic beam of X rays of intensity I_0 traversing a material homogenous screen of thickness x and of specific mass ρ. The intensity I of the emerging beam is defined by the relation:

$$I = I_0 \exp[-(\mu/\rho)\rho x] \quad \text{or} \quad \ln(I_0/I) = (\mu/\rho)\rho x$$

μ/ρ is called *mass absorption coefficient*. It is expressed in $cm^2/gram$. It has the same meaning as the extinction coefficient ε in absorption photometry. As the latter, it depends on the incident wavelength, the absorption X spectrum being defined by the curve $\mu/\rho = f(\lambda)$. The difference comes from the fact that the absorption in spectroscopy X involves electronic transitions at the level of the deep shells of the atoms. As a result, the absorption of X-rays is an atomic phenomenon independent of the bonds and of the physic state of the absorbing substance. This point is important. It means that since the inner electrons are not concerned with the state of chemical combination of the atoms (except for the lighter elements), the X-ray properties of the elements are independent of their chemical combinations.

The losses of energy of X-rays by absorption are due to two factors: the diffusion and the losses by photoelectric effect.

The Diffusion

It can be of two orders. They are always weak overall in the domain of wavelength used in analysis, i.e., for 1< lambda< 3 A°. The X photons are deviated from their trajectory:

By a Coherent Diffusion

The electrons and the atoms enter in forced vibration and reemit waves of the same frequency in all the directions with possibilities of interferences. Such a phenomenon is the basis of the diffractometry of X-rays. The coherent diffusion can be understood as considering the atoms as being resonators.

By the Compton's Effect

It is the incoherent diffusion with an increase of the wavelength due to the elastic shocks of photons X with the electrons. This phenomenon is overall observed with light elements or with X-rays of great energy $\lambda < 0.2$ Å. The Compton's effect gives rise to the corpuscular character of electrons. The image of the phenomenon is the shock between a photon and an electron with a change of directions of both particles.

The Photoelectric Effect

The losses by the photoelectric effect represent the quasi-totality of the absorption. X-rays provoke the ionization of the traversed matter by ejection of the internal electrons as in the case of interaction of cathodic rays with the matter of the anticathode. As a result, there are:

- A radiation β due to the ejected electrons;
- A radiation of X fluorescence due to the reorganization of the electronic shells of the ionized atom.

The wavelengths of this secondary emission X are then characteristic of the absorbent substance. Hence, the principle of the method of analysis by fluorometry of X-rays.

Discontinuities

Spectra X of absorption exhibit discontinuities or edges at some wavelengths. They have the form shown in Figure 4.

The phenomenon appears when an electron of one atom of the anticathode has the quantic level of energy $U_i = h\nu_i$. Hence, there exists possibility to absorb any ray of frequency $\nu_i > \nu$. When the energy of the incident photon is lower than the value $h\nu_i$, the photoelectric effect vanishes and the absorption coefficient frankly decreases. Then, one observes the discontinuities at each change of energy level K, L and M.

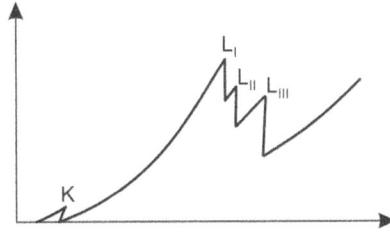

Figure 4: Discontinuities in the Absorption X-Spectra.

Analytical Applications

- The absorptiometry of X-rays offers possibilities quasi-unlimited in analysis by rule of thumb because of its simplicity, quickness and its sensitivity which may reach 0.01%. It is overall used in the domains of liquids and gases because solids and powders may contain several types of heterogeneities. Among the most remarkable examples, let us mention the determination of sulfur in hydrocarbons, that of the tetraethyllead in essence. The method is adapted for the determination of heavy metals (such as Pb) in glasses and alloys, of mineral salts in waters and of uranium in solution.

- One can apply the possibilities offered by the Beer-Lambert's law to the X-ray absorptiometry. Let us consider the determination of the centesimal composition of two elements (A and B; a and b%), the mass coefficients of which being $(\mu/\rho)A$ and $(\mu/\rho)B$ for a given wavelength.

Taking into account the additivity of the Beer-Lambert's law, one can write for the intensity of the emerging X-rays:

$$\ln(I_0/I) = [(\mu/\rho)_A.a + (\mu/\rho)_B.b]\rho x/100$$

(ρ is the specific mass and x is the thickness of the material screen). The calculation of a and b is straightforward after having measured the intensities of the incident and emerging beams with b = 100 − a and knowing ρ and x.

• There is another possibility to apply X-ray absorption to the qualitative and quantitative analysis. It is named *the edge absorption analysis*. Let us briefly say that it consists in studying the intensity of an X-ray beam in the vicinity of the K-edge of an element. A relative error of 1% can be obtained at concentrations down to 0.1% for many elements.

Atomic Fluorescence Spectroscopy

Generalities

After the wrench of one electron K during the production of the X-spectrum, the ionized atom recovers its stability by transition of the electrons from the levels L, M and N towards the levels K, L and M with emission of the series of rays K, L, etc. This is a case of fluorescence.

The sample absorbing the primary X-rays (with the photoelectric effect) emits fluorescent rays. A very important point, evidently, is the fact, ascertained experimentally, that the intensity of an element overall depends on its concentration and quasi-proportionally. The X-fluorescence spectrum of a given substance exhibits characteristic rays identical to those of the excitation spectrum, with the same relative. There are different the excitation and the intensity which is here very weak.

The principle, the techniques and the applying of atomic fluorescence spectrometry are therefore quasi the same as in emission X spectrography. Then emission occurs at the same or often longer wavelength. The intensity of fluorescence thus produced depends on a number of factors. The most important of which are the intensity of the source of radiation, the concentration of atoms in that area of the atomizer that is irradiated by the source, the fraction of the source radiation absorbed, the efficiency of conversion of absorbed to emitted radiation and the amount of self-absorption in the atomizer.

In the X-ray fluorescence spectroscopy, the sample (a metal or a rock) is subjected to irradiation by a powerful beam of X-rays of short wavelength. The beam may displace one electron K. The ionized atom recovers its stability by transition of another electron from one of the outer shell (levels L, M and N towards the levels K, L and M). By generalization, atomic fluorescence occurs when atoms are excited by the absorption of radiations of suitable wavelengths from an external source, with emission of the series of rays K-, L-, etc. Doing so, it releases energy in the form of X-rays.

Further Information About the Origin of the Emission of the Characteristic X-Rays by the Elements

It is interesting to further study the emission of the characteristic X–rays, over all in relation with the phenomena of X-ray fluorescence. These rays consist of two or more groups of definite frequencies. The groups are designated by the K-, L-; M-, N- and O-series in order to increase wavelength. The X-ray spectra were first explained in terms of Bohr's theory that is to say in terms of energy levels of electrons. It is assumed that electrons in an atom are occurring in series of shells. The first one is called the K-shell, the second the L-shell, etc. These shells have groups of electrons possessing the same principal number n. Thus n =1 corresponds to the K-shell, n = 2 to the L-shell, etc. The electrons in the same shell have approximately the same energy. They are bond more and less equally strongly to the nucleus. The strength of the binding falls off with the increasing distance of the shell from the nucleus. Hence, it diminishes in the order K, L and M (the strength of the binding may be chosen in literature as being the inverse. This is arbitrary).

For a K-line to be excited, it can be postulated that the atom absorbs sufficient energy to eject an electron from the K-shell. The vacancy in the K-level can now be filled by an electron moving in from the L-, M- and N- shells, the involved energy differences between the levels being materialized by the emissions of the K-series. In a similar manner, the ejection of an electron from the L level results in the emission of the L-series (Figure 5).

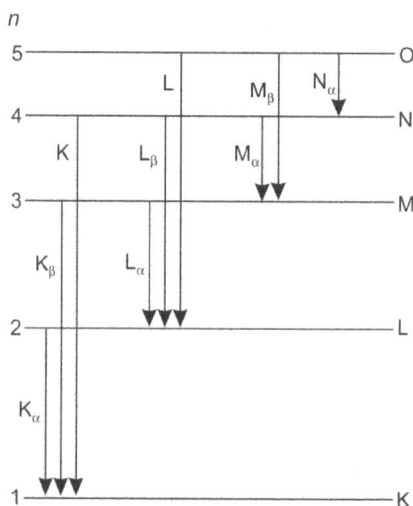

Figure 5: Characteristic X-Rays.

Analysis by Fluorometry X

X-ray fluorescence spectroscopy is one of the most powerful techniques for the elemental analysis. Several domains can be explored by this technique included artistic ones such as those of paintings, archeological, jewelry. They can be analyzed since it is not destructive. X-ray fluorescence cannot be used for the determination of light elements.

It is at least applicable to the elements the atomic number of which greater than 20 is. Perhaps, the most remarkable results obtained with this technology have been obtained in the domain of study of steels where the simultaneous quantitative of Cr, Co, Fe, Mo and Ni is possible with a precision superior to 1%. Let us also mention the determination of the elements which are difficult to dose by a chemical way. They are metals of the second and the third series of transition, uranium and thorium and lanthanides. It can be applied also to liquids and solutions.

Moreover, it applies to very little samples and to very big ones as well. It may be used in difficult conditions of work. It is endowed with an excellent precision and accuracy which is at least that of other competing techniques such as UV-visible spectroscopy or atomic absorption spectroscopy.

However, we can say that X-ray spectroscopy is often less sensitive than other optical methods. Unfortunately, the instrumentation is expensive and tends to remain localized in laboratories. Finally, one rarely uses primary emission spectrum of a sample playing the part of a target in point the local heat produced by the rays striking the anticathode may give rise to volatilizations of some components of the sample. One avoids most of these problems by operating with the fluorescence spectra emanating from samples located outside the X-ray tube. The analysis of fluorescence X is very interesting for the elements the concentrations of which are spread from a few percent until cent % since it then gives correct results with an error smaller than 1%. By way of example, it has been possible to measure until 0.03% of manganese and iron in some aluminum alloys. Let us also mention the determination of iron in the hemoglobin, sulfur in oils and so forth. The method is particularly interesting for the determination of elements for which only few sure methods of chemical analysis exist.

A very important point, evidently, is the fact, ascertained experimentally, that the intensity of an element overall depends on its concentration and quasi-proportionately.

Other Modes of Generation of X-Rays

One can essentially obtain X-rays in three ways:

- By bombardment of metal target with a beam of high-energy electrons. This possibility has been already considered.
- By exposure of a substance to a primary beam of X-rays in order to generate a secondary beam of X-ray fluorescence.
- By use of a radioactive source whose decay process results in X-ray emission.

Use of Secondary Fluorescent Sources

X-ray fluorescence spectroscopy is associated with electronic transitions. The fluorescence spectrum of an element is induced by excitation by radiations in the X-ray region either from an X-ray tube or via the use of a radioactive nuclide.

One can also operate with fluorescence spectra of samples located at the outside of the X-rays tube. The characteristic spectra can still be obtained. The incident radiation need not be of a specific frequency. They must be sufficiently energetic to cause excitation of the electrons located usually in the K-shell. The primary X-rays emitted by the anticathode (target) strike the sample which is able to give a secondary beam which through a slit penetrates into one Geiger-Muller or scintillation counters.

X-ray fluorescence spectra are measured by instruments based on the energy dispersive principle. The X-ray spectrophotometers utilize polychromatic X-ray sources, a semi-conductor detector and various electronic devices for the energy discrimination. All of the frequencies constituting the fluorescence spectrum are measured simultaneously. In these instruments, the X-ray tube (the target) and the sample are placed in a judicious position, one with respect to the other. The origin of the name energy dispersive instruments or X-ray spectrometers comes, of course, from the well-known relation:

$$E\lambda = hc.$$

hc is a constant. It shows that it is possible to draw an X-ray spectrum with respect to energy instead to wavelength.

This methodology depends on the availability of detectors to respond linearly to the energy content of incident photons. (This condition must be distinct from the other one according to which the detector should respond to total incident energy which is function of the energy per photon and also of the number of photons. There exist such required detectors in the domain of X-rays. They are the scintillation counters, gases counters, lithium-drifted silicon or germanium detectors.

Radioactive Sources Decays Producing X-Rays

Some radioactive substances are used as sources of X-rays. Some of them produce simple spectra whereas others also produce the continuum. Some elements are more suitable for absorption studies, others for excitation of fluorescence. Phenomena involved are the β-decay and the electron capture decay.

Electron capture is a process which involves the capture of a K electron by the nucleus and formation of another element. It is the next lower which has lost a K-electron. As a result, an electronic transition into the vacated orbital can occur in the former atom. More rarely, a capture of L or M electron may occur. An interesting example to mention is that of $^{55}_{26}$Fe. It undergoes a K-capture with a half-live of 2.6 years

$$^{55}\text{Fe} \rightarrow \ ^{54}\text{Mn} + h\nu$$

The result manganese K_α line is located at 2.1 Å. It is used for absorption and fluorescence X-ray spectroscopy. Let us also mention $^{57}_{27}$Co $^{109}_{48}$Cd, $^{125}_{53}$I and $^{210}_{90}$Pb.

Some Components of the Apparatus

X-rays may participate to the following phenomena – absorption, emission, fluorescence and diffraction. Therefore, the material elements which permit to handle them possess the components which are necessary to these participations. Analogous, but not identical ones, also participate for measurements in optical spectroscopy.

Notably, some components of systems producing X-rays deserve further commentaries.

Monochromator Sources

Narrow bands can be obtained by three methods:
- By choosing characteristic emission lines and by isolating them with the aid of filters.
- By means of a monochromator in which a crystal of known spacing acts as a diffraction grating.
- By using a radioactive source.

An interesting methodology of filtering consists of using an element or its compound which has a critical absorption edge judiciously located on the scale of the wavelengths with respect to the element to study. It is possible to find elements which have a strong absorption for the K_β X-ray of the target in question, while it presents a rather weak absorption for the corresponding ray K_α. Figure 7 gives an example of this methodology. It is the filtrating action of the nickel on the X-rays of copper (Figure 6).

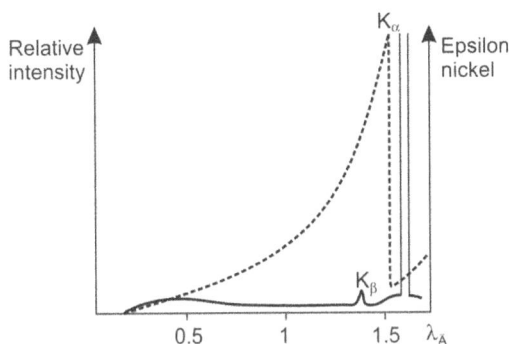

Figure 6: Filtrating Action of a Nickel Foil on the X-Rays Emanating From Copper.

Grating monochromators give greater monochromaticity than that obtained with filters. The former ones are used for experiments of X-ray diffraction.

Radioactive Sources

X-ray may be emitted from some radioactive elements by either of these two mechanisms. The first one involves intranuclear energy levels. The other is by K capture. This process has already been described. We confine ourselves to saying, here, that there is a finite probability of the capture of a K electron by the nucleus in many atoms. They have the advantage of not requiring Coolidge's tube with limited lifetime, but they have the disadvantage that they cannot be turned off.

X-Ray Detectors

X-rays were first detected by a photographic way, but it is no longer used for quantitative applying. Some detectors are grounded on the fact that X-rays produce flashes of light (scintillations) in certain materials. They may also cause ionization in other materials. Some of the latter ones have the property of emitting a tiny flash of light when an X-ray photon is absorbed. It is a case of fluorescence. An example is provided by crystalline NaI into which a small amount (1 to 2 percent) of Tl(II) (thallium (1) iodide) has been added. Observation of these flashes with a photomultiplier gives a reliable measure of the number of photons incident on the crystal.

Gas-Ionization Detectors

Since X-rays ionize gases through which they pass, their presence can be detected by the conductivity of the gas. This can be done with an ionization chamber which can be considered as being a simple metallic container filled with dry gas. A voltage of 100 V or more is impressed across the container and an insulated central electrode. The resulting current is measured with an electrometer.

Solid-State Ionization Detectors

Germanium and silicon as free elements which can be sensitized to ionizing radiation, such as X-radiation, by the addition of lithium. Lithium is allowed to diffuse into the crystalline material, scavenging impurities as it penetrates. An X-ray photon entering the crystal dislodges electrons from the lattice, leaving vacancies commonly called holes. Holes are equivalent in effect to mobile positive charges. The number of charge separation events is directly related to the energy content of the photon. Therefore, it is also the case of the conductivity. The electrical output from these solid-state detectors is much smaller than the corresponding signal.

Diffraction X

X-Ray Diffraction

Since X-rays are electromagnetic waves such as lights, they can be diffracted in a similar manner. The applications of X- rays in this domain are very numerous and it is impossible to summarize the principles of these radio-crystallographic methods and their applying here. This is the reason why we confine ourselves to only giving a brief account concerning them.

The X-rays diffused by the ions and molecules located at the nodes of the crystalline gratings give rise to diffraction phenomena.

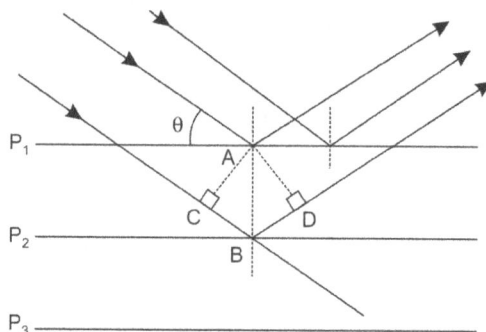

Figure 7: Diffraction of a Beam of X-Rays by a Family of Reticular Planes.

Firstly, it must be said that the X-rays diffused by the nodes of the same reticular plane that are not in concordance of phases except those which are in the direction symmetric of that of the incident one with respect to the normal to the plane (Figure 7). But, since the X-rays are penetrating, the parallel reticular planes P2, P3, etc., also participate to the reflection. In order to make sure that the waves diffracted by two successive planes P1 and P2 should be in phase, it is necessary that the path-length Δ should be a multiple of the wavelength λ. Hence, one must have:

$$\Delta = CBD - n\lambda$$

Since

$$CB = AB \sin \theta$$

$$2d\sin\theta = n\lambda \quad \text{Bragg's law}$$

This is Bragg's law.[52] The d is the distance between two successive planes (inter-reticular distance).

With the measurement of the angles of diffraction, one can determine the inter-reticular distances of the crystals by knowing the wavelength of the incident rays. It is the principle of the radio-crystallographic analysis. In this domain, the simplest technique (and the most used) is the powder method of Debye-Scherrer. In this method, the diffracted beam of X-rays is detected photographically in a special apparatus called a *Debye-Scherrer powder camera* (Figure 8).

In qualitative analysis, the powders method permits identifying the crystalline phases of a sample quickly by simple comparison with the reference spectra of the pure substances which are the components.

---

[52] William-Laurence BRAGG, Australian physicist, (1890–1971), Nobel prize in 1915.

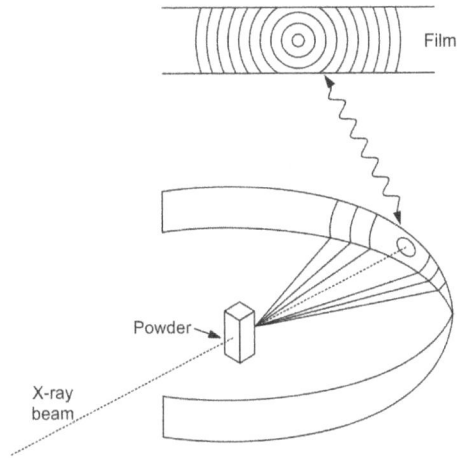

Figure 8: Schematic Apparatus Used for the Powder Method.

In quantitative analysis, the expression of the intensity of the rays of the spectrum of its component X as a function of its concentration is easy to handle:

$$I = I_0 Ax$$

I_0 is the intensity of the incident beam. A is a proportionality factor the value of which depending on several factors. One operates with a calibration curve.

CHAPTER 37

Radiochemical Methods of Analysis

Radioisotopes are detected and identified due to their radioactivity. The dose (content) of radioelements can be determined since the intensity of the radioactive emission is linearly related to their concentration. This is the basic principle of the radiochemical methods of analysis, the number of which has considerably grown because of the measurement facilities that furnish artificial radioisotopes. In this chapter, we successively:

- Recall some points concerning the disintegration of the radioisotopes.
- Study the detection and measurement of the radioactivity.
- Give the principles of the radiochemical methods of analysis.

Radioactivity

Radioactive Decay

Nearly all known elements exist in several isotopic forms. Few of them are natural. The most known are probably ^2_1H, $^{13}_6C$, $^{15}_7N$ and $^{18}_8O$. Many of these isotopes are formed artificially from suitable isotopes of the same element or of other elements. Most artificial isotopes and even many natural ones are unstable. The result is that their nuclei disintegrate spontaneously. The disintegration of isotopes is called *radioactive decay*. They eject energetic particles and sometimes, together with this ejection, emit radiant energy. At the end of this phenomenon, it remains a nucleus slightly lighter in mass than before. The whole of this phenomenon is named the phenomenon of *radioactivity*.

Different types of particles are ejected by radioactive decay. Those which are important for purpose of chemical analysis are the electron, the positron (positive electron) and the alpha particle. The emission of these particles is often accompanied by the radiation of energy as gamma rays. Table 1 mentions some characteristics of particles produced in radioactive decay:

Table 1: Particles Produced in Radioactive Decay.

Particle	Symbol	Mass[1]	Charge[2]
Electron	β^-	5.439×10^{-4}	-1
Positron	β^+	5.439×10^{-4}	$+1$
Alpha particle	α	3.9948	$+2$
Photon (gamma ray)	γ	0	0

[1] in units of $1.673 \times 10^{-27}\,kg$

[2] in units of $1.60240 \times 10^{-19}\,C$

- Rays α, also called α particles, are the nucleus of helium atom $^4_2He^{2+}$ is specific of the natural radioisotopes. As an example of decay:

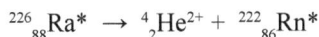

$$^{226}_{88}Ra* \rightarrow {}^4_2He^{2+} + {}^{222}_{86}Rn*$$

- Rays β⁻ (electrons)

$$^{40}_{19}K* \rightarrow {}^0_{-1}e + {}^{40}_{20}Ca$$

- Rays β⁺ (positrons)

$$^{30}_{15}P* \rightarrow {}^0_1e + {}^{30}_{14}Si$$

These rays are only encountered in artificial radioactivity.

- Rays γ (electromagnetic rays), very energetic, accompanying the emission of particles α, β⁺, β⁻

Another particle also deserves mention, it is the neutron together with another mode of radioactive decomposition which is the spontaneous capture by the nucleus of an electron from the K or L level. This is the process of *electron capture* (see preceding chapter).

Half-Life of One Isotope

In all cases, the rate of disintegration depends only on the number n of radioactive atoms at any moment. Hence, we can write:

$$-dn/dt = \lambda n$$

λ is the constant of disintegration of the radioactive isotope, $\lambda n = A$ its activity (1) which is also named rate of disintegration at the moment t. This is a differential equation of first order at separate variables. It leads with $n = n_0$ for $t = 0$ at the decreasing exponential law of radioactivity:

$$n = n_0 e^{-\lambda t}$$

The radioactive period T is defined as being the time necessary for the initial activity to be reduced by one half, that is to say:

$$n_0/2 = n_0 e^{-\lambda T}$$

$$\ln(1/2) = -\lambda T$$

$$\text{and} \quad T = 0{,}693/\lambda$$

Hence, the radioactive period is a characteristic constant of the isotope via the value λ. It is determined by following the activity as a function of the time. Effectively, according to the preceding relations:

$$\log A = \log \lambda n_0 - (\lambda/2.3)t = cte - (0.3/T)t$$

The disintegration curve $\log A = f(t)$ is a straight line, the slope of which permits the calculation of T and the identification of the radioactive isotope (Figure 1).

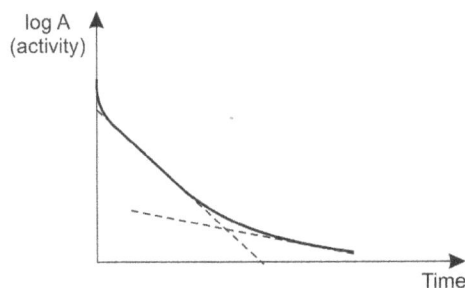

Figure 1: Example of Disintegration Curve with Formation of Manganese 58.

Detection and Measurement

The detection and the measurement of the radioactivity are grounded to the three fundamental properties of the nuclear rays, i.e., photographic impression, ionization of encountered substances and excitation of their fluorescence spectra.

Photographic Measurements

One takes plates on which the sensitive emulsion is thick and rich in silver bromide. If one sets down a solution of a salt of a radioactive element on the surface, the emitted particles amble along through the emulsion, make an impression on it and finally will be revealed by the development of the plate. β-rays are difficult to reveal but α-rays ionize silver bromide very strongly and give large traces easily visible with a microscope. This process permits measuring the energy of the emitted particle in some experimental conditions. (It is due to photography that radioactivity has been discovered – Becquerel).

Ionization of Gases and Measurements Apparatus

The particles α or β are absorbed through the matter with formation of ions, the emitted electrons possessing a sufficient energy to ionize other atoms or molecules along their path. Such ionization is still efficient in the case of γ-rays, which are absorbed by the photoelectric effect or by Compton effect (see the preceding chapter). The ionizing power increases with the mass, the charge and the energy of the nuclear species.

Therefore, for an indicative viewpoint, the number of ions pairs formed per cm of traversed air: for α-rays (1 to 70); for β-rays (100 to 700), γ-rays (30,000 to 50,000). The penetrating power follows the progression α → β → γ.

Geiger-Müller Counter

This is the most used counter, since among other reasons, it is well-adapted the detection of β-rays which are emitted by the most radioisotopes (Figure 2).

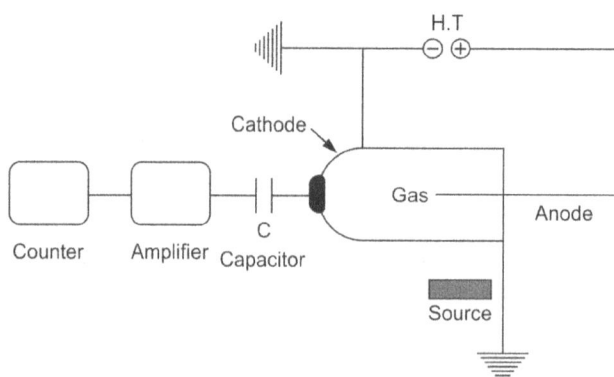

Figure 2: Scheme of a Geiger-Müller Counter.

It is constituted by a metallic tube (a cylinder of perhaps 2 cm in diameter and 10 cm in length) filled with a gas (argon and helium) under reduced pressure. It plays the part of a cathode. It is also constituted by a central wire passing along the axis of the metallic tube. The central wire is brought to a voltage of 500 to 2,500 V through a high resistance R to the positive terminal of a variable-voltage of direct current supply, while the tube is held at ground potential, as the negative point of the power supply. The central wire plays the part of the anode. The latter (positive anode) is connected through a capacitor C to the input of an amplifier (operational amplifier). The radioactive substance located face to the window (mica-10 microns thickness) related to the ground.

The rays emitted by the substance traverse the wall of the tube and provoke the formation of a small discharge between the central wire and the wall. This discharge induces a brief and sharp decrease of the potential V_B; it is transmitted by the capacitor C to an amplifier and then to a numerator. Therefore, the very intense field created thus as it is described by the anode and permits an acceleration of the electrons a succession of ionizations. One can say that the current produced through the resistance R corresponds to an important ohmic fall. In these conditions, the voltage of the anode reaches a critical value for which the successive ionizations can no longer occur. The discharge stops by itself.

With this device, it is easy to count the particles β and the rays γ. Particles α are not sufficiently penetrating. It is interesting to mention that the numerator of impulsions permits counting the ionizing particles having penetrated into the counter, hence to determine the intensity of the radiations and the activity of the radioactive preparation. The "dead-time" between two impulsions is on the average 10^{-4} s. This result permits a measurement which can go up to 10,000 kicks/second.

Scintillation Counters

A scintillation counter is built as follows. A big transparent Crystal C of a particular substance is placed near the radioactive source S, in obscurity (Figure 3). The rays emitted by S go into C and provoke a luminescence, that is to say a small luminous trail. A sensitive Detector D (a multiplier of electrons) is impressed by the small trail. The current produced by this detector is amplified and then a special apparatus called a numerator, records and counts the number of pulses given by the amplifier. Therefore, one can count the number of particles which have a radioactive origin and which are emitted by the source S.

When the matter of the subject is the occurrence of rays β, one uses a crystal of sodium iodide which contains traces of thallium (in some modalities and crystals of anthracene). In order to reveal particles α, one uses a screen built in zinc sulfide containing traces of copper. In this general presentation of scintillation counters, there is the apparatus in which an excitation of the fluorescence in the domain of visible of some crystals takes place.

In these methodologies, the "dead times" (10^{-6}–10^{-7} s) are more reduced than with Geiger Müller counters. As a result, there is a better precision on the high values of counting.

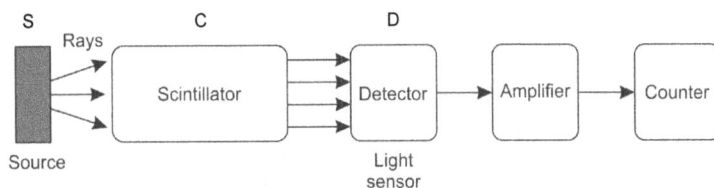

Figure 3: Scheme of a Scintillation Counter.

Other possibilities of counting exist. They are the use of electroscope, if needed the Wilson's chamber and solid-state ionization detectors (see the preceding chapter).

Unities

Activity is given in units of second^{-1}. The becquerel (Bq) corresponds to 1 decay per second. That is:

$$1 \text{ Bq} = 1 \text{ s}^{-1}$$

An older unit of activity is the Curie (Ci) which was originally defined as the activity of 1 g of radium 236.1 curie is equal to 3.70×10^{10} Bq, i.e., 1 curie corresponds to 3.70×10^{10} disintegrations per second. Activities of analytes range usually from a nanocurie or less to a few microcuries.

Precision of the Measurements

It depends on several factors. Here, we only consider two aspects of the precision of radiochemical methods.

Losses of Radiations Before the Counter. They are the losses of radiations before the counter which depend on the location of the sample, its geometry and its thickness. Therefore, it is necessary to appeal to a standardization with the help of a source of known activity (U_3O_8 – uranium oxide which is the most stable) in the same experimental conditions in order to determine the efficacity factor. By doing that, we are eliminating the imperfections of the set of measurements which only records a part of the rays arriving to the counter; absorption through the window, transparence to the radiations and "dead time" between two pulses.

Statistical Uncertainties. The number of counts (disintegrations) is submitted to statistical fluctuations. As a result, the precision of observation increases with the number N of measured impulsions since the aleatory errors neutralize each other. Given the normal law of distribution of N the standard deviation σ_N has the expression:

$$\sigma_N = \sqrt{N}$$

If for example 10,000 disintegrations have been recorded in 10 minutes, the standard deviation is equal to 100 disintegrations. The error is hence smaller than 200 disintegrations for 95% of the cases. Thus, with the uncertainty of 5%, the precision of the measurement is 2%:

$$N = 10,000 \pm 200 \text{ pulses}$$

It is this precision of 2% that is entertained. For weaker number of pulses, the precision falls quickly.

Analysis by Radiochemical Methods

The radiochemical methods of analysis are simple and rapid. Overall, they are extremely sensitive. In order to prove that, let us calculate at what quantity of radioactive matter can correspond to a number of 10^3 pulses/minutes. One considers that the total efficacy of the measurement is 10%. Thus, the real activity is of 104 impulsions/minutes. According to the definition of the activity (see preceding sections):

$$A = \lambda n = (0.693/T)n$$

Or

$$A = (0.693/T)(6.10^{23}/M)m$$

The right term permits obtaining the result in a mass of product m and M is its atomic mass. This term results from a simple rule of steps test. $6\ 10^{23}$ is the Avogadro number. The general expression relating the analyzed mass of the radioactive compound and the number of impulsions is:

$$m = (MT/4.17.10^{23})\ A$$

In the case of the isotope ^{31}Si, emitter β of period $T = 2.6$ h, m is equal to $0.12.10^{-15}$ g! One can appreciate how low the limit of detection of the radiochemical methods of analysis is.

Direct Radiometric Determinations

They apply to the natural radioelements such as uranium and thorium. The case of potassium, element which is difficult by a chemical way, is particularly interesting. It is sufficient to draw a calibration curve activity concentration by starting from solutions of known concentrations of a potassium salt. The used (natural) isotope is ^{40}K is a β emitter and of period $1.4.10^9$ years. Its content is 0.012%.

Analysis by Activation

In this method, activity is induced in one or more elements of the sample by irradiation with suitable radiation or particles. The resulting radioactivity is then measured. The sample is submitted to the irradiation of thermal neutrons of an atomic pile and then is analyzed with the help of the formed radioisotopes. Usually, the irradiation is carried out by neutrons bombardment of some stable nuclides. Then there is release of radiations. Three sources of neutrons are employed; nuclear reactors and radionuclides sources, such as the Transuranium elements and accelerators (cyclotrons). Neutrons formed within all three sources are often too highly energetic. They require slowing down to energies of –0.04 MeV ($1.6022\ 10^{-19}$ J = 1eV).

The energy loss is realized by passing them through a moderating material containing many protons or deuterium atoms, such as water, paraffin or deuterium oxide. There are collisions with the nuclei in the moderator. Neutrons obtained in that manner are called "thermal neutrons". For some light elements such as nitrogen, oxygen, fluorine, silicon and fast neutrons (of about 14 MeV) are necessary. They are produced by accelerators which are generator of deuterium ions which are accelerated by a difference of potential.

Neutron capture is the most important reaction for activation methods. The newly formed nuclide is in a high excited state since, in the process, it acquired about 8 MeV of energy. This excess energy is finally released by a gamma-ray emission, as in this example

$$^{23}_{11}\text{Na} + {}^{1}_{0}\text{n} \rightarrow {}^{24}_{11}\text{Na} + \gamma$$

The theory of this analysis method is the following. When it is exposed to a flux of neutrons, the rate of formation of radioactive nuclei from a single isotope is:

$$dN^*/dt = N\phi\sigma$$

dN^*/dt is the rate of formation of active particles in neutrons per second. N is the number of stable target atoms, ϕ is the flux ($cm^{-2}s^{-1}$) of neutrons and σ is the capture cross section in cm^2/target atom. σ is a measure of the probability of the nuclei reacting with a neutron at the particle energy employed (σ is measured in barns; 1 b = 10^{-24} cm^2/target atom and metric unit of area). Once formed, the radioactive nuclei decay at the rate:

$$-dN^*/dt = k\ N^*$$

During the irradiation with a uniform flux of neutrons, there is at the same time formation of active particles via N* and decay of the radioactive nuclei symbolized by the term λN^*. At the beginning of irradiation, the equilibrium is not reached. One can represent the phenomenon by the differential equation:

$$dN^*/dt = N\phi\sigma - \lambda N^*$$

The solution of this differential equation is (*t* from 0 to *t*)

$$N^* = N\phi\sigma/\lambda[1 - \exp(-\lambda t)]$$

By replacing λ by its expression:

$$N^* = N\phi\sigma/\lambda[1 - \exp(-0.693t/T)]$$

And by introducing the activity A

$$A = \lambda N^* = N\phi\sigma[1 - \exp(-0.693t/T)]$$

$$A = N\phi\sigma S$$

Here S, which is the term between the braces, is called the saturation factor. This relation can be written in terms of experimental rate measurements by introducing the parameter c. The parameter c is given by the relation;

$$R = cA = c\lambda N$$

R is named the counting rate and c is a constant called the detection coefficient.

$$R = N\phi\sigma c[1 - \exp(-0.693t/T)] = N\phi\sigma Sc \tag{1}$$

c depends on the nature of the detector, the efficiency of counting disintegrations and the geometric arrangement of the sample and detector. Of course, according to what is preceding:

$$R = R_0\, e^{-\lambda t}$$

In the majority of cases, irradiation of the sample and standards is carried out for a long period enough to reach saturation. In these conditions, all the terms except N on the right-hand side of (1) are constant. As a result, the number of analyte radionuclides is directly proportional to the counting rate. Usually, it is interesting to irradiate a standard (although it is not obligatory) with very similar characteristics alongside the sample. If one admits the assumption that the specific activities of the analyte and the standard are identical, we may write the indices x and s; then we may write:

$$Rx = kw_x$$

$$Rs = kw_s$$

Here, k is a proportionality constant and x and s represent the analyte and the standard. Dividing one equation by the other:

$$w_x = (Rx/Rs)w_s$$

All the isotopes of the same element have the same chemical properties. If one of them is radioactive, one can follow its behavior thanks to the radioactivity. The radioisotope plays the part of a radioactive indicator or a radioactive tracer. The studied element is called the marked element.

The applications of the tracers have considerably increased in number since the development of artificial isotopes which permit marking any element. If we only consider the problems of analysis, let us notice that all questions of efficiency of separation (precipitation, extraction and distillation) can be elegantly treated in this way. It is also on the use of tracers that one of the most powerful methods of the radioactive ones, the analysis by isotopic dilution, is based.

Let us consider a compound C, the concentration of which is x. It must be dosed in a mixture of which we take p grams. For this sample, one adds m grams of a radioactive preparation of the compound C of specific activity a = A/m (the specific activity corresponds to the activity per gram of substance). The mixture contains at this point of the experiment (m + xp) of the corps C pure of specific activity a':

$$a' = A/(m + xp) = am/(m + xp)$$

Therefore:

$$x = m/p(a/a' - 1)$$

In order to know x, it is sufficient to know the specific activity a', to separate a quantity m' of C and to measure its activity A', with:

$$a' = A'/m'.$$

Let us remark that in this kind of determination, the separations must not obligatorily be complete. It is sufficient to isolate sufficiently pure substance in order to measure its activity. This is the reason why the method is advantageous in the cases of mixtures of neighboring compounds.

Applications

- Activation methods are particularly advantageous for several reasons. Among them, we can mention:
  - The fact is that often this is a non-destructive method. This property permits the application of it to the analysis of delicate samples.
  - It has high sensitivity.

They suffer from two disadvantages:

- They need large and expensive equipment.
- They require long times for complete analysis.

The domain of applying the neutron activation analysis is probably very large since, potentially, it applies to the determination of about 70 elements.

The isotope dilution technique has been employed in different and sometimes very different domains. It has been used also for the determination of organic compounds. It is particularly interesting when concentrations are very weak.

Part III
Various Methods

Introduction
Definitions and Classification of the Thermal Methods of Analysis

According to IUPAC (International Union for Pure and Applied Chemistry) and ICTAC (International Confederation for Thermal Analysis and Calorimetry) the general title "Thermal Methods" or "Thermoanalytical and Enthalpimetric Methods" refers to two groups of analytical methods, which are called "Thermal Analysis" or "Enthalpimetric Analysis".

The term *thermal analysis* regroups the methods in which a physical property (fusion, crystallization, calorific (thermal) capacity, glass transition and solid-solid transition) or a chemical property (dehydration, oxidation, decomposition and desolvation) of the studied substance is measured as a function of its temperature that the operator may change according to a programmed mode. There exist several methods of this kind. We confine ourselves to only mentioning those we shall study here, that is to say:

- The thermogravimetry (TG)
- The thermal differential analysis (ATD)
- The calorimetric differential analysis or "Differential Scanning Calorimetry or DSC".

The term *enthalpimetric analysis* regroups the methods in which the change of enthalpy accompanying one reaction of the studied substance by which a substance is directly or indirectly determined quantitatively. Let us mention in this group the thermometric titrimetry.

CHAPTER 38

Thermogravimetry

Thermogravimetry (TG, TGA or thermogravimetric analysis) is a method in which the mass of the substance to analyze is measured as a function of its temperature, the change of which is planned. We shall successively:

- Go into more details concerning its principle;
- Give some definitions in order to correctly infer the results from the experimental curves;
- Describe the apparatus.
- Give some examples of applications and mention others.

Principle

The principle of the method is based on the fact that under the influence of the heating, the substance under study is decomposed or, another alternative, its volatile components disappear.

The measurement of the change of the mass of the analyte as a function of the temperature implicates that some conditions should be satisfied:

- The increase or decrease at a constant rate of the temperature of the sample must be possible.
- It is the same for the possibility of keeping the temperature of the sample constant and for the possibility of recording the mass temperature as a function of time.

Likewise:

- A change of the rate of heating must in some conditions be null.
- One judicious adaptation of the rate of heating to the changes of masses in order to optimize the resolution must be feasible. The rate of heating can be high when there is no reaction from the sample. It can be slowed down once the mass evolves. Here is the matter of the high resolution thermogravimetry.
- Finally, the superposition of a sinusoidal change of the temperature to its linear variation is interesting. It is modulated thermogravimetry (MTG or MTGA). This method offers possibilities for the determination of kinetic constants of the studied processes.

Remark: In the case of loss of mass by vaporization, one may notice that the principle of thermogravimetry obeys the same concept as the conventional method of the pharmacopeia of different countries entitled "loss to desiccation" or "loss to calcination" which consists in determining the loss of mass by desiccation in well-defined experimental conditions.

Definitions

The apparatus of thermogravimetry provide thermogravimetric curves. They are diagrams mass of the sample (written in ordinates) as a function of its temperature T (or of the time *t*) spent since the founding of the change in temperature if the latter is carried out linearly (this is usually the case) (Figure 1). The axis of ordinates can also be graduated in percentages of the total mass lost.

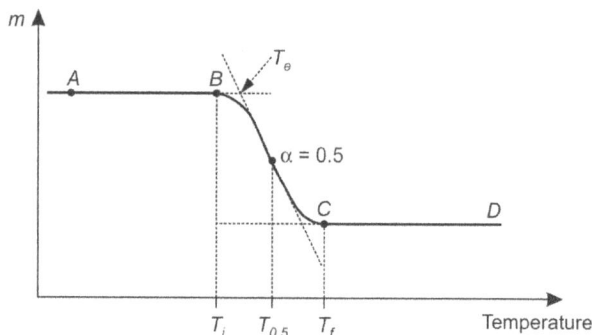

Figure 1: A Thermogravimetric Curve. (Reprinted with Permission from Gwenola and Jean-Louis Burgot, Paris, Lavoisier, Tec et Doc, 2017, 510).

The curve is constituted of plateaus when the mass of the sample is constant (plateaus AB and CD). In the example just above, the part BC corresponds to a loss of mass. The initial temperature T_i (point B) is the temperature (°C or K) from which the apparatus is able to detect a change in the mass. The final temperature T_f (point C) is that for which the loss of mass is maximal. The interval of reaction is the temperature interval $T_f - T_i$. T_i also depends on the sensitiveness of the apparatus and on the rate with which the reaction is carried out. Therefore, other approaches are envisaged, among which the extrapolation of the initial temperature T_e. When it is difficult to carry out (for reasons of slow decomposition for example), one rather determines T_α which is measured for a fraction α of mass lost such as $T_{0.5}$ when the half-quantity of the sample has disappeared.

Apparatus

The apparatus of thermogravimetry comprises:
- An analytical balance
- A furnace (a drying room at a controlled temperature)
- A device for the introduction or the purge of the atmosphere of the system
- A computer for the control of the apparatus and of the treatment of data

The thermobalances constituent the sensitive device. They can give information concerning samples the mass of which extends from 1mg to 100 mg; usually in the range of 5 to 20 mg. One estimation of the precision of present thermobalances is given by the following figures; 10 mg of the sample can be weighed with a precision of 0.1%. The whole balance, except the sample and its holder, must be thermically isolated from the furnace. The holder of the sample (the crucible) can (usually) contain 1 to 20 mg.

(*Remark*: For this purpose, it must be known that an empty crucible has its mass which increases when it is heated from the ambiance temperature up to about 1,000°C. Hence, it is necessary to proceed to corrections. They can be done automatically by the apparatus.)

The furnaces generally procure changes of temperature which linearly vary from −160 to 1,600°C. The scanning rates of the temperature must be reproducible. Generally, they are located in the range of quasi-zero until 200°C/minutes, a value considered as being high. Placing the intermediary compounds in a prominent position by thermogravimetry depends, in effect, much on the rate of heating. However, when the activated reaction is reversible and rapid, it practically plays no part.

A device of introduction or purge of the atmosphere is necessary since the nature of this latter may exercise a strong effect on the temperature of decomposition and the nature of the processes. Nitrogen or argon are often used in order to purge the atmosphere and to avoid the decomposition of the sample by oxidation by the dioxygen of air. The inverse operation can be carried out with some apparatus. For example, one can introduce the dioxygen at a given moment of the experiment in order to precisely study the eventual oxidation of the sample.

The measurement of the sample temperature is often carried on with the help of thermocouple platinum/platinum-rhodium. Let us recall that the principle of functioning of a thermocouple is based on the fact that when two wires of different metals are welded together only by their extremity and that extremity is placed at temperature T_1, the other is at one temperature T_2 which has a difference of potential settles between the two extremities. It is proportional to the difference $(T_2 - T_1)$. This fact permits the measurement of the temperatures with the help of a millivoltmeter.

Some apparatus permits the recording of the diagrams dm/dt or dm/dT as a function of the temperature T. It is the differential thermogravimetry. It permits evidence of the changes which spontaneously, and successively occur at temperatures that are close to each other when a slow reaction is followed by a rapid reaction. Moreover, the basic line comes automatically back to zero when the process is achieved since the slope is then null.

Finally, the thermogravimetry can be coupled with other methods, such as, the spectrometries of mass or infrared. The origin of the observed phenomenon can be then more easily cleared up.

Examples

A classic example is provided by the copper (2) sulfate pentahydrate $CuSO_4$, $5H_2O$ (Figure 2). The successive transformations occurring under the effect of the heat are given in the following balance sheets:

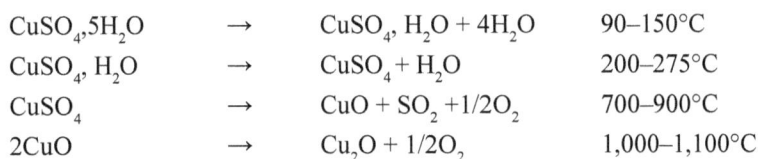

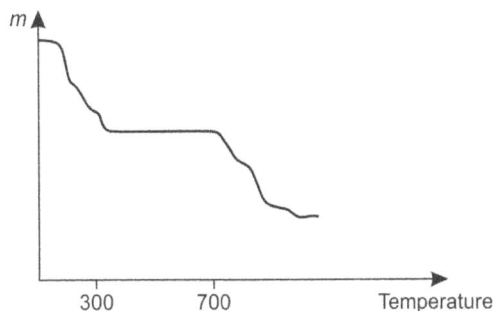

$CuSO_4, 5H_2O$	$\rightarrow$	$CuSO_4, H_2O + 4H_2O$	90–150°C
$CuSO_4, H_2O$	$\rightarrow$	$CuSO_4 + H_2O$	200–275°C
$CuSO_4$	$\rightarrow$	$CuO + SO_2 + 1/2O_2$	700–900°C
$2CuO$	$\rightarrow$	$Cu_2O + 1/2O_2$	1,000–1,100°C

Figure 2: Thermogravimetric Curve of the Copper (2) Sulfate $CuSO_4$ and $5H_2O$. (Reprinted with Permission from Gwenola and Jean-Louis Burgot, Paris, Lavoisier, Tec et Doc, 2017, 512).

Another example is provided by the tetracycline base (Figure 3). In the conditions of experience, the tetracycline base exhibits two plateaus of desolvation; a "true" plateau which indicates the occurrence of one hexahydrate, besides appears at whatever the sample is. The other plateau is what corresponds to the partial drying of the product. Its height varies with the sample.

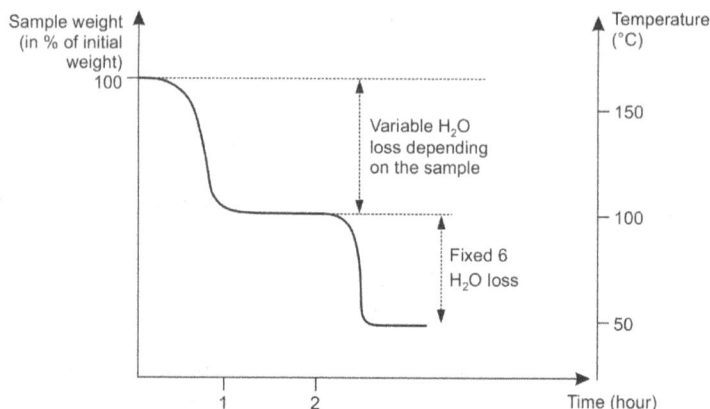

Figure 3: Thermogravimetric Curve of the Tetracycline Base. (The Abscissa Points Out the Intervals of Time in Hours and Ordinates. On the Left is the Weight of the Sample in the Percentage of the Initial Weight and On the Right is the Temperatures in Degrees Celsius). (Reprinted with Permission from Gwenola and Jean-Louis Burgot, Paris, Lavoisier, Tec et Doc, 2017, 513).

Applications of Thermogravimetry

The thermogravimetry is evidently used for the recognition of the processes which are shown by a loss of mass of the sample. This can be the volatilization of some impurities, the transformation of the sample and even, at the limit, the degradation of the studied compound.

Placing in a Prominent Position the Volatile Impurities

The thermogravimetry permits knowing the quantity of water contained in the sample (see the section on 'Examples'). Another interesting example is that provided by the dihydrate of the complexon II, Na_2 edta and $2H_2O$ which tends to absorb slightly more than 2 water molecules. The method shows that it can be dry by starting from 80°C for the excess water, variable beyond its two molecules of water. The loss of these latter ones involves a stronger heating. The salt begins to dehydrate by starting from 110–120°C.

Water is not the sole volatile impurity. Let us mention the residual solvents the importance of which is notable in the domain of the pharmacy. After their separation, the active principles regularly contain more and less solvents used in the last steps of their preparations. They are undesirable and their quantities are governed by assigned limits. Hence, their identifications and their quantifications are very important since some are toxic. When they are free, solvents vaporize in some domain of temperature. In contrast, the solvents which are incorporated in crystals are only discovered during the fusion, that is to say, sometimes at very high temperature.

The advantage of thermogravimetry in this domain is its rate of use which is relatively high, its simplicity of handling and the weak mass of sample necessary for one analysis. Thus, some pharmacopeias prescribe the use of thermogravimetry for the determination of the assay "loss of the desiccation". In USP 24, it is the case of the sulfate of vincristine, the sulfate of vinblastine and the mesylate of bromocryptine. In the European pharmacopeia (10th Edition), it is the case of some biological products, such as meningococcal vaccine, typhoidal or Hemophilus influenzae vaccines. It is estimated that the presence of volatile impurities can be judged up to a percentage as weak as 0,1% in a sample.

Study of the Transformation or the Decomposition of a Compound Under the Action of Heat

Some derivatives may undergo one dehydration under the action of the heat with rupture and formation of chemical bonds. For example, the formation of a cyclic amide is possible. Another possibility is the study of the loss of mass or of the gain of mass in an atmosphere of dioxygen or of another reactive gas. It is in a such manner that the study of the quality of vegetal oils has been carried out.

It is easily conceivable that the thermogravimetry is a very valuable auxiliary method of the gravimetry. The study of the thermogravimetric curve procures essential information concerning the stability of precipitates and also on their conditions of drying and eventually on their kinetic of decomposition.

However, the thermogravimetry is less useful in this domain than are the thermal differential analysis (ATD) and the calorimetric differential analysis (DSC) because the latter ones can apply to other processes than those which solely involve a change of mass (see Chapter 39). Moreover, it has the disadvantage not to permit the identification of the fragments or species which are forming. This is the reason why it is sometimes coupled with other instrumental methods.

CHAPTER 39

Thermal Differential Analysis and Calorimetric Differential Analysis

The thermal differential analysis and calorimetric differential analysis are methods of thermal analysis. They permit to studying the transitions of phases and the evolving of chemical reactions when the compound under study is heated or cooled to a temperature close to that of transition. Transitions of second-order and glass transitions may also be investigated with these methods. We shall successively consider:

- The definitions of these methods
- The physical phenomena which are of interest to them
- Some theoretical aspects in order to explain the possibilities of quantitative determinations with these methods
- The apparatus
- Their applications

Definitions

• The *Differential Thermal Analysis (ATD)* is the method in which the difference in temperature between the sample to study and one compound of reference is measured as a function of the temperature when both compounds are submitted to the same program of variation of it.

• The *Calorimetric Differential Analysis* or *Differential Scanning Calorimetry (DSC)* is the method in which one measures the difference of energy necessary to keep the sample and the reference at the same temperature when they are submitted to the same program of variation of temperature.

Physical Basic Phenomena

The comparison of the curves obtained in ATD and of the individual curves of changes of temperature of the product to study and of the reference permits explaining the obtained diagrams (Figure 1). For the chosen example, one supposes that the sample is submitted to a transformation of one allotropic form to another.

The curves of the temperature changes of the sample and of the reference coincide from point A up to the point B. At the point B, the transformation of an allotropic form into another is realized. Then the temperature of the sample becomes weaker than that of the reference (it is an endothermic phenomenon). The transformation is continuing until the point D by passing by the point C. From E up to F, both compounds again exhibit the same temperature.

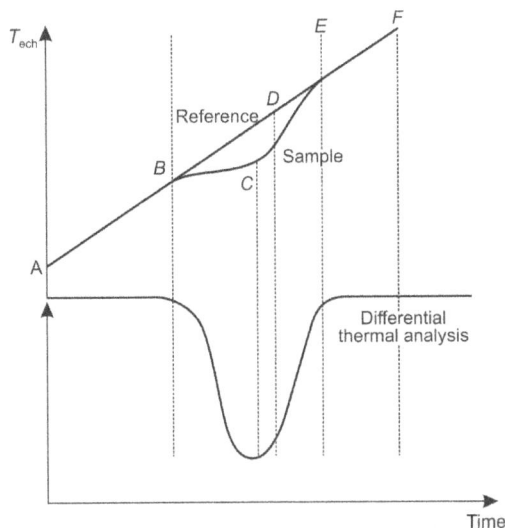

Figure 1: Comparison of the Curves of ATD and Changes of Temperature of the Sample and the Reference. T_{ech} is the Temperature of the Sample and ΔT is Difference of Temperature Between the Sample and the Substance of Reference. (Reprinted with Permission from Gwenola and Jean-Louis Burgot, Paris, Lavoisier, Tec et Doc, 2017, 516).

(*Remark*: It is interesting to highlight in passing that the transformation does not evolve at constant temperature as it should be required by the rule of phases since the phenomenon evolves at constant pressure. The reason lies in the fact that it is submitted to an unceasing change of temperature, according to a forced manner and the principle of the method.)

Let us consider the corresponding curve of ATD. From A to B, the two compounds are at the same temperature. This is expressed by the fact that there is no electromotive force to appear in the circuit where are placed the sensor (see the section on 'Apparatus'). As a result, the slope of the ATD curve is null between A and B. Between B and D, the transformation evolves and the temperature of the sample is weaker than that of the reference. As a result, there appears a potential difference, both compounds being submitted to an increasing temperature. The studied compound shows a slight increase of its temperature, although its transformation takes place simultaneously. The difference of the maxima ΔT is reached at the point C. From point C to point D which represents the end of the phenomenon of fusion, the quantity of sample transformed is weak. Finally, they are at the same temperature from E to F. No electromotive force appears.

Naturally, if an exothermic chemical reaction concerning the sample affecting the sample could evolve in place of the endothermic (preceding) physical phenomenon, its temperature would be higher than that of the reference. The differences of temperature and of potential would be of opposite signs to those of the preceding case. The thermogram would present a peak in the opposite sign.

Theoretical Aspects

The ATD and the DSC permit not only qualitative determinations but also quantitative measurements. We give the grounds of these ones in this paragraph.

From a general viewpoint, the position, slope, height and area of a peak are characteristic of the transition or the reaction responsible for the obtained curve. As a result, the study of a thermogram permits obtaining three types of information. They concern:

- The temperature at which the transformation or the reaction evolves
- The heat of transformation
- Its rate

The quantitative studies are based on the fact that the heat of transformation or of reaction is related with the Area A of the peak of thermogram. If the heat of transformation or of the reaction is known, the area of the peak can be used in order to determine the quantity of product which has been transformed or which has reacted, that is to say the quantity of products responsible for the peak.

Determination of the Mathematical Relation Between the Heat of Transformation or Reaction and the Area Under the Peak.

Let the temperatures of the sample and the reference T_e and T_f at a given moment and T_o that of the external ambient surrounding (temperature of the furnace; see later). The calorimetric balance-sheet is for each product:

$$Cp_e dT_e = dH + k_e(T_o - T_e)dt \qquad (1)$$

$$Cp_r dT_r = k_r(T_o - T_r)dt \qquad (2)$$

For the sample, the gain of enthalpy (the thermal effect that it receives under constant pressure). $Cp_e dT_e$ is equal to the heat evolved by the process at constant pressure dH (dH is negative) to which it is necessary to add that it receives from the surroundings $k_e(T_o - T_e)$ during the time dt. For reference, only plays a part in the exchange with the surroundings. k_e and k_r are the coefficients of heat transfer of cells where are respectively the sample and the reference (dimension: joule/degree x temps). Some hypotheses concerning these basic relations can be done reasonably:

- The temperature is uniform in both compartments (sample and reference). This condition is not obligatorily easily obeyed in the case of solids.
- The heat is transferred to the sample and to the reference only by conduction.
- The coefficients of heat transfer are equal:

$$k_e = k_r = k';$$

The calorific capacities of both products can be considered as being identical:

$$Cp_e = Cp_r = Cp$$

With these hypotheses, according to (1) and (2)

$$CpdT_e = dH + k'(T_0 - T_e)dt$$

$$CpdT_r = k'(T_0 - T_r)dt$$

By substracting these two relations and by setting $T_0 - T_r = \Delta T$ and hence $dT_e - dTr = d\Delta T$, we obtain:

$$dH = Cpd\Delta T + k'\Delta tdt \qquad (3)$$

$$\Delta H = Cp(\Delta T - \Delta T_0) + k' \int_0^\infty \Delta Tdt$$

Since $\Delta T = 0$ pour $t = 0$ and for $t \to \infty$ and since the sum is equal to the area A under the peak of the thermogram, we can write:

$$\Delta H = k'A \quad \text{with} \quad A = \int_0^\infty \Delta Tdt$$

The area under the curve is, as it has already been said, proportional to the heat evolved during the process at constant pressure. The constant k' is characteristic of the used instrument. Its value is obtained by studying standard compounds, that is to say compounds giving rise to processes, the heat effect of which is known.

Since, in DSC, the energy required to maintain the sample and the reference exactly at the same temperature is measured as a function of the temperature, one conceives that it is better adapted than ATD for the determination of the variation of enthalpies accompanying a physical process or a chemical reaction.

The relation $\Delta H = k'A$ also permits making kinetic measurements since the heat evolved at constant pressure dH during a very brief moment dt, is proportional to the number of moles of the sample transformed or having reacted (during the same time).

These kinetic measurements, as all the kinetic measurements, permit calculating of the rate constants, the orders of reactions and the activation energies by study of thermograms. The preceding considerations constitute the justification of the quantitative measurements by DSC.

(*Remark*: However, to be put into practice, it is necessary to proceed to corrections of the obtained thermograms in order to obtain accurate data. There exists, a slight shifting of temperatures struck up by the apparatus and the true one. This fact is due to the changes in temperatures that cannot be instantaneous. The corrections are carried out graphically by following the thermal evolution of one substance the fusion point of which is known very exactly. Usually, one chooses Indium the point of fusion of which is 156.3°C. Moreover, indium is one metal. It is a very good conductor of heat and the slope of its endotherm is a satisfactory approach to the rate of the exchanges of heat. It is reported on the thermogram of the studied sample by the graph of parallels (obliques in Figures 2a and 2b). Point G indicates the temperature to which the fraction ADE has melted.)

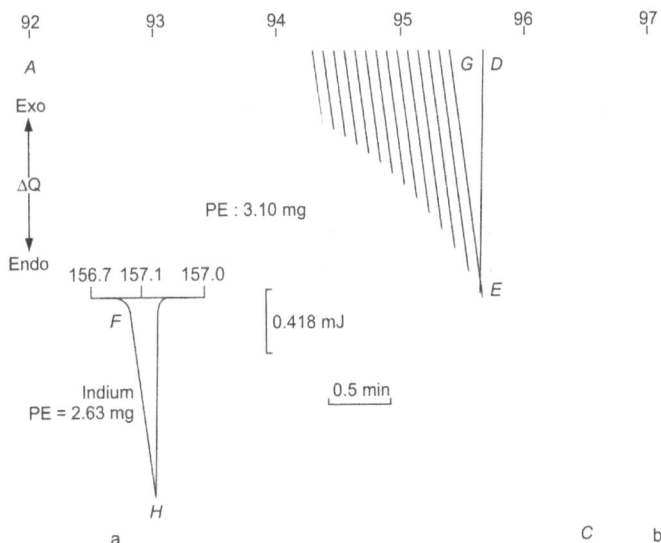

Figure 2: Graphical Determination of the True Temperature in Quantitative Analysis by DSC. (According to Masse and Chauvet, 1983) A) Endotherm of Fusion of Indium. B) Graphical Corrections (Notice that the Straight Lines GE and FH Must be Parallel) (Example of a Pharmaceutical Substance).

Apparatus

A scheme of one apparatus of ATD is given Figure 3. The heart of the apparatus is constituted by a heating block (furnace, drying-room) the temperature of which varies linearly with time according to a scheduled sequence. The block contains two identical cavities, one containing the sample, the other the reference (Figure 3). The latter must be thermally inert in the explored domain of temperature. In each cavity also plunges a sensor with thermocouple which is in immediate contact with the product. Both thermocouples are mounted in opposition in the same electrical circuitry. If the sample does not undergo a transformation, the tensions given by both thermocouples are equal

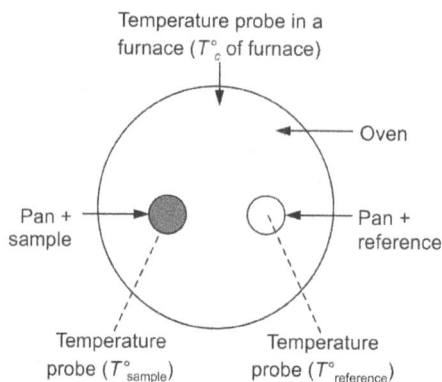

Figure 3: Scheme of an Apparatus of ATD. (Reprinted with Permission from Gwenola and Jean-Louis BURGOT, Paris, Lavoisier, Tec et Doc, 2017, 520).

and the voltages of both couples and the difference of potential at the terminals of the circuit is null. During a reaction or a process endo or exothermic, the lack of thermal equilibrium which occurs is objectivized by a potential difference which is proportional to the difference of temperature existing between the two cavities.

In some apparatus, there also exist devices which ensure the checking of pressure in the apparatus together with other devices permitting the change of the atmosphere. Usually, the weight of the samples is located in the range spreading from 0.1 mg to 10 mg. The rates of heating are of the order of 10 to 20°C/minutes.

In DSC, the sample and the standard reference material are thermally isolated from each other. Each of the two cavities is equipped by a temperature probe and by a heating device. In each cavity, each product is deposited in a small pan. The temperature of the enclosure, the sample and of the reference are varying linearly with time. The temperatures of the sample and of the reference are compared in continuous and the power of each heating device is regulated in such a manner that their temperature is maintained equal. A signal proportional to the difference of powers delivered at the two devices as a function of the temperature is registered. A scheme of an apparatus of DSC is given Figure 4.

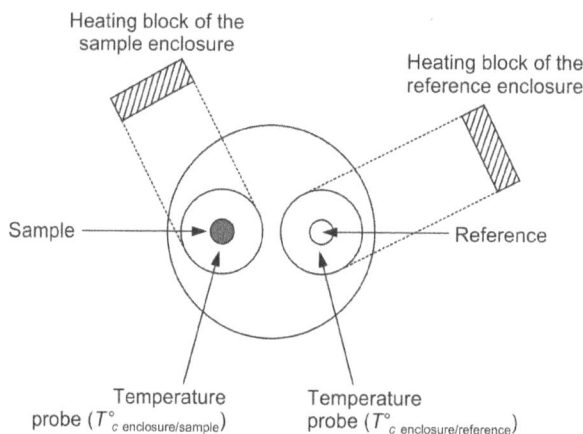

Figure 4: Apparatus of Differential Scanning Calorimetry. (Reprinted with Permission from Gwenola and Jean-Louis Burgot, Paris, Lavoisier, Tec et Doc, 2017, 521).

Samples in the range 1 to 20 mg are currently studied. Some samples of 0.1 mg can even be worked out.

Applications of ATD and DSC

It is difficult to classify the applications of ATD and DSC given their great number and given their different natures. They concern numerous inorganic and organic compounds. These methods involve several types of analyses such as:

- The proximate and quantitative analysis with the determination of the purity of a compound. Such an analysis can be the determination of its optical purity.
- The study of some chemical reactions.
- ATD and DSC also exhibit a great interest in some domains of the physical chemistry.

Proximate, Qualitative and Quantitative Analysis

Determination of the Fusion Point

Let us recall that the measurement of the fusion point of a solid constitutes an identity criterium and a purity criterium of the sample. The determination of a fusion point is an operation less harmless than it looks like at first sight. For example, there exist controversies concerning numerous substances, with among them, reference substances. Most often, they are due to the existence of several crystalline forms.

In the simplest case of a pure compound which does not suffer any modification before the fusion and which does not decompose during the fusion, the thermogram shows an endothermic peak (Figure 5). The part AB corresponds to the heating of the product in the solid state. In B, it begins to melt and as the phase rule foresees it, the temperature must remain constant, whence the vertical BC which is located precisely at the fusion temperature T_f. The part CD corresponds to the return to the thermal equilibrium from the moment C when the last crystal has disappeared. The fusion is a process endothermic and the area of the peak, in this occurrence called endotherm, is proportional to the enthalpy of fusion (Figure 5).

The form of the thermograms permits to detect the case for which the fusion is accompanied by a decomposition. Such a state of affairs does not correspond to what is called fusion *stricto sensu*. The substance can begin by decompose before the beginning of the liquefaction or decompose in liquid phase. In the first case, one follows the fusion of the eutectic substances/products of their decomposition. In the second case, one is faced with a mixture of a solid and of a liquid containing several constituents. Anyway, the thermograms show peaks which are neat or curves like bells which are perfectly reproducible in well-defined experimental conditions.

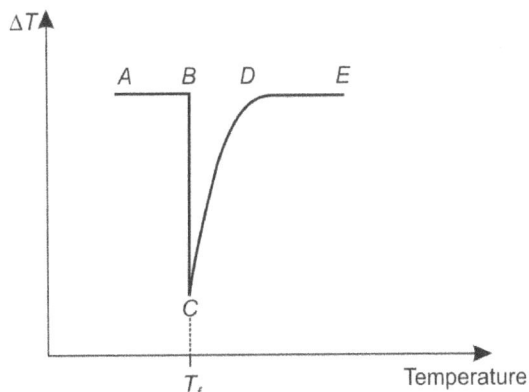

Figure 5: Thermogram Obtained During a Simple Fusion. (Reprinted with Permission from Gwenola and Jean-Louis Burgot, Paris, Lavoisier, Tec et Doc, 2017, 522).

Determination of the Purity of a Compound

ATD and DSC permit to appreciate the purity of a compound. Already, the actual form of the fusion endotherm one principal characteristic of which must be its verticality, permits very quickly to answer the question. It is the same for the presence of supplementary peaks indicating the presence of impurities.

The quantitative determination is possible. It is grounded on the presence of eutectics. We begin by recalling some characteristics of them.

EUTECTICS

Qualitative Aspects

Here, we only consider one case of system liquid-solid in which two compounds are completely miscible in the liquid state whereas the solid phases are constituted only by pure compounds.

When the liquid phase is cooled, a solid begins to separate at a definite temperature, just the temperature of congelation (or fusion). At this temperature, there are two phases (liquid + solid). Since there are two components, according to the rule of phases, the system does possess two degrees of liberty. Usually, one arbitrarily fixes the pressure. In these conditions, either the temperature or the composition of the liquid phase is sufficient to completely define the system. Therefore, by determining the freezing point of liquid solutions of different compositions in A and B as a function of these compositions, one obtains the two curves AC and BC which are useful for establishing the following binary diagram.

Figure 6a represents the phases diagram. Let us mention that the abscissa axis is graduated in molar fractions x of the compound B. The molar fraction $x = 0$ corresponds to the pure product A whereas $x = 1$ corresponds to the product B pure. Let us also recall that for a binary mixture containing n_B and n_A, the molar fractions of A and B are given by the expressions $x_A = n_A/(n_A + n_B)$ and $x_B = n_B/(n_A + n_B)$.

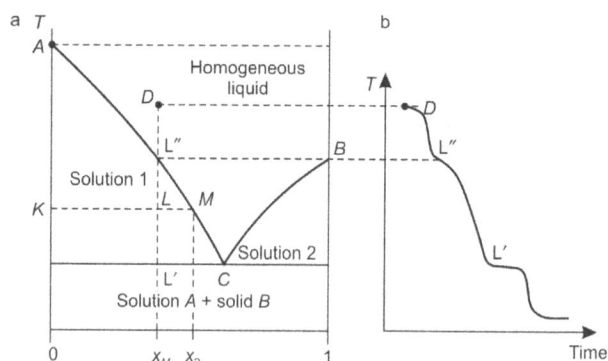

Figure 6: Diagram of an Eutectic (See the Text for the Characteristics of the System). (Reprinted with Permission from Gwenola and Jean-Louis Burgot, Paris, Lavoisier, Tec et Doc, 2017, 523).

When a rich liquid in A (point D) is cooled, the compound A separates in solid state. For an example in L, the mixture in global composition xm (in molar fraction) is formed of the liquid of composition x_2 (point M) and of the solid A (point K) in the proportion $L\bar{K}/L\bar{M}$ (general rule of moments). At the point k ($x = 0$) since A is pure (analogous considerations would apply to the branch BC, but it would be the product B which would settle).

At the temperature corresponding to the point L', the point which represents the composition of the liquid is C. The curve of the deposit of B passes also through C. At this point, with two solids immiscible and one liquid phase in presence, the system has become monovariant. At chosen pressure, the temperature corresponding to point C is fixed. It is unique. Point C is the eutectic point. Its temperature is at the lowest one it can reach.

As a result of these considerations that the existence of impurities lowers the fusion point of a pure substance. This property is used in the identification of organic compounds which melt at law temperatures.

If the temperature of the system is recorded (Figure 5b) as a function of time, that is to say as a function of the extent of the temperature lowering according to the vertical DL', one obtains the curve of Figure 5b. From D to L" there is a simple lowering of the temperature of the liquid. From L" to L' appears the compound A solid. There is liberation of the heat of fusion which lessens the cooling of the system. When L' is reached, the temperature remains constant as long it remains liquid. Then, the curve corresponds to the simple cooling of both solid phases.

Quantitative Aspects

From a theoretical standpoint, the molar fraction x_A of the compound, the purity of which is studied in the liquid of fusion, is given by the relation of Schröder-Van Laar which is an extension of the equation of Clapeyron which applicates to the phase changes. It is written:

$$\ln x_A = (\Delta H_A f / R)(1/T'_A - 1/Tf) \tag{4}$$

Here, ΔH_{Af} is the enthalpy of fusion of the pure compound A, T_{Af} is its temperature when it is pure and T_f is the reached temperature at the end of the fusion of the mixture. Let us recall that the fusion enthalpy is always positive (endothermic phenomenon). The relation (4) immediately explains why the increase of the content of impurities (x_A decreasing) provokes the decrease of the temperature of fusion of the mixture. $\ln x_A$ becomes, indeed, more and more negative in proportion as T_f decreases. The relation (4) is transformed easily in (5) in which intervenes the molar fraction x_B of the impurity in the liquid phase. It is written:

$$T_f = T_{Af} - [(T_{Af})^2 R/\Delta H_{Af}] x_B \tag{5}$$

It is obtained from the preceding one by considering that $x_B = 1 - x_A$ is very small versus 1. This point permits to write $\ln(1 - x_B) = -x_B$. Moreover, one makes the hypothesis that the product TT_{Af} is close to $(T_{Af})^2$ since the depression of the fusion point is very weak. The relation (5) can be expressed in terms of *total purity X* of the substance by starting from its definition. Let us name:

- n_B is the number of moles of B in the liquid of fusion
- n_f the total number of melted moles
- n_T the total number of moles of the sample
- $x_B = n_B/n_f$

x_B the ratio (number of moles B in the liquid of fusion)/(total number of melted moles) which can be also written: x_B = [number of moles of B in the liquid of fusion/total number of moles in the sample]x[total number of moles in the sample]/[total number of melted moles].

The first term n_B/n_T is the molar fraction X of the impurity in the sample. The second n_T/n_f is the inverse of the fraction f melted of the melted sample. F is obtained by starting from the following reasoning.

The straight line T_f as a function of 1/f has for slope $-[R(T_{Af})^2/\Delta H_{Af}]X$ and for intercept T_{Af}. ΔH_{Af} is given by the area of the endotherm of fusion, that is to say directly by DSC. The melted fraction at a given temperature is the ratio of the obtained curve since the beginning of the fusion until the given temperature and the total area under the curve. Thus, the curve of DSC gives all the information necessary to calculate X. It is one of the interests that presents the DSC versus the ATD: that is to furnish directly the energies.

Optical Purity of a Crystallized Compound

The existence of eutectics sometimes permits determining the optical purity of a crystallized compound and thus to follow its duplication by ATD by working with very weak quantities and of the order of some mg. One immediately conceives the interest that can have this possibility in the pharmaceutical field. One knows, indeed, that enantiomers may exhibit pharmacologic, therapeutic and pharmacokinetic properties different. There exist guidelines that demand that the pharmaceutical principles should be wholly described. They also demand that when an impurity isomer (included an enantiomer) present in the pharmaceutical formulation, its content and its other properties are specified.

The principle of the determination of the optical purity is rigorously the same as that described previously. It is also founded on the existence of eutectics. Let us take the example of the determination of the excess of one enantiomer in a racemic.

(*Remark*: Let us already mention the fact that the methodology exposed under cannot be applied to the all types of racemic. It can be applied only to the conglomerates (e) and to the racemic compounds.)

Recall there exist three kinds of racemic compounds:

- The conglomerates of crystals of two enantiomers. Each crystal contains only one enantiomer.
- The racemic compounds. The crystals are formed of two enantiomers in the ratio of 1/1. They are solid compounds.
- The pseudo-racemates which are solid solutions of both enantiomers in the stoichiometric ratio of 1/1.

The case of a conglomerate is shown by the existence of a eutectic state (Figure 7).

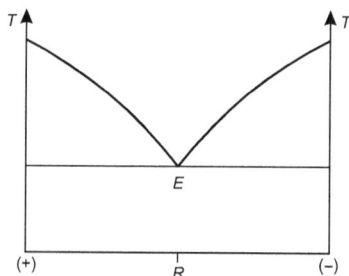

Figure 7: Phases Diagram in the Case of a Conglomerate. (Reprinted with Permission from Gwenola and Jean-Louis Burgot, Paris, Lavoisier, Tec et Doc, 2017, 526).

However, the present binary diagram is slightly different from the preceding. Both enantiomers, indeed, possess the same freezing point. Moreover, it is symmetric. The racemic is located for $x+ = x- = 0.5$. As previously, there is no mutual miscibility of the two enantiomers in the solid state and the Eutectic E consists of a mixture of the two kinds of crystals and of the liquid. When the enantiomers are in unequalled proportions, both disappear during the fusion of the eutectic mixture by giving a racemic liquid. During this fusion, the temperature remains constant and equal to TE. After the disappearing of all the racemic, it does not remain (for the chosen example) the enantiomer (–) in the solid phase and the remaining enantiomer (+) melts gradually until the temperature Tf. The obtained thermogram is represented (Figure 8).

The thermal energy necessary to liquefy all the eutectic is proportional to the mass of the racemic present. Therefore, the comparison of the areas obtained with the racemic (that is to say with the eutectic) and with a product having suffered a partial duplication permits to fix the concentration of the latter. Let us also notice that the equation of Schröder-van Laar is still applicable as just above.

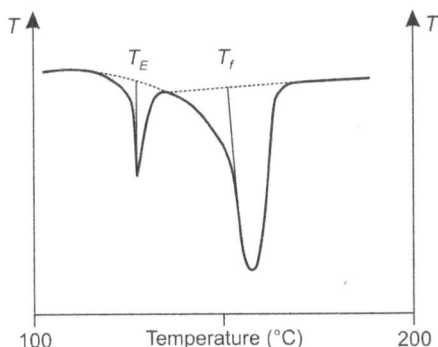

Figure 8: Thermogram (DSC) Obtained with an Excess of the Enantiomer (+). (Reprinted with Permission from Gwenola and Jean-Louis Burgot, Paris, Lavoisier, Tec et Doc, 2017, 526).

Let us also draw attention to the fact that one can determine the optical purity of a racemic compound by DSC but not that of pseudo-racemates.

Possibilities of Application in the Domain of the Physical Chemistry

Study of the Crystalline "Polymorphism"

ATD and DSC permit to detect the crystalline "polymorphism". Numerous organic compounds exist under two or several crystalline forms (the phenobarbital does possess eight forms). By confining ourselves solely to the case of dimorphism, the passing from one form to the other in the solid state is made known by a simple endotherm corresponding to the crystalline rearrangement followed by the endotherm due to the melting of the formed compound. It is the case of the sulfathiazole (Figure 8). It must be highlighted for this purpose the case of compounds with double melting points. For example, the testosterone acetate melts at 129°C gives at 132°C, an exotherm of crystallization of a second form which melts at 143°C.

When one solely disposes of the second form, only the endotherm of fusion appears (Figure 9).

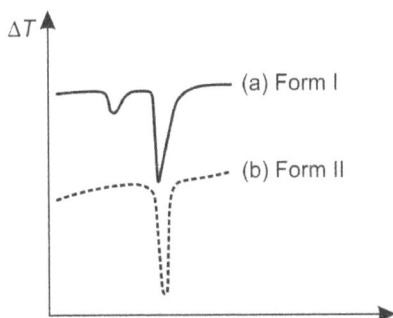

Figure 9: Polymorphism of the Sulfathiazole. (Reprinted with Permission from Gwenola and Jean-Louis Burgot, Paris, Lavoisier, Tec et Doc, 2017, 527).

Smectic and Nematic States

The smectic state (liquid crystals) is the state in which the liquid of fusion is not isotropic. The molecules have their axis parallel and can move only inside parallel shares regularly located far apart. In the nematic state, the molecules move according to any manner but keep parallel their axis. ATD and DSC permit to following of the transitions between these states and the state liquid isotropic.

Desolvation

A solvent may be retained by the crystals by simple adsorption, but it can be also retained because it is an integrant part of the crystal and, hence, participates to its reticular organization. The study of crystals often shows a different structure from that of the desolvated compound. This can fallaciously orient the phenomenon towards a dimorphism. The phenomenon is sometimes called pseudo-polymorphism. For example, the β-estradiol gives crystals of pseudo-polymorphs with the totality of about 30 usual solvents. Usually, the presence of solvent is made known by the presence of one endotherm in form of bell preceding the liquefaction the abscissa of which is function of the pressure under which one works. It may arrive that both solvents are retained simultaneously. The isotherm of fusion is then preceded by two others, the locations of which are function of the pressure.

Studies of Complex Products

ATD and DSC also permit the study of complex products:

- Fatty matters
- Soaps
- Numerous polymers such as the poly-olefines, polyamides, polyesters and polycarbonates
- Filters
- The big biochemical molecules (such as dextrans, poly-glucosans, starches) proteins (such as the gluten of corn, albumin of eggs, gelatin, etc.)
- Some catalyzers

Let us mention the fact for the purpose of the polymers that these methods permit obtaining "the degree of crystallinity" of a polymer. For example, the heats of fusion of polyethylenes depend on their degree of crystallinity.

Studies of Chemical Reactions

Generally, ATD and DSC permit studying the behavior of organic and inorganic compounds in different conditions. For the mineral compounds, the reactions which are the most often followed are those of decomposition. For the organic compounds, let us also mention their decomposition but also other ones, such as for example the formation of anhydrides, amides with formation of water, etc. They also permit studying isomerization with or without catalyzers and even in some cases, the corresponding activation heats. It is in this way that the isomerization Z – E of stilbene has been studied.

In absence of catalyzer, the conversion begins at 300°C and the summit of the exotherm is located near 385°C. With a catalyzer (black-palladium), the maximum of the exotherm is located close 290°C. The energies activation energies have been determined during the same experiences. They are of 46 ± 2 kcal/mole and 26 ± 1.6 kcal/mole with and without catalyzer. Another example is that of the conversion of the oleic acid (Isomer Z) into its Isomer E, the elaidic acid. The exotherm of isomerization appears near 500°C. Due to ATD and DSC, numerous isomerization in steroid series have been evidenced. These methods also permit the study of the aromatization of 3-keto-Δ-5-steroids with the formation of the corresponding phenols. Finally, let us mention the study of molecules such as the complexes of insertion urea-paraffines.

ATD, DSC and Pharmaceutical Technology

The examples given preceding already show the interests that present ATD and DSC in the pharmaceutic field. But there is another supplement. It is located in the domain of the pharmaceutical technology, the ultimate goal of which is to prepare the medicines as active as possible.

The same chemical species can occur in the solid state under several crystalline forms so-called polymorph. Inside the crystals, the molecular arrangements are different in such a way that two polymorphs of the same compound differ from each other from the standpoint of their physical properties as far as their crystals differ from each other. In particular, the fusion points, the solubilities, the vapor tensions in the solid state, the electrical and optical properties, the diagrams of diffraction of X-rays and the spectra IR differ. The crystalline edifices disaggregate by fusion or by dissolution. As a result, it appears that different polymorphs lead to identical liquid and gaseous states. The phenomenon of polymorphism is frequently encountered in the great series of medicines. For example, about 67% of steroids, 40% of sulfonamides and 63% of barbiturates present it.

CHAPTER 40

Thermometric Titration

Thermometric titration is a method of enthalpimetric analysis. It is endowed with interesting applications in the domain of analytical chemistry and in that physical chemistry as well. We shall successively consider:

- Its principle;
- The apparatus;
- Its theoretical possibilities;
- The true thermograms;
- Its analytical possibilities;
- Some examples of these possibilities;
- Some applications in the domain of physical chemistry.

Principle of the Method

According to IUPAC, thermometric titration is a method of titrimetric analysis that consists in recording the changes in temperature of the solution to titrate in the proportion that the titrating solution is added. The obtained records (thermograms) are diagrams, such as variations of temperature ΔT/volume v of added titrant solution (Figure 1):

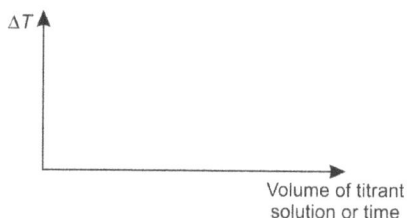

Figure 1: Coordinates of the Obtained Thermograms in Thermometric Titration. (Reprinted with Permission from Gwenola and Jean-Louis Burgot, Paris, Lavoisier, Tec et Doc, 2017, 531).

Frequently, it is time t which is retained as a coordinate in abscissa since the addition of the titrant solution is often carried out at a constant rate. More precisely, the solution to be titrated is placed in a reaction enclosure as adiabatic as possible. It is an isoperibolic calorimeter. At the beginning of the addition of the titrating solution, the solution to be titrated, which is in the calorimeter, is at the same temperature T_0 as the surroundings and as the solution to titrate. In proportion as the titrating solution is added, the titrating reaction generates a thermal effect q which is a function of the variation of enthalpy ΔH accompanying the titration reaction since the pressure is constant. The thermal effect dissipated in the bulk calorimeter makes its temperature changing and reaching the value T. It is the difference $\Delta T = T - T_0$ which is recorded during the titration. The final point of the titration is perceived by a singularity of the curve (ΔT/v or t).

It is because a change in enthalpy (the reactional enthalpy) can be determined by this analytical method that thermometric titration is also called *enthalpimetric titration.*

(*Remark*: In all scientific rigor, a calorimeter cannot be perfectly adiabatic because of thermal unavoidable leakages. This is an experimental fact that thermal leakages are due to the existence of a gradient of temperature between the points where they appear. In order for a calorimeter to be strictly adiabatic, it is necessary that the temperature be the same inside and outside the calorimeter. This cannot be the case in thermometric titrimetry since there is only such equality before any addition of titrant solution and before the beginning of the solution to be titrate. This is the reason why the calorimeter is said to be isoperibolic.)

Apparatus

The apparatus necessary to perform thermometric titrations is relatively simple. They are variants of that represented (Figure 2).

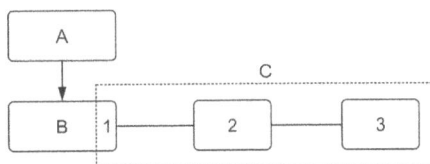

Figure 2: Block-Diagram of an Apparatus of Thermometric Titration. (Reprinted with Permission from Gwenola and Jean-Louis Burgot, Paris, Lavoisier, Tec et Doc, 2017, 532).

There are three principal types of components:

- The automatic burette A
- The isoperibolic calorimeter B
- The ensemble of measurement and recording of the temperature C (1: lead of temperature or thermistor; 2: Wheatstone-bridge; 3; recorder).

The automatic burette must deliver the titrating solution at a constant rate. If it is not the case, fluctuations on the found values of reactional enthalpies appear (see the section on 'Theoretical Possibilities of the Thermometric Titration). The titrating solution winds in a thermostatically controlled water bath at constant temperature T_o. The thermostatically controlled bath also plays the part of the surroundings of the calorimeter at constant temperature.

The reactional device must obey a lot of conditions. It must be as adiabatic as possible. The heat coming from the titration reaction must only serve to increase (or to decrease) the temperature of the calorimeter. There must be also a thermal capacity very weak in order to have a change of temperature as important as possible at equal thermal effect. Finally, it must exhibit a time of equilibration as short as possible. These calorimeters are particular Dewar's flasks. There are supplied with a head which involves the entering of the titration solution apparatus and a mechanical stirring responsible for a motor rotating at constant speed.

In Ensemble C, are found in:

- The sound of temperature 1, which is a thermistor.
- A Wheatstone bridge 2.
- Eventually an amplifier and a recorder 3.

It is the tension of disequilibrium at the terminals of the bridge, amplified or not which is recorded as a function of v. The most performing apparatus are equipped with a sensitive lead of temperature with a quick answer and of weak thermal capacity. Thermistors comply with these exigences. Doubtless, what is now called the renewal of thermometric titration is due to their employment. They are semi-conductors in ceramic. Their behavior is that of resistances with great negative coefficients of change of temperature. Their resistance decreases with the increase of temperature. It is important to know that the resistance of thermistors is not a linear function of the temperature. This is not an important drawback since for a change of temperature of 0.1°C, the gap between the result of this fact and the linearity is of the order of 0.2%. Also, the gap is reduced for changes of weaker temperature.

Moreover, this is an extreme case since a change of temperature of 0.1°C is truly a maximal value in thermometric titration. Finally, the answer of the bridge is also not linear with the change of temperature, but here the gap with the linearity is of the same order as the preceding and yet in the inverse sense.

Concerning the changes encountered in thermometric titrimetric for the variations of temperature, 5 cm of vertical deviation of the pen of the recorder correspond perceptibly to a change of temperature of 10^{-3} K.

Theoretical Possibilities of the Thermometric Titration

Let us titrate a solution of the species A by an antagonist solution of B according to the total and immediate reaction:

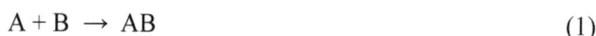

$$A + B \rightarrow AB \tag{1}$$

Let ΔH_{AB} the change in molar enthalpy accompanying it. The thermal effect q due to it (joules) is related to the molar enthalpy and the change of temperature of the system ΔT through the relations:

$$q = -n\Delta H_{AB} \tag{2}$$

$$q = C_S \Delta T \tag{3}$$

Here, n is the number of moles of product formed AB and Cs the thermal capacity of the system. The combination of these two equations gives the relation (4):

$$\Delta T = -(1/Cs)\Delta H_{AB}n \tag{4}$$

This relation permits to deduce some fundamental characteristics of the thermometric titration.

It is explicit the form of the titration curves. If we suppose the enthalpy ΔH_{AB} and the thermal capacity C_S constant all along the titration, the relation (4) becomes a linear relation between the change in temperature ΔT and the number n of moles of product AB formed. The thermometric titration is hence, fundamentally, a *linear method of titration*. It brings to play the measurement of a variable (the change in temperature) which is in linear relation with the number of moles of the species to titrate, that is to say with its concentration.

It is important to highlight the importance of the reactional enthalpy among the parameters which condition the success of a thermometric titration, as it is shown by relation (4).

(*Remark*: From the viewpoint of linearity, the thermometric titration is quite comparable for example to amperometry, conductometry, to the photometric titrations for which the measured variables are the limit current of diffusion, the conductance and the absorbance. For the same reason, it is quite different from potentiometric methods in which the measured variable is a logarithmic function of concentrations. In order to prove on this subject, the reactional enthalpy plays the part of the equivalent conductivity in conductometry, of the molar extinction coefficients, etc. It is well-known however that the molar enthalpy, as any thermodynamic function, does possess a physical significance of much higher importance than the other parameters already cited concerning the other linear methods of titration.)

The relation (4) indicates that a change of temperature appears so long as molecules AB are formed. According to the reaction of dosage (1), for one molecule B added, there is one molecule of AB formed immediately, provided there remain molecules A in the medium, since by hypothesis, the reaction is total and immediate. One sees that when n = 0 (before the beginning of the titration) the change in temperature is null. It is constant. The titration curve is represented by the segment AB horizontal (Figure 3). The segment BC represents the change in temperature during the reaction. The horizontal segment CD corresponds to a value n constant, that is to say when no molecule A (to titrate) subsists. Point C is the equivalent point. As in every linear method of titration, the equivalent point (C) is the point of intersection of two segments of lines. The thermometric titration is hence, also, an indicative method of the end of the titration reaction.

a) The curve of titrimetric titration (exothermal and total titration reaction-ideal conditions).

b) The curve of titrimetric titration (same conditions as above but reaction endothermal).

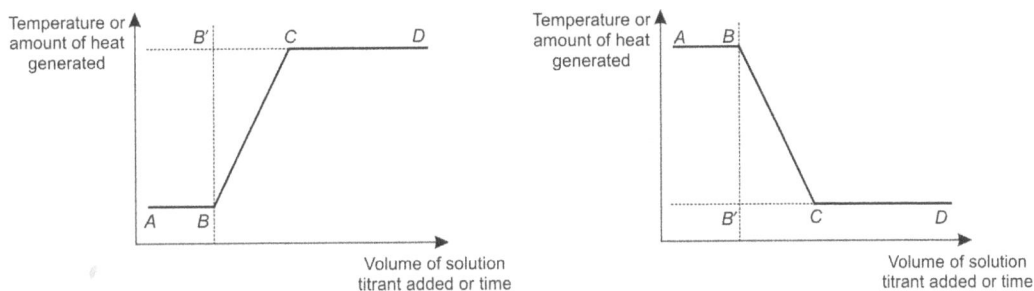

Figure 3: Two Examples of Thermometric Titration Curves. (Reprinted with Permission from Gwenola and Jean-Louis Burgot, Paris, Lavoisier, Tec et Doc, 2017, 535).

The relation (4) indicates that if the change in temperature, the thermal capacity and the number of moles of product AB are known, the measurement of the molar reactional enthalpy is possible. In particular, the recorded curve permits to immediately see if the reaction is exothermic or endothermic.

The molar enthalpy is not the sole thermodynamic grandeur accessible by thermometric titration. When the titration reaction (1) is equilibrated, more precisely when its equilibrium constant is located in some limits, the thermogram presents a curvature in the neighboring of the equivalent point. With the help of the study of this curvature, it is possible to determine graphically or by some calculations, the value of the equilibrium constant K and, through it, the variation of the free standard enthalpy of the titration reaction; (ΔG_{AB} must rather be called now; the reactional Gibbs energy).

$$\Delta G^{\circ}_{AB} = -RT\ln K$$

R is the constant of perfect gases. Usually, in the calorimeter, the solutes are sufficiently diluted for thinking that the change in the Gibbs energy ΔG_{AB} is practically equal to the change in the Gibbs energy standard ΔG_{AB}° (see *thermodynamics*). With this possibility, one can immediately calculate the change in the entropy standard accompanying the titration reaction, due to the fundamental relation:

$$\Delta H^{\circ}_{AB} = \Delta G^{\circ}_{AB} + T\Delta S^{\circ}_{AB}$$

Therefore, the three thermodynamic standard fundamental functions of the titration reaction can be obtained by starting from only one titration and even better, in the same experimental conditions. No other method, up to now, is endowed with such possibilities in solution.

Still remaining in the domain of the measurement of physical data, the method offers another important possibility. It is the study of the kinetics of the titrating reaction. One has seen that when it is equilibrated, the thermogram presents a curvature. It is a consequence of the fact that n does no longer vary linearly with the volume added. It is the same when the titration reaction is too slow with respect to the rate of addition of the titrating solution.

In conclusion in these two paragraphs, one can say that thermometric titrimetry possesses the very interesting characteristic that it is governed both by the enthalpy and the Gibbs energy of the titration reaction, that is to say by the enthalpy and the entropy of the titration reaction.

True Thermograms

Although it is convenient to introduce the method, the relation (4) is only a very simplified and incomplete of the reality of one thermometric titration. Nevertheless, the general conclusions to which it leads remain sound provided that during the study of thermograms one takes into account the supplementary following considerations.

If we refer ourselves to the titrations represented Figure 3, one understands that the diagram registered is not the same. It is that represented (Figure 4).

The segments AB and DE are not horizontal since they result from a parasitic gain of heat by the system, that coming from the stirring of the solution (see Joule's experiment). The part CD corresponds to the thermal effect of dilution (here exothermal) of the titrating reactant. In order to keep the thermal capacity quasi-constant during the titration, that is to say during the addition of the titrating solution, the latter is as concentrated as possible in order to avoid the addition of a too big volume of titrating solution. The molar enthalpy of the titrant is not the same in the calorimeter where it is very diluted. The difference between both is the enthalpy of dilution. The thermal effects of dilution and stirring add to the reactional one in the part BC. However, despite these summing up thermal effects, the final point remains perceived by a slope rupture at the point C.

For the obtention of the calorimetric data, graphical processes are used to get rid of the parasitic thermal effects.

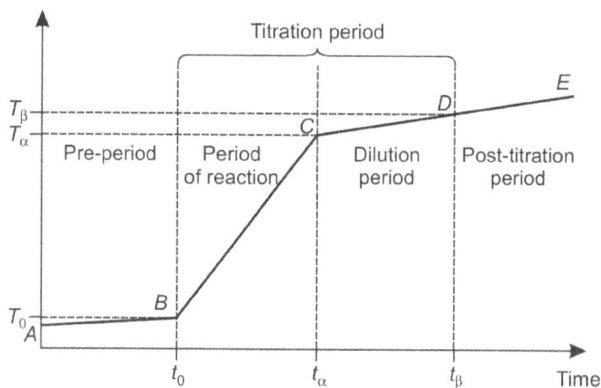

Figure 4: Real Thermogram (Titration Reaction Complete and Immediate). (Reprinted with Permission from Gwenola and Jean-Louis Burgot, Paris, Lavoisier, Tec et Doc, 2017, 537).

Analytical Possibilities of the Thermometric Titration

Its domain of applications is very large. This is due to the facts that:

– It can be practiced in any solvent since the nature of the thermistors makes their answer independent from the media surrounding them. One is not limited for example in some electrochemical methods by the ionizing character or not of the solvent or by the existence or not of adapted electrodes. Another example is the media of great ionic strength are not particularly inconvenient. Thermometric titration has been already used in media such as fused salts and poly-phased media.

– It permits to follow every kind of reaction in solution: reactions acid-base, redox, complexation, of precipitation. In the last cases, one again finds the fact that the method is insensible to the presence of several phases.

- It can permit some sequential titrations impossible to carry out with other titrimetric methods. Their success is conditioned by the conjunction of three factors: the linearity of the method, the values of the successive reactional enthalpies and the values of the equilibrium constants of the successive titration reactions. The conjunction of the first two factors is characteristic of the method.
- The precision and the accuracy of the thermometric titrations are quite acceptable. They depend, indeed, on a certain number of parameters inherent to the chemical system of titration. The most important are the values of the reactional enthalpies and the initial concentration of the compound determined. When the values of these two factors are sufficiently high, the precision is that of the burette. It is no longer imputable to the calorimetric system.
- One of its advantages (which inversely can be considered as being in some circumstances a disadvantage) is that it is a universal method. This means that it is not selective, given the nature of the variable measured (change of temperature). Nothing is more common than a thermal effect and nothing is more anonymous than a calorie.
- Its detection limit (or limit of sensitivity) is of the same order than those shown by other methods of titration, as for example the potentiometry. Acceptable titrations can be still carried out with concentrations of the order of $5 \cdot 10^{-3}$ mol l^{-1}, even 10^{-4} mol l^{-1} of the solution to titrate. These limits, of course, are function of the reactional enthalpy. The factor limiting the "sensibility" of the method seems to be the thermal effect of agitation which is when the concentration of the solution to titrate is too weak. They can become of the same order of grandeur.

Some Examples of Analytical Applications

The preceding considerations justify the fact that the analytical applications of thermometric titration are numerous. We focus ourselves on examples for which the method offers greater possibilities than other usual methods. One can mention:

- The possibility of titration of boric acid $B(OH)_3$ ($pK_a = 9.20$) by a solution of sodium hydroxide. It is impossible by potentiometry or by using colored indicators because it is not sufficiently acidic for its reaction with the hydroxide ion being sufficiently complete. The change in pH at the equivalent point is not sufficient. Contrarily, the thermometric titration is possible (Figures 5a and 5b) because on one hand the neutralization enthalpy is high (of the same order of grandeur as those obtained with strong acids); on the other one, the thermometric titration is a linear method of titration.

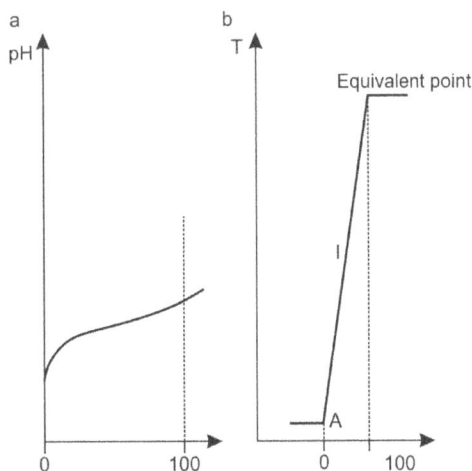

Figure 5: a) Potentiometric Titration of Boric Acid. b) Thermometric Titration. (Reprinted with Permission from Gwenola and Jean-Louis Burgot, Paris, Lavoisier, Tec et Doc, 2017, 539).

- The possibility of sequential titration of the ions Ca^{2+} and Mg^{2+} by the tetrasodium salt of edtaH$_2$Na$_2$ (Figure 6). It is possible because, for a first reason, the reactional enthalpies are of opposite signs and, for a second reason, the thermometric titration is a linear method of titration.

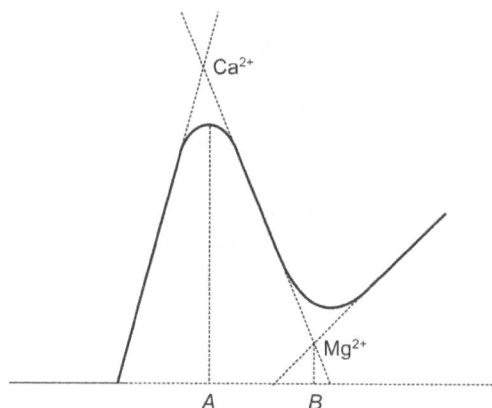

Figure 6: Sequential Titrations of Ions Ca^{2+} and Mg^{2+} by the Tetrasodium Salt of edtaH$_2$Na$_2$. A) Equivalent Point for Ions Ca^{2+}. B) Equivalent Point for the Ions Mg^{2+}. (Reprinted with Permission from Gwenola and Jean-Louis Burgot, Paris, Lavoisier, Tec et Doc, 2017, 539).

(*Remark*: One notice that Ca^{2+} is titrated the first. This is normal since the conditional constant of formation of the complex [CaY^{2-}] is one hundred times higher than that of the complex [MgY^{2-}] ($10^{10.7}$ and $10^{8.7}$ respectively) in the required pH conditions. The ratio of the two equilibrium constants seems to be weak in order to obtain a satisfactory sequential titration. This is true for the logarithmic methods of titration for which a ratio of 10^4 is necessary. But it is at this level precisely, that is the interest of the linear methods of titration The deep reason of this matter of fact lies in the property that have the linear methods to be more complete at the beginning and at the end of the titration. It is into these regions that interfere the less the other substances which can be titrated, so that one can take one's stand on it in order to draw the segments of straight lines which constitute the titration curve.)

- The possibility of titrations in polyphasic media. We have already begun to discuss this subject. A very interesting one is provided by the thermometric titration in the presence of two non-miscible phases. The advantage of processing according in this manner is to ascertain a good solubility of all the solutes. A second advantage is in the possibility to displace the equilibrium of the titration reaction by making it more complete. A typical example is that of the titration of protonated rather strong organic bases, the pka of which are statistically located between 9 and 10.5. In a purely aqueous media only, the titration reaction is:

$$BH^+ + OH^- \rightarrow B\downarrow + H_2O$$

In these conditions, titration suffers from two drawbacks. The reaction is not sufficiently complete. This is redhibitory by the principle of titrations. Moreover, there is frequent precipitation of the organic base. This disturbs the measurements. In presence of another solvent, immiscible with water, the titration reaction is symbolized by:

$$BH^+ + OH^- \rightarrow B_{org} + H_2O$$

The preceding drawbacks disappear. The acid BH$^+$ is more dissociated in these conditions. It is hence stronger in these conditions. Its new acidity constant in these conditions (which is an apparent constant) K_a' is equal to its true constant in water K_a multiplied by the partition coefficient P of the conjugated base B $(P = [B_{org}]/[B_{aq}])$.

Applications in the Domain of Physical Chemistry

The applications proceed from the previous considerations. The thermometric titration permits to determine the reactional enthalpies in dilute solutions. That is to say to approach the standard reactional enthalpies. It also permits the determination of the dilution enthalpies that are interesting values for the study of solvation phenomena and finally, in judicious conditions of equilibrium, the reactional Gibbs energies and hence the corresponding equilibrium constants. As a result, in the same conditions of equilibrium, thermometric titration also permits the determination of the reactional entropies.

Otherwise, we have seen some examples of analytical applications. They are difficult to forecast because their success is dependent on several parameters.

CHAPTER 41

Mass Spectrometry

We arbitrarily present the study of mass spectrometry in three chapters. In the first one, we present some generalities. This permits us to introduce the subject. In the second chapter, we study some different types of mass spectrometry and describe the corresponding mass spectrometers. In the last chapter, we give some applications of the method.

In this chapter, we begin by a description of the historical apparatus of Aston.[53] It permits to master the principle of the method and with the fact that there exist several kinds of apparatuses according to the nature of the components constituting them. Some generalities are added.

Definition of Mass Spectrometry and Mass Spectrometer

The mass spectrometry is grounded on the measurement of the deviation imposed to the trajectory of ionized atoms under the actions of an electric field and of a magnetic field.

A mass spectrometer is an apparatus which quickly separates moving ions according to the value of their ratio mass/electric charge (m/z). Most often, these ions are monocharged so that, when it is the case, they are finally separated as a function of their molecular mass. Most often, these ions are positively charged, but this is not obligatory.

We shall begin by giving a description of one of the first apparatus which is that of Aston.

A Brief Description of Aston's Apparatus

Aston used a modification of J.J. Thomson's[54] apparatus to increase the precision of his method. In Thomson's apparatus, the magnetic and the electric fields produced deflections in planes at right angles. In Aston's apparatus, the deflections were in the same plane, but in opposite directions.

This spectrometer of mass was constituted by a great enclosure under vacuum in which we could distinguish (Figure 1):

- An *inlet system* that permits the introduction of the sample into the ion source.
- The *ion source*: Only gaseous ions can be treated by the mass analyzer (compounds having normally appreciable vapor pressure values can also be worked). The positive rays observed by J.J. Thomson are various fragments coming from the breaking of the interatomic bonds of the species under study. They were obtained in a discharge tube at low pressure. (Besides, in mass spectrometry, the output of ion sources is a stream of positively, more rarely negatively, charged ions, whatever the spectrometer is. There are several other possible ion sources).

---

[53] Aston, Francis William, 1877–1945, English physicist, Nobel prize (physics) 1922.
[54] Thomson, Joseph John, English physicist, 1856–1940, Nobel Price (physics), 1906.

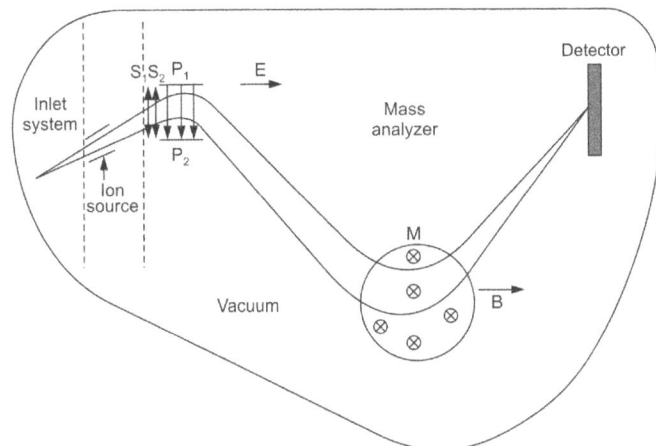

Figure 1: Aston's Mass Spectrograph.

- The *mass analyzer* into which the fine ions rays (fragments) were made to pass. The mass analyzer is called sometimes a "filter of masses". In the analyzer, the fragments are accelerated by an electrical field and dispersed by a magnetic field and then finally allowed to fall onto a photographic plate.

In Thomson's apparatus, the magnetic and electric fields produced deflections in planes at right angles. This is due to the fact that between the plates P_1 and P_2 where occurs the electrical field **E**, the particles are deviated in a plane whereas, because of their passing between the two poles of the electromagnet, the particles become deviated in a plane perpendicular to the preceding. This is because the force developed by the magnetic field is normal to the plane defined by the intensity and by the vector magnetic induction.

In Aston's arrangement, the deflections are in the same plane but in opposite direction. The positive particles first passed through two narrow parallel slits S_1 and S_2. The resulting fine beam was spread out by means of an electric field applied across the plates P_1 and P_2. A section of the beam passed between the magnetic poles (in M) of an electromagnet. The magnetic field was disposed so that the rays were bent in the opposite direction of their initial one. The mass spectrum is represented by a sharp line for each type of particle present, that is to say for each particle possessing its own mass/charge (e/m) ratio, where m is the mass of the ion in atomic mass units and z its charge.

- The *detector* was a photographic plate.

Different Components of a Mass Spectrometer

This description is simplified because there are several components of the same kind. They have nearly the same function in each kind of mass spectrometer and they are interchangeable. But we cannot describe all the possible combinations of these components here. We shall describe some of them at greater length in more details, along with the description of some kinds of mass spectrometry, in the next chapter. However, we decided to already begin to study a kind of analyzer, the double focusing system owing to its importance.

The different components of a mass spectrometer are classified according to the different functions the apparatus must carry out. They are:

- The introduction of the sample into the ion source. This is done by the *inlet system*.

- The ionization of the sample. It is done in the ion source. Often, the inlet system contains a means for volatilizing solid or liquid samples. Moreover, the inlet system and the ion source can be combined into a single component.

- The dispersion of the ions according to their mass-to-charge ratio (see the following paragraph entitled the double focusing *mass analyzer*).

- Their detection and the production of a corresponding signal (the components are the *ion transducer* and the signal processing.

In any case, the pressure in a mass spectrometer, from ion source to detector, must not be greater than about 10^{-5} torr to avoid collisions and air molecules (1 torr = 1 mm Hg or 133.322 Pa). This condition requires pumping with, for example, oil diffusion pumps trapped with liquid nitrogen.

Other components of mass spectrometers will be briefly described later with the apparatuses themselves we have retained for consideration (see next chapter).

The Double Focusing Mass Analyzer

The first point concerning the Aston's apparatus to stress its mass analyzer. It is a double focusing mass analyzer. The apparatus consists of the passage of charged particles through an electrostatic sector and then through a magnetic one. The electromagnet is used to focus ions with similar mass-to-charge ratios in a direction along a common trajectory. The electrostatic sector renders the ions isoenergetic.

Let us first consider the case for which the analyzer is constituted only by the magnetic device called a *magnetic sector analyzer*. The beam of ions coming from the ion source through two slits is forced to travel according to a circular path of 180, 90 or 60 deg. It is shown that a homogeneous beam of ions diverging from a slit can be brought to focus by a magnetic field in the shape of a sector. The ions enter into an electromagnet in such a way that the magnetic field **B** is perpendicular to the trajectory of the ions. Therefore, each ion suffers from a force $\mathbf{F}_{B\ perpendicular}$ to the direction of B. The value of its module is zvB where z is the electric charge on the ion, V is the applied potential at the exit of the source, v is the speed of the ion. Under the effect of this force, each ion takes a circular trajectory of radius r_B. Then, they are obligatorily submitted to the corresponding centripetal force $\mathbf{F}_c$:

$$Fc = mv^2/r_B$$

Here m is the mass of the ion. (Let us recall that a centripetal force occurs when a mobile of mass m describes a circle of center O and radius R according to a uniform motion of speed v. The centripetal force from which the latter suffers is constant, continuously directed toward the center O and equal to mv^2/r). The two forces to which are submitted the ions must be equal:

$$mv^2/r_B = zvB$$

Taking into account all these equations, we obtain:

$$m/z = r_B^2 B^2/2V$$

The radius r_B of the trajectory in the magnet of one ion of ratio m/z depends on B and V. Thus, ions of different ratios m/z possess different circular trajectories. If a collector of ions of very fine thickness is well located at the exit of the magnetic sector, only the ions possessing the requires ratio m/z will be detected.

Unfortunately, the focalization of the ions of the same ratio m/z on the same slit is not perfect because the ions, at the outset of the source, do not have the same kinetic energy. The explanation of this phenomenon is as follows.

The distribution of the kinetic energy of ions living in the source obeys Boltzmann's relation. There is, indeed, some inhomogeneity both in the formation of the ions in the source and the magnetic field. They are the cause of a broadening of the beam of ions reaching the transducer with, as a result, a loss of resolution. This may be extremely inconvenient for the measurement of some grandeurs such as the atomic and molecular masses. For that, dispositive double-focusing is applied. With it, the directional and energetic inhomogeneities are diminished. In a double-focusing apparatus, the ion beam is first passed through an electrostatic analyzer (ESA) consisting of two smooth *curved* metallic planes across which a dc potential E is applied. This difference of potential gives rise to an electrostatic force that is perpendicular to the surface of the electrodes, the intensity of which must be equal to the centripetal force according to:

$$zE = mv^2/r_E$$

Here r_E is the radius of the circular trajectory of the ion in the electrostatic sector. (It is interesting to notice that the fact that electrodes are curved makes the trajectory of the ion circular in the electrostatic sector, as is the case in the magnetic sector).

By substituting mv^2 by $2zV$ according to the equation linking the tension V of the source and the kinetic energy of the accelerated ions, one finds:

$$r_E = 2V/E$$

This relation shows that the radius r_E is independent of the mass and the charge of the ions. It only depends on the ratio V/E. Since V/E is itself constant, the changes in E commanded by the operator permits the selection of the ions as a function of their kinetic energy. This potential has for effect to diminish the kinetic energy of the ions.

In summary, one can say that the focalization in kinetic energy is provided by the electrostatic sector whereas that in direction is provided by the magnetic sector. Hence, in order to obtain a good resolution, it is necessary to add an electric sector to the magnetic one. Such analyzers are entitled double focusing analyzers. They have the property to give spectra finer than the previous ones because they have the property of concentrating the ions on the target.

Elementary Description of a Mass Spectrum—Fragmentation

Let us consider the mass spectrum of a simple organic compound. The mass spectrum has been obtained with an electron impact ion source, which consists of the bombardment of molecules by energetic electrons coming from the source. There is then the formation of positive ions. They are extracted from the region in which they are formed by an electrical field, go through the analyzer and, finally, fall onto the detector. The mass spectrum is recorded in such a way that the energy of the electrons of the source is increased during the recording.

In some way, the recorded mass spectrum represents the sequence of events occurring since the ion source. The first ionization potential of an organic compound is of the order of 8 to 12 eV.

The ionization potential of a compound or element is the energy measured in electron volts necessary to remove the least strongly held electron to an infinite distance. In electron-impact ionization, this may be taken as the accelerating potential in the electrode gun.

So, when this energy is reached, the first ion deriving from the studied compound begins to appear. It results from the removal of an electron from the initial molecule. For this reason, it is called a *molecular ion* or *parent ion*. As the potential is increased, other bonds of the molecule are broken. Each bond is characterized by an appearance potential. Hence, mass spectrometry can provide information about ionization potentials and relative bond strengths. The majority of mass spectra are obtained with a 70 eV electron beam, sufficiently energetic to break any bond of the molecule. Therefore, each molecule yields a series of fragments and the phenomenon is called *fragmentation*. Mass spectra are often represented by a listing or by a bar-graph presentation. Let us consider the case of methanol for example. The molecular ion is formed by the reaction:

$$M + e- \rightarrow M^{+\cdot} + 2e-$$

One finds the following fragments:

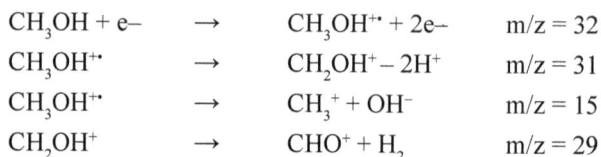

$$CH_3OH + e- \quad \rightarrow \quad CH_3OH^{+\cdot} + 2e- \quad m/z = 32$$
$$CH_3OH^{+\cdot} \quad \rightarrow \quad CH_2OH^+ - 2H^+ \quad m/z = 31$$
$$CH_3OH^{+\cdot} \quad \rightarrow \quad CH_3^+ + OH^- \quad m/z = 15$$
$$CH_2OH^+ \quad \rightarrow \quad CHO^+ + H_2 \quad m/z = 29$$

In the occurrence, the mass spectrum is the diagram: ordinates (height of the fragment peak expressed in percentage of the peak the highest itself, expressed at 100%)/abscissa (m/z) value of the fragment. In these conditions the mass spectrum of methanol is represented (Figure 2).

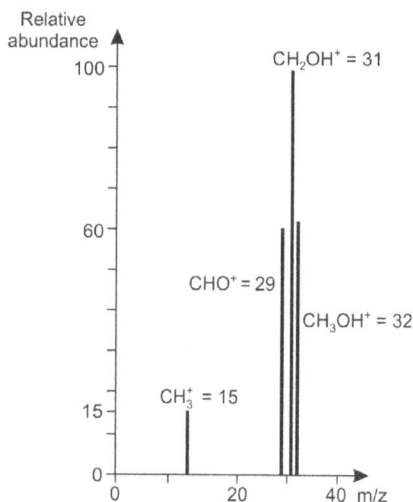

Figure 2: Mass Spectrum of Methanol.

Its bar-graph presentation is represented in Table 1.

Table 1: Presentation of the Mass Spectrum of the Methanol.

	m/z	Percentage
$CH_3OH^{+\cdot}$	32	62
CH_2OH^+	31	100
CHO^+	29	60
CH_3^+	15	15

We can notice that the highest peak is not obligatorily the peak parent. It is the case here. The highest peak is the base peak. It must be also noticed that all the peaks accessible are not retained in this example. Besides, their number is a function of the conditions of ionization. In this simplified spectrum, the peaks corresponding to the different isotopes are not present.

The phenomenon of fragmentation is of great importance in organic chemistry (see 'Applications').

Technical Characteristics of Analyzers

It is interesting to regard the technical characteristics of analyzers. One considers that they are four.

Range of Worked Ratios m/z

This range can reach values of 10,000th (Thomson) even well above. (The Thomson is the unity of the ratio m/z). For an analyzer of the kind TOF, it is illimited. In coupling GC-MS (see under), it is rather limited to about 1,000. Working with a range too broad may have, as a consequence, a loss of sensibility.

Transmission

By definition, the transmission is the ratio of the number of detected ions and the number of produced ions. It gives an idea of the loss of ions between their formation and their detection. It is difficult to evaluate because one does not exactly know the number of formed ions in the source. The transmission is never equal to 100%. It depends evidently on the threshold of detection.

Scan Rate

It is the rate with which the analyzer of can obtains some range of data m/z. A great scan rate permits working with a great registering rate. This parameter is of utmost importance for the matching of the coupling gas or liquid chromatography/mass spectrometry.

Resolution

There are two most widely adopted definitions.

First, resolution in mass spectrometry is the following; the resolution is the ratio $R_s = M/\Delta M$ where ΔM is the difference in mass numbers which will give a valley of no greater than 10 percent of the peaks of mass numbers M and $M + \Delta M$ when the two peaks are of equal heights. The resolution is considered satisfactory if:

$$\Delta M > 1$$

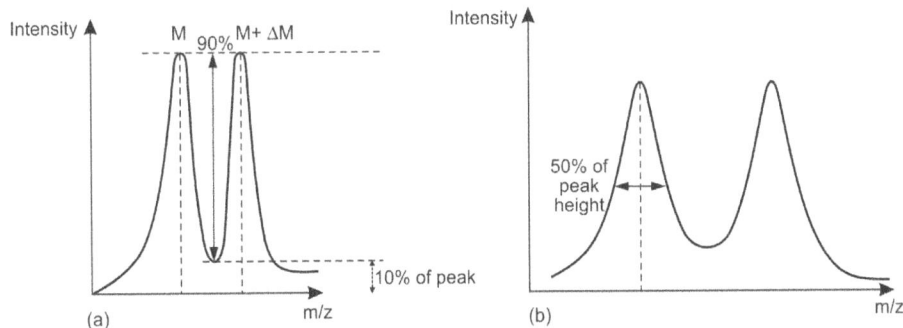

Figures 3a and 3b: Definitions of Resolution in Mass Spectrometry.

The second approach valuable for TOF is the following ΔM is defined as the full width at half maximum (FWHM), that is to say, the width of the peak at half its height (Figures 3a and 3b). Generally, the higher the resolution, the better the separation.

CHAPTER 42

Some Kinds of Mass Spectrometry and Some Mass Spectrometers and Components

In this chapter, we make arbitrarily the choice to describe some apparatus instead of others. We decide to privilege some kinds of mass spectrometry which are particularly useful, notably in bioanalytical chemistry. Given this choice, we cannot describe all the components which, possibly, constitute mass spectrometers.

This is the reason why we give some complements on the structures of other spectrometers in the following paragraphs.

Inductively Coupled Plasma Mass Spectrometry—ICPMS

ICPMS is a very important technique for the analysis of the elements. There are several reasons at the origin of this assertion. Its exhibits low detection limits for most elements, high degree of selectivity and good precision and accuracy as well. An ICP torch serves as an atomizer and as an ionizer.

The starting point for mass spectrometric analysis is the formation of gaseous analyte ions. The ion sources can be classified into two categories: the gas-phase sources and the desorption sources. In the former, the sample is first vaporized and then ionized. In the latter a solid or a liquid is converted directly into gaseous ions. It is a matter of fact that ion sources are classified as hard and soft sources. *Hard sources* impart sufficient energy to analyte molecules so that they are left in a highly excited energy state. Relaxation then involves rupture of bonds, producing fragment ions that have a mass-to-charge ratios less than that out the molecular ions. *Soft sources* cause little fragmentation. Consequently, the resulting mass spectra often consists of the molecular ion peak and eventually a few other ones, contrary to the results obtained with the hard sources.

Plasma

A plasma is an electrical conducting gaseous mixture containing a significant concentration of cations and electrons. The concentrations of both types of particles are such that the net charge approaches zero. It is very energetic. Its temperature may reach 10^5 K. It requires a single nonflammable gas. It can be produced artificially by radio frequency induction or by a d.c. discharge.

Inlet and Ion Source

Let us notice that now plasma sources have become the most important and widely used sources for *atomic* emission spectrometry. In ICPMS, the ionization and atomization are carried out in a

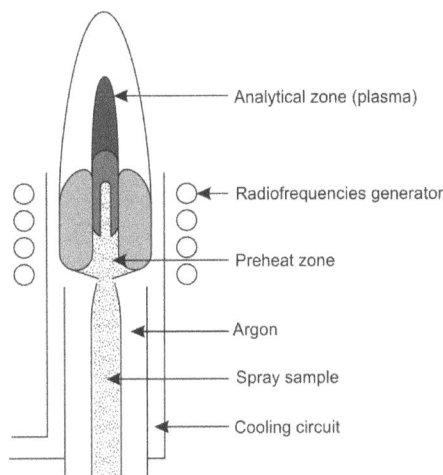

Figure 1: A Typical ICP Source. (Reprinted with the Permission from of the Society Perkin-Elmer Instrument, 12 Avenue de la Baltique 91140 Villebon sur Yvette).

plasma torch. The sample is introduced in an argon flux into a special device through a nebulizer (Figure 1).

The argon flux is at hot temperature, between 6,000 K and 8,000 K. Ionization of the flowing argon is initialized by a spark from a coil. The resulting ions and the associated electrons interact with the fluctuating magnetic field produced by the inducing coil. These interactions are the origin of the movements of ions and electrons and of the ohmic heating. The torch is rather a hard gas phase source.

Interface Torch/SM

This the delicate point of the coupling because the ICP torch operates at atmospheric pressure whereas the spectrometer can work only at a pressure lesser than 10^{-4} torr. The problem is surmounted by a coupler, interface through which the hot plasma is transmitted, with the help of a very small orifice, into a region where the pressure is of the order of 1 torr. The rapid expansion of the gas cools it.

The Quadrupole Analyzer

The quadrupole analyzer is often the analyzer used in the ICPMS. It is constituted of four cylindric or semi-cylindric metallic rods. They are strictly parallel and straight. They have been positioned in such a manner that the ion beam passes through the center of the array (Figure 2).

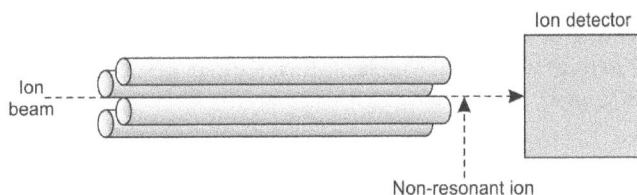

Figure 2: Principle of the Quadrupole Analyzer – Scheme (1).

Diagonally opposite rods are connected electrically as it is indicated in the Figure 3. Hence, a direct current (d.c.) and an alternating current (a.c.) are applied to these electrodes. Neither the d.c current nor the ac fields have any influence on the longitudinal motions of the ions.

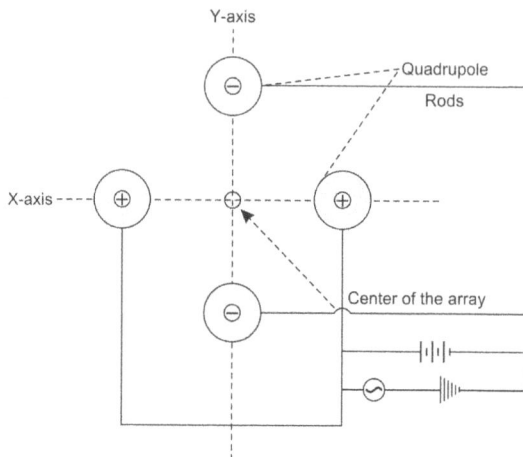

Figure 3: Description of a Quadrupole Analyzer, Geometry and Charges of the Rods.

One can write for the differential equations of motion of an ion of mass m and charge *e* brought into this device:

$$d^2x/dt^2 + (2e/r^2m)(V_{dc} + V_0 \cos \omega t) \, x = 0$$

$$d^2y/dt^2 + (2e/r^2m)(V_{dc} + V_0 \cos \omega t) \, y = 0$$

$$m(d^2z/dt^2) = 0$$

Here, V_{dc} is the applied direct potential, V_0 is the amplitude of the alternating voltage of frequency ω radians per second and r is the distance between the summit of each demi-cylinder and the central point between the four rods. They indicate that the motions of the ions have a periodic component of frequency ω and are dependent on the ratio e/m. The ions that can pass till the detector are called the resonant ions; the other is the non-resonant ones. By varying the parameters, all ions become sequentially resonant.

The quadrupole analyzers are of low sensitivity. They are well-adapted to the couplings with gas-chromatography instruments.

Double focusing analyzers are also combined with the ICP.

Detectors

The most used transducers in the case of ICPMS are the electrodes multipliers. They are of several types.

- The discrete dynode electrode multiplier. They are metallic plates recovered by an alloy lead/ lead oxide rich in electrons. The shock due to the arrival of an ion, the alloy emits electrons. A potential difference accelerates the latter ones up to a second dynode where again new electrons are emitted and so forth with new Dynodes, which are kept at successively higher voltages than the preceding (Figure 4).

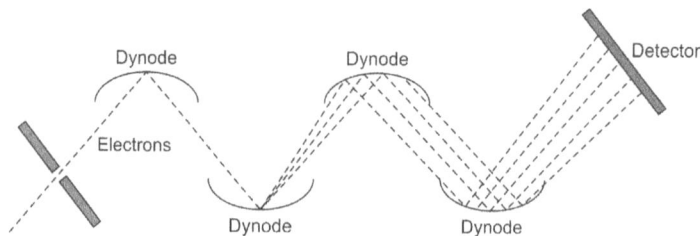

Figure 4: A Discrete Dynode.

- The continuous dynode electron multiplier or "chaneltron" is a single dynode. It has the form of a large cornucopia. It is doped with lead. A voltage is applied between its entry and its exit. When an ion hits the internal side, electrons are emitted and give rise to bursts from the surface. These, in turn, give rise to cascades, etc. (Figure 5).

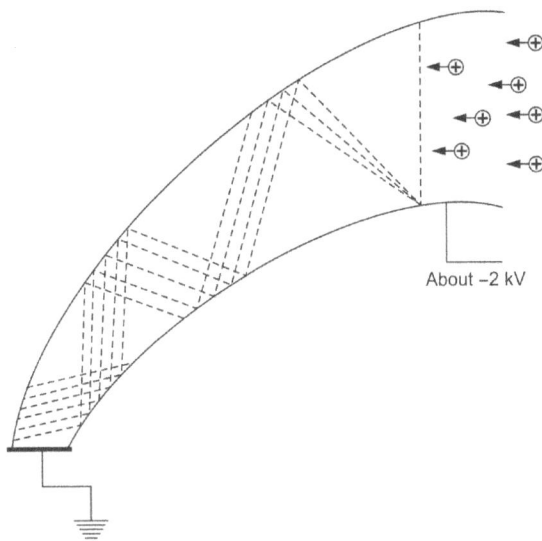

About −2 kV

Figure 5: A Continuous Dynode.

Applications of ICPMS

ICPMS can carry out analysis of numerous elements in samples of matter, such as qualitative, semiquantitative and quantitative determinations. Generally mass spectra are easier to interpret than optical emission spectra. This property is particularly important for the analysis of complicated mixtures. ICPMS exhibits very low detection limits. They are in the same order as those obtained, for example, with electrothermal atomic absorption spectroscopy, that is 10^{-2} ng.mL^{-1}.

Matrix-Assisted Laser Desorption Ionization—Time-of-Flight Mass Spectrometry: (MALDI-TOF/MS)

Matrix-assisted laser desorption ionization (MALDI) is usually combined with time-of- flight (TOF) analyzer. MALDI is a soft ionization method. It leads to molecular ions [M]$^+$ and to the quasi-molecular ion, such as [M + H]$^+$. Soft sources cause a weak fragmentation.

As a result, the resulting mass spectrum often consists of the molecular ion peak and some other peaks. Soft ionizations methods have one considerable importance in bioanalytical chemistry as they allow the analysis of whole proteins or DNA molecules. M. Karas et al. were awarded the Nobel prize for their invention in 2002.

Principle

The principle of the method lies in the ionization. The sample is co-crystallized with a matrix and then ions can be generated by exposure to photons.

Ionization

The ionization is based on the soft desorption of the solid sample molecules from a surface into the vacuum and subsequent ionization. It applies to organic molecules with poorly volatile, thermolabile and high molecular mass. First, the sample is co-crystallized with 1,000–10,000 excess of suitable matrix on a metallic plate. Small organic UV absorbing molecules like 2,5-dihydroxybenzoic acid (DHBA), sinapinic acid (SA) (dihydroxy3,5-4-methoxybenzoic acid, nicotinic acid, α-cyano-4-hydroxycinnamic acid (α-CHCA), 4-hydroxypicolinic acid, succinic acid and glycerin are used as matrix materials. An electric field is applied between the simple plate and the entrance to the time-of-flight analyzer. A pulsed laser beam (Appendix 4), whose pulses are very brief (of the order of the nano or even of the picosecond) is focused onto the crystal. The matrix is chosen such that it absorbs readily at the laser wavelength. The sample must not, however, absorb at this wavelength. When the matrix molecules are bombarded with the photons from the laser pulses, they are excited rapidly and transferred to the neighboring molecules. Some matrix and analyte molecules become ionized during this process (Figure 6).

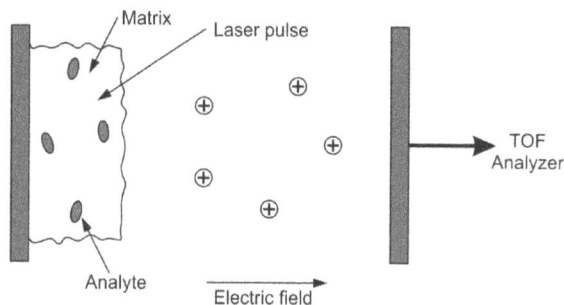

Figure 6: Principle of Matrix-Assisted Laser Desorption.

Once in the gaseous phase, the ions are accelerated towards the TOF analyzer by the applied electrical field (Figure 7).

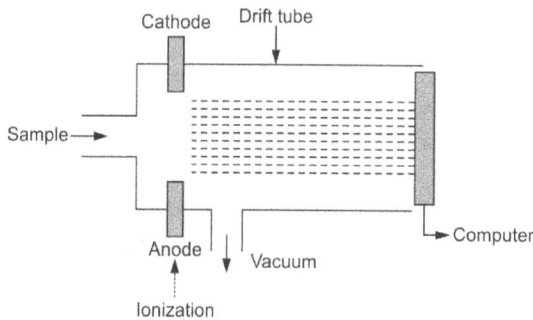

Figure 7: Linear TOF Analyzer.

MALDI being a soft ionization method, it leads to molecular ions $[M]^+$ and to the quasi-molecular ion such as $[M + H]^+$. The matrix molecules can also take and donate protons or electrons to the analyte molecules and transfer ionization energy. The analyte ions obtained are, hence, predominantly molecular ions $[M^+]$ or quasi-molecular ions like $[M + H^+]^+$ as well. They can also be some adducts with alkali metal ions from buffer solutions like $[M + Na]^+$. Multiply-charged molecular ions also occur. Due to the high matrix concentration, the analyte ions are prevented from interacting with each other. Nitrogen lasers with a wavelength of $\lambda = 337$ nm are most commonly used for matrices that absorb in the UV domain. The wavelengths of lasers can be located up to the infra-red region. The lasers used are carbon dioxide lasers or colorings lasers.

Mass Analyzer

The mass analyzer used with MALDI is usually a time-of-flight (TOF) analyzer. Its use is pertinent for the mass determination of high molecular weight species. The time-of-flight spectrometers work by sequentially monitoring when and how ions of differing mass-to-charge reach a detector. The measurement follows the injection of a single burst of ion from an ion source. The desorbed ions are accelerated by an electric field (10^3–10^4 V) to a kinetic energy of several keV. Then, they enter a tube devoted from any field. They drift along it with different speeds according to their mass/charge ratios. Ions are dispersed and monitored via the differing times by ions to travel the length of the tube.

At the end of the tube, the ions strike the detector and their drift time is measured electronically. It is well-known that light ions are accelerated more than heavier ones. Because of that, they reach the detector first. The kinetic energy of the drifting ion is defined as:

$$E_{kin} = 1/2mv^2$$

$$E_{kin} = zeV$$

Here, *m* is the mass of the ion, *v* is its velocity at the end of the range of acceleration, *z* is its ion charge, *e* is the elementary one and V is the voltage of the applied electrical field. The velocity *v* can also be defined as the length of the field-free tube L over the term of flight *t*:

$$v = L/t$$

Grouping these three equations permits us to write:

$$m/z = (2eV/L^2)t^2$$

The m/z ratio of the ion is, indeed, proportional to the square of the drift time. Hence, the mass of an ion can be determined by measuring the drift time once the analyzer is calibrated with substances of known weight and charge. The time measurement makes that even heavy ions can be detected accurately. Typically, the flight tubes have a length of about two meters, resulting in flight times in the order of microseconds. Very good sensitivities can be obtained, in comparison to other analyzers. In this result, also plays the fact that all the ions that have suffered the action of the analyzer reach the detector. This is because all these ions have passed across a pin hole before entering the analyzer and are hence canalized.

Because not all the molecules get desorbed at the same time and place, slightly different velocities are obtained for identical ions. The result is the occurrence of broad peaks and poor resolution. Using a corrector device, whose name is "reflector TOF", rectifies the situation. Good sensitivities can be obtained. This method is above all used for the analysis of proteins and peptides.

Detection of Ions: "Secondary Electron Multipliers"

A mass spectrometer contains a transducer that converts the beam of ions into an electrical signal that can be processed. In general, electron multipliers are capable of providing high current gains. For the detection of ions in time-of-flight analyzers, secondary electron multipliers are used. More precisely, one distinguishes, as we have already seen:

- The system of discrete dynode electrode multipliers.
- The system of the continuous dynode electron multiplier.
- They can typically provide a current gain of 10^7–10^5.
- They are also used for the detection of ions in the case of MALDI.
- The Faraday cup is a faraday cage surrounding a collector electrode. The Faraday cage and the collector electrode are electrically connected. Ions striking any part of this device can give rise to a signal or the emission of secondary.
- Electrons which, in turn, may enhance the signal.

- Scintillation ion detectors whose recording of ions is sensed via the emission of visible light following collision of the ions or secondary electrons with a phosphor-coated surface.
- Photographic detection which is now rarely used but which has formerly been of great utility.

MALDI demands a preparation of the sample, but it is straightforward.

Applications of MALDI

It is interesting to recall that no method was available to transfer large biomolecules with molecular weights of more than 1,000 Da into the vacuum without fragmenting them, before the discovery of this methodology. This method with the method ESI (see the following section) makes the mass spectrometry among the most powerful tool for the study of proteins and DNA. With MALDI-TOF, molecular weights above 500,000 Da can be determined with sensitivities as low as 1 fmol and mass accuracies as high as 0.1–0.01%. MALDI is mainly used for the analysis of proteins and peptides and their mixtures. It is possible to determine the molecular weights and to obtain structural information. Molecular weights of proteins and peptides can be determined accurately with only a small amount of sample.

The strong point in favor of MALDI includes the fact that a very low amount of sample is only necessary for analysis. The spectra obtained are simple so that mixtures can be analyzed without the need to separate the components prior to MALDI analysis. However, in the contrary to ESI (following paragraph), MALDI cannot be directed coupled to liquid chromatography (LC) or capillary electrophoresis (CE) since it is not a continuous but a batch ionization method.

Often, in typical MALDI spectra, little or no fragmentation occurs. The molecular peak does not appear obligatorily and on the contrary, a major peak at $[M + H]^+$, and minor peaks $[M + 2H]^{2+}$ and the agglomerate ones $[2M + H]^+$ may be observed. MALDI-TOF can be regarded as a very fast separation method. In many ways, it is considered as being more powerful than electrophoresis or chromatography.

Finally, MALDI spectra are of some utility in the identification of proteins by measuring its "peptid fingerprint" and comparing it to a database. The protein is reacted with an enzyme which cleaves the amino acid chain in specific places. The obtained fragment is the "peptide fingerprint". It is very specific of a given protein. The measurement is done by comparing the data obtained from the MALDI spectrum to a data base.

Electrospray Ionization Mass Spectrometry (ESI-MS)

Electrosprays are generated by dispersing a liquid under the form of small droplets generated via an electric field. This method has been known for a long time and is used for a variety of tasks. ESI for mass spectrometry, as used in modern instruments today, was developed by John Fenn in the 1980s. In 2002, Fenn was awarded the Nobel Prize for his invention.

Ionization Principle

Electrospray ionization is based on the dispersion of a liquid due to an electric field. The method takes place under atmospheric pressure. The sample solution, containing analyte solution, is pumped into a heated chamber through a capillary or needle. A potential difference of several kilovolts is applied between the capillary and the opposite chamber walls. It creates an intense electric field at the capillarity exit. If the capillary has a positive potential, negative ions are held back and positive ions are drawn away from the capillary towards the opposing chamber wall. This leads to the formation of a liquid cone at the end of the capillary and droplets with positively charged analyte ions form at the tip of the cone. These are dragged through the chamber by the electric field whilst continuously losing solvent due to evaporation. At the end of the drag, droplets shrink. This leads to an increase of charge density on the droplet surface. The repulsive forces on the droplet surface move eventually so close together, that the droplet bursts into a mist of finer droplets. The process is repetitive; shrinking and bursting occurs repeatedly until eventually the analyte is completely desolvated and transferred into the mass analyzer (Figure 8).

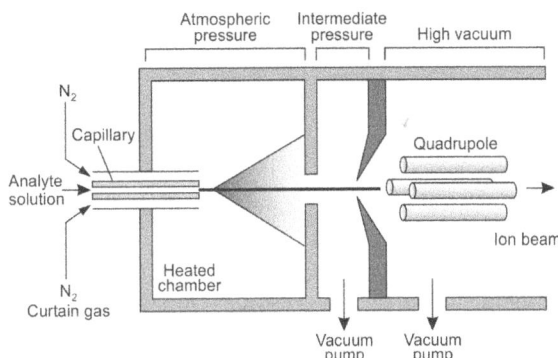

Figure 8: Scheme of Electrospray Ionization Interface Connecting the Ionization Chamber at Atmosphere Pressure to the Mass Analyzer Reproduced with the Permission of A. Manz et al., 2004. Bioanalytical Chemistry. p. 100, copyright@2022, World Scientific Publishing Ldt., Imperial College Press.

A typical characteristic of electrospray ionization is the formation of ions which can be several times charged. For example, large biomolecules give rise to signals corresponding to the peaks $[M + H]^+$, $[M + 2H]^{2+}$, $[M + 3H]^{3+}$ to $[M + nH]^{n+}$. These highly charged ions appear at relatively low m/z values in the mass spectrum so that this method permits the study of very high molecular weights. They are not accessible by other techniques.

The potential difference between the capillary and the electrospray chamber wall can be applied in two ways, depending on whether cations or anions are to be analyzed. In the positive ion mode, the capillary has a positive potential. Negatively charged ions are blocked back by the capillary and cations are drawn through the chamber into the mass analyzer and detected. Often low pH values of the sample solution are used to promote formation of cations. In the negative ion mode, the potential difference is reversed and the capillary is negative. Cations are blocked back whereas anions are drawn towards the analyzer. In this mode, high pH values, pH > pI are employed.

ESI—Source and Interface

Electrospray ionization is achieved at *atmospheric pressure* but the mass analyzer, however, operates under high vacuum. A special interface is, therefore, necessary to transfer the ions from the ionization chamber into the mass spectrometer. Such an interface is shown in Figure 7. Usually, a zone of intermediate pressure separates the ionization chamber and the analyzer. The liquid sample together with a nebulizing gas is introduced into the heated ionization chamber. An electrospray is generated by applying a potential difference between the needle and the opposite interface plate. A small proportion of the desolvated analyte ions exits the ionization chamber through a submillimeter orifice and enter the zone of intermediate pressure. The analyte ion then passes via another small orifice into the mass analyzer. It is usually a quadrupole which is operated under high vacuum.

A characteristic of ESI is that the sample can be pumped into the mass analyzer continuously. MALDI, on the other hand, is a pulsed method requiring a dry sample. Thus, ESI-MS can be coupled directly with liquid separation methods such as RP-HPLC and CE. As the sample emerges from the separation column it is directly pumped into the electrospray chamber. As outlined earlier, MALDI-TOF is capable of separating sample components directly from the sample mixture. ESI-MS must be coupled to LC or CE for separation of sample components.

Depending on the amount of sample available, different flow rates are used. A low flow rate allows for long measurement times to optimize instrument parameters. When the pneumatically assist electrosprays as it can be the case, rather large capillaries and flow rates are used. Other values for capillaries and flow rates are employed for micro-electrospray. For bioanalysis, often only a limited amount of sample is available, requiring very low flow rates.

Quadrupole Analyzer

The mass analyzer most commonly used with ESI is the quadrupolar analyzer whose functioning principle has been described.

Applications of ESI-MS

ESI enables the production of molecular ions directly from samples in solution. It can be used for small as well as large biopolymers up to about 200,000 Da, including peptides, proteins, carbohydrates, DNA fragments and lipids. Unlike MALDI, ESI is a continuous ionization method like HPLC or CE.ESI is suitable for almost all types of biomolecules as long as they are polar and soluble in a solvent that can be used for spraying. Molecular weight determination is one of its main applications. Moreover, sequencing of peptides and DNA fragments is possible with ESI connected to a tandem mass spectrometer (ESI-MS/MS). Samples must be soluble and stable in solution and need to be relatively clean. Ion formation in the spray is hindered by buffers, salts and detergents. A problem with electrospray ionization is its low tolerance for impurities or additives. Buffer and salt concentrations of more than 0.1 m M can prevent sufficient ion formation in the electrospray process. In almost every case, it is necessary to clean the sample from salt and impurity contents prior to introduction into the electrospray chamber. Commonly used techniques for desalting include micro-dialyze and solid-phase microextraction. Reversed phase liquid chromatography (RP-LC) can be used for preconcentrating and isolating the compounds of interest. RP-LC can be coupled directly to ESI/MS as the organic solvents used in RP-LC are compatible with electrospray ionization. At low flow rates, the sample can be injected directly from the column into the ionization chamber. At higher flow rates, the sample stream is split and only a fraction is directed into the mass spectrometer.

ESI is a soft ionization technique. A few fragments are formed. However, ESI promotes the formation of multiply charged ions. Peptides and proteins thus give a series of signals with $[M + H]^+$, $[M + 2H]^{2+}$ to $[M + nH]^{n+}$. The number of peaks depends on the size of the molecule as well as the number of acidic and basic groups. Larger proteins can have a signal series going up $[M + 100\,H]^{100+}$. There is a possible difficulty. How is it possible to determine which peak refers to the $[M + H]^+$ ion and the molecular weight of the analyte molecule? The presence of isotope peaks complicates the problem. The molecular weight can be then calculated with algorithms, which is called a deconvoluted spectrum. In this case, the molecular weight is given in the form of a peak.

Some Other Components of Mass Spectrometers

Possible components of a mass spectrometer are numerous. This is the reason why everything has not been evoked during the previous description of some mass spectrometers. Now, we briefly mention some of them. Here, the choice to evoke one component rather another one is arbitrary.

Sample Inlet Systems

These devices must permit the introduction of a sample into the spectrometer. The majority of these apparatus are equipped at least with two inlet systems in order to permit the introduction of gaseous, liquid and solid samples. Inlet systems generally contain a nebulizer or atomizer and a heater to facilitate the vaporization of the sample before the ionization. The most employed systems are:

- The batch inlet systems. They constitute the simplest types. They permit the samples to leak through a microporous metallic or glass diaphragm into the ionization chamber;
- The direct probe inlet systems. With these devices extremely small samples (of the order of the nanogram) can be introduced directly into the ionization chamber. Their ionization chamber is isolated from the entrance by a vacuum lock;
- Chromatographic inlet systems (see the following chapter).

Ion Sources

It must not be forgotten that sometimes the inlet system and ionization sources may be combined, thus forming only one unit.

One knows that the samples must be in the gaseous state or be vaporized prior to ionization. There are two types of sources for achieving the ionization process:

- The gas-phase sources;
- The desorption sources.

For the process of the gas-phase sources, the sample is first vaporized and then ionized. Functioning according to this type, one finds as ion-sources:

- The *electron impact (EI)* or *electron impact ionization* is widely used. Ionizing agents are energetic electrons. This is a hard source. The gaseous stream of sample pass directly through a pass of electrons accelerated towards an anode. The historical Aston's apparatus was of this kind.
- The *chemical ionization (CI)*. In chemical ionization, gaseous atoms of the sample (coming from either a batch inlet or a heated probe) are ionized by collision with ions produced by electron bombardment of an excess of a reagent gas.

Usually, positive ions are formed. The gaseous reagent is introduced into the ionization part of the spectrophotometer in an amount such that the concentration ratio of reagent to sample is 10^3–10^4. Because of this large value, the electron beam reacts nearly exclusively with reagent molecules. Methane is a reagent of choice since it gives important amounts of CH_4^+, CH_3^+ and CH_2^+. The first two react rapidly with additional methane molecules as follows:

$$CH_4^+ + CH_4 \rightarrow CH_5^+ + CH_3$$

$$CH_3^+ + CH_4 \rightarrow C_2H_5^+ + H_2$$

Collisions between the sample molecule MH and CH_5^+ or $C_2H_5^+$ involve proton or hydride transfer according to:

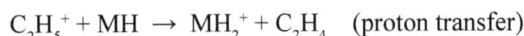

$$CH_5^+ + MH \rightarrow MH_2^+ + CH_4 \quad \text{(proton transfer)}$$

$$C_2H_5^+ + MH \rightarrow MH_2^+ + C_2H_4 \quad \text{(proton transfer)}$$

$$C_2H_5^+ + MH \rightarrow M^+ + C_2H_6 \quad \text{(hydride transfer)}$$

- The field ionization (FI). In this method, ions are formed under the influence of a large electric field (10^8 V/cm). Such fields are produced by applying high voltages (10 to 20 kV) to specially emitters consisting of numerous fine tips having diameters of less than 1 μm.

For the process of desorption sources, the sample in the liquid or solid state is converted directly into gaseous ions. As in the case of field ionization, there are several types of ion sources to achieve the process.

- The field desorption (FD). As in the case of field ionization, there are production of electrostatic gradients close to the tip of the electrode.
- The electrospray ionization (ESI) or (EI) has already been described.
- The matrix- assisted desorption/ionization (MALDI) has already been described.
- The plasma desorption (PD) or plasma desorption ionization (PDI). In this methodology, the sample is first adsorbed on a small, aluminized nylon foil before being bombarded with highly energetic fission fragments of a radionuclide to induce the desorption and the ionization of the analyte. This method is still used for the ionization of analytes with relative masses of 10,000 Da.
- The fast atom bombardment (FAB). In this methodology, the sample in a condensed state, often in a non-volatile solvent (in order to help subsequently its desorption and its liberation) is ionized by bombardment with energetic argon or xenon atoms. Both positive or negative (analyte) are sputtered from the surface of the solution. It is the process of desorption. There is a very rapid heating. The beam of fast atoms is obtained by passing accelerated argon or xenon ions from an ion source through a chamber containing argon or xenon atoms at a pressure of about 10^{-5} torr. A phenomenon of electrons exchange reaction between the ions and the atoms with a loss of translational energy is produced. It gives the beam of energy atoms. The fast bombardment usually produces significant amounts of molecular ions and of ion fragments even for high-molecular and thermally instable samples. Molecular weights over 10,000 have been determined. It is a hard ionization method.
- Secondary ion mass spectrometry (SIMS) presents a strong analogy with FAB spectrometry. (Sims) is devoted to the analysis of solid surfaces. Within this methodology, surface solid is bombarded by high-energy ions, such as Ar^+, Cs^+, N_2^+ and O_2^+, formed with an electron impact ion source. These primary ions provoke the desorption of atoms. A few of them are further ionized. They are known as secondary ions. The latter are electronically accelerated and passed to the analyzer for determination.
- The thermospray ionization (TS). In this method, a solution of the analyte is passing through a thin steel capillary tube at very high (supersonic) velocities into a vacuum chamber to produce a fine spray of analyte ions. There is formation of droplets that enter a vacuum chamber. Ions are focused, extracted and accelerated from the vacuum chamber and into the mass analyzer by applying an electric field.
- Atmospheric pressure electron impact ionization sources. Usually, electron impact ionization sources operate under high vacuum conditions. However, ionization via an electron impact may be carried out under normal atmospheric conditions with an efficiency greater than that obtainable under high vacuum conditions. One difficulty comes from the fact the ions must leave the source at atmospheric pressure and enter the analyzer under vacuum. This may be achieved by allowing ions to leak through a diaphragm of very narrow aperture.

Analyzers

Recall that mass analyzers separate ionized samples according to differences in their mass-to-charge ratios as they emerge from the ionization source.

Other Analyzers

We have already encountered:
- The magnetic sector analyzers;
- The double focusing mass analyzers;
- The quadrupole mass analyzers;
- The time-of-flight analyzer (TOF).

Ion trap analyzer can be classified in this category.

- An *ion trap analyzer* is a device in which ions may be stored for some interval of time. A simple instrument possesses a doughnut-shaped ring electrode together with a pair of end-cap electrodes. A radio frequency voltage is applied to the ring electrode. It may be changed to vary the radius of orbit of ions with differing m/z ratios. As the voltage is swept, ions of different m/z ratios become stabilized and may leave the ring electrode cavity via openings in the endcaps to make contact with detector.

- There is another type of ion trap. It is an *ion-cyclotron-based* instrument. Ions are generated via an electron impact ion source. Then, they are injected into an enclosure known as a cyclotron. Strong electromagnets cause the ions to rotate in a circular path in a plane of motion that is perpendicular to the direction of the field. For a given magnetic Field B, the angular motion of an ion depends on its charge z and is inversely dependent upon its mass *m*. The angular frequency of an ion (under a particular set of conditions) is referred to as its cyclotron frequency. An ion trapped in a circular path in a magnetic field is capable of absorbing energy from an electric field provided the frequency of the field matches the cyclotron resonance. The absorbed energy then increases the velocity of the ion. In these conditions, the velocity of the ion increases. It is the same for the radius of the pass. Mass spectra can be obtained by sequentially accelerating ions of differing mass-to-charge ratios by applications of an electrical fields of different frequencies (Figure 10).

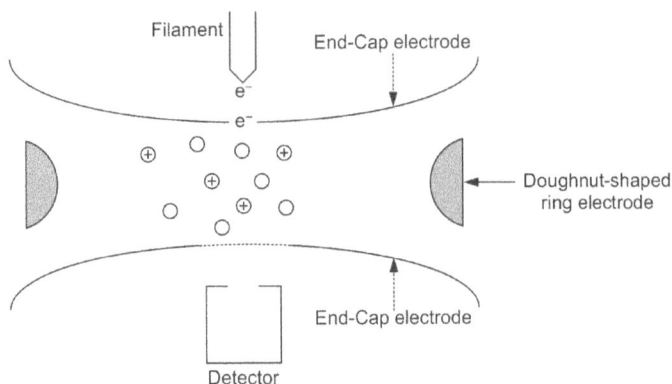

Figure 9: Scheme of an Ion Trap Analyzer.

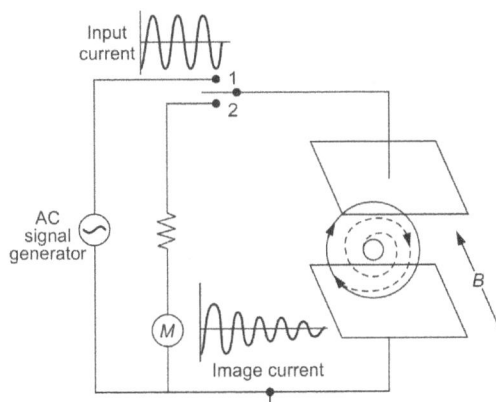

Figure 10: Ion Cyclotron Analyzer: Path of an Ion in a Strong Magnetic Field B. Inner Solid Line Represents the Original Circular Path of the Ion. Dashed Line Shows Spiral Path when Switch is Moved Briefly to Position 1. Outer Solid Line is New Circular Path when Switch is Again Opened. Reproduction with the Permission of D.A. Skoog et al., 2018, 7th ed., p. 522, Fig. 20.19, Cengage Learning.

In practice, the mass spectra from ion cyclotron analyzers are obtained via Fourier transform signals.

Detectors

We have already encountered the different dynodes devices on which are based the *electron multiplier detectors*. Let us also mention:

- The Faraday cup detectors. It is a Faraday cage which surrounds a collector electrode. Ions exiting the analyzer strike the collector electrode. The cage prevents the escape of reflected ions and ejects secondary electrons. The Faraday cage and the collector electrode are electrically connected to the ground potential. Ions striking any part of this ensemble can give rise to a signal or to an emission of secondary electrons which, in turn, may enhance the signal if these strike a further part of the Faraday cup. The charge of the positive ions striking the collector is neutralized by a flow of electrons from ground through the resistor. The resulting potential drop across the resistor is amplified by a high impedance amplifier.

- Scintillation detectors. They operate by recording ions via the emission of visible light following collision of the ions (or secondary electrons) with a phosphor-coated surface.

- Photographic detection of ions in mass spectrometry.

Photographic plates or films are very rarely used to day. This type of detection, however, is still encountered in spark instruments because these detection systems are well suited to the simultaneous observations of a wide range of m/z values.

Fourier Transform Mass Spectrometers

These apparatuses do generally possess an ion trap analyzer cell. Gaseous sample molecules are ionized in the center of the molecule by electrons coming from an electron impact ion source. A pulsed voltage applied at the grid of the cell allows to switch the electron beam on and off periodically. The ions are held in the cell by a 1 to 5 V voltage applied between the trap plates. The ions are accelerated by a radiofrequency signal applied to the transmitter plate. (The plates are the rectangular sides of the cell). I gen such of way, storage times may reach several minutes. After have been generated by an electron beam, the ions are trapped and are subjected to short radio-frequency pulses. During each pulse, the frequency increases linearly from 0.070 to 3.6 MHz, After the frequency sweep is discontinued, the image current, induced by the various ion packets, is amplified, digitized and stored in memory. The time domain decay signal is, then, transformed to yield a frequency domain signal that can be converted to a mass-domain signal.

This kind of instrument is characterized by a great resolution power. The signal treatment by Fourier transform, by accumulating the signals produced by the ions in weak quantities permits to reach mass domains of 10,000 Da.

Mass Spectrometers and Computers

A mass spectrum contains a great number of data that a human cannot treat seriously without the help of microcomputers.

CHAPTER 43

Applications of Mass Spectrometry

They are very numerous. We confine ourselves:

- To somewhat describe the identification of pure compounds;
- To briefly recall the biological applications of the method already quoted at the time of the description of the methods used in the biological analysis (see Chapter 42), to analyzing mixtures by hyphenated mass spectral methods;
- And to consider quantitative applications of mass spectrometry.

Identification of Pure Compounds

A mass spectrum of a pure compound provides several kinds of data that are useful for its identification. They are its molecular weight, the determination of the molecular formula from exact molecular weights by the isotopic ratios, by the presence or the absence of various functional groups and by study of its fragmentation patterns and the comparison of its spectrum with compounds of close formula.

This determination of molecular weights from mass spectra requires the identification of the molecular ion peak or in some cases the identification of the (M + 1)+ and (M − 1)+ peaks. The location of the peak (on the abscissa) gives the molecular weight with the highest possible accuracy. With electron-impact sources, an error may occur because the ion peak may be absent or it may be so diminished so the confusion on the identity of it molecular peak with another one of its neighbor peaks is possible.

Identification by Molecular Formulas From Exact Molecular Weights

Let us first recall that, within the course of an electron-impact mass spectrum, a molecular compound yields a characteristic series of fragments called its fragmentation and that each fragment can be referenced on the spectrum by its bar-graph (see Chapter 41). Rules have been derived to determine the elemental composition from the mass spectrum of a compound. This can be carried out only with high resolution spectroscopy. One operates by observation of *mass defects*. They are the differences (of mass) of true atomic or molecular weights and their nominal values. Obtaining the elemental composition necessitates being able to distinguish between two ratios m/z to four decimal places

Molecular Formulas From Isotopic Ratios

As a rule, it is possible to deduce molecular formula by calculating the ratios of the height of the peaks M, (M + 1)+ and (M + 2) whose exact values for a given compound are known. This methodology is only possible when the molecular ion peak is sufficiently intense.

Structural Information From Fragmentation Patterns

Today, there exist guidelines and rules which predict fragmentation mechanisms for pure compounds. They can help to interpret the mass spectrum of a new product. Generally, the fragmentations tend to give fragments and radicals more stable than the ion from which they come from. There is sometimes formation of metastable ions, the occurrence of which can help to the elucidation of the structure of the studied compound.

Biological Applications of Mass Spectrometry

We have already quoted:
- The analysis of peptides, polypeptides and proteins;
- The peptide sequencing;
- The analysis of oligonucleotides;
- The analysis of oligosaccharide;
- The analysis of lipides.

Hyphenated Mass Spectral Methods

Although mass spectrometry is a very powerful tool for the identification of pure compounds, its usefulness may be limited for analysis of some mixtures because they give rise to a very great number of fragments and it becomes impossible to interpret the totality of the spectrum. This is the reason why several methodologies have been developed. They result from the coupling of mass spectrometry and various separation methods. They are called "hyphenated mass spectral methods".

Tandem Mass Spectrometry

In tandem mass spectrometry, three quadrupoles are arranged in series (Figure 1).

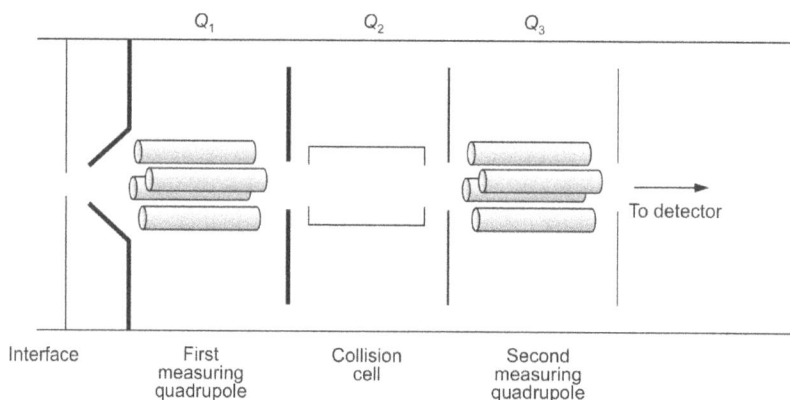

Figure 1: A Tandem Mass Spectrometer.

Q1 is the measuring quadrupole for determining the m/z of the introduced sample. The second quadrupole Q2 acts as a reaction zone. It is a cell filled with an inert gas such as nitrogen, helium or argon. The analyte ions collide with the gas molecules and become fragmented. This process is called collision induced dissociation (CID). These fragments are then introduced into the third quadrupole Q3 for mass analysis. Several combinations are possible.

The measuring quadrupoles can be run in a static or scanning mode. In the static mode, the electric fields are kept constant and only ions with one defined m/z value can pass. In the scanning mode, the quadruple sequentially allows ions within a define m/z to pass through. Within these two options several modes are possible. The daughter ion analysis method is the most commonly used analysis. It is set in a static mode. The first quadrupole is set in a static mode and only ions with one specific m/z-value can pass. They are, then reacted and fragmented in the collision cell. Ions resulting from this fragmentation, the so-called daughter ions are then analyzed in the third quadrupole.

Structural analysis of peptides, nucleic acids and oligosaccharides can be performed with a tandem mass spectrometer as for example an ESI-MS/MS. Tandem mass spectrometry can be applied for analysis of peptide mixtures. The first quadrupole only passes on one specific peptide ion which is then fragmented in the collision chamber, that is to say amino acids are cleaved from the peptide chain. In the third quadrupole, the difference between mass peaks gives information about the amino-acid sequence in the peptide. Oligonucleotides and oligosaccharides can be analyzed in a similar fashion.

Chromatography/SM

Prior to any study of these couplings, let us notice some difficulties encountered by the realization of these couplings. They are:

- The minimum quantities being into the chromatographic flood;
- The difference of pressure occurring between the chromatograph and the spectrometer;
- The presence of the eluting phase;
- The harmonization of the emergence time of the chromatographic peak with the registering of the mass spectrum.

Let us already say that the problem is, in part solely, solved by repetitive scans.

Gas Chromatography and Mass Spectrometry GC/MS

The coupling of both methods gas chromatography and mass spectrometry constitutes the method now called gas chromatography/mass spectrometry. It is one of the most powerful for the analysis of complex organic or biochemical mixtures. With this apparatus, the spectra of the solutes to separate are collected at their exit of the chromatographic column. It is interesting to notice that, in this method, the mass spectrometer plays the part of the detector of the gaseous phase, a part finally rarely played. The technique GC/SM seems to be now well dominated, although several experimental protocols remain difficult to put in perfect running. The difficulties come from the following points:

- The stationary phase must possess very weak volatility together with great thermal stability.
- The vector gas must be eliminated easily. For example, it is possible to use dihydrogen and helium, but it is not the case with nitrogen.
- The difference of pressure between the column and the spectrometer must be lowered. The goal is to reduce the pressure of the eluting phase and to enrich the vapor with the compound to analyze. In the occurrence, this is done by elimination of a great part of the carrier gas.

Liquid Chromatography/Mass Spectrometry (LC/MS)

The essential problem is the presence of the mobile phase which must enter into the spectrometer which is more difficult to eliminate than one gas. Its presence modifies the functioning of the ionization chamber. One solution of the problem is the pumping of a great part of the solvent and to make the remaining present solvent play a part in the ionization of the solutes.

There is a notable difference between the two kinds of couplings. The coupling with the liquid chromatography may also concern polar or thermolabile molecules.

Capillary Electrophoresis/Mass Spectrometry

This coupling proves to be a powerful and important tool in the analysis of large biopolymers such as proteins, polypeptides and DNA species. The capillary effluent is passed directly into an electrospray ionization device and the products then enter a quadrupole mass filter.

Quantitative Applications of Mass Spectrometry

One can say that there are two main types of analyses that are practiced by mass spectrometry. They are:

- The quantitative determination of molecular species is essentially organic, biological and rarely inorganic species. We shall call this type of analysis "organic analysis".
- The quantitative determination of the concentration of elements essentially in inorganic samples. We shall call this type of analysis "inorganic analysis".
- In organic analysis, all the types of ionization processes are used. In inorganic analysis, a harder source is used. This is essentially the inductively coupled plasma source although some other sources are sometimes used.

In a first methodology used in organic analysis, the analyte concentrations are obtained directly from the heights of the spectral peaks. It may be possible to find a peak at one own peak, that is to say at m/z characteristic values of the element m/z. One can also work by incorporating a fixed value of an internal standard in both the sample and in the curve of calibration. The ratio of the peak intensity of the analyte species in the two media is then plotted as a function of the analyte concentration. Not surprisingly, a good internal standard is a stable isotope analog of the analyte. It is assumed that the labeled molecules behave chemically as their unlabeled counterparts. The spectrometer makes the difference. Some other methodologies analogous to those used in some cases for the spectrophotometric determinations of mixtures are used.

It is clear, evidently, that analyses can be performed by passage of the sample through a chromatograph or a capillary electrophoresis instrument.

CHAPTER 44

Criteria of Purity

When the organic chemist has succeeded in isolating an organic species with the help of one of the means of the proximate analysis, it is imperative to control its homogeneity of purity by the determination of some of its physico-chemical constants. These constants are called *criteria of purity*.

Case of Solids

Crystalline Aspect

Organic crystallized compounds at ambient temperature T_0 are some good candidates for an examination of the homogeneity of their crystals. Sometimes, the occurrence of a strong proportion of impurities modifies slightly the physical constants whereas the examination of several successive "crystallizats" together with that of the residue shows the heterogeneity. The examination of the crystalline aspect can be carried out:

- With the naked eye.
- With the microscope. One deposes some drops of a solution of the compounds on a slide. After evaporation, the crystallization must be regular and the size and the form of crystals must be homogeneous. The crystals are not always neat with sharp limits. When the crystals are grouped giving compact masses, they are impure.
- With the polarizing microscope. In recto-linear polarized light, the crystallized systems give characteristic phenomena. In order to enhance the security of the examination, one compares the obtained results to those obtained with pure standards. The difficulty of this study is:
  - The polymorphism of a pure species,
  - The isomorphism. In principle, different chemical substances give unique crystalline forms.

It can be the inverse in neighboring chemical series. Some examples are provided by the phenol and the metacresol on one hand (Figure 1).

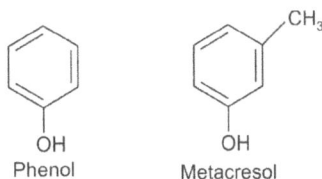

Figure 1: Structures of Phenol and Metacresol.

And another example is provided by the zinc, iron and calcium carbonates CO_3Zn, CO_3Fe, CO_3Ca on the other.

Melting Point

Definition

It is the temperature at which the solid and liquid phases coexist in a stable equilibrium under their own vapor tension. At constant pressure, every pure substance is melting at a constant temperature. This temperature remains constant all along the phenomenon of melting. There is never late in the phenomenon of fusion. From the practical standpoint, the fusion point is determined at the pressure for which its variations have a negligible influence on the fusion temperature.

The origin of this property lies in the phase rule which is represented by the general relation:

$$v = C - \varphi + 2$$

Here, v is the variance of the system. v is the number of intensive variables that can be changed independently without disturbing the number of phases in equilibrium. C is the number of components and φ the number of phases at equilibrium. When two phases are in equilibrium, $v = 1$ which implies that temperature is not freely variable if the pressure is fixed. Therefore, freezing (or melting) or any other phase transition occurs at a definite temperature at a given pressure.

Influence of the Impurities

Recalls of Curves of Binary Systems. As we shall see it, the presence of an impurity modifies the freezing point. Generally, it decreases it. If one draws the points of beginning and finishing solidification as a function of the proportions of a mixture of two compounds, A and B, on the same diagram, one obtains two kinds of curves, already encountered in this book.

• Curves without maximum or minimum

F_A and F_B are the temperatures of fusion of the compounds A and B (Figure 2).

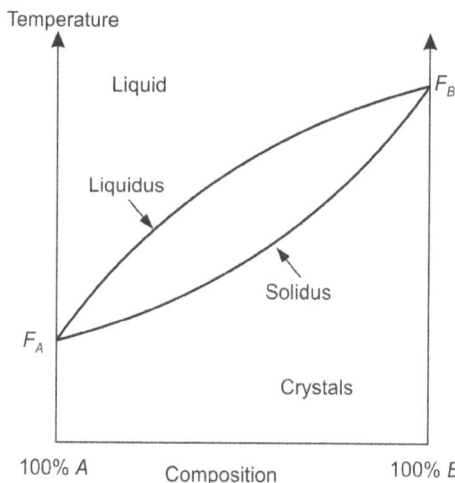

Figure 2: Curves of Liquidus and Solidus of the Mixture of Compounds A + B.

The whole of the points of beginning solidification is called liquidus and that of finishing is called solidus. In a certain domain of temperature, for every proportion of the mixture A + B corresponds a different mixture. The freezing point is not net (clear-cut) as it is in the case in which one passes easily from the solid phase to the liquid phase (Figure 2). In the case of the mixture, there exists an intermediary step with the coexistence of the two phases liquid and solid (Figure 3). As a result, there is an imprecision on the measurement of the fusion point.

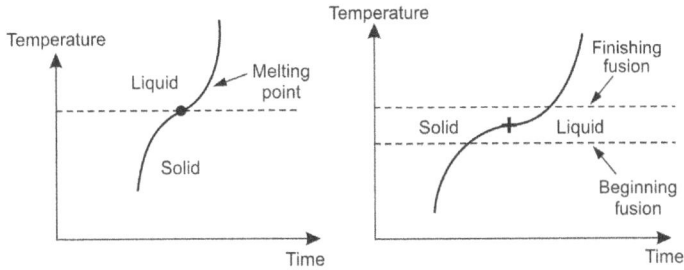

Figure 3: Curve Temperature/Time During the Fusion of a Pure Compound.

• Curves with maximum or minimum

They constitute azeotropic systems. (Analogous curves have been already encountered and studied in this book.) We confine ourselves to reproducing the cases where there exists a minimum or a maximum (Figures 4 and 5).

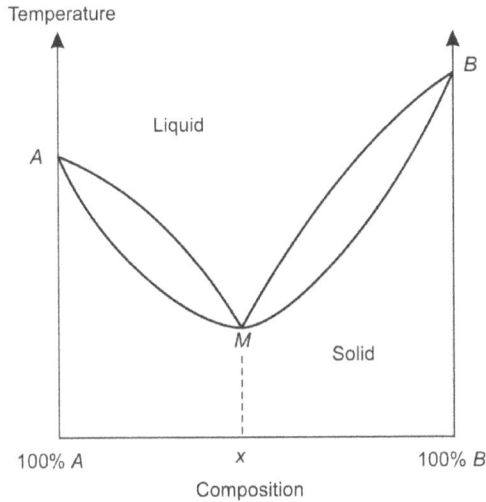

Figure 4: Curve of Liquidus-Solidus in the Case of Positive Azeotropism.

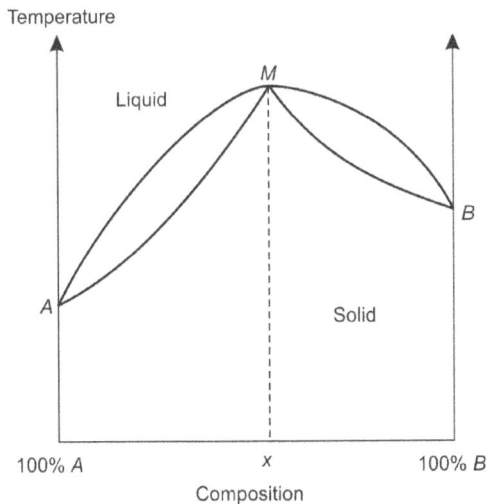

Figure 5: Curve Liquidus-Solidus in the Case of Negative Azeotropism.

In both cases, the two curves intersect at the maximum or minimum of the points. The corresponding mixtures have the behavior of a pure compound. They have a constant freezing point and the composition of their mixture at their extremum is invariable.

Clapeyron's Equation: Calculation of the Quantities of Impurities. During the purification (crystallization, for example), the fusion point of an organic compound increases until it becomes the fusion point of the pure product. It can be interesting to determine the number of impurities occurring in a mixture. Let us consider a Compound A containing a small quantity of the impurity B, then that the quantity $\Delta T = T - T_0$ the difference between the temperature T of fusion of the mixture and T_0 that of product A pure (the difference $T - T_0$ is always negative) and Q the molecular heat of fusion of A, the Clapeyron equation is:

$$At = -(RTT°/Q) \log N_A$$

In this relation, N_A is the molar fraction of A in the mixture ($N_A = n_A/n_A + n_B$) where n_A and n_B are the numbers of moles of A and B in the mixture and at the sought of impurities. As $T \approx T°$, we can write:

$$\Delta T = f\,(T^2)/Q$$

The interest of this relation is to show that ΔT is weak when the fusion heat of the compound is great and its fusion point low. Therefore, 10% of palmitic acid in stearic acid only induces a lowering of 2 degrees of its fusion point.

Applications to the Identification. The identification of a compound can be carried out by the determination of the fusion point of its mixture with a reference sample melting at the same temperature T. Let us consider two compounds A which is known and B which is unknown. They melt (per accident) at the same temperature.

When one mixes them in equal quantities and when one determines the fusion point, if the latter has not varied, one can say that both compounds are identical. If the fusion point of the mixture A + B is lowered, one can say that both products are different. This behavior is not formal proof, but it is a serious suspicion.

Techniques of Determination

Technique by Slow-Blow. The substance is slowly heated until its fusion is reached. There is a risk of alteration of the substance. It is as great as the heating prolongs is and the decomposition products lower the true fusion point.

- Method of the simple capillary tube.
- Method of the Thiele's tube.
- Method of the fusion under a microscope.

Techniques by Instantaneous Fusion. They consist in determining the temperature at which a particle of substance begins to melt immediately. This eliminates errors due to adulterations.

- Maquenne bloc
- Köffler bench

Purification Methods by Fusion

It is the method of *zone refining*.

Principle. It is a fractionated crystallization method carried out by moving a small melting zone alongside the sample to purify. Let us consider a solid A containing an impurity B under the form of a solid solution (Let us recall that if the two solid solutions are completely soluble in each other, only one solid phase can exist, this is a solid solution.) The sample is in the form of a narrow cylinder of

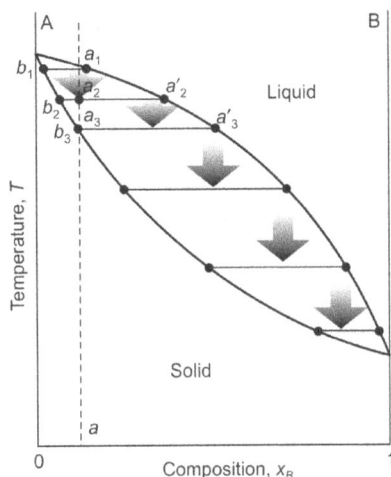

Figure 6: A Device for the Refining Zone (from Atkins's Physical Chemistry, 2002, p. 211 with the permission of P. Atkins and J. Paula, 2002, Oxford Publishing Limited).

sufficient length which is heated by zones that are displaced repeatedly due to a heating ring from one end of the sample to the other (Figure 6).

At the end of the process, the zone at the end of the sample contains all the initial quantity of B and cools to a dirty solid.

Explanation. Let us consider the mixture of A and B. Its molten zone is represented in Figure 6. Let us suppose that at a given instant, the concentration of A of the liquid phase is a_1. Let the system is cooled to a_2, a solid of composition b_2 is deposited and the remaining liquid (the zone where the heater has moved on is at a'_2. The process is equivalent to follow the vertical passing by a_1, a_2, a_3. The process continues until the last drop of liquid to solidify is heavily contaminated with B ($x_B \approx 1$).

This process permits very powerful purifications.

Case of Liquids

Ebullition Point

Definition

The ebullition point for the characterization of liquids is endowed with the same importance as the freezing point has for the characterization of solids. When one heats a liquid, the temperature increases and then stabilizes at the value of the ebullition point. It is reached when its vapor tension is equal to the pressure exerting on its surface.

Determination

By a Macro-Technique. One carries out a micro-distillation provided one has a sufficient quantity of liquid. It is necessary that there is a sufficient volume of vapor so that the distillation should be achieved with a constant speed.

By the Micro-Technique of Perkins. Some drops of the liquid under study are poured into a small tube of the type "hemolysis tube". It is plunged by its open extremity into the liquid of the bath the temperature of which is controlled by the nature of the solvent which filled it. The bath is heated quickly. The heating is continued until its temperature exceeds that of the ebullition of the studied compound. Then, there is a release of small bubbles coming from the compound, whereas the liquid of the bath tends to go into the capillary. The accuracy of the method is 0.5 degrees.

Refractive Index

Definition

When radiation passes at an angle through the interface between two transparent media that have different densities, a brutal change in direction of the beam happens. This phenomenon is called *refraction*. The angle that it forms with the normal surface changes. The ratio of the sinus of the incidence i and refraction r angles is equal to the inverse of the refractive indexes:

$$\sin i/\sin r = n_2/n_1$$

This is Snell's law (Figure 7).

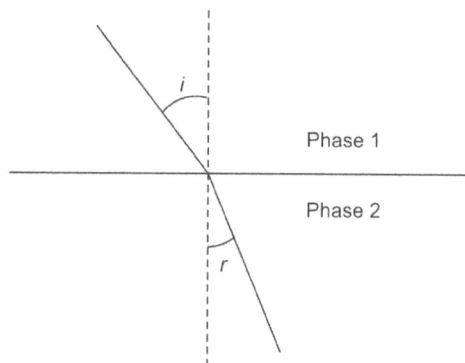

Figure 7: Snell's law: Definition of the Index of Refraction.

When the medium is air, $n_1 \approx 1$. The ratio is equal to n_2. Then, the ratio of the sinus depends on the refractive index of the substance. It changes with temperature and also with wavelength. One always specifies the temperature of the measurement of the refractive index in the exponent and the wavelength in subscript. For example, we find in literature, symbols of the kind such as n^{20}_D. It is the refractive index of a liquid at 20°C for the ray D of sodium.

One can reach the accuracy of the fourth decimal with this method. It is more accurate than the ebullition point method.

Apparatus

One uses Abbes and Pulfrich refractometers.

Molecular Refraction and Other Methods

Molecular Refraction

According to Lorenz, the molecular refraction R of a substance depends on the refractive index n, the density of the liquid and the molecular mass. R is given by the relation:

$$R = (n^2 - 1)/(n^2 + 2) \times M/d$$

This molecular refraction R can be calculated by the addition of the atomic refractions which depend on the functional part played. By the atom in the molecule and of the increments corresponding at some bonds. For example, let us take the case of acetic acid: CH_3COOH, i.e., $C_2H_4O_2$.

$$R = 2R_C + 4R_H + 1R_O(carbonyl) + 1R_O(hydroxyl)$$

Addition of the atomic refractions:

$$Rcalc = (2 \times 2.418) + (4 \times 1.100) + (2.211) + (1.575) = (13.022)$$

The refraction calculated according to this method must be in accordance with the observed molecular refraction. There are other methods, the general principle being the same as that followed for the refraction. They differ by the fact that there are other molecular properties that are calculated and compared with the observed. This kind of calculation can also bring an argument in favor or not of the purity of the compound.

The method of the capillary tube is simple. In a liquid bath, one plunges a thermometer. To the reservoir is welded a capillary tube containing the test material. As a liquid bath one uses H_2SO_4, H_3PO_4, phthalate esters and silicones.

Method of the Thiele's Tube

In order to equalize the temperature of the bath, Thiele utilizes an apparatus in which a circulation establishes itself by thermosiphon due to the current of convection.

Method of Fusion Under a Microscope

The microscope with a constant platin permits to determine the fusion point when one only has a very small sample. The method gives some on the phenomena accompanying the passage to the liquid state. With the use of a polarizing microscope, the fusion points are determined with a precision of 5% of degree. The birefringence of the crystals disappears during the fusion.

Techniques by Instantaneous Fusion

They consist in determining the temperature at which a part of a substance enters in immediate fusion. This eliminates the errors due to the alterations:

• Block Maquenne
It is a parallelopiped block, electrically heated or with a ramp to gas. One proceeds by repeated projections of substances finely divided on a heated block progressively. One increases by 5 degrees by min and then by 1° by minute just before the fusion.

• Köffler Bench
When we must do several determinations, it makes more sense to use a block that shows continuous variations of temperature between its two extremities.

Appendices

Appendix 1: Obtention of the van't Hoff Relation of Osmotic Pressure (Chapter 5)

The simplest demonstration, probably to obtain, is the one that involves the notion of fugacity. The fugacity f_B of substance B, which can be gas, liquid or solid and is a kind of fictitious pressure. It is related to its chemical potential μ_B by the relation:

$$\mu_B = \mu_B^\circ + RT \ln f_B/f_B^\circ$$

Here, f_B° is its fugacity in its standard state. μ_B° is its chemical potential in its standard state. They are constant in the conditions of the experience. They depend on the temperature of the system, of the nature of the substance. When the solute is non-added, the two compartments are under the same pressure P_o. Since the fugacity of a species, 1 depends on the total pressure and since it is the same, we can say that the fugacity of the solvent f_1 is the same in both compartments:

$$f_1^\cdot \text{ (left)} = f_1^\cdot \text{(right)} \quad \text{(Before addition of solute)}$$

The black superscript point means that the compound considered is pure. After the addition of some solute into the left compartment, the chemical potential of the liquid in the right compartment has not changed since its thermodynamic state has not varied.

The corresponding chemical potential $f_1^\cdot$ remains the same. Since there has been a spontaneous transfer from the right to the left, one can conclude that:

$$f_1 \text{ (left)} < f_1^\cdot \text{(right)}$$

For the equilibrium to be brought back to an identical level of liquids in both compartments or, equivalently, after having counterbalanced the osmotic pressure previously formed, we must have again:

$$f_1 \text{ (left)} = f_1^\cdot \text{ (right)}$$

At constant temperature, f_1 depends on the pressure and concentration of the solute, therefore we can write:

$$d \ln f_1 = (\partial \ln f_1/\partial P)_{T,x2} dP + (\partial \ln f_1/\partial x_2)_{T,P} dx_2$$

P is the total pressure above the left compartment. When an identical level is brought back there is no longer a change in f_1, this means that:

$$d \ln f_1 = 0$$

Hence:

$$(\partial \ln f_1/\partial P)_{T,x2} dP = -(\partial \ln f_1/\partial x_2)_{T,P} dx_2$$

According to the following general thermodynamic relation:

$$(\partial \ln f/\partial P)_T = \overline{V}_m/RT$$

In the present case, it is written:

$$(\partial \ln f_1 / \partial P)_T = \bar{V}_{m1}/RT$$

Here, $\bar{V}_{m1}$ is the partial molar volume of solvent 1. From these relations, one can write:

$$\bar{V}_{m1}/RT = -(\partial \ln f_1/\partial x_2)_{T,P}dx_2$$

The partial derivative on the right-hand term can be simplified as follows if one admits that the solution on the left remains sufficiently dilute. Then, it obeys Raoult's law. It is an ideal solution:

$$f_1 = f_1 {}^\bullet x_1$$

Since the solution is binary:

$$f_1 = f_1 {}^\bullet (1 - x_2)$$

$$d\ln f_1 = d\ln(1 - x_2)$$

$$d\ln f_1 = d(1 - x_2)/(1 - x_2)$$

$$d\ln f_1 = -dx_2(1 - x_2)$$

Since the solution is diluted:

$$(1 - x_2) \approx 1$$

$$(\partial \ln f_1/\partial x_2)_{T,P} = -1$$

Transferring this relation in the one where the partial derivative intervenes with $dP = P - Po$, we find:

$$P - Po = (RT/\bar{V}_{m1})x_2$$

In very dilute solution, the partial molal volume $\bar{V}_{m1}$ does not differ appreciably from its molar volume in the pure state $V_{m1}{}^\bullet$. As a result:

$$P - Po = (RT/V_{m1}{}^\bullet)x_2$$

$$\pi = (RT/V_{m1}{}^\bullet)x_2$$

This is the true van't Hoff relation. We can go further into approximations by considering again that the solution is sufficiently diluted so that the following approximation is valid:

$$x_2 \approx n_2/n_1$$

$$\pi = (RTV_{m1}{}^\bullet n_2/n_1)$$

Now, taking, into account the total solvent in the left compartment $V = n_1 V_{m1}{}^\bullet$,

$$\pi = n_2(RT/V)$$

Appendix 2: (Chapter 14)

Hence, it remains to demonstrate that:

$$< z^2 > = 2Dt$$

This is the proof of the above identity.

The reasoning is the following. Let us consider the situation in which the concentration of the solute which diffuses is constant in the plan xy normal to the z-axis but changes in the direction of z. In order to study the diffusion, let us consider an area of surface unity in the plane xy called the *transit plane*. There exists an aleatory motion of the solute through this plane from the left towards the right and from the right towards the left.

On the left and the right, let us imagine two planes parallel to that of transit. They are called the left plane(l) and the right plane (r). They are located at the distance $\sqrt{<z^2>}$ from the transit plane. The two compartments delimited in such a way on the left and on the right of the transit plane contain the concentrations of solute C_l and C_r (Figure 1).

During an interval of time t, a solute covers a square of average distance $<z^2>$, that is an average distance of $\sqrt{<z^2>}$. By choosing the plane l at the distance $\sqrt{<z^2>}$ from the transit plane, one is certain that statistically all the molecules of solute of the left compartment cross over the transit plane from the left to the right during the time t, provided they are going from left to right. The number of molecules in the left compartment is equal to its volume multiplied by its concentration:

$$\sqrt{<z^2>} \times 1 \times C_l$$

The number of moles that cross the transit plane from the left to the right during the time t is:

$$1/2\sqrt{<z^2>}\, C_l/t$$

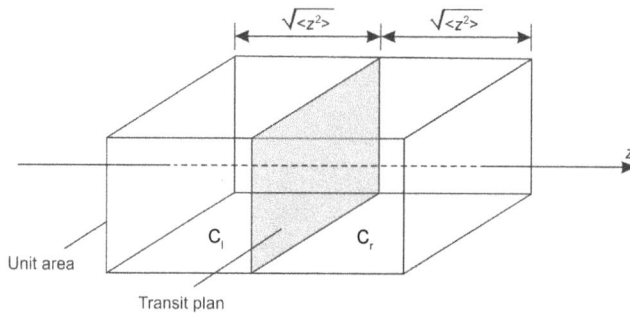

Figure 1: Schematic Diagram Permitting the Derivation of the Relation $<z^2> = 2Dt$. (Reprinted with Permission from Gwenola and Jean-Louis Burgot, Paris, Lavoisier, Tec et Doc, 2017, 167).

The factor ½ comes from the fact that the solute has equal opportunities to migrate from the left to the right that it has from the right to the left. The number of moles that cross the transit plane from the right to the left is:

$$1/2\sqrt{<z^2>}\, C_r/t$$

Therefore, the net flux J of diffusion of the solute across the transit plane, that is to say if the net of a number of ions that cross the transit plane by a unit of surface and time from the left to the right are:

$$J = \tfrac{1}{2}\,[\sqrt{<z^2>}/t\,](C_l - C_r)$$

The gradient de concentration along the z-axis by going from left to right is:

$$(dC/dz) = -(C_l - C_r)/\sqrt{<z^2>}$$

By substituting in the preceding relation:

$$J = -\tfrac{1}{2}\,[<z^2>/t\,](dC/dz)$$

By comparing with the first law of Fick, one obtains:

$$D = <z^2>/2t \quad \text{or} \quad <z^2> = 2Dt$$

(This relation is named the relation of Einstein-Smoluchowsky).

Otherwise, one demonstrates with the help of reasoning by induction that $<z^2>$ is proportional to the number of steps n and to the square of each distance l when a step is done:

$$<z^2> = knl^2 \quad (k \text{ proportionality factor})$$

Let us set up k = 1, n the number of theoretical plates and l the plate height, one obtains the equality:

$$< z^2 > = nH^2$$

$$\sigma^2 = nH^2$$

This relation has been already founded in the plates theory and used by Van Deemter et al. (1956) to find their equation. One must not forget that during a chromatographic separation, several aleatory walks can take place simultaneously with different values for *n* and for the square of the length of a theoretical plate. The effects on the widening of peaks accumulate to give a global variance equal to the sum of variances of the different mechanisms contributing to the widening of the bands:

$$\sigma^2 = \sigma_1^2 + \sigma_2^2 + \ldots \sigma_i^2$$

As a result, the variances are additive. In connection with this, it is important to notice that some experimental factors, exterior to the column itself, can contribute to the widening of the peaks. It is notably the case when the solute cannot be introduced as a band infinitely thin. The consequence is the occurrence of variance of injection σ^2 inj. Significantly, some authors have cut the H into portions related to one mechanism or another given the additivity of the variances and the relation:

$$H = \sigma^2 \, inj/L$$

Let us recall that in the integral solution ((6) of Chapter 14), it appeared two global parameters σ_1^2 and σ_2^2 could be considered as being different variances linked to the displacements by diffusion and by transfer of masses, the latter, could be themselves independent.

Thus, the stochastic theory of chromatography exhibits an undeniable coherence with the other theories of chromatography, including that of plates. It is in this context that some authors have estimated that the stochastic theory reconciles all the other theories of chromatography.

Appendix 3: Solid Angle (Introduction Spectral Methods)

We are interested in a cone whose origin is O (Figures 1a and b). The portion of space delimited by the generator of the cone corresponds to the solid angle Ω. In Figure 1, Σ is the intersection area of some spheres of center O and of radius R and the portion of space characterized by Ω.

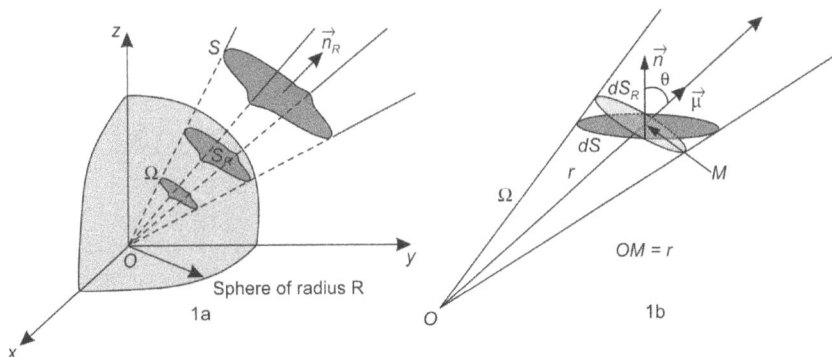

Figures 1a and b: Definition of a Solid Angle Ω.

The solid angle Ω is defined by the relation:

$$\Omega = \Sigma/R^2$$

Σ does not depend on the considered sphere. Let us consider, indeed, the surface Σ' determined by a sphere of radius R'. One can write:

$$\Sigma/R^2 = \Sigma'/R'^2$$

Since the two spheres are homothetic. Ω is expressed in steradians (sr). The sphere of unit radius corresponding to the whole space gives a solid angle.

$$\Omega = 4\pi \ (sr)$$

Let us notice dS_R the elementary area obtained by orthogonal projection of dS on the sphere of center O and radius r. The elementary algebraic solid angle is given by the expression:

$$d\Omega = dS_R \cos\theta / r^2$$

It is the solid angle under which we see the elementary surface dS from a distance r located at the point O oriented by its normal **n** (vector), which makes an angle θ with the radial vector **u** in M.

Appendix 4: Lasers (Chapter 19)

Owing to their importance, lasers deserve some particular consideration. The word *laser* is an acronym for *light amplification by stimulated emission of radiation.* Laser action results from two quasi-simultaneous phenomena:

- One, in which the population of atoms in an excited electronic state, is made exceeding in a lower state. This is the *population inversion.*
- The other in which there is *stimulation* of a radiative emission between both states.

The excited state, under the stimulation, emits a photon by interacting with a radiation of the same frequency that it has absorbed previously. The more present photons of that frequency, the greater the number of photons the excited states are stimulated to emit.

The simultaneous absorption and stimulated emission have a strong consequence. This is the reinforcement of the number of photons that the excited states are stimulated to emit. This process is named the *gain* of the laser medium.

The lasing medium may be a solid crystal, a solution of an organic dye or a gas. It may be also a semi-conductor. As a solid crystal let us cite a crystal of ruby (Al_2O_3) mixed with Cr_2O_3 as a minor constituent. Lasers can also be made with other materials, such as a semi-conductor, let us cite gallium arsenide and as a gas argon or krypton. Dye lasers have become particularly important because they can provide narrow bands of radiations at any chosen wavelength. They are fluorescent organic compounds such as fluoresceine or rhodamine G in solution. Their emission can be made to occur anywhere within the fluorescent emission spectrum (see Chapter 22) and can be selected by means of a prism or of a grating. Dye lasers permit to overcome a disadvantage of early lasers, that is to say, a somewhat limited choice of wavelengths to use.

Figure 1 shows the important components of a laser source. The lasing medium is often activated by the energy of pumping directed through the side of the source called the system. On one end of the cavity, a mirror is placed so that all light approaching it from the interior of the source is reflected. The mirror at the other end is coated with a thin layer of, for example, silver so that only a fraction (80 to 90 percent) of the incident light is reflected, the remainder escaping.

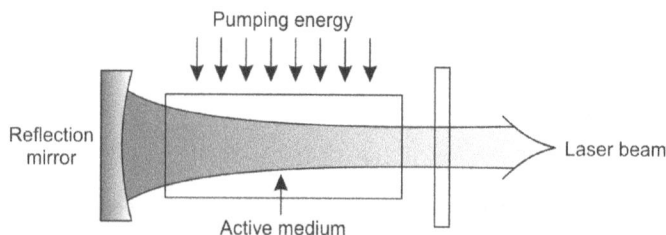

Figure 1: Scheme of a Laser System.

The population inversion is created from the *pumping energy*. In turn, the pumping energy may come from intense lamps, electric discharges or from another laser. The pumping energy is necessary because Boltzmann's law says that the population is greater in the lowest energy state than in an upper one, at thermal equilibrium.

We witness the results of the laser because the presence of the radiant energy at the exact required energy (that is to say corresponding to photons whose energy exactly spans the two states-ground and excited ones) stimulates emission. As a result, a radiant flux appears and increases rapidly because, at every reflection from the end mirror, some light passes through (this is illustrated schematically by the different arrows in Figure 1). Finally, a great quantity of monochromatic light is emitted during a very brief interval of time (of the order of 0.5 ms). Power of the order of the megawatt may be reached.

There are several kinds of laser systems. They are classified according to the levels of energy and the transitions that occur during the phenomena. In the three-level system, the transition laser, that is to say, the one which is characterized by light amplification by stimulated emission, is between the excited state E_{ex} and the ground state E_0 (Figure 2). In the four-level system, the laser transition terminates in another excited state E_{ex}, easier to attain than the ground state (Figure 3).

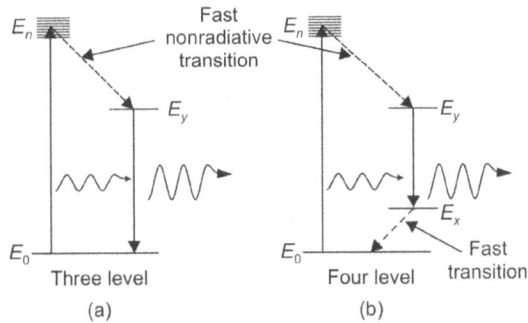

Figure 2: Laser Transitions in the Three-Level System.

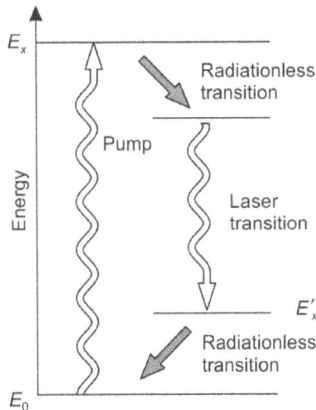

Figure 3: Laser Transitions in the Four-Level System.

The principal advantages of the lasers are the following:

- They are sources of monochromatic radiations.
- The band-with is located in the range of 10^{-2} nm, as a result, the light coming from a laser is highly monochromatic.
- They possess high power at one wavelength.

These two points are probably the two principal advantages of lasers. In particular, their strong luminous intensity permits them to excite very high densities of particles. This implies that the "sensitivity limits" are very good with that methodology.

- Their rays are parallel. They are collimated.
- They are polarized.
- They are coherent, that is to say, all waves are in phase.

Their disadvantages are they can be of limited wavelengths and are expensive. The ray spectra, indeed, encountered in atomic spectroscopy, require laser sources perfectly well-adapted to the metallic element from the double standpoint of the emission and of the narrowness of the rays. Fortunately, the existence now of dye lasers permits to reach the wavelengths spreading from U-V to infrared.

They are important devices as sources of radiations, inter alia, in Raman, UV-visible, infra-red spectroscopies and emission spectroscopy. Lasers play also a part in a method of introducing solid samples into atomizers, called ablation (see Chapters 34 and 35).

Appendix 5: Thermocouples-Bolometers (Chapter 29)

- Thermocouples

Electrons can be used as the working material of heat engines or heat pumps. An electric current flowing through a junction of one conductor with another absorbs or evolves depending on the direction of current flows. This effect, named "the Peltier heat", is related by thermodynamics principles to the "Seebeck effect" which is the potential of a thermocouple. (Besides, the German physicist Johann Seebeck is considered as being the inventor of the thermocouple).

A thermocouple is a device that permits the measurement of the temperature of an object. It is constituted of two different metals soldered by one extremity (Figure 1). This point of the probe called the "hot probe" is put in contact with the piece whose temperature must be determined. The other extremity "the cold solder" is maintained at a reference temperature, such as 0°C. Then, it appears a voltage between both solders which is related to the temperature via a known relation.

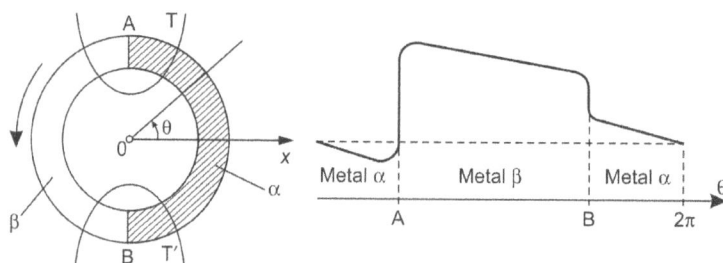

Figure 1: Scheme of a Thermocouple.

According to the Seebeck effect, when both metals are used to form an open circuit (there is no circulating current), a voltage between the two terminals if a temperature difference does exist (Figure 2). For the determination of the temperature, the two different metals are soldered at their extremity which is the hot solder. The two other extremities are related to the terminals of a voltmeter (Figure 3). The potential difference measured with the voltmeter is a (known) function of the temperatures of the hot and cold sources.

A well-designed thermocouple transducer is capable of responding to temperature differences of 10^{-6} K. As the goal, when one uses a thermocouple is to transform the radiant energy into, thermal energy, the hot soldered joint is generally covered by a black substance. The thermocouple is overall useful in the domain of the great wavelengths (wavelengths).

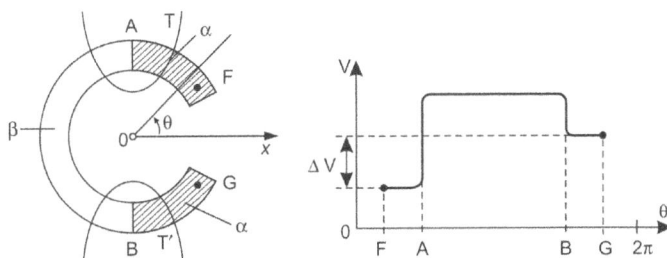

Figure 2: Thermocouple, Open Circuit.

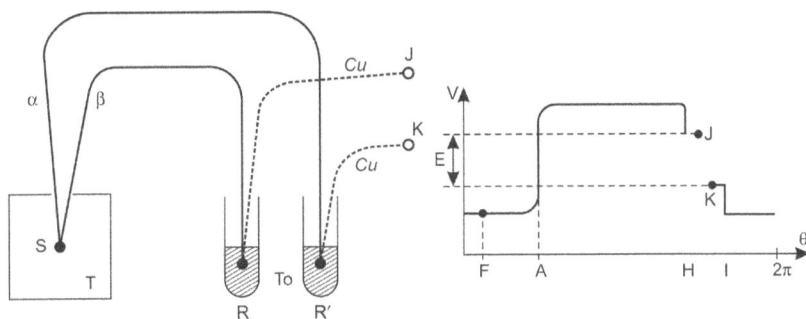

Figure 3: Other Representation of Thermocouple Apparatus.

• Bolometer

It is a tiny resistance thermometer. Its sensitive element is a thermistor. It is about five times more sensitive than platinum. The apparatus is constituted by two principal elements. One element is exposed to the radiation whereas its other element is protected from it.

The resistance of the element exposed to the radiation differs from that of the other element. The disequilibrium is registered by a Wheatstone bridge and treated as usual. Often it is prolongated by an amplifier circuitry.

Appendix 6: Homotopic, Enantiotropic and Diastereotopic Substituents (Chapter 25)

Some substituents that are equivalent in a molecule do not exhibit the same reactivity. This is an experimental fact. For example, they differently react with some enzymes and, for another example, have different signals in N.M.R. If they are equivalent, they are said to be homotopic, if not heterotopic. In the latter ones, one distinguishes the enantiotropes and the diastereotopes.

Homotopes

The substituents homotope comply with the following criterium: two identical substituents of the same molecule are homotopes or homotopic if the replacement of the first and, then, in a different experiment of the second by the same substituent, leads to the same structure.

It is the case of the two substituents H of the dichloromethane which affords the bromodichloromethane after substitution by one atom of bromine. H_A and H_B are homotopes (Figure 1).

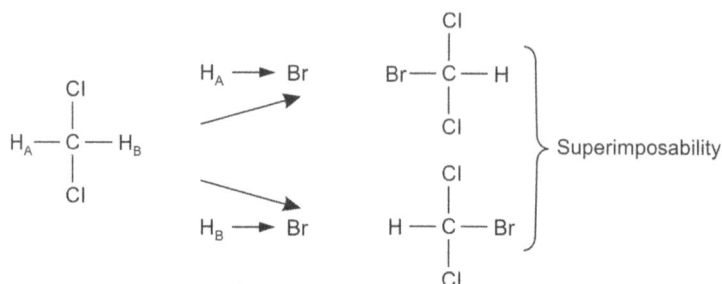

Figure 1: Example of Two Substituents Homotopes. (Reprinted with Permission from Gwenola and Jean-Louis Burgot, Paris, Lavoisier, Tec et Doc, 2017, 460).

Enantiotropes

The two substituents are called enantiotropes or enantiotropic if the replacement of the two by the same achiral substituent leads to two enantiomers:

For example, the bromochloromethane substituted by a fluorine atom (Figure 2).

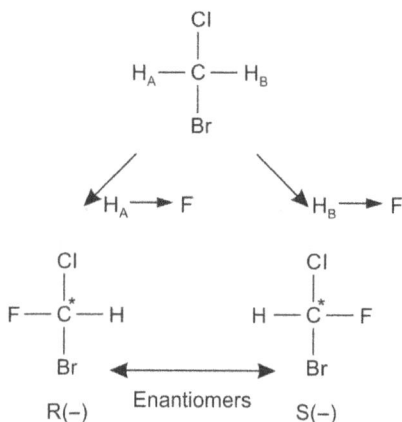

Figure 2: Example of Two Substituents Enantiotropes. (Reprinted with Permission from Gwenola and Jean-Louis Burgot, Paris, Lavoisier, Tec et Doc 2017, 460).

Diastereotopes

Replacing the two substituents with another one achiral leads to two diastereoisomers. It is the case for example of the 2-bromobutane (Figure 3). The achiral substituent is the chlorine atom. H_A and H_B are two diastereotopes and are said diastereotopic. The diastereotopic nuclei, in principle, exhibit different chemical shifts, included in achiral media. They are different in any media, chiral or achiral; They also lead to different coupling constants. They are anisochrones. The cause of the isochronous character is the magnetic field felt by the two protons which are not the same. Sometimes, the differences in the resonance frequencies of the two diastereotopes are so weak that they are not perceptible, except in the case for which the study is made in high fields N.M.R.

The anisochronicity is perceptible with several nuclei, i.e., ^1H, ^{13}C and ^{19}F.

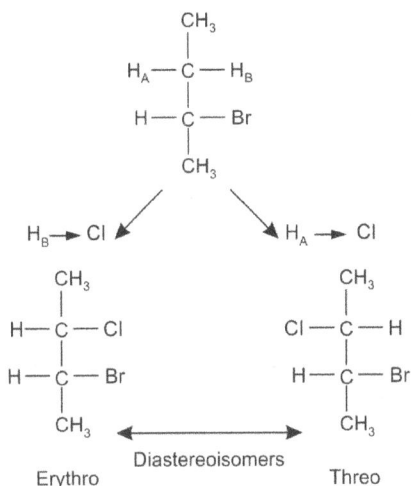

Figure 3: Examples of Two Diastereotopic Substituents. (Reprinted with Permission from Gwenola and Jean-Louis Burgot, Paris, Lavoisier, Tec et Doc, 2017, 461).

Appendix 7: Michelson's Interferometer IR (Chapter 29)

Light from a source S is divided into two equal parts by the splitter BS. The two beams are reflected by mirrors M1 and M2 and portions of them, finally, strike the detector. If mirror M2 is moved to the right along the optical axis, the two phases of the two beams striking the detector differ and interference results. It is the part played by interferometer the retardation p of the phase for a given increase in path length depends on the wavelength (and hence of the time) of the radiation. It is observed by the detector as a series of successive maxima and minima of intensity.

The mirror M2 is made to move at a constant speed for a distance long compared to the wavelength (of the order of the millimeter). Ideally, monochromatic radiation should produce a cosine wave as the two optical paths deviate from equality. If both frequencies enter together into the interferometer, the detected signal oscillates as the two components come into and out of phase as p changes.

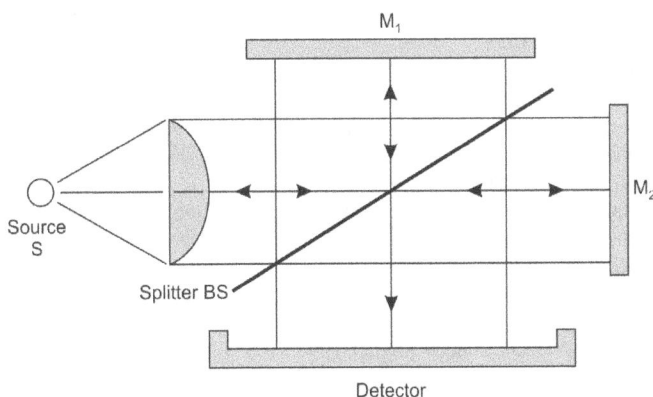

Figure 1: Michelson's Interferometer.

Bibliography

Albani, R. 2001. Absorption et fluorescence. Éditions Tec et Doc, Paris.

Atkins, P.W. and J. de Paula. 2002. Physical Chemistry. 7th edition. Oxford University Press, Oxford.

Bertoncini, F., C. Vendeuvre, D. Espinat and D. Thiebaut. 2005. Apport de la chromatographie en phase gazeuse bidimensionnelle pour la caractérisation de matrices. Spectra Analyse 247: 26–31.

Bertrand, P. 2010. La spectroscopie de résonance paramagnétique électronique-Fondements. EDP sciences, Les Ulis.

Billy, M. 1973. Introduction à la chimie analytique. Dunod, Paris.

Boas, M.L. 2005. Mathematical Methods in the Physical Sciences. Wiley, New York, 3rd edition.

Bosch Ojeda, C., F. Sanchez Rojas and J.-M. Cano Pavon. 1995. Recents developments in derivative ultraviolet/visible absorption spectrophotometry. Talanta 42: 1195–1214.

Bouchonnet, S. 2009. La spectrométrie de masse en couplage avec la chromatographie en phase gazeuse. Éditions Tec & Doc, Lavoisier, Paris.

Bouchoux, G. and M. Sablier. 2005. Spectrométrie de masse, principe et appareillage, P2645v2 Techniques de l'Ingénieur, Paris, 2017 et applications P2646, Techniques de l'Ingénieur, Paris.

Bourdon, R. and J. Yonger. 1977. Cours de chimie analytique. Centre de Documentation Universitaire, Paris.

Brothier, F. 2014. Développements d'outils bioanalytiques miniaturisés: greffage de biomolécules sur monolithes en capillaire couplés à la nanochromatographie pour l'analyse d'échantillons complexes. Thèse de Doctorat de l'Université Pierre et Marie Curie, Paris.

Caude, M. and N. Bargmann-Leyder. 2001. Séparations chirales par CPL, CPS et CPG. Techniques de l'Ingénieur, P1470, Paris.

Cazes, J. and R.P.W. Scott. 2002. Chromatography Theory. Marcel Dekker, INC, New York, Basel.

Chapuis, F., V. Pichon and M.C. Hennion. 2005. Méthode de préconcentration par extraction en phase solide. Oil and Gas Science and Technology 60(6): 899–912.

Chatten, L.G. 1969. Pharmaceutical chemistry. Marcel Dekker, Taylor and Francis, New York.

Delaunay, N. 2021. Électrophorèse capillaire: Principes, Techniques de l'Ingénieur, Traité Analyse et Caractérisation, Paris, P. 3365V2.

Delaunay, N. 2021. Électrophorèse capillaire: applications, Techniques de l'Ingénieur, Traité Analyse et Caractérisation, Paris, P. 3367V2.

Denbigh, K. 1989. The Principles of Chemical Equilibrium. 4th edition. Cambridge University Press, Cambridge.

Djerassi, C. 1960. Optical Rotatory Dispersion. Mc Graw-Hill Book Company, New York.

Dyer, J.R. 1970. Spectroscopie d'absorption appliquée aux composés organiques. Dunod, Paris.

Einstein, A. 1905. Uber einen die Erzeugung and Verwandlung des Lichtes betreffenden heuristischen Gesichtspunkt. Annalen der Physik. 322(6): 132–148.

Ewing, G.W. 1985. Instrumental Methods of Chemical Analysis. 5th edition, McGraw-Hill Education, New York.

Fernandez, X., J.-J. Filippi and M. Jeanville. 2011. Chromatographie en phase gazeuse à deux dimensions: GC-GC et GCxGC. P1489, Techniques de l'Ingénieur, Paris.

Fifield, F.N. and D. Kealey. 2000. Principles and Practice of Analytical Chemistry. 5th ed. Blackie Academic and Professional, Glasgow.

Gaillard, O., N. Kapel, J. Galli, J. Delattre and D. Meillet. 1994. Fluorimétrie en temps retardé: principe et applications en biologie clinique. Ann. Biol. Clin. 52: 751–755.

Gareil, O. 1990. L'électrophorèse de zone et la chromatographie électrocinétique capillaire, Partie I: Principes et notions fondamentales. Analusis 18: 221.

Gareil, O. 1990. L'électrophorèse de zone et la chromatographie électrocinétique capillaire, Partie II: Mise en œuvre expérimentale et applications, Analusis 18: 447.

Gareil, O. and G. Peltre. 1995. Électrophorèse. Techniques de l'Ingénieur, Traité Analyse et Caractérisation, Paris P. 1815.

Giddings, J.C. 1961. The role of lateral diffusion as a rate-controlling mechanism in chromatography. Journal of Chromatography A, 5: 46–60.

Giddings, J.C. 1961. Lateral diffusion and local nonequilibrium in gas chromatography. Journal of Chromatography A, 5: 61–67.

Glasstone, S. 1948. Textbook of Physical Chemistry. 2nd ed. Sixth printing. D. Van Nostrand Company Inc, New York, London, Toronto.

Guernet, M. and M. Hamon. 1981. Abrégé de Chimie Analytique, Tome 1. 2nd edition. Masson, Paris.

Greaves, J. and J. Roboz. 2014. Mass Spectrometry for the Novice, CRC Press, Boca Raton.

Greenwood, N.N. and A. Earnshaw. 1997. Chemistry of the Elements. Butterworth Heinemann, 2nd edition, Oxford.

Guillarme, D. 2014. Les récentes évolutions en chromatographie liquide. Techniques de l'Ingénieur, Traité Analyse et Caractérisation, Paris, P. 1494.

Hagege, A. and T.N.S. Huynh. 2013. Electrophorèse capillaire: applications, Techniques de l'Ingénieur, Traité Analyse et Caractérisation, P. 3366. Paris.

Hamon, M., F. Pellerin, M. Guernet and G. Mahuzier. 1990. Abrégé de Chimie Analytique, Tome 3, 2nd édition, Masson, Paris.

Harris, D.C. 2015. Quantitative Chemical Analysis. 9th edition, Freeman, New York.

Harvey, D. 2000. Modern Analytical Chemistry, McGraw-Hill Education, New York.

Hollas, J. and D.J. Hollas. 1998. Spectroscopie, Cours et Exercices, Dunod, Paris.

Holme, D.J. and H. Peck. 1998. Analytical Biochemistry, 3rd edition, Longman, Harlow.

Inczedy, J., T. Lengyel and A.M. Ure. 1998. Compendium of Analytical Nomenclature, Blackwell Science, Oxford.

Kellner, R., J.M. Mermet, M. Otto and H.M. Widmer. 1998. Analytical Chemistry. Wiley-VCH, Weinheim.

Klotz, I.M. and T.F. Young. 1964.Chemical Thermodynamics Basic Theory and Methods. W.A Benjamin Inc., New York.

Kolthoff, I.M. and P.J. Elving. 1968. Treatise on analytical chemistry. Part I, vol. 8, Intersciences Publisher, New York.

Kusnetz, J. and H.P. Mansberg. 1978. Optical consideration nephelometry in automated immunoanalysis. Part I, Intersciences Marcel Dekker, New York.

Levillain, P. and D. Fompeydie. 1986. Spectroscopie dérivée: intérêts, limites et applications. Analusis 14(1): 1.

Linden, G. 1981. Techniques d'analyse et de contrôle dans les industries agroalimentaires, tome 2. Techniques et Documentation, Paris.

Mahuzier, G., M. Hamon, D. Ferrier and P. Prognon. 1999. Abrégé de Chimie Analytique, Tome 2. 3rd edition. Masson, Paris.

Manz, A., N. Pamme and D. Lossifidis. 2004. Bioanalytical Chemistry. Imperial College Press, London.

Marin, C. 2000. Apport de la chimie analytique pour le contrôle de l'authenticité des arômes d'origine naturelle: aspects techniques et économiques. Ann. Fals. Exp. Chim. 93: 950; 95–110.

Masse, J. and A. Chauvet. 1983. Application of thermal analysis to the determination of the purity of chemical drug, chlorprothixene. Annales Pharmaceutiques Françaises 41(6): 579–589.

McCabe, W.L., J.C. Smith and P. Harriott. 2004. Unit Operations of Chemical Engineering. McGraw-Hill Education, New York.

Metais, H.P., J. Agneray, G. Ferard, J.C. Fruchard, J.-C. Jardiller, A. Revol, G. Siest and A. Stahl. 1977. Biochimie clinique, Tome 1: Biochimie analytique. Simep, Villeurbanne.

Meyer, V. 1998. Practical High-performance Liquid Chromatography. Wiley, Chichester.

Miller, J.M. 2004. Chromatography—Concepts and Contrasts. 2nd edition, Wiley, Interscience New Jersey.

Moore, W.J. 1965. Chimie Physique. 2nd edition. Dunod, Paris.

Munier, R.L. 1972. Principes des méthodes chromatographiques. Azoulay, Paris.

Nelson, D.L. and M.M. Cox. 2013. Lehninger: Principles of Biochemistry. McMillan, New York.

Nerst, W. 1891. Verteibing eines stoffes zwischer zwei Losungsmitteln und zwischen Losungsmittel und dampfraum. Zeitschift fur Physikalische Chemie 8U: 110–139.

Nyiredi, S. 2001. Planar chromatography a retrospective view for the third Millenium. Springer Scientific Publisher, Budapest.

Paucot, G. and M. Potin-Gautier. 2015. ICP-MS: couplage plasma induit par haute fréquence – spectrométrie de masse. P. 2720V3, Techniques de l'Ingénieur, Traité Analyse et Caractérisation, Paris.

Penicaut, B., C. Bonnefoy, C. Moesch and G. Lachatre. 2006. Spectrométrie de masse à plasma couplé par induction (ICP-MS), potentialités en analyse et en biologie. Ann. Pharm. Fr. 64: 312–327.

Pichon, V. 2006. Extraction sur phase solide pour l'analyse de composés organiques. Techniques de l'Ingénieur, P1420.

Plotka-Wasylka, J., N. Szczepanka, M. de la Guardia and J. Namiesnik. 2016. Modern trends in solid phase extraction: New sorbent media. Trends in Analytical Chemistry 77: 23–43.

Pradeau, D. 1992. Analyse pratique du médicament. Éditions médicales internationales, Cachan.

Randerath, R. 1971. Chromatographie sur couches minces. Gauthier-Villars, Paris.

Richardin, P. 2001. La chromatographie en phase gazeuse: en route vers un nouveau millénaire. Spectra Analyse 222: 19–26.

Rosset, R., M. Caude and A. Jardy. 1991. Chromatographies en phases liquide et supercritique, Masson, Paris.

Rouessac, F. and A. Rouessac. 2019. Analyse chimique: méthodes et techniques instrumentales. 9th edition, Dunod, Paris.

Sablier, M. 2021. Couplage CG/SM/SM, Techniques de l'Ingénieur, P1487v2.

Schwedt, F. 1998. Atlas de poche des méthodes d'analyse, 2e édition, Médecine-sciences-Flammarion, Paris.

Serway, R.A. and J.W. Jewett. 1992. Physics for Scientists and Engineers with Modern Physics, volume 3. 3rd edition, Editions Etudes Vivantes, Laval.

Skoog, D.A., D.M. West, F.J. Holler and S.R. Crouch. 2013. Fundamentals of Analytical Chemistry. 8th edition. Cengage Learning Inc, Boston.

Skoog, D.A., F.J. Holler and S.R. Crouch. 2018. Principles of Instrumental Analysis. 7th edition, Cengage Learning Inc, Boston.

Touchstone, D.J. 1992. Practice of Thin Layer Chromatography. 3rd edition. John Wiley, Hoboken.

Tranchant, J. 1997. Manuel pratique de chromatographie en phase gazeuse. 4th edition, Dunod, Paris.

Van Deemter, J.J., F.J. Zuiderweg and A. Klinkenberg. 1956. Longitudinal diffusion and resistance to mass transfer is causes of non ideality in chromatography. Chem. Eng. Sci. 5(6): 271–280.

Van Degans, J., A.-M. de Kersabiec and M. Hoenig. 1997. Spectrométrie d'absorption atomique. Techniques de l'Ingénieur, Traité Analyse et Caractérisation, P. 2825V2, Paris.

Veluz, L., M. Legrand and M. Grosjean. 1965. Optical circular dichroism. Verlag chemie G.M.B.H. Weinheim/Bergstrasse, Academic press, New York-London.

Vogel's, J. Mendham, R.C. Denney, J.D. Barnes and M.J.K. Thomas. 1999. A. Textbook of Quantitative Inorganic Analysis. 6th edition, Longman, Upper Saddle River.

Wainer, I.W. and D.E. Drayer. 1988. Drug Stereochemistry. Marcel Dekker, New York.

Wayne, R.P. 1994. Chemical Instrumentation. Oxford Chemistry Printers, Oxford Sciences Publication, Oxford.

Wheatley, P.J. 1959. The Determination of the Molecules Structure. Oxford and Clarendon Press, Oxford.

Whittaker, A.G., A.R. Mount and M.R. Heal. 2000. Physical Chemistry. BIOS, Oxford.

Willard, H.H., L.L. Merritt and J.A. Dean. 1965. Méthodes physiques de l'analyse chimique, Dunod, Paris.

Wilson and Wilson's. 2011. Green Analytical Chemistry: Theory and Practices. vol. 57, Elsevier, Oxford.

Index

For Product Safety Concerns and Information please contact our EU
representative GPSR@taylorandfrancis.com
Taylor & Francis Verlag GmbH, Kaufingerstraße 24, 80331 München, Germany